Smith's
Elements of Soil Mechanics

Smith's
Elements of Soil Mechanics

Smith's Elements of Soil Mechanics

10TH EDITION

Ian Smith

WILEY Blackwell

This tenth edition first published 2021
© 2021 John Wiley & Sons Ltd

Edition History
© 2014 by John Wiley & Sons, Ltd
© 2006 Ian Smith

Registered Offices
John Wiley & Sons, Inc., 111 River Street, Hoboken, NJ 07030, USA
John Wiley & Sons Ltd, The Atrium, Southern Gate, Chichester, West Sussex, PO19 8SQ, UK

Editorial Office
9600 Garsington Road, Oxford, OX4 2DQ, UK

For details of our global editorial offices, customer services, and more information about Wiley products, visit us at www.wiley.com.

Wiley also publishes its books in a variety of electronic formats and by print-on-demand. Some content that appears in standard print versions of this book may not be available in other formats.

Library of Congress Cataloging-in-Publication Data

Names: Smith, G. N. (Geoffrey Nesbitt), author.
Title: Smith's elements of soil mechanics / Ian Smith.
Other titles: Elements of soil mechanics for civil and mining engineers |
    Elements of soil mechanics
Description: 10th edition. | Hoboken, NJ, USA : Wiley-Blackwell, 2021. |
    Originally published under the title: Elements of soil mechanics for
    civil and mining engineers. | Includes bibliographical references and
    index.
Identifiers: LCCN 2021003404 (print) | LCCN 2021003405 (ebook) | ISBN
    9781119750390 (paperback) | ISBN 9781119750406 (adobe pdf) | ISBN
    9781119750413 (epub)
Subjects: LCSH: Soil mechanics.
Classification: LCC TA710 .S557 2021 (print) | LCC TA710 (ebook) | DDC
    624.1/5136–dc23
LC record available at https://lccn.loc.gov/2021003404
LC ebook record available at https://lccn.loc.gov/2021003405

Cover Design: Wiley
Cover Image: © Monty Rakusen/Getty Images

Set in 9/11.5pt Avenir by Straive, Pondicherry, India

10 9 8 7 6 5 4 3 2 1

Contents

Preface

When the ninth Edition of this book was published in 2014, I was pleased to receive positive feedback for making the understanding of the new geotechnical design code, Eurocode 7, straightforward and easy to follow. At that same time, a review of all Eurocodes commenced with the intention of producing a 'second generation' of the codes containing improvements and changes to the design procedures. The second generation of the Eurocodes will be published throughout the 2020s and Eurocode 7, in particular, is set to be significantly different from the first generation of the code.

The changes in the new Eurocode 7 affect many aspects of the existing design procedure, and I describe these and explain how to perform the new geotechnical design process in this tenth edition of the book. Many new worked examples help to illustrate the new procedures. Since the first generation of Eurocode 7 is still very much in use in geotechnical design, I have retained the content on it so that readers of this book can gain a thorough understanding of the design procedures used in both generations of the code.

As with the ninth edition, to help the reader fully understand the stages of a Eurocode 7 design, I have arranged the sequence of chapters in the book appropriately: Chapter 6 describes the design methods (aligning mainly to Eurocode 7 Part 1) and Chapter 7 describes the ground investigation aspects (aligning to Part 2). The new Part 3 of the second generation of the code is then introduced in the appropriate later chapters in the book (e.g. retaining structures, shallow and deep foundations).

In recognition of the growing coverage of geomechanical modelling in university degrees, I have introduced a brand-new chapter on constitutive modelling in geomechanics (Chapter 16). The content of this chapter was provided by my friend and colleague Dr Rodaina Aboul Hosn from the *École Spéciale des Travaux Publics*, Paris. What Rodaina does not know about constitutive modelling is not worth knowing, and I owe her an immense amount of thanks for her excellent work. I have embedded Rodaina's work into my style of writing so that the chapter sits neatly alongside all the other chapters.

In addition to the above new content, I have also updated all sections on laboratory and field testing to align the descriptions of the procedures with the new international standards for these tests (Chapters 1, 4 and 7). Another update introduced in this edition is highway pavement foundation design (Chapter 15) – brought about as a result of revisions to the UK Design Manual for Roads and Bridges (DMRB). The early chapters of the book continue to cover the fundamentals of the behaviour of soils. To ease understanding of critical state soil mechanics, I have rearranged and expanded this subject and this is now in a new chapter alongside description of stress paths (Chapter 5).

As with previous editions, I have provided many worked examples throughout the book that illustrate the principles of soil mechanics and the geotechnical design processes. To help the reader further, I have produced a suite of spreadsheets and documents to accompany the book that match up against many of the worked examples. These can be used to better understand the analysis being adopted in the examples. In addition, I have produced the solutions to the exercises at the end of the chapters, a suite of video animations of lab tests and geotechnical processes, and various other teaching resources to accompany the book. All of these files can be freely downloaded from the companion website.

The content of the book aligns to the subjects typically covered in all university geotechnics courses. Teachers should find the content and teaching resources helpful for the own needs, and students will find that the book covers all courses in all years of their degree – all written in an easy-to-follow style.

In addition to Dr Aboul Hosn, I must also express my thanks to Dr Andrew Bond (past Chair TC250/SC7 – Eurocode 7) and to Dr Daniel Barreto (Edinburgh Napier University) for answering the various questions I posed to them as I wrote this edition.

Professor Ian Smith
Edinburgh, January 2021

About the Author

Ian Smith is a freelance Geotechnical and Educational consultant, and Professor of Geotechnical Engineering and Leader of the Built Environment Education at Heriot Watt University, Edinburgh. He has taught Geotechnical Engineering for 25 years at various universities across the globe, having spent some years beforehand working in the site investigation industry. He was Head, then Dean, of the School of Engineering and the Built Environment at Edinburgh Napier University before leaving to set up his own consultancy in 2017. He is an authority on the use of Eurocode 7 in geotechnical design, and has instructed designers and academics in the use of the code throughout the UK, Europe and in China. He is also Visiting Professor at three universities in China and is regularly invited to teach geotechnical engineering at universities across Europe and Asia.

Notation Index

The following is a list of the more important symbols used in the text.

Notation specific to the second generation of Eurocode 7 are indicated where appropriate.

A	Area, pore pressure coefficient
A'	Effective foundation area
A_b	Area of base of pile
A_s	Area of surface of embedded length of pile shaft
B	Width, diameter, pore pressure coefficient, foundation width
B'	Effective foundation width
C	Cohesive force, constant
C_a	Area ratio
C_C	Compression index, soil compressibility, coefficient of curvature
$C_{d,SLS}$	Eurocode 7 serviceability limit state limiting design value
C_N	SPT overburden correction factor
C_r	Static cone resistance
C_s	Constant of compressibility
C_u	Uniformity coefficient
C_v	Void fluid compressibility
C_w	Adhesive force
D	Diameter, depth factor, foundation depth, embedded length of pile
D_w	Depth of groundwater table
D_r	Relative density
D_1, D_2	Cutting shoe diameters
D_{10}, D_{30}, D_{60}	Effective particle sizes (10, 30, 60%)
E	Modulus of elasticity, efficiency of pile group
E'_0	One-dimensional modulus of elasticity
E_d	Eurocode 7 design value of effect of actions
$E_{dst;d}$	Eurocode 7 design value of effect of destabilising actions
$E_{stb;d}$	Eurocode 7 design value of effect of stabilising actions
E_M	Pressuremeter modulus
E_m	Eurocode 7 design value of modulus of elasticity
E_r	SPT energy ratio
F	Factor of safety
F_b	Factor of safety on pile base resistance
$F_{c;d}$	Eurocode 7 design axial compression load on a pile
F_d	Eurocode 7 design value of an action
F_{rep}	Eurocode 7 representative value of an action
F_s	Factor of safety on pile shaft resistance

$G;\ G'$	Shear modulus; effective shear modulus
$G_k;\ G_d$	Eurocode 7 (2nd G) characteristic permanent action; design permanent action
$G_{dst;d}$	Eurocode 7 design value of destabilising permanent vertical action (uplift)
G_s	Particle specific gravity
$G_{stb;d}$	Eurocode 7 design value of stabilising permanent vertical action (uplift)
$G'_{stb;d}$	Eurocode 7 design value of stabilising permanent vertical action (heave)
G_{wk}	Eurocode 7 (2nd G) characteristic value of G_w
$G_{w,rep}$	Eurocode 7 (2nd G) representative value of G_w
$G_{wk;inf}$	Eurocode 7 (2nd G) inferior (lower) characteristic value of G_w
$G_{wk;sup}$	Eurocode 7 (2nd G) superior (upper) characteristic value of G_w
H	Thickness, height, horizontal load
I	Index, moment of inertia
I_D	Density index
I_L	Liquidity index
I_P	Plasticity index, immediate settlement coefficient
I_σ	Vertical stress influence factor
K	Boussinesq Influence factor, ratio of σ_3/σ_1, bulk/volumetric modulus
K'	Effective bulk/volumetric modulus
K_a	Coefficient of active earth pressure
K_F	Eurocode 7 (2nd G) action consequence factor
K_M	Eurocode 7 (2nd G) material consequence factor
K_0	Coefficient of earth pressure at rest
K_p	Coefficient of passive earth pressure
K'_{ps}	Effective plane strain bulk/volumetric modulus
K_s	Pile constant
L	Length
L'	Effective foundation length
M	Moment, slope projection of critical state line, mass, mobilisation factor
M_s	Mass of solids
M_w	Mass of water
N	Number, stability number, uncorrected blow count in SPT
N_{60}	Number of blows from the SPT corrected to energy losses
$(N_1)_{60}$	Number of blows from the SPT corrected to energy losses, rod length and normalised for effective vertical overburden stress
$N_c,\ N_q,\ N_\gamma$	Bearing capacity coefficients
N_p	Immediate settlement coefficient
P	Force
P_a	Thrust due to active earth pressure
P_p	Thrust due to passive earth pressure
P_w	Thrust due to water or seepage forces
Q	Total quantity of flow in time t, applied surface line load
Q_b	Ultimate soil strength at pile base
$Q_k;\ Q_d$	Eurocode 7 (2nd G) characteristic variable action; design variable action
Q_s	Ultimate soil strength around pile shaft
Q_u	Ultimate load carrying capacity of pile
R	Radius, reaction
$R_{b;cal}$	Eurocode 7 calculated value of pile base resistance
$R_{b;k}$	Eurocode 7 characteristic value of pile base resistance
$R_{b,rep}$	Eurocode 7 (2nd G) representative base resistance

R_c	Eurocode 7 compressive resistance of ground against a pile at ultimate limit state
$R_{c;cal}$	Eurocode 7 calculated value of R_c
$R_{c;d}$	Eurocode 7 design value of R_c
R_{cd}	Eurocode 7 (2nd G) design value of R_c
$R_{c;k}$	Eurocode 7 characteristic value of R_c
$R_{c;m}$	Eurocode 7 measured value of R_c
$R_{c,rep}$	Eurocode 7 (2nd G) representative total pile resistance
R_d	Eurocode 7 design resisting force
R_o	Overconsolidation ratio (one-dimensional)
R_p	Overconsolidation ratio (isotropic)
$R_{s;cal}$	Eurocode 7 calculated value of pile shaft resistance
$R_{s;k}$	Eurocode 7 characteristic value of pile shaft resistance
$R_{s,rep}$	Eurocode 7 (2nd G) representative shaft resistance
S	Vane shear strength
$S_{dst;d}$	Eurocode 7 design value of destabilising seepage force
S_r	Degree of saturation
S_t	Sensitivity
T	Time factor, tangential force, surface tension, vane torque
T_d	Eurocode 7 design value of total shearing resistance around structure
U	Average degree of consolidation
U_z	Degree of consolidation at a point at depth z
V	Volume, vertical load
V_a	Volume of air, percentage air voids
$V_{dst;d}$	Eurocode 7 design value of destabilising vertical action on a structure
V_s	Volume of solids
V_v	Volume of voids
V_w	Volume of water
W	Weight
W_s	Weight of solids
W_w	Weight of water
X_d	Eurocode 7 design value of a material property
X_k	Eurocode 7 representative value of a material property
	Eurocode 7 (2nd G) characteristic value of a material property
X_{rep}	Eurocode 7 (2nd G) representative value of a material property
Z	Section modulus
a	Area, wall adhesion, intercept of MCV calibration line with w-axis
b	Width, slope of MCV calibration line
c	Unit cohesion
c'	Unit cohesion with respect to effective stresses
c_b	Undisturbed soil shear strength at pile base
c'_d	Eurocode 7 design value of effective cohesion
c_r	Residual value of cohesion
c_u	Undrained unit cohesion
$\overline{c_u}$	Average undrained shear strength of soil
$c_{u;d}$	Eurocode 7 design value of undrained shear strength
c_v	Coefficient of consolidation
c_w	Unit cohesion between wall and soil
d	Pile penetration, pile diameter, particle size
d_c, d_q, d_γ	Depth factors

x	Horizontal distance
y	Vertical, or horizontal, distance
z	Vertical distance, depth
z_a	Depth of investigation points
z_o	Depth of tension crack
z_w	Depth below water table
α	Angle, pile adhesion factor, slope of K_f line
β	Slope angle
Γ	Eurocode 7 over-design factor, specific volume at $\ln p' = 0$
γ	Unit weight (weight density), shear strain
γ'	Submerged, buoyant or effective unit weight (effective weight density)
$\gamma_{A;dst}$	Eurocode 7 partial factor: accidental action – unfavourable
γ_b	Bulk unit weight (bulk weight density), Eurocode 7 partial factor: pile base resistance
$\gamma_{c'}$	Eurocode 7 partial factor: effective cohesion
γ_{cu}	Eurocode 7 partial factor: undrained shear strength
γ_d	Dry unit weight (dry weight density)
γ_E	Eurocode 7 partial factor for effects of actions, Eurocode 7 (2nd G) partial factor for unfavourable effects of actions
$\gamma_{E,fav}$	Eurocode 7 (2nd G) partial factor for favourable effects of actions
γ_F	Eurocode 7 partial factor for an action
γ_G	Eurocode 7 (2nd G) partial factor for permanent unfavourable/destabilising action
$\gamma_{G;dst}$	Eurocode 7 partial factor: permanent action – unfavourable
$\gamma_{G,fav}$	Eurocode 7 (2nd G) partial factor for permanent favourable action
$\gamma_{G;stb}$	Eurocode 7 partial factor: permanent action – favourable
$\gamma_{G,w}$	Eurocode 7 (2nd G) partial factor for unfavourable/destabilising water action
$\gamma_{G,w,stb}$	Eurocode 7 (2nd G) partial factor for stabilising water action
γ_M	Eurocode 7 partial factor for a soil parameter
γ_Q	Eurocode 7 (2nd G) partial factor for variable (unfavourable) action
$\gamma_{Q;dst}$	Eurocode 7 partial factor: variable action – unfavourable
$\gamma_{Q,w}$	Eurocode 7 (2nd G) partial factor for unfavourable water action
γ_{qu}	Eurocode 7 partial factor: unconfined compressive strength
γ_R	Eurocode 7 partial factor for a resistance
γ_{Rb}	Eurocode 7 (2nd G) pile design base resistance factor
γ_{Rc}	Eurocode 7 (2nd G) pile design total resistance factor
γ_{Rd}	Eurocode 7 (2nd G) pile design model factor
γ_{Re}	Eurocode 7 partial factor: earth resistance
γ_{Rh}	Eurocode 7 partial factor: sliding resistance
γ_{Rs}	Eurocode 7 (2nd G) pile design shaft resistance factor
γ_{Rv}	Eurocode 7 partial factor: bearing resistance
γ_s	Eurocode 7 partial factor: pile shaft resistance
γ_{sat}	Saturated unit weight (saturated weight density)
γ_t	Eurocode 7 partial factor: pile total resistance
$\gamma_{\tan\delta}$	Eurocode 7 (2nd G) partial factor: ground/structure interface friction
$\gamma_{\tan\phi}$	Eurocode 7 (2nd G) partial factor: angle of shearing resistance
γ_{ts}	Eurocode 7 (2nd G) partial factor: effective shear strength
γ_w	Unit weight of water (weight density of water)
γ_γ	Eurocode 7 partial factor: weight density
$\gamma_{\phi'}$	Eurocode 7 partial factor: angle of shearing resistance
δ	Ground–structure interface friction angle

ε	Strain
η	Dynamic viscosity of water, conversion factor
θ	Angle of failure plane to major principal plane, angle subtended at centre of slip circle
κ	Slope of swelling line
λ	Slope of normal consolidation line, SPT rod length correction factor
μ	Settlement coefficient, one micron
ν	Poisson's ratio
ξ_m	Eurocode 7 (2nd G) pile design correlation factor
ξ_1, ξ_2	Eurocode 7 correlation factors to evaluate results of static pile load tests
ξ_3, ξ_4	Eurocode 7 correlation factors to derive pile resistance from ground investigation results
ρ	Density, settlement
ρ'	Submerged, buoyant or effective density
ρ_b	Bulk density
ρ_c	Consolidation settlement
ρ_d	Dry density
ρ_i	Immediate settlement
ρ_s	Particle density
ρ_{sat}	Saturated density
ρ_w	Density of water
σ	Total normal stress
σ'	Effective normal stress
σ'_a	Total, effective axial stress
σ'_e	Equivalent consolidation pressure (one-dimensional)
σ_r, σ'_r	Total, effective radial stress
$\sigma_{stb;d}$	Eurocode 7 design value of stabilising total vertical stress
σ'_v	Effective overburden pressure
$\overline{\sigma'_v}$	Average effective overburden pressure
$\sigma_1, \sigma_2, \sigma_3$	Total major, intermediate and minor stress
$\sigma'_1, \sigma'_2, \sigma'_3$	Effective major, intermediate and minor stress
τ	Shear stress
ϕ'	Angle of shearing resistance with respect to effective stresses
ϕ_{cv}	Critical state, or constant volume, angle of shearing resistance
$\phi_{cv;d}$	Design value of critical state angle of shearing resistance
ϕ'_d	Design value of ϕ'
ϕ_u	Angle of shearing resistance with respect to total stresses (=0)
ψ	Angle of back of wall to horizontal, Eurocode 7 correlation factor

About the Companion Website

The book's companion website www.wiley.com/go/smith/soilmechanics10e provides you with resources and downloads to further your understanding of the fundamentals of soil mechanics and the use of Eurocode 7:

- A suite of editable spreadsheets which map onto the worked examples in the book, showing how they are solved.
- Solutions to the end-of-chapter exercises, including the full workings and accompanying spreadsheets.
- Convenient tables with useful data and formulae.
- Animations to demonstrate some of the more complex laboratory testing and geotechnical procedures.

Part I
Fundamentals of Soil Mechanics

Part 1

Fundamentals of Soil Mechanics

Chapter 1

Classification and Physical Properties of Soils

Learning objectives:

By the end of this chapter, you will have been introduced to:
- the formation of rocks and soils;
- clay soils and the field identification of soils;
- the classification of soils;
- various physical properties of soils and the relationships between them.

In the field of civil engineering, nearly all projects are built on to, or into, the ground. Whether the project is a structure, a roadway, a tunnel, or a bridge, the nature of the soil at that location is of great importance to the civil engineer. *Geotechnical engineering* is the term given to the branch of engineering that is concerned with aspects pertaining to the ground. *Soil mechanics* is the subject within this branch that looks at the behaviour of soils in civil engineering.

Geotechnical engineers are not the only professionals interested in the ground; soil physicists, agricultural engineers, farmers and gardeners all take an interest in the types of soil with which they are working. These workers, however, concern themselves mostly with the organic topsoils found at the soil surface. In contrast, geotechnical engineers are mainly interested in the engineering soils found beneath the topsoil. It is the engineering properties and behaviour of these soils, which are their concern.

1.1 Agricultural and engineering soil

If an excavation is made through previously undisturbed ground the following materials are usually encountered (Fig. 1.1).

Topsoil

A layer of organic soil, usually not more than 500 mm thick, in which humus (highly organic partly decomposed vegetable matter) is often found.

Smith's Elements of Soil Mechanics, 10th Edition. Ian Smith.
© 2021 John Wiley & Sons Ltd. Published 2021 by John Wiley & Sons Ltd.
Companion website: www.wiley.com/go/smith/soilmechanics10e

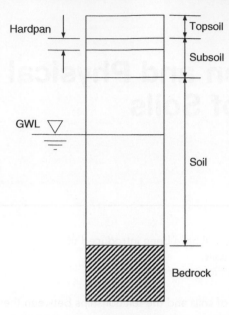

Fig. 1.1   Materials encountered during excavation.

Subsoil

The portion of the Earth's crust affected by current weathering and lying between the topsoil and the unweathered soil below.

Hardpan

In humid climates, humic acid can be formed by rainwater causing decomposition of humus. This acid leaches out iron and alumina oxides down into the lower layers where they act as cementation agents to form a hard, rock-like material. Hardpan is difficult to excavate and, as it does not soften when wet, has a high resistance to normal soil drilling methods. A hardpan layer is sometimes found at the junction of the topsoil and the subsoil.

Soil

The soft geological deposits extending from the subsoil to bedrock constitute soils. In some soils, there is a certain amount of cementation between the grains, which affects the physical properties of the soil. If this cementation is such that a rock-hard material has been produced, then the material must be described as rock. A rough rule is that if the material can be excavated by hand or hand tools, then it is a soil.

Bedrock

Beneath the soil, rock is encountered. This rock is often referred to as *bedrock* and the horizon at which the soil meets the rock is known as the *rockhead*.

Groundwater

A reservoir of underground water. The upper surface of this water may occur at any depth and is known as the *water table* or *groundwater level* (GWL).

1.2   The rock cycle

Rocks and soils are formed within a geological process known as the *rock cycle* (Fig. 1.2). The process is continuous and has lasted for millions of years.

1. Magma (molten rock) rises towards the surface, cooling and solidifying along the way.
2. The magma crystallises beneath or above the Earth's surface forming igneous rocks.
3. At the surface, rocks undergo physical and chemical weathering which break down the parent rock into particles.
4. The rock particles (sediments) are moved downslope and transported by glaciers, rivers and wind. The combined processes of weathering and transportation are referred to as *erosion*.
5. Eventually the sediments are deposited in oceans and floodplains, where they undergo *lithification*: the process where the particles are compressed under great pressure over time to form rock.
6. Sedimentary rocks are formed.
7. If the sedimentary rock is subjected to further great pressures and heat, it will react and change to a metamorphic rock.
8. Metamorphic rocks subjected to very high pressure or temperature, liquefy into magma that eventually crystallises into igneous rock, and the cycle starts again.

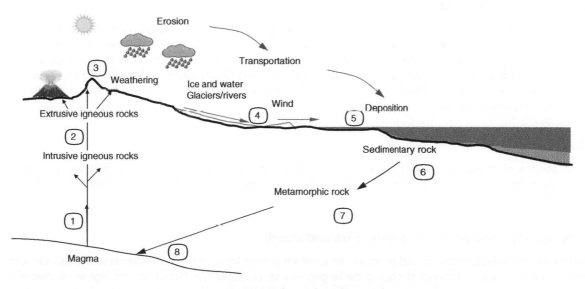

Fig. 1.2   The rock cycle.

1.2.1 Rock

Rocks are made from various types of minerals. Minerals are substances of crystalline form made up from a particular chemical combination. The main minerals found in rocks include quartz, feldspar, calcite and mica. We can classify all rocks into three basic groups: *igneous*, *sedimentary* and *metamorphic*. The position of each within the rock cycle is shown in Fig. 1.2.

Igneous rocks

These rocks have become solid from a melted liquid state. *Extrusive* igneous rocks are those that arrived on the surface of the Earth as molten lava and cooled. *Intrusive* igneous rocks are formed from magma that forced itself through cracks into the rock beds below the surface and solidified there.

Examples of igneous rocks: *granite, basalt, gabbro.*

Sedimentary rocks

Weathering reduces the rock mass into fragmented particles, which can be more easily transported by wind, water and ice. When dropped by the agents of weathering, they are termed *sediments*. These sediments are typically deposited in layers or beds called strata and when compacted and cemented together through lithification, they form sedimentary rocks.

Examples of sedimentary rocks: *shale, sandstone, chalk.*

Metamorphic rocks

Metamorphism through high temperatures and pressures acting on sedimentary or igneous rocks produces metamorphic rocks. The original rock undergoes both chemical and physical alterations.

Examples of metamorphic rocks: *slate, quartzite, marble.*

Identification of rocks

The identification of rocks may, initially, be considered quite a tricky thing to get right. With practice and experience however, engineers and geologists can rapidly identify features, that enable the identification to be made swiftly. Features, which assist in the identification and description of the rock type, include colour, grain size, mineralogical composition, structure and void content. Guidance on the identification and description of rock types is given in BS EN ISO 14689 (BSI, 2018b) and in the *Code of practice for ground investigations*, BS 5930 (BSI, 2015).

1.2.2 Soil

The actions of frost, temperature, gravity, wind, rain and chemical weathering are continually forming rock particles that eventually become soils. There are three types of soil when considering modes of formation.

Transported soil (gravels, sands, silts and clays)

Many soils have been transported by water. As a stream or river loses its velocity, it tends to deposit some of the particles that it is carrying, dropping the larger, heavier particles first. Hence, on the higher reaches of a river, gravel and sand are found whilst on the lower or older parts, silts and clays predominate, especially where the river enters the sea or a lake and loses its velocity. Ice, in the form of huge slow-moving yet enormously powerful glaciers, is another important transportation agent, and large deposits of *boulder clay* and *moraine* are formed from *glaciation*.

In arid parts of the world, wind is continually forming sand deposits in the form of ridges. The sand particles in these ridges have been more or less rolled along and are invariably rounded and fairly uniform in size. Light brown, wind-blown deposits of silt-size particles, known as *loess*, are often encountered in thin layers, the particles having sometimes travelled considerable distances.

Residual soil (topsoil, laterites)

These soils are formed *in situ* by chemical weathering and may be found on level rock surfaces where the action of the elements has produced a soil with little tendency to move. Residual soils can also occur whenever the rate of break-up of the rock exceeds the rate of removal. If the parent rock is igneous or metamorphic the resulting soil sizes range from silt to gravel.

Laterites are formed by chemical weathering under warm, humid tropical conditions when the rainwater leaches out of the soluble rock material leaving behind the insoluble hydroxides of iron and aluminium, giving them their characteristic red-brown colour.

Organic soil

These soils contain large amounts of decomposed animal and vegetable matter. They are usually dark in colour and give off a distinctive odour. Deposits of organic silts and clays have usually been created from river or lake sediments. *Peat* is a special form of organic soil and is a dark brown spongy material, which almost entirely consists of lightly to fully decomposed vegetable matter. It exists in one of three forms:

- *Fibrous*: non-plastic with a firm structure only slightly altered by decay.
- *Pseudo-fibrous*: peat in this form still has a fibrous appearance but is much softer and more plastic than fibrous peat. The change is due more to prolonged submergence in airless water than to decomposition.
- *Amorphous*: with this type of peat, decomposition has destroyed the original fibrous vegetable structure so that it has virtually become an organic clay.

Peat deposits occur extensively throughout the world and can be extremely troublesome when encountered in civil engineering work.

1.2.3   Granular and cohesive soils

Geotechnical engineers classify soils as either *granular* or *cohesive*. Granular soils (sometimes referred to as *cohesionless* soils) are formed from loose particles without strong inter-particle forces, e.g. sands and gravels. Cohesive soils (e.g. clays, clayey silts) are made from particles bound together with clay minerals. The particles are flaky and sheet-like and retain a significant amount of adsorbed water on their surfaces. The ability of the sheet-like particles to slide relative to one another, gives a cohesive soil the property known as *plasticity*.

1.3   Clay soils

Rock fragments can be reduced by mechanical means to a limiting size of about 0.002 mm, so that a soil containing particles above this size has a mineral content similar to the parent rock from which it was created.

For the production of particles smaller than 0.002 mm, some form of chemical action is generally necessary before breakdown can be achieved. Such particles, although having a chemical content similar to the parent rock, have a different crystalline structure and are known as clay particles. An exception is rock flour, rock grains smaller than 0.002 mm, produced by the glacial action of rocks grinding against each other.

1.3.1   Classes of clay minerals

The minerals constituting a clay are invariably the result of the chemical weathering of rock particles and are hydrates of aluminium, iron or magnesium silicate generally combined in such a manner, as to create sheet-like structures only a few molecules thick. These sheets are built from two basic units, the tetrahedral unit of silica and the octahedral unit of the hydroxide of aluminium, iron or magnesium. The main dimension of a clay particle is usually less than 0.002 mm and the different types of minerals have been created from the manner in which these structures were stacked together.

The three main groups of clay minerals are as follows.

Kaolinite group

This mineral is the most dominant part of residual clay deposits and is made up from large stacks of alternating single tetrahedral sheets of silicate and octahedral sheets of aluminium. Kaolinites are very stable with a strong structure and absorb little water. They have low swelling and shrinkage responses to water content variation.

Illite group

Consists of a series of single octahedral sheets of aluminium sandwiched between two tetrahedral sheets of silicon. In the octahedral sheets, some of the aluminium is replaced by iron and magnesium; and in the tetrahedral sheets, there is a partial replacement of silicon by aluminium. Illites tend to absorb more water than kaolinites and have higher swelling and shrinkage characteristics.

Montmorillonite group

This mineral has a similar structure to the illite group but, in the tetrahedral sheets, some of the silicon is replaced by iron, magnesium and aluminium. Montmorillonites exhibit extremely high water absorption, swelling and shrinkage characteristics. Bentonite is a member of this mineral group and is usually formed from weathered volcanic ash. Because of its large expansive properties when it is mixed with water it is much in demand as a general grout in the plugging of leaks in reservoirs and tunnels. It is also used as a drilling mud for soil borings.

Readers interested in this subject of clay mineralogy are referred to the publication by Murray (2006).

1.3.2   Structure of a clay deposit

Macrostructure

The visible features of a clay deposit collectively form its macrostructure and include such features as fissures, root holes, bedding patterns, silt and sand seams or lenses and other discontinuities.

A study of the macrostructure is important as it usually has an effect on the behaviour of the soil mass. For example, the strength of an unfissured clay mass is much stronger than along a crack.

Microstructure

The structural arrangement of microscopic sized clay particles, or groups of particles, defines the microstructure of a clay deposit. Clay deposits have been laid down under water and were created by the settlement and deposition of clay particles out of suspension. Often during their deposition, the action of Van der Waals forces attracted clay particles together and created flocculant, or honeycombed, structures, which, although still microscopic, are of considerably greater volume than single clay particles. Such groups of clay particles are referred to as *clay flocs*.

1.4   Field identification of soil

Gravels, sands and peats are easily recognisable, but difficulty arises in deciding when a soil is a fine sand or a coarse silt or when it is a fine silt or a clay. The following rules may, however, help:

Fine sand	Silt	Clay
Individual particles visible	Some particles visible	No particles visible
Exhibits dilatancy	Exhibits dilatancy	No dilatancy
Easy to crumble and falls off hands when dry	Easy to crumble and can be dusted off hands when dry	Hard to crumble and sticks to hands when dry
Feels gritty	Feels rough	Feels smooth
No plasticity	Some plasticity	Exhibits plasticity

The dilatancy test involves moulding a small amount of soil in the palm of the hand; if water is seen to recede when the soil is pressed, then it is either a sand or a silt.

Organic silts and clays are invariably dark grey to blue-black in colour and give off a characteristic odour, particularly with fresh samples.

The condition of a clay very much depends upon its degree of *consolidation*. At one extreme, a soft normally consolidated clay can be moulded by the fingers whereas, at the other extreme, a hard overconsolidated clay cannot. Overconsolidation is defined in Section 4.4 and the subject of degree of consolidation is covered in Chapter 13. Guidance on the identification of soils may be found in the international standard for *Geotechnical investigation and testing*, BS EN ISO 14688-1 (BSI, 2018a) and in the *Code of practice for ground investigations*, BS 5930 (BSI, 2015).

Common types of soil

In the field, soils are usually found in the form of a mixture of components, e.g. silty clay, sandy silt, etc. Local names are sometimes used for soil types that occur within a particular region. e.g. London clay.

Boulder clay, also referred to as *glacial till*, is an unstratified and irregular mixture of boulders, cobbles, gravel, sand, silt and clay of glacial origin. In spite of its name, boulder clay is not a pure clay and contains more granular material than clay particles.

Moraines are gravel and sand deposits of glacial origin. Loam is a soft deposit consisting of a mixture of sand, silt and clay in approximately equal quantities.

Fill is soil excavated from a 'borrow' area, which is used for filling hollows or for the construction of earth-fill structures, such as dams or embankments. Fill will sometimes contain man-made materials such as crushed concrete or bricks from demolished buildings.

Soil classification (see Section 1.6) enables the engineer to assign a soil to one of a limited number of groups, based on the properties and characteristics of the soil. The classification groups are then used as a system of reference for soils. Soils can be classified in the field or in the laboratory. Field techniques are usually based upon visual recognition as described above. Laboratory techniques involve several specialised tests and, in Europe, these are described in different parts of BS EN ISO 17892: *Geotechnical investigation and testing – Laboratory testing of soil* (BSI, 2014–2019). In much of North America and some other nations, specification D2487-17, published by ASTM International, is used. The procedures are largely the same across all the testing standards.

1.5   Soil classification laboratory testing

1.5.1   Drying soils

Soils can be either oven or air dried. It is standard practice to oven dry soils at a temperature of 105 °C but it should be remembered that some soils can be damaged by such a temperature. Oven drying is necessary for water content, sieve analysis and a few other tests, but air drying can be used whenever a test does not require a fully dry sample, e.g. compaction tests, described in Chapter 15.

1.5.2   Determination of water content, w

The most common way of expressing the amount of water present in a soil is the *water content*. The water content, also called the *moisture content*, is given the symbol w and is the ratio of the amount of water to the amount of dry soil.

$$w = \frac{\text{Weight of water}}{\text{Weight of solids}} = \frac{W_w}{W_s} \text{ or } w = \frac{\text{Mass of water}}{\text{Mass of solids}} = \frac{M_w}{M_s} \tag{1.1}$$

w is usually expressed as a percentage and should be quoted to two significant figures.

Example 1.1: Water content determination

A sample of soil was placed in a water content tin of mass 19.52 g. The combined mass of the soil and the tin was 48.27 g. After oven drying the soil and the tin had a mass of 42.31 g.

Determine the water content of the soil.

Solution:

$$w = \frac{M_w}{M_s} = \frac{48.27 - 42.31}{42.31 - 19.52} = \frac{5.96}{22.79} = 0.262 = 26\%$$

1.5.3   Granular soils – particle size distribution

A standardised system helps to ensure consistency between engineers in the classification of granular soils. The usual method is based on the determination of the *particle size distribution* (PSD) by shaking an oven dried sample of the soil (usually after washing the sample over a 63 μm sieve) through a set of sieves. The aperture size of each succeeding sieve is smaller than the one above. By weighing the mass of soil retained on each sieve, we can obtain the particle size distribution for the soil.

The particle size scale is based on the limits for each fraction listed in Table 1.1.

Table 1.1   Particle size limits for different soil fractions.

Particle size fractions and symbol	Upper size (mm)	Lower size (mm)
Gravel, Gr	60	2
Sand, Sa	2	0.06
Silt, Si	0.06	0.002
Clay, Cl	0.002	–

These fractions can be subdivided on the basis of the 6-2, 6-2 pattern:

e.g. gravel:

coarse, c: 60–20 mm
medium, m: 20–6 mm
fine, f: 6–2 mm.

Similar subdivisions can be made for sand and silt.

The symbols are combined to aid classification, e.g.

cGr – coarse gravel; fSa – fine sand; mSi – medium silt.

The particle size fractions listed above refer to the classification system adopted in Europe, see also BS EN ISO 14688-1:2018 (BSI, 2018a). ASTM International standard D2487-17 (2017) offers an alternative system (*United Soil Classification System, USCS*) which adopts different sizes as the boundaries between the particle size fractions.

The results of the sieve analysis are plotted with the particle sizes horizontal and the summation percentages vertical. As soil particles vary in size from molecular to boulder it is necessary to use a log scale for the horizontal plot so that the full range can be shown on the one sheet.

The smallest aperture generally used in soils work is that of the 0.063 mm size sieve. Below this size (i.e. silt and clay sizes) the distribution curve must be obtained by sedimentation (following either a pipette or hydrometer method of analysis). Unless a centrifuge is used, it is not possible to determine the range of clay sizes in a soil, and usually it is adequate to just obtain the total percentage of clay sizes present. The procedures for sieve, pipette and hydrometer analyses are given in BS EN ISO 17892-4:2016 (BSI, 2016).

Examples of particle size distribution, or *grading*, curves for different soil types are shown in Fig. 1.10. From these grading curves it is possible to determine for each soil, the total percentage of a particular size, and the percentage of particle sizes larger or smaller than any particular particle size.

The effective size of a distribution, D_{10}

An important particle size within a soil distribution is the *effective size*, which is the largest size of the smallest 10%. It is given the symbol D_{10}. Other particle sizes, such as D_{30} and D_{60}, are defined in the same manner.

Grading of a distribution

For a granular soil, the shape of its grading curve indicates the distribution of the soil particles within it.

If the shape of the curve is not too steep and is more or less constant over the full range of the soil's particle sizes, then the particle size distribution extends evenly over the range of the particle sizes within the soil and there is no deficiency or excess of any particular particle size. Such a soil is said to be *well graded*.

If the soil has any other form of distribution curve, then it is said to be *poorly graded*. According to their distribution curves, there are two types of poorly graded soil:

- if the major part of the curve is steep, then the soil has a particle size distribution extending over a limited range with most particles tending to be about the same size. The soil is said to be *closely graded* or, more commonly, *uniformly graded*;
- if a soil has large percentages of its bigger and smaller particles and only a small percentage of the intermediate sizes, then its grading curve will exhibit a significantly flat section or plateau. Such a soil is said to be *gap graded*.

The uniformity coefficient C_u and the coefficient of curvature, C_c

The grading of a soil is best determined by direct observation of its particle size distribution curve. This can be difficult for those studying the subject for the first time, but some guidance can be obtained by the use of a grading parameter known as the *uniformity coefficient* or by the *coefficient of curvature*.

$$C_u = \frac{D_{60}}{D_{10}} \tag{1.2}$$

$$C_c = \frac{(D_{30})^2}{(D_{60} \times D_{10})} \tag{1.3}$$

The shape of the grading of the soil can then be established through the guidance given in Table 1.2. Other statistical data can also be obtained from the psd;

e.g. median – size at which 50% of sample is finer, D_{50}.

Table 1.2 Shape of grading curve.

Term	C_u	C_c
Uniformly graded	<3	<1
Poorly graded	3–6	<1
Medium graded	6–15	<1
Well graded	>15	1–3
Gap graded	>15	<0.5

Example 1.2: Particle size distribution

The results of a sieve analysis on a soil sample were:

Sieve size (mm)	Mass retained (g)
10	0
6.3	5.5
2	25.7
1	23.1
0.600	22.0
0.300	17.3
0.150	12.7
0.063	6.9

2.3 g passed through the 63 µm sieve.

Plot the particle size distribution curve and determine the uniformity coefficient of the soil.

Solution:

The aim is to determine the percentage of soil (by mass) passing through each sieve. To do this, the percentage retained on each sieve is determined and subtracted from the percentage passing through the previous sieve. This gives the percentage passing through the current sieve.

Calculations may be set out as follows:

Sieve size (mm)	Mass retained (g)	Percentage retained	Percentage passing
10	0	0	100
6.3	5.5	5	95
2	25.7	22	73
1	23.1	20	53
0.600	22.0	19	34
0.300	17.3	15	19
0.150	12.7	11	8
0.063	6.9	6	2
Pass 0.063	2.3	2	
	Σ115.5 g		

e.g. 2 mm sieve:

$$\text{Percentage retained} = \frac{25.7}{115.5} \times 100 = 22\%$$

Percentage passing $= 95 - 22 = 73\%$

The particle size distribution curve is shown in Fig. 1.3. Using the vertical axis, we can easily see that the soil has approximate proportions of 30% gravel and 70% sand.

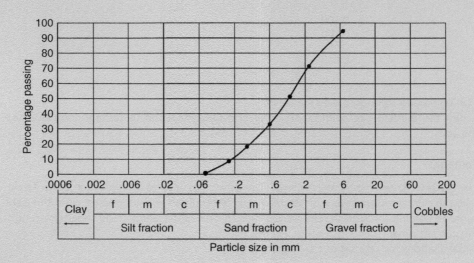

Fig. 1.3   Example 1.2.

$$D_{10} = 0.18 \text{ mm}; \quad D_{60} = 1.5 \text{ mm}; \quad C_u = \frac{D_{60}}{D_{10}} = \frac{1.5}{0.18} = 8.3$$

It is also seen that the grading curve has a regular slope and therefore contains roughly equal percentages of particle sizes. The soil is a medium graded, gravelly SAND.

1.5.4   Sedimentation analysis

The fraction of soil smaller than 0.06 mm cannot be separated by sieves, so a *sedimentation analysis* is used to establish the proportions of silt and clay fractions. The procedure is only considered necessary if more than 10% of the soil passes the 63 μm sieve. In the test, a sample of the dry soil passing the 63 μm sieve is placed into suspension with water in a sedimentation cylinder and allowed to settle over a period of time. Measurements of the percentage of particles remaining in suspension at set time intervals are established by either using a pipette or a hydrometer. Details of the test procedures are given in BE EN ISO 17892-4:2016 and Head (2006). The pipette procedure is described below.

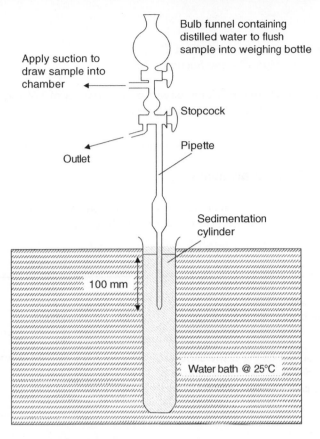

Fig. 1.4   Pipette analysis arrangement.

The dry soil of known mass is placed into a sedimentation cylinder of volume 500 ml and distilled water, containing a small amount of dispersing agent, is added. A dispersing procedure (e.g. end over end shaking) is adopted to separate all the particles in the suspension. The test begins by placing the upright cylinder into a water bath at 25 °C and a stopwatch is started. Three separate sampling dips are made at a depth of 100 mm using the pipette (Fig. 1.4) at pre-determined times. Different particle sizes will be captured in each sample, since the larger particles fall to the bottom of the cylinder before the smaller ones.

The pipette has a standard volume of 10 ml. This equals the volume of the solution sampled. That sample is flushed from the pipette into a small glass weighing bottle. By weighing the bottle before sampling and again after oven drying, the mass of the sampled soil is determined. By scaling the mass retained in the pipette sample (10 ml), less the mass of the dispersant, to that of the cylinder (500 ml), the mass of soil in suspension is determined. That, together with the determination of the particle size at the time of sampling, allows us to establish the percentage of particles passing that size.

The particle size, d (mm), sampled at depth, h (mm), at time, t (min) is established by:

$$d = 0.005531 \sqrt{\frac{\eta \times h}{(\rho_s - 1) \times t}}$$

(1.4)

where

η is the dynamic viscosity of water (mPa s) (=0.89 at 25 °C)
ρ_s = particle density of the soil (Mg/m^3) – see Section 1.7.3.

If samples are taken at 4, 46 and 414 minutes for a soil with particle density of 2.65 Mg/m^3, the particle sizes in each sample will be 0.02, 0.006 and 0.002 mm respectively.

Example 1.3: Pipette analysis

The results of a standard pipette analysis, carried out on a sample of soil passing the 63 μm sieve of dry mass 27.25 g and particle density = 2.65 Mg/m^3, were:

	Sample 1	Sample 2	Sample 3
Time (min)	4	46	414
Mass of empty sample bottle (g)	5.1926	5.3710	5.2983
Mass of sample bottle plus dry soil (g)	5.6927	5.8052	5.6898

During a control test on the dispersant/water only solution (500 ml), it was established that the mass of dry dispersant in a sample of 10 ml was 0.0173 g.
   Determine the percentages of fine silt and clay in the soil.

Solution:

Standard test, therefore:

Volume of cylinder = 500 ml;
Volume of pipette = 10 ml;
Water bath temperature = 25 °C;
Sampling depth = 100 mm

	Sample 1	Sample 2	Sample 3
Particle size, d (mm)	0.02	0.006	0.002
Mass of soil in sample (g)	0.5001	0.4342	0.3915
Equivalent mass in cylinder (g)	$= (0.5001 - 0.0173) \times \left(\dfrac{500}{10}\right) = 24.15$	8.0	5.5
Percentage passing d (%)	$= \dfrac{24.15}{27.25} \times 100 = 88.6$	76.5	68.7

From the results it is seen that the percentage of fine silt (0.006–0.002 mm) is 76.5 – 68.7 = 7.8% and the percentage of clay (<0.002 mm) is 68.7%.

1.5.5 Cohesive soils – liquid and plastic limit tests

The results of the grading tests described above can only classify a soil with regard to its particle size distribution. They do not indicate whether the fine grained particles will exhibit the plasticity generally associated with fine grained soils. Hence, although a particle size analysis will completely define a gravel and a sand, it is necessary to carry out plasticity tests in order to fully classify a clay or a fine silt.

These tests were evolved by Atterberg (1911) and determine the various values of water content at which changes in a soil's strength characteristics occur. As an introduction to these tests, let us consider the volume of a soil as the amount of water within it is varied (Fig. 1.5).

As the water is added to the dry clay, there is no immediate increase in volume. However, as the amount of water is gradually increased, the volume increases too. Also, as the water content changes, the soil behaves differently: when dry, the soil is brittle but when extremely wet, the soil is a liquid (referred to as a *slurry*). There is a stage in between the brittle and liquid states and the soil in this range of water contents behaves as a plastic material.

The boundaries to the four states in which a soil may exist are defined:

liquid limit, w_L: the boundary between the liquid and the plastic state;
plastic limit, w_P: the boundary between the plastic and the semi-solid state;
shrinkage limit, w_S: the boundary between the semi-solid and the solid state.

The limits are defined as the water contents of the soil at each boundary, as indicated in Fig. 1.5.

Liquid limit (w_L) and plastic limit (w_P)

The water content at which the soil stops acting as a liquid and starts acting as a plastic solid is known as the *liquid limit* (w_L). If the soil is dried from the liquid limit, it passes through the plastic state. When plastic, the soil can be moulded in the fingers akin to working with modelling clay. But as further water is driven from the soil, the soil no longer behaves as a plastic material and instead acts as a brittle solid. The limit at which plastic behaviour changes to brittle failure is known as the *plastic limit* (w_P).

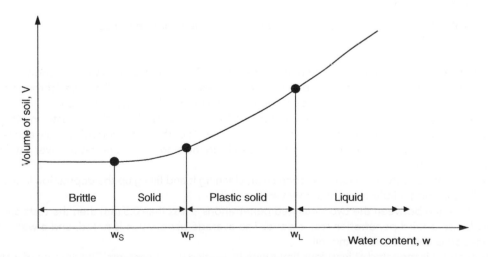

Fig. 1.5 Changes in volume against water content.

Plasticity index (I_P)

The *plasticity index* is the range of water content within which a soil is plastic; the finer the soil the greater its plasticity index.

Plasticity index = Liquid limit − Plastic limit

$$I_P = w_L - w_P \qquad\qquad (1.5)$$

Liquidity index

The *liquidity index* enables a comparison to be made of a soil's plasticity with its natural water content (w).

$$I_L = \frac{w - w_P}{I_P} \qquad\qquad (1.6)$$

If $I_L = 1.0$ the soil is at its liquid limit. If $I_L = 0$ the soil is at its plastic limit.

Shrinkage limit (w_S)

If the drying process is prolonged after the plastic limit has been reached, the soil will continue to decrease in volume until a certain value of water content is reached. This value is known as the *shrinkage limit* and at values of water content below this level the soil is partially saturated. In other words, below the shrinkage limit the volume of the soil remains constant with further drying, but the weight of the soil decreases until the soil is fully dried.

Determination of liquid and plastic limits

The test procedures are given in BS EN ISO 17892-12 (BSI, 2018).

Liquid limit test

BS EN ISO 17892-12 specifies two methods for determining the liquid limit of soil.

(1) Fall cone method (preferred method)
    This is typically referred to as the *cone penetrometer test*. Details of the apparatus are shown in Fig. 1.6. The soil to be tested is air dried and thoroughly mixed. At least 200 g of the soil is sieved through a 425 μm sieve and placed on a glass plate. The soil is then mixed with distilled water into a paste.
    A metal cup, approximately 55 mm in diameter and 40 mm deep, is filled with the paste and the surface struck level. The cone, of mass 80 g, is next placed at the centre of the smoothed soil surface and level with it. The cone is released so that it penetrates into the soil and the amount of penetration, over a time period of five seconds, is measured.
    The test is now repeated by lifting the cone clear, cleaning it and filling up the depression in the surface of the soil by adding a little more of the wet soil.
    If the difference between the two measured penetrations is less than 0.5 mm then the tests are considered valid, else a third penetration test is done. The average penetration is noted, and a water content determination is carried out on the soil.
    The procedure is repeated at least four times with increasing water contents. The amount of water used throughout should be such that the penetrations obtained lie within a range of 15–25 mm.

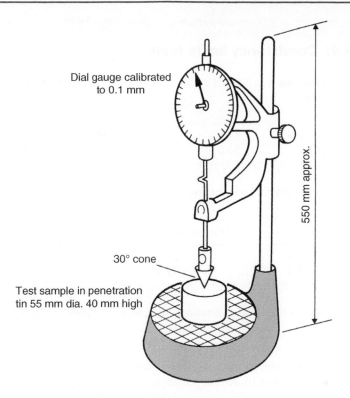

Dial gauge calibrated
to 0.1 mm

550 mm approx.

30° cone

Test sample in penetration
tin 55 mm dia. 40 mm high

Fig. 1.6 Fall cone apparatus.

To obtain the liquid limit, the variation of cone penetration is plotted against water content and the best straight line is drawn through the experimental points. The liquid limit is taken to be the water content corresponding to a cone penetration of 20 mm, expressed as a whole number.

(2) *Casagrande apparatus*

Although still used worldwide, the Casagrande test has now been largely superseded by the fall cone method because the latter achieves more repeatable results. In the UK, the fall cone test is adopted as the first-choice test method.

Plastic limit test

About 20 g of soil prepared as in the liquid limit test is used. The soil is mixed on the glass plate with just enough water to make it sufficiently plastic for rolling into a ball, which is then rolled out between the hand and the glass to form a thread. The soil is judged to be at its plastic limit when it just begins to crack at a thread diameter of 3 mm. At this stage, several pieces of the thread are taken for water content determination. The test is repeated until two water content tins contain an adequate mass of soil threads – approximately 10–15 g in each tin.

It is interesting to note that in some countries, the fall cone penetrometer is used to determine both w_L and w_P. The apparatus uses a 30° angle cone of mass 76 g. The procedure is the same as the fall cone described above, only this time w_L is taken at a penetration of 17 mm and w_P is taken as the water content at a penetration of 2 mm.

Example 1.4: Consistency limits tests

A cone penetrometer test was carried out on a sample of the clay from Example 1.3, with the following results:

Cone penetration (mm)	16.1	17.6	19.3	21.3	22.6
Water content (%)	50.0	52.1	54.1	57.0	58.2

The results from the plastic limit test were:

Test no.	Mass of tin (g)	Mass of wet soil + tin (g)	Mass of dry soil + tin (g)
1	8.1	20.7	18.7
2	8.4	19.6	17.8

Determine the liquid limit, plastic limit and the plasticity index of the soil.

Solution:

The plot of cone penetration to water content is shown in Fig. 1.7. The liquid limit is the water content corresponding to 20 mm penetration, i.e. $w_L = 55\%$.

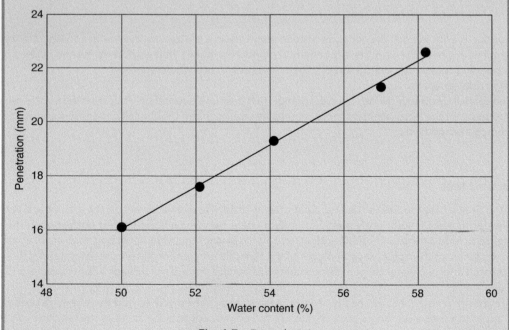

Fig. 1.7   Example 1.4.

The plastic limit is determined thus:

$$w_P(1) = \frac{20.7 - 18.7}{18.7 - 8.1} \times 100 = 18.9\%$$

$$w_P(2) = \frac{19.6 - 17.8}{17.8 - 8.4} \times 100 = 19.1\%$$

Average $w_P = 19\%$

The plasticity index is the difference between w_L and w_P i.e.

$$I_P = 55 - 19 = 36\%$$

1.5.6   Activity of a clay

In addition to their use in soil classification, the w_L and w_P values of a plastic soil also give an indication of the types and amount of the clay minerals present in the soil.

It has been found that, for a given soil, the plasticity index increases in proportion to the percentage of clay particles in the soil. Indeed, if a group of soils is examined and their I_P values are plotted against their clay percentages, a straight line, passing through the origin, is obtained.

If a soil sample is taken and its clay percentage artificially varied, a relationship between I_P and clay percentage can be obtained. Each soil will have its own straight line because, although in two differing soils the percentages of clay may be the same, they will contain different minerals.

The relationship between montmorillonite, illite, kaolinite and the plasticity index is shown in Fig. 1.8. The plot of London clay is also shown on the figure and, from its position, it is seen that the mineral content of this

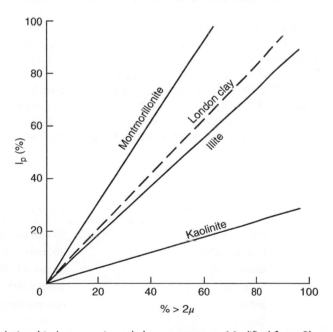

Fig. 1.8   Relationship between I_P and clay percentage. Modified from Skempton (1953).

soil is predominantly illite. London Clay has a clay fraction of about 46% and consists of illite (70%), kaolinite (20%) and montmorillonite (10%). The remaining fraction of 54% consists of silt (quartz, feldspar and mica: 44%) and sand (quartz and feldspar: 10%). In Fig. 1.8, the slope of the line is the ratio:

$$\frac{I_P}{\%clay}$$

Skempton (1953) defined this ratio as the *activity* of the clay. Clays with large activities are called active clays and exhibit plastic properties over a wide range of water content values.

1.6   Soil classification and description

1.6.1   Soil classification

Soil classification systems have been in use for a very long time with the first recorded use being in China over 4000 years ago. In 1896 a soil classification system was proposed by the Bureau of Soils, United States Department of Agriculture in which the various soil types were classified purely on particle size and it is interesting to note that the limiting sizes used are more or less the same as those in use today. Further improved systems allowed for the plasticity characteristics of soil, and a modified form of the system proposed by Casagrande in 1947 was the basis of the soil classification system used in the UK for many years. With the widespread adoption of BS EN ISO 14688:2018 (BSI, 2018a,c), the classification of soils across Europe now follows a standardised approach.

In order to classify soils, the system considers natural soils as falling into one of the following categories: *very coarse*, *coarse*, *fine* and *organic*. The majority of soils are inorganic, and particle sizes and plasticity characteristics of these soils are used to identify the *primary soil fraction*. Developing the notion of fraction sizes listed in Table 1.1, BS EN ISO 14688-2:2018 (BSI, 2018c) offers the principles of classification listed in Table 1.3.

Since soils are usually composite (i.e. contain various amounts of different particle sizes) it will be the case that secondary and tertiary fractions will also exist in the soil. In soil classification and description, these additional fractions are used as adjectives and the primary fraction is the noun. The primary fraction is written in capitals: e.g. sandy, silty CLAY; gravelly, coarse SAND; clayey SILT.

For fine soils, the results of the liquid and plastic limits tests are used to classify these soils. Classification is done through the use of a plasticity chart (Fig. 1.9). To use the plasticity chart, a point is plotted whose coordinates are the liquid limit and the plasticity index of the soil. The soil is then classified by observing the position of the point relative to the sloping straight line drawn across the diagram. This line, known as the *A-line*, is

Table 1.3   Principles of soil classification.

Soil group	Primary function	Criteria
Very coarse	Boulders (Bo)	50% of particles >200 mm in size
	Cobbles (Co)	200 mm > 50% > 63 mm
Coarse	Gravel (Gr)	63 mm > 50% > 2 mm
	Sand (Sa)	2 mm > 50% > 0.06 mm
Fine	Silt (Si)	Low plasticity or non-plastic
	Clay (Cl)	Plastic

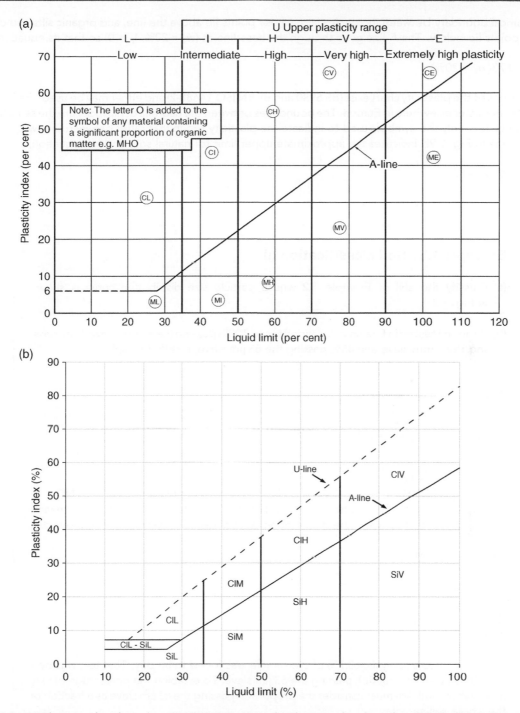

Fig. 1.9 (a) Plasticity chart based on BS 5030 (BSI, 2015) (b) Plasticity chart based on BS EN ISO 14688-2 (BSI, 2018c).

an empirical boundary between inorganic clays, whose points lie above the line, and organic silts and clays whose points lie below. The A-line goes through the base line at (w_L = 20%, I_P = 0) so that its equation is:

$$I_P = 0.73(w_L - 20\%)$$

Two versions of the plasticity chart exist (BS 5930 and BS EN ISO 14688-2) and either may be used: the former in the UK, the latter elsewhere in Europe. The boundaries between the degrees of plasticity are the same from both charts. It is only the symbols used to indicate the plasticity group that differ between the two.

The U-line in Fig. 1.9b indicates the approximate upper limit for natural soils and has equation,

$$I_P = 0.9(w_L - 8\%)$$

Example 1.5: Soil classification (i)

(a) Classify the soil of Example 1.2 whose particle size distribution curve is shown in Fig. 1.3.

(b) If the soil tested in Example 1.3 experienced 100% passing the 63 mm sieve, 96% passing the 2 mm sieve and 85% passing the 63 μm sieve, classify the soil.

Solution:

(a) $C_u = \dfrac{D_{60}}{D_{10}} = \dfrac{1.5}{0.18} = 8.3;$

$C_c = \dfrac{(D_{30})^2}{(D_{60} \times D_{10})} = \dfrac{0.51^2}{1.5 \times 0.18} = 0.96$

Using Tables 1.2 and 1.3, the soil is classified a medium graded gravelly SAND.

(b) The PSD may be plotted to visualise the results (see spreadsheet example_1.3_and_1.5b.xls) or the classification may be made by reviewing the data directly:

Particle fractions:

Gravel: 100 − 96 = 4%
Sand: 96 − 85 = 11%
Silt: 85 − 58* = 27%
Clay: 58*%

*In the pipette analysis, 68.7% of the sample was found to be clay. That test was performed only on the fraction passing the 63 μm sieve. To establish the percentage of clay in the whole soil, we must consider the proportion passing the 63 μm sieve as a fraction of the whole soil sample.

i.e. %clay = 0.687 × 85 = 58%

Based on the percentages of each fraction, the soil may therefore be classified as a slightly sandy, silty CLAY.

Since it is a clay, it must also be classified in terms of its plasticity. To do this we use the results of Example 1.4 together with the plasticity chart (Fig. 1.9). It is seen that the soil therefore is a slightly sandy, silty CLAY of high plasticity.

Example 1.6: Soil classification (ii)

A set of particle size distribution analyses on three soils, A, B and C, gave the following results:

Sieve size (mm)	Percentage passing		
	Soil A	Soil B	Soil C
20	90	–	–
10	56	–	–
6.3	47	–	–
2	43	–	–
0.6	39	93	–
0.425	–	78	–
0.300	28	16	–
0.212	–	5	–
0.150	–	–	100
0.063	5	2	92

Soil C: Since more than 10% passed the 63 µm sieve, a pipette analysis was performed. The results were:

Particle size (mm)	Percentage passing
0.04	79
0.02	62
0.006	47
0.002	40

Soil C was found to have a liquid limit of 48% and a plastic limit of 29%. Plot the particle size distribution curves and classify each soil.

Solution:

The particle size distribution curves for the three soils are shown in Fig. 1.10. The curves can be used to obtain the following particle sizes for soils A and B.

Soil	D_{10} (mm)	D_{30} (mm)	D_{60} (mm)
A	0.1	0.31	12.0
B	0.26	0.36	0.38

Soil A: From the grading curve it is seen that this soil consists of 57% gravel and 43% sand and is therefore predominantly gravel. The curve has a horizontal portion indicating that the soil has only a small percentage of soil particles within this range. It is therefore gap graded. Also, $C_u = 120$.

The soil is a gap graded sandy GRAVEL.

Soil B: From the grading curve, it is immediately seen that this soil is a sand with most of its particles about the same size. Also, $C_u = 1.5$.

The soil is a uniformly graded SAND.

Soil C: From the grading curve, it is seen that the soil is a mixture of 10% sand, 50% silt and 40% clay so it is a slightly sandy, very clayey SILT. The liquid limit of the soil = 48% and the plasticity index, $(w_L - w_P) = 19\%$. Using Fig. 1.9, it is seen that the soil is a silt with the group symbol MI (BS 5930) or SiM (BS EN ISO 14688-2).

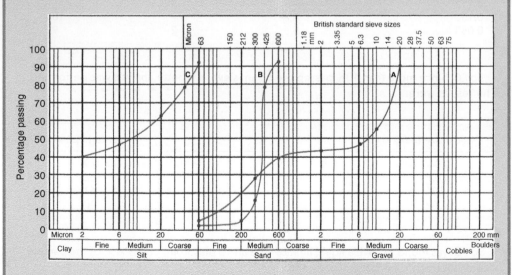

Fig. 1.10   Example 1.6.

1.6.2   Description of soils

Classifying and describing a soil are two operations, which are not necessarily the same. An operator who has not even visited the site from which a soil came can classify the soil from the information obtained from grading and plasticity tests carried out on disturbed samples. Such tests are necessary if the soil is being considered as a possible construction material and the information obtained from them must be included in any description of the soil.

Further information regarding the colour of a soil, the texture of its particles, etc., can be obtained in the laboratory from disturbed soil samples but a full description of a soil must include its *in situ*, as well as its laboratory, characteristics. Some of this latter information can be found in the laboratory from undisturbed samples of the soil collected for other purposes, such as strength or permeability tests, but usually not until after the tests have taken place and the samples can then be split open for proper examination. Other relevant information such as bedding details, gravel particle shapes (e.g. angular, rounded, elongated), clay consistency (e.g. soft, firm, stiff) and site observations can also be included in the soil's description.

1.7   Soil properties

From the foregoing it is seen that soil consists of a mass of solid particles separated by spaces, or *voids*. A cross-section through a granular soil may have an appearance similar to that shown in Fig. 1.11a.

In order to study the properties of such a soil mass, it is advantageous to adopt an idealised form of the diagram as shown in Fig. 1.11b. The soil mass has a total volume V and a volume of solid particles equal to V_s. The volume of the voids, V_v is obviously equal to $V - V_s$.

1.7.1   Void ratio and porosity

From a study of Fig. 1.11, the following may be defined:

Void ratio, e

$$e = \frac{\text{Volume of voids}}{\text{Volume of solids}} = \frac{V_v}{V_s} \tag{1.7}$$

Porosity, n

$$n = \frac{\text{Volume of voids}}{\text{Total volume}}$$

$$n = \frac{V_v}{V} = \frac{V_v}{V_s + V_v} = \frac{e}{1 + e} \tag{1.8}$$

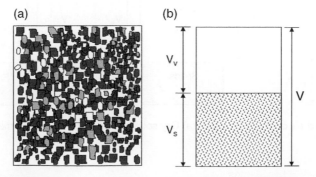

(a)   (b)

Fig. 1.11   Cross-section through a granular soil. (a) Actual form. (b) Idealised form.

1.7.2   Degree of saturation, S_r

The voids of a soil may be filled with air or water or both. If only air is present the soil is dry, whereas if only water is present the soil is *saturated*. When both air and water are present the soil is said to be partially saturated. These three conditions are represented in Fig. 1.12a–c.

The degree of saturation is simply:

$$S_r = \frac{\text{Volume of water}}{\text{Volume of voids}} = \frac{V_w}{V_v} \tag{1.9}$$

(usually expressed as a percentage)

For a dry soil, $S_r = 0$

For a saturated soil, $S_r = 1.0$

1.7.3   Particle density, ρ_s and specific gravity, G_s

The specific gravity of a material is the ratio of the weight or mass of a volume of the material to the weight or mass of an equal volume of water. In soil mechanics the most important specific gravity is that of the actual soil grains and is given the symbol G_s.

From the above definition it is seen that for a soil sample with volume of solids, V_s, mass of solids, M_s and weight of solids, W_s,

$$G_s = \frac{M_s}{V_s \rho_w} = \frac{W_s}{V_s \gamma_w} \tag{1.10}$$

where ρ_w is the density of water (=1.0 Mg/m^3 at 20 °C) and γ_w is the unit weight of water (=9.81 kN/m^3).

The density of the particles ρ_s is defined as:

$$\rho_s = \frac{M_s}{V_s}$$

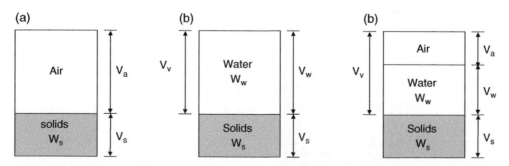

Fig. 1.12   Water and air contents in a soil. (a) Dry soil. (b) Saturated soil. (c) Partially saturated soil.

therefore,

$$G_s = \frac{\rho_s}{\rho_w}$$

If ρ_s is measured in units of Mg/m^3 and the water temperature is assumed to be 20 °C, it follows that ρ_s and G_s are numerically equal. G_s, however is dimensionless whereas ρ_s has the units of density, Mg/m^3.

Particle density is determined in the laboratory through a well-established and reliable testing procedure (described in BS EN ISO 17892-3:2015, BSI, 2015). A mass of dry soil is mixed with distilled water in a standard glass vessel known as a *pycnometer* to separate all the particles and to enable all the air to be removed from the soil. In the test the difference in the volumes of water required to fill the pycnometer, both with and without the soil present, is determined. The particle density is equal to the dry mass of the soil divided by that volume difference.

For sands and fine soils, a sample of the soil (minimum 10 g) is oven dried, weighed and placed into the pycnometer of minimum volume 50 ml along with distilled water at room temperature (Fig. 1.13). Coarse soils can be tested in a larger pycnometer or the gravel portion can be mechanically broken down to pass a 4 mm sieve and placed in the smaller pycnometer. Formerly, in the UK, a 1 l gas jar was used for coarse soils – see earlier editions of this book for details of the gas jar procedure.

The pycnometer and contents are shaken steadily but vigorously to remove all air bubbles. The glass stopper is then removed, the jar topped up carefully to full capacity with further distilled water and the stopper

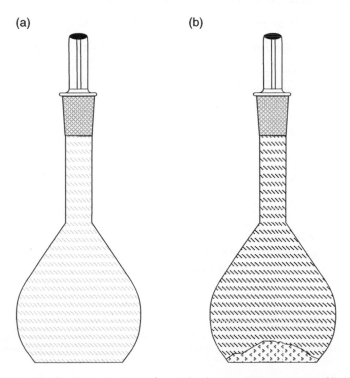

Fig. 1.13 Pycnometer used in the determination of particle density. (a) Pycnometer filled with water only. (b) Pycnometer filled with soil and water.

replaced to seal the vessel without trapping any air inside. The stopper has a capillary tube through it, which permits the water to flow out its top as it is inserted into pycnometer. From various weighings that are made, the particle density of the soil can be calculated (see Example 1.7).

Example 1.7: Particle density

The mass of an empty pycnometer, together with its glass stopper, was 178.0 g. When completely filled with water and the stopper fitted the mass was 228.2 g. An oven dried sample of soil was placed in the pycnometer and the total mass, including the stopper, was 191.2 g. Water was added to the soil and, after a suitable period of shaking, was topped up until the vessel was brim full. The stopper was fitted and the total mass was found to be 236.4 g.

Determine the particle density of the soil.

Solution:

Mass of soil + water = 236.4 − 178.0 = 58.4 g

Mass of dry soil = 191.2 − 178.0 = 13.2 g

Mass of water present with soil = 58.4 − 13.2 = 45.2 g

Mass of water when pycnometer full = 228.2 − 178.0 = 50.2 g.

Therefore, mass of water of same volume as soil = 50.2 − 45.2 = 5.0 g

$$\rho_s = \frac{\text{Mass of soil}}{\text{Mass of same volume of water}} = \frac{13.2}{5.0} = 2.64$$

The particle density can be quickly found from a formula thus:

$$\rho_s = \frac{m_s}{(m_1 - m_2) + m_s}$$

where

m_s = mass of dry soil (g)
m_1 = mass of pycnometer + water (g)
m_2 = mass of pycnometer + water + soil (g)

$$\rho_s = \frac{13.2}{[228.2 - 236.4] + 13.2} = 2.64$$

1.7.4   Density and unit weight

The amount of material in a given volume, V, may be expressed in two ways:

the amount of mass, M, in the volume, or the amount of weight, W, in the volume.

If we consider unit volume, the two systems give the *mass density* and the *weight density* of the material respectively. The *mass density* is usually simply referred to as *density* and the *weight density* is routinely referred to as the *unit weight*:

$$\text{Density}, \rho = \frac{\text{Mass}}{\text{Volume}} = \frac{M}{V} \tag{1.11}$$

$$\text{Unit weight}, \gamma = \frac{\text{Weight}}{\text{Volume}} = \frac{W}{V} \tag{1.12}$$

$$\text{Weight} = \text{mass} \times 9.81$$

As an example, consider water at 20 °C:

Density of water, $\rho_w = 1000 \text{ kg/m}^3 = 1.0 \text{ Mg/m}^3$

Hence the unit weight of water, $\gamma_w = 1.0 \times 9.81 = 9.81 \text{ kN/m}^3$.

Soil densities are usually expressed in Mg/m^3 to the nearest 0.01.

Soil weights are usually expressed in kN/m^3.

   In soils work, it is generally more convenient to measure the density of a soil through test (e.g. Example 1.8) then to perform the geotechnical analysis using the unit weight, or weight density, derived from the density value.

Density of soil

Bulk density

The bulk density of a soil is a frequently requested, easily determined, geotechnical property. By sampling a soil using a sampler (e.g. a steel tube) of known volume, V, the mass, M can easily be measured on laboratory scales and the bulk density determined:

$$\text{Bulk density}, \rho = \frac{M}{V}$$

The bulk density is the density of the wet soil, as it considers the mass of water in the soil as well as the soil particles.

Dry density

The dry density is the density of the soil particles within the same volume, V as the bulk density.

$$\text{Dry density}, \rho_d = \frac{M_s}{V}$$

The dry density can be determined from the bulk density and the water content (in %):

$$\rho_d = \frac{\rho \times 100}{100 + w} \tag{1.13}$$

Unit weight, or weight density, of soil

As mentioned, the unit weight of a material is its weight per unit volume. In soils work the most important unit weights are:

Bulk unit weight, γ

This is the natural *in situ* unit weight of the soil. Referring to Fig. 1.12 and Section 1.7.3:

$$\gamma = \frac{\text{Total weight}}{\text{Total volume}} = \frac{W}{V} = \frac{W_s + W_w}{V_s + V_v}$$

$$\gamma = \frac{G_s V_s \gamma_w + V_v \gamma_w S_r}{V_s + V_v} = \gamma_w \left[\frac{G_s + e S_r}{1 + e} \right] \tag{1.14}$$

Saturated unit weight, γ_{sat}

$$\text{Saturated unit weight,}\ \gamma_{sat} = \frac{\text{Saturated weight}}{\text{Total volume}}$$

Recall, when soil is saturated, $S_r = 1$. Therefore,

$$\gamma_{sat} = \gamma_w \frac{G_s + e}{1 + e} \tag{1.15}$$

Dry unit weight, γ_d

$$\gamma_d = \frac{\text{Dry weight}}{\text{Total volume}}$$

$$\gamma_d = \frac{\gamma_w G_s}{1 + e} \quad (\text{as } S_r = 0) \tag{1.16}$$

Effective, or buoyant, unit weight, γ'

When a soil is below the water table, part of its weight is balanced by the buoyant effect of the water. This upthrust equals the weight of the volume of the water displaced.

Hence, considering unit volume:

$$\text{Effective unit weight} = \text{Saturated unit weight} - \text{Unit weight of water}$$

$$\gamma' = \gamma_w \frac{G_s + e}{1 + e} - \gamma_w = \gamma_w \frac{G_s - 1}{1 + e} \tag{1.17}$$

The effective unit weight is also referred to as the *buoyant unit weight* or the *submerged unit weight*.

Note: The two terms *weight density* and *unit weight* are synonymous, and both are in common use. Both terms are used throughout this book.

Additional expressions for densities

As with the unit weights, similar expressions can be obtained for densities:

$$\text{Bulk density,} \, \rho_b = \rho_w \frac{(G_s + eS_r)}{1 + e} \tag{1.18}$$

$$\text{Saturated density,} \, \rho_{sat} = \rho_w \frac{(G_s + e)}{1 + e} \tag{1.19}$$

$$\text{Dry density,} \, \rho_d = \rho_w \frac{G_s}{1 + e} \tag{1.20}$$

$$\text{Effective density,} \, \rho' = \rho_w \frac{G_s - 1}{1 + e} \tag{1.21}$$

Relationship between density and unit weight values

In the previous expressions, G_s, e, S_r and the number 1 are all dimensionless.

Hence, a particular unit weight = γ_w times a constant.

The corresponding density = ρ_w times the same constant.

Example 1.8: Dry unit weight

A sample of wet soil was extruded from a sampling tube of diameter 100 mm in a soil test-ing laboratory. The length of extruded sample was 200 mm. The mass of the wet soil was 3.15 kg. Following a water content determination, the mass of the dry soil was found to be 2.82 kg.

Determine the bulk density, water content, dry density and dry unit weight of the soil.

Solution:

$$\text{Volume of sample} = \frac{\pi \times 0.1^2}{4} \times 0.2 = 0.0016 \, m^3$$

$$\rho = \frac{M}{V} = \frac{3.15}{0.0016} = 1969 \, kg/m^3 - 1.97 \, Mg/m^3$$

$$w = \frac{3.15 - 2.82}{2.82} = 12\%$$

$$\rho_d = \frac{\rho \times 100}{100 + w} = \frac{197}{112} = 1.76 \, Mg/m^3$$

$$\gamma_d = \rho_d \times 9.81 = 17.3 \, kN/m^3$$

Relationship between w, γ_d and γ

$$\gamma = \frac{W_w + W_s}{V} \tag{1.22}$$

$$\gamma_d = \frac{W_s}{V} \tag{1.23}$$

$$w = \frac{W_w}{W_s} \tag{1.24}$$

Rearranging (1.24) and substituting into (1.22) gives:

$$\gamma = \frac{W_s}{V}(1 + w)$$

And substituting into (1.23) gives:

$$\gamma_d = \frac{\gamma}{1 + w}$$

Thus, to find the dry unit weight from the bulk unit weight, divide the latter by $(1 + w)$ where w is the water content expressed as a decimal.

Relationship between e, w and G_s for a saturated soil

$$w = \frac{W_w}{W_s} = \frac{V_w \gamma_w}{V_s \gamma_w G_s} = \frac{V_v}{V_s G_s} = \frac{e}{G_s} \quad (V_w = V_v \text{ if the soil is saturated})$$

i.e.

$$e = wG_s \tag{1.25}$$

Relationship between e, w and G_s for a partially saturated soil

$$w = \frac{W_w}{W_s} = \frac{V_w \gamma_w}{V_s \gamma_w G_s} = \frac{V_v S_r}{V_s G_s} = \frac{eS_r}{G_s}$$

i.e.

$$e = \frac{wG_s}{S_r} \tag{1.26}$$

Example 1.9: Physical properties determination

In a bulk density determination, a sample of clay with a mass of 683 g was coated with wax. The combined mass of the clay and the wax was 690.6 g. The volume of the clay and the wax was found, by immersion in water, to be 350 ml.

The sample was then broken open and water content and particle specific gravity tests gave respectively 17% and 2.73.

The specific gravity of the wax was 0.89. Determine the bulk density, unit weight, void ratio and degree of saturation of the soil.

Solution:

Mass of soil = 683 g

Mass of wax = 690.6 − 683 = 7.6 g

$$\Rightarrow \text{Volume of wax} = \frac{7.6}{0.89} = 8.55 \text{ ml}$$

$$\Rightarrow \text{Volume of soil} = 350 - 8.6 = 341.4 \text{ ml}$$

$$\rho_b = \frac{683}{341.4} = 2 \text{ g/ml} = 2.0 \text{ Mg/m}^3$$

$$\gamma_b = 2 \times 9.81 = 19.6 \text{ kN/m}^3$$

$$\rho_d = \frac{2}{1.17} = 1.71 \text{ Mg/m}^3$$

Now,

$$\frac{\rho_w G_s}{1 + e} = 1.71$$

$$\Rightarrow e = \frac{2.73 - 1.71}{1.71} = 0.596$$

Now,

$$\rho_b = 2.0 = \rho_w \frac{(G_s + eS_r)}{1 + e}$$

$$\Rightarrow 1.596 \times 2.0 = 2.73 + 0.596 \times S_r$$

$$\Rightarrow S_r = 78.0\%$$

1.7.5   Density index, I_D

A granular soil generally has a large range into which the value of its void ratio may be fitted. If the soil is vibrated and compacted the particles are pressed close together and a minimum value of void ratio is obtained, but if the soil is loosely poured a maximum value of void ratio will result.

These maximum and minimum values can be obtained from laboratory tests and it is often convenient to relate them to the naturally occurring void ratio of the soil. This relationship is expressed as the *density index*, I_D or *relative density*, of the soil:

$$I_D = \frac{e_{max} - e}{e_{max} - e_{min}} \tag{1.27}$$

The theoretical maximum possible density of a granular soil must occur when $e = e_{min}$, i.e. when $I_D = 1.0$. Similarly, the minimum possible density occurs when $e = e_{max}$ and $I_D = 0$. In practical terms this means that a loose granular soil will have an I_D value close to zero whilst a dense granular soil will have an I_D value close to 1.0.

1.7.6 Summary of soil physical relations

A summary of the relationships established in Section 1.7 is given below:

$$\text{Water content} \quad w = \frac{W_w}{W_s} = \frac{M_w}{M_s}$$

$$\text{Void ratio} \quad e = \frac{V_v}{V_s}$$

$$e = wG_s \quad \text{(saturated)}$$

$$e = \frac{wG_s}{S_r} \quad \text{(partially saturated)}$$

$$\text{Porosity} \quad n = \frac{V_v}{V} = \frac{e}{1 + e}$$

$$\text{Degree of saturation} \quad S_r = \frac{V_w}{V_v}$$

$$\text{Particle specific gravity} \quad G_s = \frac{M_s}{V_s \rho_w} = \frac{W_s}{V_s \gamma_w}$$

$$\text{Bulk density} \quad \rho_b = \rho_w \frac{(G_s + eS_r)}{1 + e}$$

$$\text{Dry density} \quad \rho_d = \frac{\rho_w G_s}{1 + e} = \frac{\rho \times 100}{100 + w} \quad (w \text{ as a percentage})$$

$$\text{Saturated density} \quad \rho_{sat} = \rho_w \frac{(G_s + e)}{1 + e}$$

$$\text{Effective density} \quad \rho' = \rho_w \frac{G_s - 1}{1 + e}$$

$$\text{Bulk unit weight} \quad \gamma = \gamma_w \frac{G_s + eS_r}{1 + e}$$

$$\text{Dry unit weight} \quad \gamma_d = \frac{\gamma_w G_s}{1 + e} = \frac{\gamma}{1 + w} \quad (w \text{ as a decimal})$$

$$\text{Saturated unit weight} \quad \gamma_{sat} = \gamma_w \frac{G_s + e}{1 + e}$$

$$\text{Effective unit weight} \quad \gamma' = \gamma_w \frac{G_s - 1}{1 + e}$$

$$\text{Density index} \quad I_D = \frac{e_{max} - e}{e_{max} - e_{min}}$$

Exercises

Exercise 1.1

The results of a sieve analysis on a soil were:

Sieve size (mm)	Mass retained (g)
50	0
37.5	15.5
20	17.0
14	10.0
10	11.0
6.3	33.0
3.35	114.5
1.18	63.3
0.6	18.2
0.15	17.0
0.063	10.5

The total mass of the sample was 311 g. Plot the particle size distribution curve and, from the inspection of this curve, determine the effective size and uniformity coefficient. Classify the soil.

Answer $D_{10} = 0.7$ mm; $D_{60} = 5.2$ mm. $C_u = 7.4$. 70% gravel, 30% sand. Well graded sandy GRAVEL.

Exercise 1.2

Plot the particle size distribution curve for the following sieve analysis, given the sieve sizes and the mass retained on each. Classify the soil.
   Sample mass = 642 g.
   Retained on 425 µm sieve – 11 g, 300 µm sieve – 28 g, 212 µm sieve – 77 g, 150 µm sieve – 173 g, 63 µm sieve – 321 g.

Answer   By inspection of grading curve soil is a uniform SAND. This is confirmed from the value of $C_u = 2.3$.

Exercise 1.3

A BS cone penetrometer test carried out on a sample of boulder clay gave the following results:

Cone penetration (mm)	15.9	17.1	19.4	20.9	22.8
Water content (%)	32.0	32.8	34.5	35.7	37.0

Determine the liquid limit of the soil.

Answer w_L = 35%

Exercise 1.4

A liquid and plastic limit test gave the following results:

Test no.	1	2	3	4	PL	PL
Wet mass (g)	33.20	32.10	28.20	31.00	11.83	15.04
Dry mass (g)	28.20	26.50	22.40	23.90	11.25	14.07
Tin (g)	7.02	7.04	7.10	7.02	7.04	7.25
Penetration (mm)	14.5	17.0	20.9	22.7	–	–

Determine the plasticity index of the soil and classify the soil.

Answer   22, CI

If the natural water content was 28%, determine the liquidity index in the field.

Answer   0.64

Exercise 1.5

A sand sample has a porosity of 35% and the specific gravity of the particles is 2.73. What is its dry density and void ratio?

Answer   e = 0.54, ρ_d = 1.77 Mg/m³

Exercise 1.6

A sample of silty clay was found to have a volume of 14.88 ml, whilst its mass at natural water content was 28.81 g and the particle specific gravity was 2.7. Calculate the void ratio and degree of saturation if, after oven drying, the sample had a mass of 24.83 g.

Answer   e = 0.618, S_r = 70%

Exercise 1.7

A sample of moist sand was cut from a natural deposit by means of a sampling cylinder. The volume of the cylinder was 478 ml and the mass of the soil was 884 g before drying, and 830 g after drying. The volume of the dried sample, when rammed tight into a graduated cylinder, was 418 ml and its volume, when poured loosely into the same cylinder, was 616 ml. If the particle specific gravity was 2.67, determine the density index and the degree of saturation of the deposit.

Answer I_D = 69%, S_r = 32%

Exercise 1.8

In order to determine the density of a clay soil, an undisturbed sample was taken in a sampling tube of volume $0.001\,664\,m^3$.

The following data were obtained:

Mass of tube (empty) = 1.864 kg

Mass of tube and clay sample = 5.018 kg

Mass of tube and clay sample after drying = 4.323 kg

Calculate the water content, the bulk, and the dry densities.

If the particle specific gravity was 2.69, determine the void ratio and the degree of saturation of the clay.

Answer $w = 28\%$, $\rho_b = 1.90\,Mg/m^3$, $\rho_d = 1.49\,Mg/m^3$, $e = 0.82$, $S_r = 93\%$

Chapter 2

Permeability and Flow of Water in Soils

Learning objectives:

By the end of this chapter, you will have been introduced to:
- water in soils and the differences between aeration and saturation zones;
- the groundwater table and groundwater flow;
- hydraulic head, the hydraulic gradient, Darcy's Law, and saturated flow;
- the laboratory and field determination of the coefficient of permeability;
- the use and construction of flow nets to determine flow quantities;
- measurement of suction pressures in unsaturated soils;
- flow of water through earth dams and deposits of different permeabilities.

2.1 Subsurface water

This is the term used to define all water found beneath the Earth's surface. The main source of subsurface water is rainfall, which percolates downwards to fill up the voids and interstices. Water can penetrate to a considerable depth, estimated to be as much as 12 000 m, but at depths greater than this, due to the large pressures involved, the interstices have been closed by plastic flow of the rocks. Below this level, water cannot exist in a free state, although it is often found in chemical combination with the rock minerals, so that the upper limit of plastic flow within the rock determines the lower limit of subsurface water.

Subsurface water can be split into two distinct zones: *saturation zone* and *aeration zone*.

2.1.1 Saturation zone

This is the depth throughout which all the fissures and voids are filled with water under pressure. The upper level of this water is known as the *groundwater table (GWT)*, *phreatic surface* or *groundwater level (GWL)*, and the water within this zone is called phreatic water or, more commonly, *groundwater*.

The water table tends to follow, in a more gentle manner, the topographical features of the surface above (Fig. 2.1). At groundwater level, the hydrostatic pressure is zero, so another definition of water table is the level to which water will eventually rise in an unlined borehole.

Smith's Elements of Soil Mechanics, 10th Edition. Ian Smith.
Companion website: www.wiley.com/go/smith/soilmechanics10e

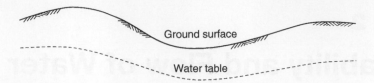

Fig. 2.1 Tendency of the water table to follow the earth's surface.

The water table is not constant in depth but rises and falls with variations of rainfall, atmospheric pressure, temperature, and proximity to tree roots. In coastal regions, the GWL can be affected by tides and is said to be tidal. At locations where the water table reaches the ground surface, springs, lakes, swamps, and similar features can be formed.

2.1.2   Aeration zone

Sometimes referred to as the *vadose zone*, this zone occurs between the water table and the surface, and can be split into three sections.

Capillary fringe

Owing to capillarity, water is drawn up above the water table into the voids of the soil. Water in this fringe can be regarded as being in a state of negative pressure, i.e. at pressure values below atmospheric. The minimum height of the fringe is governed by the maximum size of the voids within the soil. Up to this height above the water table, the soil will be sufficiently close to full saturation to be considered as such. The maximum height of the fringe is governed by the minimum size of the voids. Between the minimum and maximum heights, the soil is *partially* saturated.

Terzaghi and Peck (1948) give an approximate relationship between the maximum height and the grain size for a granular soil:

$$h_c \approx \frac{C}{eD_{10}} \, mm$$

where C is a constant depending upon the shape of the grains and the surface impurities (varying from 10.0 to 50.0 mm^2), and D_{10} is the effective size expressed in millimetres.

Intermediate belt

As rainwater percolates downward to the water table, a certain amount is held in the soil by the action of surface tension, capillarity, adsorption, and chemical action. The water retained in this manner is termed held water and is deep enough not to be affected by plants.

Soil belt

This zone is constantly affected by precipitation, evaporation, and plant transpiration. Moist soil in contact with the atmosphere either evaporates water or condenses water into itself until its vapour pressure is equal to atmospheric pressure. Soil water in atmospheric equilibrium is called *hygroscopic water* and its water content (which depends upon relative humidity) is known as the *hygroscopic water content*.

The various zones are illustrated in Fig. 2.2

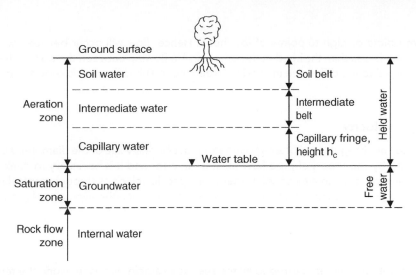

Fig. 2.2 Types of subsurface water.

2.2 Flow of water through soils

The voids of a soil (and of most rocks) are connected together and form continuous passageways for the movement of water brought about by rainfall infiltration, transpiration of plants, imbalance of chemical energy, or a variation of the intensity of dissolved salts.

When rainfall falls on the soil surface, some of the water infiltrates the surface and percolates downward through the soil. This downward flow results from a gravitational force acting on the water. During flow, some of the water is held in the voids in the aeration zone and the remainder reaches the groundwater table and the saturation zone. In the aeration zone, flow is said to be *unsaturated*. Below the water table, flow is said to be *saturated*. Flow of water through soils is often referred to as *seepage*.

2.2.1 Saturated flow

The water within the voids of a soil is under pressure. This water, known as *pore water*, may be static or flowing. Water in saturated soil will flow in response to variations in hydraulic head within the soil mass. These variations may be natural or induced by excavation or construction.

2.2.2 Hydraulic head

The head of water acting at a point in a submerged soil mass is known as the *hydraulic head* and is expressed by Bernoulli's equation:

Hydraulic head = Velocity head + Pressure head + Elevation head

$$h = \frac{v^2}{2g} + \frac{p}{\gamma_w} + z \tag{2.1}$$

In seepage problems, atmospheric pressure is taken as zero and the velocity is so small that the velocity head becomes negligible; the hydraulic head is therefore taken as:

$$h = \frac{p}{\gamma_w} + z \tag{2.2}$$

Excess head

Water flows from points of high to points of low head. Hence, flow will occur between two points if the hydraulic head at one is less than the head at the other, and in flowing between the points, the water experiences a head loss equal to the difference in head between them. This difference is known as the *excess head*.

2.2.3 Seepage velocity

The conduits of a soil are irregular and of small diameter – an average value of the diameter is $D_{10}/5$. Any flow quantities calculated by the theory of pipe flow must be in error and it is necessary to think in terms of an average velocity through a given area of soil rather than specific velocities through particular conduits.

If Q is the quantity of flow passing through an area A in time t, then the average velocity, v is:

$$v = \frac{Q}{At} \tag{2.3}$$

This average velocity is sometimes referred to as the *seepage velocity*. In further work, the term velocity will imply average velocity.

2.2.4 Hydraulic gradient

If we install two vertical standpipes, a distance apart and with their lower ends embedded into a zone of saturated soil, groundwater will rise in the pipes to different levels as shown in Fig. 2.3. (A standpipe is simply an open-ended tube, perforated at its lower end, which permits groundwater to flow into the tube and to rise to a final position. See Section 7.5.1 for more details.)

Akin to the equation of a straight line in mathematics, the hydraulic gradient is defined as the difference in vertical values divided by the difference in horizontal values, at the two points considered.

i.e.

$$\text{hydraulic gradient} = \frac{\text{change in head}}{\text{length of flow path}}$$

$$i = \frac{\Delta h}{l} \tag{2.4}$$

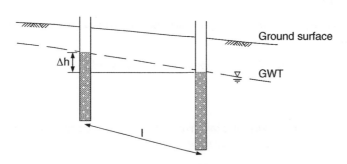

Fig. 2.3 Difference in hydraulic head between two points.

2.3   Darcy's law of saturated flow

In 1856, Darcy showed experimentally that a fluid's velocity of flow through a porous medium is directly related to the hydraulic gradient causing the flow, i.e.

$v \propto i$

where i = hydraulic gradient (the head loss per unit length), or

$v = Ci$

where C = a constant involving the properties of both the fluid and the porous material.

2.4   Coefficient of permeability, k

In soils, we are generally concerned with water flow, and the constant C is determined from tests in which the permeant is water. The particular value of the constant C obtained from these tests is known as the *coefficient of permeability* and is given the symbol k. The unit for k is m/s.

It is important to realise that when a soil is said to have a certain coefficient of permeability, this value only applies to water (at 20 °C). If heavy oil is used as the permeant, the value of C would be considerably less than k. Temperature causes variation in k, but in most soils work this is insignificant.

Provided that the hydraulic gradient is less than 1.0, as is the case in most seepage problems, the flow of water through a soil is linear and Darcy's law applies, i.e.

$v = ki$

or

$$Q = Atki \qquad (2.5)$$

or

$q = Aki \quad \left(\text{where } q = \text{quantity of unit flow} = \dfrac{Q}{t} \right)$

From this latter expression, a definition of k is apparent: the coefficient of permeability is the rate of flow of water per unit area of soil when under a unit hydraulic gradient.

2.5   Determination of permeability in the laboratory

Laboratory tests can be performed to establish the coefficient of permeability for both granular and cohesive soils, and the testing procedures are described in BS EN ISO 17892-11:2019. The tests involve placing the soil in a cylindrical *permeameter*, which can take different forms:

- Rigid wall permeameter: standard cylindrical vessel (see Sections 2.5.1 and 2.5.2) or oedometer ring within oedometer cell. (The oedometer test is described in Chapter 12.)
- Flexible wall permeameter: rubber membrane placed around soil and tested in a triaxial cell, under required effective stress conditions. (The triaxial test is described in Chapter 4.)

In all cases, water is passed through the permeameter and the volume of water passing through the soil in a time interval, together with a measurable hydraulic gradient, can be used to establish the coefficient of permeability. Two of the most well-established tests are the *constant head test* (for granular soils) and the *falling head test* (for cohesive soils).

2.5.1   Constant head test

The apparatus is shown in Fig. 2.4. Water flows through the soil under a head which is kept constant by means of the overflow arrangement. The head loss, h, between two points along the length of the sample, distance l apart, is measured by means of a manometer (in practice there are more than just two manometer tappings).

$$\text{From Darcy's law : } q = Aki$$

$$\text{The unit quantity of flow, } q = \frac{Q}{t}$$

$$\text{The hydraulic gradient, } i = \frac{h}{l}$$

$$\text{and } A = \text{area of sample}$$

Hence, k can be found from the expression:

$$k = \frac{q}{Ai} \text{ or } k = \frac{Ql}{tAh} \tag{2.6}$$

A series of readings can be obtained from each test and an average value of k determined. The test is suitable for gravels and sands, and could be used for many fill materials.

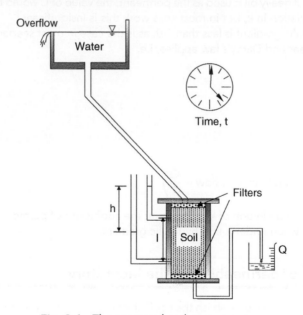

Fig. 2.4   The constant head permeameter.

Example 2.1: Constant head test

In a constant head permeameter test, the following results were obtained:

Duration of test = 4.0 min
Quantity of water collected = 300 ml

Head difference in manometer = 50 mm
Distance between manometer tappings = 100 mm
Diameter of test sample = 100 mm

Determine the coefficient of permeability in m/s.

Solution:

$$A = \frac{\pi \times 100^2}{4} = 7854\,mm^2 \quad q = \frac{300}{4 \times 60} = 1.25\,ml/s$$

$$k = \frac{ql}{Ah} = \frac{1.25 \times 1000 \times 100}{7854 \times 50} = 3.18 \times 10^{-1}mm/s = 3.2 \times 10^{-4}m/s$$

2.5.2   The falling head permeameter

A sketch of the falling head permeameter is shown in Fig. 2.5. In this test, which is suitable for silts and some clays, the flow of water through the sample is measured at the inlet. The height, h_1, in the standpipe is measured and the valve is then opened as a stop clock is started. After a measured time, t, the height to which the water level has fallen, h_2, is determined.

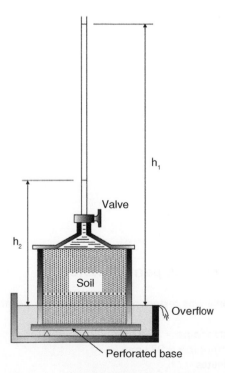

Fig. 2.5   The falling head permeameter.

k is given by the formula:

$$k = \frac{al}{At} \ln \frac{h_1}{h_2}$$

where

A = cross-sectional area of sample
a = cross-sectional area of standpipe
l = length of sample.

As just mentioned, during the test, the water in the standpipe falls from a height h_1 to a final height h_2.
Let h be the height at some time, t.
Consider a small time interval, dt, and let the change in the level of h during this time be −dh (negative as it is a drop in elevation).
The quantity of flow through the sample in time dt = −adh and is given the symbol dQ. Now

$$dQ = Aki\ dt$$

or

$$= Ak\frac{h}{l}dt = -adh$$

$$dt = -\frac{al}{Ak}\frac{dh}{h}$$

Integrating between the test limits:

$$\int_0^t dt = -\frac{al}{Ak}\int_{h_1}^{h_2}\frac{1}{h}dh$$

i.e.

$$t = -\frac{al}{Ak}\ln\frac{h_2}{h_1} = \frac{al}{Ak}\ln\frac{h_1}{h_2}$$

or

$$k = \frac{al}{At}\ln\frac{h_1}{h_2} \qquad\qquad (2.7)$$

Example 2.2: Falling head permeameter

An undisturbed soil sample was tested in a falling head permeameter. The results were:

Initial head of water in standpipe = 1500 mm
Final head of water in standpipe = 283 mm
Duration of test = 20 minutes
Sample length = 150 mm
Sample diameter = 100 mm

Stand-pipe diameter = 5 mm
Determine the permeability of the soil in m/s.

Solution:

$$a = \frac{\pi \times 5^2}{4} = 19.64 \text{ mm}^2 \quad A = \frac{\pi \times 100^2}{4} = 7854 \text{ mm}^2$$

$$k = \frac{al}{At} \ln \frac{h_1}{h_2}$$

$$k = \frac{19.64 \times 150}{7854 \times 20 \times 60} \times \ln \frac{1500}{283} = 5.21 \times 10^{-4} \text{ mm/s} = 0.521 \times 10^{-6} \text{ m/s}$$

2.5.3   The hydraulic consolidation cell (Rowe cell)

The Rowe cell (described in Chapter 12) was developed for carrying out consolidation tests. The apparatus can also be used for determining the permeability of a soil, though it is fairly rare to see this equipment in a commercial soils laboratory.

2.6   Determination of permeability in the field

2.6.1   Field pumping test

Laboratory tests can only determine k for the small sample of soil tested. To establish the permeability of a whole aquifer, a field pumping test is carried out. The test can be used to measure the average k value of a stratum of soil below the water table and is effective up to depths of about 45 m. Details of pumping test procedures are given in BS EN ISO 22282-4:2012 (BSI, 2012) and BS ISO 14686:2003 (BSI, 2003).

A casing of about 400 mm diameter is driven to bedrock or to impervious stratum. Observation wells of at least 35 mm diameter are put down on radial lines from the casing, and both the casing and the observation wells are perforated to allow easy entrance of water. The test consists of pumping water out from the central casing at a measured rate, q and observing the resulting drawdown in groundwater level by means of the observation wells.

At least four observation wells, arranged in two rows at right angles to each other should be used although it may be necessary to install extra wells if the initial ones give irregular results. If there is a risk of fine soil particles clogging the observation wells then the wells should be surrounded by a suitably graded filter material (the design of filters is discussed later in this chapter) or a geofabric filter.

It may be that the site boundary conditions, e.g. a river, canal or a steep sloping surface of impermeable subsurface rock, a fault or a dyke, do not allow the two rows of observation wells to be placed at right angles. In such circumstances, the two rows of wells should be placed parallel to each other and at right angles to the offending boundary.

The minimum distance between the observation wells and the pumping well should be 10 times the radius of the pumping well and at least one of the observation wells in each row should be at a radial distance greater than twice the thickness of the ground being tested.

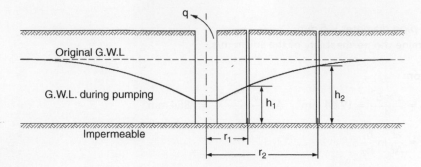

Fig. 2.6   Field pumping test.

In addition to the observation wells, an additional standpipe inside the pumping well is desirable so that a reliable record of the drawdown of the well itself can be obtained. Figure 2.6 illustrates conditions during pumping.

Consider an intermediate distance r from the centre of the pumping well and let the height of the GWL above the impermeable layer during pumping be h.

The hydraulic gradient, i, is equal to the slope of the h – r curve = $\dfrac{\partial h}{\partial r}$

and 2πrh = area of the walls of an imaginary cylinder of radius r and height h.

Now,

$$q = Aki = 2\pi rhk\frac{\partial h}{\partial r}$$

i.e.

$$q\frac{\partial r}{r} = 2\pi hk\partial h$$

and, integrating between test limits:

$$q = \int_{r_1}^{r_2} \frac{1}{r}\,\partial r = 2\pi k\int_{h_1}^{h_2} h\,\partial h$$

$$= 2\pi k\left[\frac{h_2^2 - h_1^2}{2}\right]$$

i.e.

$$q\ln\frac{r_2}{r_1} = k\pi\left(h_2^2 - h_1^2\right)$$

$$k = \frac{q\ln r_2/r_1}{\pi\left(h_2^2 - h_1^2\right)} \tag{2.8}$$

Pumping tests can be expensive as they require the installation of both the pumping and the observation wells as well as suitable pumping and support equipment. Care must be taken in the design of a suitable test programme and, before attempting to carry out any pumping test, reliable data should be obtained about the subsoil profile – by means of boreholes specially sunk for the purpose, if necessary. Suction pumps can be used where the groundwater does not have to be lowered by more than about 5 m below the intake chamber of the pump but for greater depths submersible pumps are generally necessary.

Example 2.3: Pumping test

A 9.15 m thick layer of sandy soil overlies an impermeable rock. Groundwater level is at a depth of 1.22 m below the top of the soil. Water was pumped out of the soil from a central well at the rate of 5680 kg/min and the drawdown of the water table was noted in two observation wells. These two wells were on a radial line from the centre of the main well at distances of 3.05 and 30.5 m.

During pumping, the water level in the well nearest to the pump was 4.57 m below ground level and in the furthest well was 2.13 m below ground level.

Determine an average value for the permeability of the soil in m/s.

Solution:

$$q = 5680 \, \text{kg/min} = 5.68 \, \text{m}^3/\text{min} = 0.0947 \, \text{m}^3/\text{s}$$

$$h_1 = 9.15 - 4.57 = 4.58 \, \text{m} \quad h_2 = 9.15 - 2.13 = 7.02 \, \text{m}$$

$$k = \frac{q \ln r_2/r_1}{\pi \left(h_2^2 - h_1^2 \right)} = \frac{0.0947 \times 2.3026}{28.3 \times \pi} = 2.45 \times 10^{-3} \, \text{m/s}$$

2.6.2   Borehole tests

Where bedrock level is very deep or where the permeabilities of different strata are required, borehole permeability tests can be used. (See Chapter 7 for details on borehole construction.)

In borehole permeability testing, *open* or *closed* systems can be used and procedures for all borehole hydraulic testing are given in BS EN ISO 22282:2012 (BSI, 2012).

Open system

A casing, perforated for a metre or so at its end, is driven into the ground. At intervals during the driving, the rate of flow of water placed under either a constant or a falling head within the borehole is determined. From this rate, together with knowledge of the hydraulic gradient and the cross-sectional area through which the water flows from the borehole into the adjacent soil, a measure of the soil's permeability can be determined.

Closed system

Whereas an open system simply uses a hydraulic head of water in the borehole to cause water to flow into the soil at the perforated section, a closed system uses a pump to introduce water into the soil or rock under pressure and involves the use of one or more *packers*.

Packer test

This test is performed in a partly unlined borehole and can apply to both open and closed systems. The unlined section is the part that is placed under test and, since loose sands and gravels would collapse in on themselves, the test is predominantly used for measuring the permeability of a section of fissured rock mass. An inflatable rubber membrane known as a *packer*, wrapped around a vertical, perforated hydraulic pipe, is lowered to the depth in the borehole at which the rock is to be tested. The pipe extends below the base of the packer. The packer is inflated to form a watertight seal above the point to be tested. Water is then pumped through the pipe at a controlled pressure and the volume of water injected into the rock over a period of time is measured and used to establish the permeability of the rock.

A variation on this single packer test is to have a packer both above and beneath the section to be tested. Such an arrangement is known as a *double packer test*.

2.7 Approximation of coefficient of permeability

It is obvious that a soil's coefficient of permeability depends upon its porosity, which is itself related to the particle size distribution curve of the soil (a gravel is much more permeable than a clay). It would therefore seem possible to approximate the permeability of a soil given its particle size distribution, and typical ranges of k for different soil types are given below. In addition, k can be approximated for a clean sand thus:

$$k \approx 0.01 D_{10}^2 \text{ m/s}$$

where D_{10} = effective size in mm.

Wise (1992) offered approximations for other soils based on pore size distribution but it should be remembered that no formula is as good as an actual permeability test.

Typical ranges of coefficient of permeability

Gravel $>10^{-1}$ m/s
Sands 10^{-1}–10^{-5} m/s
Fine sands, coarse silts 10^{-5}–10^{-7} m/s
Silts 10^{-7}–10^{-9} m/s
Clays $<10^{-9}$ m/s

Example 2.4: Approximation of k

Calculate an approximate value for the coefficient of permeability for the soil in Example 1.2.

Solution:

$$k \approx 0.01 D_{10}^2 = 0.01 \times 0.18^2 = 3.2 \times 10^{-4} \text{ m/s}$$

2.8 General differential equation of flow

Figure 2.7 shows an elemental cube, of dimensions d_x, d_y, and d_z, in an orthotropic soil with an excess hydraulic head h acting at its centre (an *orthotropic* soil is a soil whose material properties are different in all directions).

Let the coefficients of permeability in the coordinate directions x, y, and z be k_x, k_y, and k_z, respectively. Consider the component of flow in the x direction.

The component of the hydraulic gradient, i_x, at the centre of the element will be:

$$i_x = -\frac{\partial h}{\partial x} \tag{2.9}$$

(Note that it is of negative sign as there is a head loss in the direction of flow).

The rate of change of the hydraulic gradient i_x along the length of the element in the x direction will be:

$$\frac{\partial i_x}{\partial x} = -\frac{\partial^2 h}{\partial x^2} \tag{2.10}$$

Hence, the gradient at the face of the element nearest the origin

$$= -\frac{\partial h}{\partial x} + \left(\frac{\partial i_x}{\partial x}\right)\left(\frac{-dx}{2}\right)$$

$$= -\frac{\partial h}{\partial x} + \frac{\partial^2 h}{\partial x^2}\frac{dx}{2} \tag{2.11}$$

From Darcy's law:

$$\text{Flow} = Aki = k_x\left(-\frac{\partial h}{\partial x} + \frac{\partial^2 h}{\partial x^2}\frac{dx}{2}\right)dy.dz \tag{2.12}$$

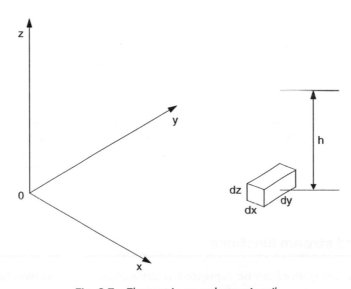

Fig. 2.7 Element in an orthotropic soil.

The gradient at the face furthest from the origin is:

$$-\frac{\partial h}{\partial x} + \left(\frac{\partial i_x}{\partial x}\right)\frac{dx}{2} = -\frac{\partial h}{\partial x} - \frac{\partial^2 h}{\partial x^2}\frac{dx}{2} \tag{2.13}$$

Therefore,

$$\text{Flow} = k_x\left(-\frac{\partial h}{\partial x} - \frac{\partial^2 h}{\partial x^2}\frac{dx}{2}\right)dy.dz \tag{2.14}$$

Equations (2.12) and (2.14) represent respectively the flow into and out of the element in the x direction, so that the net rate of increase of water within the element, i.e. the rate of change of the volume of the element, is (2.12) – (2.14).

Similar expressions may be obtained for flow in the y and z directions. The sum of the rates of change of volume in the three directions gives the rate of change of the total volume:

$$\left(\frac{k_x\partial^2 h}{\partial x^2} + \frac{k_y\partial^2 h}{\partial y^2} + \frac{k_z\partial^2 h}{\partial z^2}\right)dx.dy.dz \tag{2.15}$$

Under the laminar flow conditions that apply in seepage problems, there is no change in volume and the above expression must equal zero:

$$\frac{k_x\partial^2 h}{\partial x^2} + \frac{k_y\partial^2 h}{\partial y^2} + \frac{k_z\partial^2 h}{\partial z^2} = 0 \tag{2.16}$$

This is the general expression for three-dimensional flow. In many seepage problems, the analysis can be carried out in two dimensions, the y term usually being taken as zero so that the expression becomes:

$$\frac{k_x\partial^2 h}{\partial x^2} + \frac{k_z\partial^2 h}{\partial z^2} = 0 \tag{2.17}$$

If the soil is isotropic, $k_x = k_z = k$ and the expression is:

$$\frac{\partial^2 h}{\partial x^2} + \frac{\partial^2 h}{\partial z^2} = 0 \tag{2.18}$$

An *isotropic* soil is a soil whose material properties are the same in all directions.

It should be noted that these expressions only apply when the fluid flowing through the soil is incompressible. This is more or less the case in seepage problems when submerged soils are under consideration, but in partially saturated soils considerable volume changes may occur and the expressions are no longer valid.

2.9 Potential and stream functions

The Laplacian equation just derived can be expressed in terms of the two conjugate functions ϕ and ψ.

If we put

$$\frac{\partial \phi}{\partial x} = v_x = ki_x = -\frac{k\partial h}{\partial x} \quad \text{and} \quad \frac{\partial \phi}{\partial z} = v_z = -\frac{k\partial h}{\partial z}$$

then

$$\frac{\partial^2 \phi}{\partial x^2} = -\frac{k\partial^2 h}{\partial x^2} \quad \text{and} \quad \frac{\partial^2 \phi}{\partial z^2} = -\frac{k\partial^2 h}{\partial z^2}$$

hence

$$\frac{\partial^2 \phi}{\partial x^2} + \frac{\partial^2 \phi}{\partial z^2} = 0 \qquad\qquad (2.19)$$

Also, if we put

$$\frac{\partial \psi}{\partial z} = v_x = \frac{\partial \phi}{\partial x} \quad \text{and} \quad -\frac{\partial \psi}{\partial x} = v_z = \frac{\partial \phi}{\partial z}$$

then

$$\frac{\partial^2 \psi}{\partial z^2} = \frac{\partial^2 \phi}{\partial x \partial z} \quad \text{and} \quad \frac{\partial^2 \psi}{\partial x^2} = -\frac{\partial^2 \phi}{\partial x \partial z}$$

hence

$$\frac{\partial^2 \psi}{\partial x^2} + \frac{\partial^2 \psi}{\partial z^2} = 0 \qquad\qquad (2.20)$$

ϕ and ψ are known respectively as potential and stream functions. If ϕ is given a particular constant value then an equation of the form h = a constant can be derived (the equation of an equipotential line); if ψ is given a particular constant value then the equation derived is that of a stream or flow line.

Direct integration of these expressions to obtain a solution is possible for straightforward cases. However, in general, such integration cannot be easily carried out and a solution obtained by a graphical method in which a flow net is drawn has been used by engineers for many decades. Nowadays, however, much use is made of computer software to find the solution using numerical techniques, such as the finite difference and finite element methods. Nevertheless, the method for drawing a flow net by hand is given in Section 2.10.3 for readers interested in learning the techniques involved. The finite difference technique is described in Chapter 13 where it is applied to the numerical determination of consolidation.

2.10   Flow nets

The flow of water through a soil can be represented graphically by a *flow net*; a form of curvilinear net made up of a set of *flow lines* intersected by a set of *equipotential lines*.

Flow lines

The paths which water particles follow in the course of seepage are known as *flow lines*. Water flows from points of high to points of low head and makes smooth curves when changing direction. Hence, we can draw by hand or by computer, a series of smooth curves representing the paths followed by moving water particles.

Equipotential lines

As the water moves along the flow line, it experiences a continuous loss of head. If we can obtain the head causing flow at points along a flow line, then by joining up points of equal potential, we obtain a second set of lines known as *equipotential lines*.

2.10.1 Flow quantities

Referring back to Section 2.2.4, it is seen that the potential drop between two adjacent equipotentials divided by the distance between them is the hydraulic gradient. It attains a maximum along a path normal to the equipotentials and, in isotropic soil, the flow follows the paths of the steepest gradients so that flow lines cross equipotential lines at right angles.

Figure 2.8 shows a typical flow net representing seepage through a soil beneath a dam. The flow is assumed to be two-dimensional, a condition that covers a large number of seepage problems encountered in practice.

From Darcy's law $q = Aki$, so if we consider unit width of soil and if Δq = the unit flow through a flow channel (the space between adjacent flow lines), then:

$$\Delta q = b \times l \times k \times i = bki$$

where b = distance between the two flow lines.

In Fig. 2.8, the element ABCD is bounded by the same flow lines as element $A_1B_1C_1D_1$ and by the same equipotentials as element $A_2B_2C_2D_2$.

For any element in the net, $\Delta q = bki = bk\Delta h/l$, where

Δh = head loss between the two equipotentials
l = distance between the equipotentials (see Fig. 2.9).

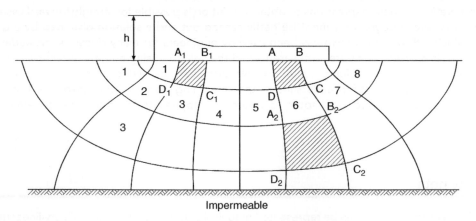

Fig. 2.8 Flow net for seepage beneath a dam.

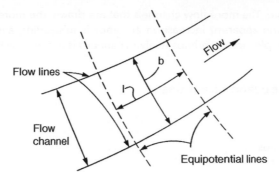

Fig. 2.9 Section of a flow net.

Referring to Fig. 2.8:

Flow through $A_1B_1C_1D_1 = \Delta q_1 = k\Delta h_1 \dfrac{b_1}{l_1}$

Flow through $A_2B_2C_2D_2 = \Delta q_2 = k\Delta h_2 \dfrac{b_2}{l_2}$

Flow through $ABCD = \Delta q = k\Delta h \dfrac{b}{l}$

If we assume that the soil is homogeneous and isotropic then k is the same for all elements and it is possible to draw the flow net so that $b_1 = l_1$, $b_2 = l_2$, $b = l$. When we have this arrangement, the elements are termed 'squares' and the flow net is a square flow net. With this condition:

$$\frac{b_1}{l_1} = \frac{b_2}{l_2} = \frac{b}{l} = 1.0$$

Since square ABCD has the same flow lines as $A_1B_1C_1D_1$,

$$\Delta q = \Delta q_1$$

Since square ABCD has the same equipotentials as $A_2B_2C_2D_2$,

$$\Delta h = \Delta h_2$$
$$\Rightarrow \Delta q_2 = k\Delta h_2 = k\Delta h = \Delta q = \Delta q_1$$

i.e.

$$\Delta q = \Delta q_1 = \Delta q_2 \text{ and } \Delta h = \Delta h_1 = \Delta h_2$$

Hence, in a flow net, where all the elements are square, there is the same quantity of unit flow through, and the same head drop across, each element.

No element in a flow net can be truly square, but the vast majority of the elements do approximate to squares in that the four corners of the element are at right angles and the distance between the flow lines, b, equals the distance between the equipotentials, l. Some relaxation is needed when asserting that a certain element is a square and some elements will be more triangular in shape but provided that the flow net is drawn with a sensible number of flow channels (generally five or six), the results obtained will be within the range of accuracy possible. The more flow channels that are drawn, the more the elements will approximate to true squares, but the apparent increase in accuracy is misleading and the extra work involved (if drawing by hand, for example, up to perhaps twelve channels) is not worthwhile.

2.10.2   Calculation of seepage quantities

Let

N_d = number of potential drops
N_f = number of flow channels
h = total head loss
q = total quantity of unit flow

Then

$$\Delta h = \frac{h}{N_d}; \quad \Delta q = \frac{q}{N_f}$$

$$\Delta q = k\Delta h \frac{b}{l} = k\Delta h \quad \left(\text{since } \frac{b}{l} = 1 \right)$$

$$\Rightarrow k\frac{h}{N_d} = \frac{q}{N_f}$$

$$\Rightarrow \text{Total unit flow per unit length, } q = kh\frac{N_f}{N_d} \tag{2.21}$$

2.10.3   Drawing a flow net

The first step is to draw in pencil the first flow line, upon which the accuracy of the final correctness of the flow net depends. There are various boundary conditions that help to position this first flow line, including:

(i)   buried surfaces (e.g. the base of the dam, sheet piling), which are flow lines as water cannot penetrate into such surfaces;

(ii) the junction between a permeable and an impermeable material, which is also a flow line; for flow net purposes, a soil that has a permeability of one-tenth or less the permeability of the other may be regarded as impermeable;

(iii) the horizontal ground surfaces on each side of the dam, which are equipotential lines.

The procedure is as follows:

(a) draw the first flow line and hence establish the first flow channel;

(b) divide the first flow channel into squares checking visually that b = l in each element;

(c) project the equipotentials beyond the first flow channel, which gives an indication of the size of the squares in the next flow channel;

(d) determine the position of the next flow line (remembering that b = l) draw this line as a smooth curve and complete the squares in the flow channel formed;

(e) project the equipotentials and repeat the procedure until the flow net is completed.

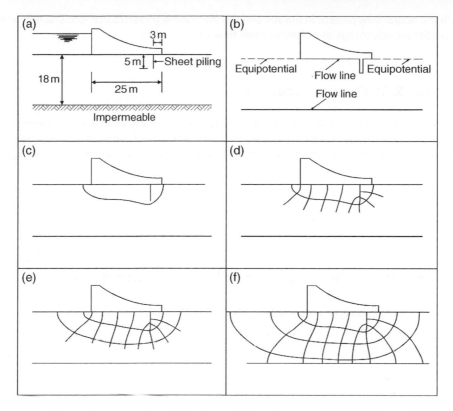

Fig. 2.10 Example of flow net construction. (a) Problem. (b) Boundary conditions. (c) Step a. (d) Step c. (e) Step d. (f) Final flow net.

As an example, suppose that it is necessary to draw the flow net for the conditions shown in Fig. 2.10a. The boundary conditions for this problem are shown in Fig. 2.10b, and the sketching procedure for the flow net is illustrated in figures c, d, e and f of Fig. 2.10.

If the flow net is correct, the following conditions will apply:

 (i) equipotentials will be at right angles to buried surfaces and the surface of the impermeable layer;
(ii) beneath the dam, the outermost flow line will be parallel to the surface of the impermeable layer.

After completing part of a flow net, it is usually possible to tell whether or not the final diagram will be correct. The curvature of the flow lines and the direction of the equipotentials indicate if there is any distortion, which tends to be magnified as more of the flow net is drawn. This gives a good indication of what was wrong with the first flow line. This line must now be redrawn in its corrected position and the procedure repeated again, amending the first flow line if necessary, until a satisfactory net is obtained.

Generally, the number of flow channels, N_f will not be a whole number, and in these cases, an estimate is made as to where the next flow line would be if the impermeable layer was lower. The width of the lowest channel can then be found (e.g. in Fig. 2.10f, $N_f = 3.3$).

Note: In flow net problems, we assume that the permeability of the soil is uniform throughout the soil's thickness. This is a considerable assumption and we see therefore that refinement in the construction of a flow net is unnecessary, since the difference between a roughly sketched net and an accurate one is small compared with the actual flow pattern in the soil and the theoretical pattern assumed. This is where numerical models have particular advantage over hand-drawn flow nets.

Example 2.5: Flow net seepage

Using Fig. 2.9f, determine the loss through seepage under the dam in cubic metres per year if $k = 3 \times 10^{-6}$ m/s and the level of water above the base of the dam is 10 m upstream and 2 m downstream. The length of the dam perpendicular to the plane of seepage is 300 m.

Solution:

From the flow net $N_f = 3.3$, $N_d = 9$
  Total head loss, $h = 10 - 2 = 8$ m

$$q/\text{metre length of dam} = kh\frac{N_f}{N_d} = 3 \times 10^{-6} \times 8 \times \frac{3.3}{9} = 8.8 \times 10^{-6}\,\text{m}^3/\text{s}$$

$$\text{Total seepage loss per year} = 300 \times 8.8 \times 60 \times 60 \times 24 \times 365 \times 10^{-6}\,\text{m}^3$$
$$= 83\,255\,\text{m}^3$$

2.11   Critical flow conditions

2.11.1   Critical hydraulic gradient, i_c

Figure 2.11 shows a sample of soil encased in a vessel of cross-sectional area A, with upward flow of water through the soil taking place under a constant head. The total head of water above the sample base = h + 1, and the head of water in the sample above the base = 1, therefore the excess hydrostatic pressure acting on the base of the sample = $\gamma_w h$.

  If any friction between the soil and the side of the container is ignored, then the soil is on the point of being washed out when the downward forces equal the upward forces:

  Downward forces = Buoyant unit weight × Volume

$$= \gamma_w \frac{G_s - 1}{1 + e} Al$$

Upward forces = $h\gamma_w A$
i.e.

$$h\,\gamma_w A = \gamma_w \frac{G_s - 1}{1 + e} Al$$

or when

$$\frac{h}{l} = \frac{G_s - 1}{1 + e} = i_c \tag{2.22}$$

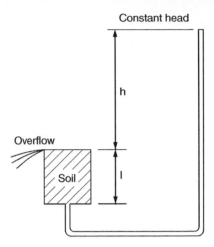

Fig. 2.11 Upward flow through a soil sample.

This particular value of hydraulic gradient is known as the *critical hydraulic gradient* and has an average value of about one for most soils. It makes a material a quicksand, which is not a type of soil but a flow condition within the soil. Generally, quicksand conditions occur in fine sands when the upward flow conditions achieve this state, but there is no theoretical reason why they should not occur in gravels (or any granular material) provided that the quantity of flow and the head are large enough. Other terms used to describe this condition are *piping* or *boiling*, but piping will not occur in fine silts and clays due to cohesive forces holding the particles together; instead there can be a heave of a large mass of soil if the upward forces are large enough.

2.11.2 Seepage force

Whenever water flows through a soil, a seepage force is exerted (as in quicksands). In Fig. 2.11, the excess head h is used up in forcing water through the soil voids over a length of l. This head dissipation is caused by friction and, because of the energy loss, a drag or force is exerted in the direction of flow.

The upward force $h\gamma_w A$ represents the seepage force, and in the case of uniform flow conditions, it can be assumed to spread uniformly throughout the volume of the soil:

$$\frac{\text{Seepage force}}{\text{Unit volume of soil}} = \frac{h\gamma_w A}{Al} = i\gamma_w$$

This means that in an isotropic soil, the seepage force acts in the direction of flow and has a magnitude = $i\gamma_w$ per unit volume.

2.11.3 Alleviation of piping

The risk of piping can occur in several circumstances, such as a cofferdam (Fig. 2.12a) or the downstream end of a dam (Fig. 2.12b).

To increase the factor of safety against piping in these cases, two methods can be adopted. The first procedure involves increasing the depth of pile penetration in Fig. 2.12a and inserting a sheet pile at the heel of the dam in Fig. 2.12b; in either case, there is an increase in the length of the flow path for the water with a resulting drop in the excess pressure at the critical section. A similar effect is achieved by laying down a blanket of impermeable material for some length along the upstream ground surface (Fig. 2.12b).

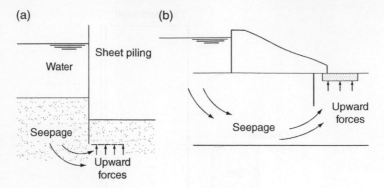

Fig. 2.12   Examples where piping can occur. (a) Cofferdam. (b) Downstream end of a dam.

The second procedure is to place a surcharge on the ground surface of the downstream side, the weight of which increases the downward forces, or to install a drainage system (e.g. perforated pipe protected by a filter) at the downstream side.

Example 2.6: Buoyant uplift

An 8 m thick layer of silty clay is overlying a gravel stratum containing water under artesian pressure. A standpipe was inserted into the gravel and water rose up the pipe to reach a level of 2 m above the top of the clay (Fig. 2.13).

The clay has a particle specific gravity of 2.7 and a natural water content of 30%. The permeability of the silty clay is 3.0×10^{-8} m/s.

It is proposed to excavate 2 m into the soil to insert a wide foundation which, when constructed, will exert a uniform pressure of 100 kPa on to its supporting soil.

Determine:

(a) the unit rate of flow of water through the silty clay in m^3 per year before the work commences;
(b) how safe the foundation will be against heaving: (i) at the end of excavation; (ii) after construction of the foundation.

Solution:

(a) Since the head of water in the gravel is greater than the depth of clay above, it follows that the GWT may be assumed to be at the ground surface.
Thus,
Head of water in clay = 8 m
Head of water in gravel = 10 m

⇒Head of water lost in clay = 2 m

q = Aki

Consider a unit area of 1 m^2 then:

$$q = 1 \times 3 \times 10^{-8} \times \frac{2}{8} = 7.5 \times 10^{-9}\,m^3/s$$

$$= 7.5 \times 10^{-9} \times 60 \times 60 \times 24 \times 365 = 0.237\,m^3/year/m^2\,of\,surface\,area$$

(b) (i) $e = wG_s = 0.3 \times 2.7 = 0.81$

$$\gamma_{sat} = \gamma_w \frac{G_s + e}{1 + e} = 9.81 \frac{3.51}{1.81} = 19.0\,kN/m^3$$

Height of clay left above gravel after excavation = 8 − 2 = 6 m

Upward pressure from water on base of clay = 10 × 9.81 = 98.1 kPa

Downward pressure of clay = 6 × 19 = 114 kPa.

It is clear that the downward pressure exceeds the upward pressure and thus, on the face of it, the foundation will not be lifted by the buoyant effect of the upward-acting water pressure, i.e. it is *safe*. We can quantify how 'safe' the foundation is against buoyancy by introducing the term *factor of safety*, F:

$$\text{Factor of safety,}\quad F = \frac{\text{Downward pressure}}{\text{Upward pressure}} = \frac{114}{98.1} = 1.16$$

(ii) Downward pressure after construction = 114 + 100 = 214 kPa

$$\text{Factor of safety,}\quad F = \frac{214}{98.1} = 2.18$$

i.e. the factor of safety against buoyant uplift is higher after construction.

We can also assess the safety against buoyancy using the limit state design approach defined in Eurocode 7 (see Chapter 6). The solution to Example 2.6 when assessed in accordance with Eurocode 7 is available for download from the companion website.

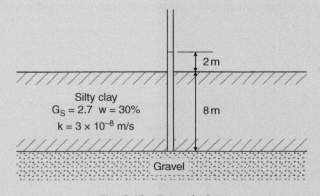

Fig. 2.13 Example 2.6.

Units of pressure

The pascal is the stress value of one newton per square metre, $1.0\,N/m^2$, and is given the symbol Pa. In the example above, pressure has been expressed in kilopascals, kPa. Pressure could have equally been expressed in kN/m^2 as the two terms are synonymous.

$$1.0\,kN/m^2 = 1.0\,kPa$$
$$1.0\,MN/m^2 = 1.0\,MPa$$

2.12   Design of soil filters

As seen above, water seeping out of the soil can lead to piping and therefore drainage should be provided in such situations to ensure ground stability. To prevent soil particles being washed into the drainage system, soil filters can be provided as the interface between base material and drain. The design procedure for a filter is largely empirical, but it must comprise granular material fine enough to prevent soil particles being washed through it and yet coarse enough to allow the passage of water.

The following formulae are used in the specification of the filter material, based initially on the work of Terzaghi and developed through the experimental research of Sherard *et al.* (1984a, b):

$$D_{15}\ \text{filter} > 5 \times D_{15}\ \text{of base material}$$
$$D_{15}\ \text{filter} < 5 \times D_{85}\ \text{of base material}$$

The first equation ensures that the filter layer has a permeability several times higher than that of the soil it is designed to protect. The requirement of the second equation is to prevent piping within the filter. The ratio D_{15} (filter)/D_{85} (base) is known as the *piping ratio*.

The required thickness of a filter layer depends upon the flow conditions and can be estimated with the use of Darcy's law of flow. The filter material should be well graded, with a grading curve more or less parallel to the soil. All material should pass the 75 mm size sieve and not more than 5% should pass the 0.063 mm size sieve (see Example 2.7 and Fig. 2.14).

Protective filters are usually constructed in layers, each of which is coarser than the one below it, and for this reason they are often referred to as *reversed filters*. Even when there is no risk of piping, filters are often used to prevent erosion of foundation materials and they are extremely important in earth dams.

Example 2.7: Filter material limits

Determine the approximate limits for a filter material suitable for the material shown in Fig. 2.14.

Solution:

From the particle size distribution curve:

$$D_{15} = 0.01\,mm; \quad D_{85} = 0.2\,mm$$

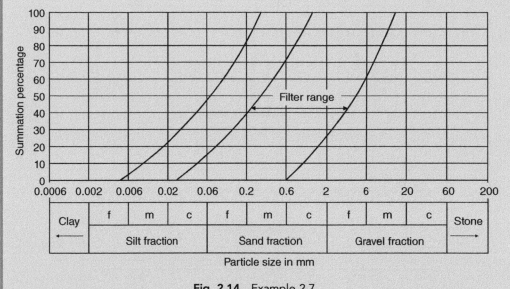

Fig. 2.14   Example 2.7

Using Terzaghi's method:

Maximum size of D_{15} for filter = 5 × D_{85} of base = 5 × 0.2 = 1.0 mm

Minimum size of D_{15} for filter = 5 × D_{15} of base = 5 × 0.01 = 0.05 mm

This method gives two points on the 15% summation line. Two lines can be drawn through these points roughly parallel to the grading curve of the soil, and the space between them is the range of material suitable as a filter (Fig. 2.14).

2.13   Capillarity and unsaturated soils

The behaviour of unsaturated soils is a relatively specialised subject area and readers interested in gaining a good understanding of the topic are referred to the publications by Fredlund *et al.* (2012) and Ng and Menzies (2007). Simple coverage of some of the key aspects involved are offered in the following sections.

2.13.1   Surface tension

Surface tension is the property of water that permits the surface molecules to carry a tensile force. Water molecules attract each other, and within a mass of water, these forces balance out. At the surface, however, the molecules are only attracted inwards and towards each other, which creates surface tension. Surface tension causes the surface of a body of water to attempt to contract into a minimum area: hence, a drop of water is spherical.

The phenomenon is easily understood if we imagine the surface of water to be covered with a thin molecular skin capable of carrying tension. Such a skin, of course, cannot exist on the surface of a liquid, but the analogy can explain surface tension effects without going into the relevant molecular theories.

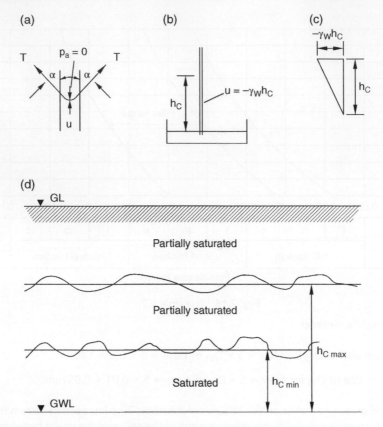

Fig. 2.15 Capillary effects.

Surface tension is given the symbol T and can be defined as the force in Newtons per millimetre length that the water surface can carry. T varies slightly with temperature, but this variation is small, and an average value usually taken for the surface tension of water is 0.000 075 N/mm (0.075 N/m).

The fact that surface tension exists can be shown in a simple laboratory experiment in which an open-ended glass capillary tube is placed in a basin of water subjected to atmospheric pressure; the rise of water within the tube is then observed. It is seen that the water wets the glass and the column of water within the tube reaches a definite height above the liquid in the basin.

The surface of the column forms a meniscus such that the curved surface of the liquid is at an angle α to the walls of the tube (Fig. 2.15a). The arrangement of the apparatus is shown in Fig. 2.15b.

The base of the column is at the same level as the water in the basin and, as the system is open, the pressure must be atmospheric. The pressure on the top surface of the column is also atmospheric. There are no externally applied forces that keep the column in position, which shows that there must be a tensile force acting within the surface film of the water.

Let

Height of water column = h_c
Radius of tube = r
Unit weight of water = γ_w

If we take atmospheric pressure as datum, i.e. the air pressure = 0, we can equate the vertical forces acting at the top of the column:

$$(T \times 2\pi r \cos\alpha) + u\pi r^2 = 0 \qquad (2.23)$$

$$\Rightarrow u = \frac{-2T\cos\alpha}{r} \qquad (2.24)$$

Hence, as expected, we see that u is negative which indicates that the water within the column is in a state of suction. The maximum value of this negative pressure is $\gamma_w h_c$ and occurs at the top of the column. The pressure distribution along the length of the tube is shown in Fig. 2.15c. It is seen that the water pressure gradually increases with loss of elevation to a value of 0 at the base of the column.

An expression for the height h_c can be obtained by substituting $u = -\gamma_w h_c$ in the above expression to yield:

$$h_c = \frac{2T\cos\alpha}{\gamma_w r} \qquad (2.25)$$

From the two expressions, we see that the magnitudes of both $-u$ and h_c increase as r decreases.

A further interesting point is that, if we assume that the weight of the capillary tube is negligible, then the only vertical forces acting are the downward weight of the water column supported by the surface tension at the top and the reaction at the base support of the tube. The tube must therefore be in compression. The compressive force acting on the walls of the tube will be constant along the length of the water column and of magnitude $2\pi T \cos\alpha$ (or $\pi r^2 h_c \gamma_w$).

It may be noted that for pure water in contact with clean glass which it wets, the value of angle α is zero. In this case, the radius of the meniscus is equal to the radius of the tube and the derived formulae can be simplified by removing the term $\cos\alpha$.

With the use of the expression for h_c we can obtain an estimate of the theoretical capillary rise that will occur in a clay deposit. The average void size in a clay is about 3 μm and, taking $\alpha = 0$, the formula gives $h_c = 5.0$ m. This possibly explains why the voids exposed when a sample of a clay deposit is split apart are often moist. However, capillary rises of this magnitude seldom occur in practice as the upward velocity of the water flow through a clay in the capillary fringe is extremely small and is often further restricted by adsorbed water films, which considerably reduce the free diameter of the voids.

2.13.2   Capillary effects in soil

The region within which water is drawn above the water table by capillarity is known as the *capillary fringe*. A soil mass, of course, is not a capillary tube system, but a study of theoretical capillarity enables the determination of a qualitative view of the behaviour of water in the capillary fringe of a soil deposit. Water in this fringe can be regarded as being in a state of negative pressure, i.e. at pressure values below atmospheric. A diagram of a capillary fringe is shown in Fig. 2.15d.

The minimum height of the fringe, $h_{c,min}$, is governed by the maximum size of the voids within the soil. Up to this height above the water table the soil will be sufficiently close to full saturation to be considered as such.

The maximum height of the fringe $h_{c,max}$, is governed by the minimum size of the voids. Within the range $h_{c,min}$ to $h_{c,max}$, the soil can be only partially saturated.

We saw in Section 2.1.2 that Terzaghi and Peck (1948) give an approximate relationship between $h_{c,max}$ and grain size for a granular soil:

$$h_c \approx \frac{C}{eD_{10}} \text{ mm} \qquad (2.26)$$

Owing to the irregular nature of the conduits in a soil mass, it is not possible, even approximately, to calculate water content distributions above the water table from the theory of capillarity. This is a problem of importance in highway engineering and is best approached by the concept of soil suction.

2.13.3 Soil suction

The capacity of a soil above the groundwater table to retain water within its structure is related to the prevailing suction and to the soil properties within the whole matrix of the soil, e.g. void and soil particle sizes, amount of held water, etc. For this reason, it is often referred to as *matrix* or *matric* suction.

It is generally accepted that the amount of matric suction, s, present in an unsaturated soil is the difference between the values of the air pressure, u_a, and the water pressure, u_w.

$$s = u_a - u_w \tag{2.27}$$

If u_a is constant, then the variation in the suction value of an unsaturated soil depends upon the value of the pore water pressure within it. This value is itself related to the degree of saturation of the soil.

2.13.4 The water retention curve

If a slight suction is applied to a saturated soil, no net outflow of water from the pores is caused. However, as the suction is increased, water starts to flow out of the larger pores within the soil matrix. As the suction is increased further, more water flows from the smaller pores until at some limit, corresponding to a very high suction, only the very narrow pores contain water. Additionally, the thickness of the adsorbed water envelopes around the soil particles reduces as the suction increases. Increasing suction is thus associated with decreasing soil wetness or water content. The amount of water remaining in the soil is a function of the pore sizes and volumes, and hence a function of the matric suction. This function can be determined experimentally and may be represented graphically as the *water retention curve*, such as the examples shown in Fig. 2.16.

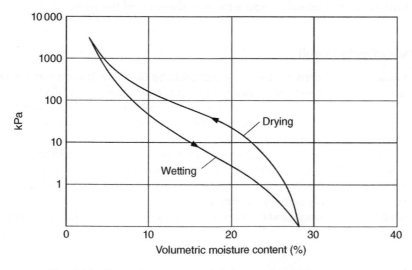

Fig. 2.16 Example wetting and drying water retention curves.

The amount of water in the pores for a particular value of suction will depend on whether the soil is wetting or drying. This gives rise to the phenomenon known as *hysteresis*, and the shape of the water retention curve for each process is shown in Fig. 2.16. A full descriptive text on water retention curves and hysteresis is given by Fredlund *et al.* (2012).

2.13.5   Measurement of soil suction

From a geotechnical point of view, there are two components of soil suction as follows:

1. *matric suction*: that part of the water retention energy created by the soil matrix;
2. *osmotic suction*: that part of the water retention energy created by the presence of dissolved salts in the soil water.

It should be noted that these two forms of soil suction are completely independent and have no effect on each other.

The total suction exhibited by a soil is obviously the summation of the matric and the osmotic suctions.

If a soil is granular and free of salt, there is no osmotic suction and the matric and total suctions are equal. However, clays contain salts and these salts cause a reduction in the vapour pressure. This results in an increase in the total suction, and this increase is the energy needed to transfer water into the vapour phase (i.e. the osmotic suction).

There are several types of equipment available which can be used to measure soil suction values. Amongst them are psychrometers, porous blocks, filter papers, suction plates, pressure plates, and tensiometers, the last being the most popular for *in situ* measurements. A useful survey was prepared by Ridley (2015).

The psychrometer method

A psychrometer is used to measure humidity and is therefore suitable to measure total soil suction, i.e. the summation of the matric and the osmotic components. The equipment and its operation have been described by Fredlund *et al.* (2012).

A sample of the soil to be tested is placed in a plastic container. A hole is then drilled to the centre of the specimen, a calibrated psychrometer inserted and the drilled hole is backfilled with extra soil material. The whole unit is finally sealed with plastic sheeting and placed in an airtight container, where it is left for three days with its temperature maintained at 25 °C. After this time, the soil sample is deemed to have achieved both thermal and vapour pressure equilibrium and relative humidity measurements can be taken.

The filter paper method

With this technique, described by Campbell and Gee (1986), both total and matric suctions can be measured. In a typical test, the soil specimen is prepared in a cylindrical plastic container and a dry filter paper disc is placed over its upper surface (Fig. 2.17). This filter will measure the matric suction.

A perforated glass disc is placed over the filter paper and a further filter paper is then placed over the glass. As this top filter paper is not in actual contact with the soil sample, it can only measure the total suction.

The assembled specimen/filter paper is left for at least a month, at a temperature of 25 °C, to obtain thermal and vapour equilibrium. At the end of this time, the assembly is dismantled, and the water contents of the specimen and the two filter papers are determined. The water contents obtained for the filter papers can be converted into the required suction values using a suction/water content curve for the filter paper material.

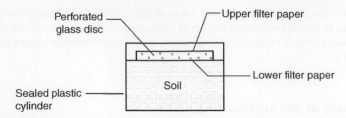

Fig. 2.17   Soil suction measurement – an arrangement for the filter paper method.

The tensiometer

Stannard (1992) presented a review of the standard tensiometer and covered the relevant theory, its construction, and possible uses. The apparatus is mainly used for *in situ* measurements and consists of a porous ceramic cup placed in contact with the soil to be tested.

A borehole is put down to the required depth and the ceramic filter lowered into position. Water is then allowed to exit from a water reservoir within the tensiometer and to enter the soil. The operation continues until the tensile stress holding the water in the tensiometer equals the stress holding the water in the soil (i.e. the total soil suction).

The tensile stress in the water in the tensiometer is measured by a pressure measuring transducer and is taken to be the value of the total soil suction.

The tensiometer must be fully de-aired during installation if accurate results are to be obtained. The response time of the type of apparatus described is only a few minutes but it has the disadvantage, until recently, that it could only be used to measure suctions up to about 100 kPa.

2.14   Earth dams

2.14.1   Seepage patterns through an earth dam

As the upper flow line is subjected to atmospheric pressure, the boundary conditions are not completely defined, and it is consequently difficult to sketch a flow net until this line has been located.

Part of such a flow net is shown in Fig. 2.18. It has already been shown that the hydraulic head at a point is the summation of velocity, pressure, and elevation heads. As the top flow line is at atmospheric pressure, the

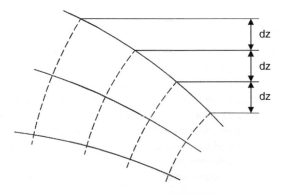

Fig. 2.18   Part of a flow net for an earth dam.

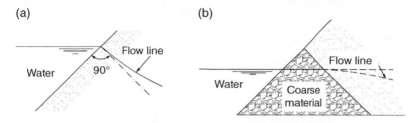

Fig. 2.19 Conditions at the start of an upper flow line.

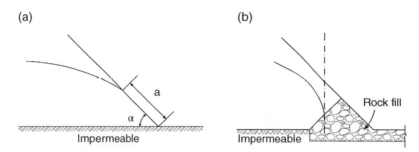

Fig. 2.20 Conditions at the downstream end of an upper flow line. (a) Flow line tangential to downstream slope. (b) Flow line vertical at exit.

only type of head that can exist along it is elevational, so that between each successive point where an equipotential cuts an upper flow line, there must be equal drops in elevation. This is the first of three conditions that must be satisfied by the upper flow line.

The second condition is that, as the upstream face of the dam is an equipotential, the flow line must start at right angles to it (see Fig. 2.19a), but an exception to this rule is illustrated in Fig. 2.19b where the coarse material is so permeable that the resistance to flow is negligible and the upstream equipotential is, in effect, the downstream face of the coarse material. The top flow line cannot be normal to this surface as water with elevation head only cannot flow upwards, so that in this case the flow line starts horizontally.

The third condition concerns the downstream end of the flow line where the water tends to follow the direction of gravity and the flow line either exits at a tangent to the downstream face of the dam (Fig. 2.20a) or, if a filter of coarse material is inserted, takes up a vertical direction in its exit into the filter (Fig. 2.20b).

2.14.2 Types of flow occurring in an earth dam

From Fig. 2.20, it is seen that an earth dam may be subjected to two types of seepage: when the dam rests on an impermeable base, the discharge must occur on the surface of the downstream slope (the upper flow line for this case is shown in Fig. 2.21a), whereas when the dam sits on a base that is permeable at its downstream end, the discharge will occur within the dam (Fig. 2.21b). This is known as the *underdrainage case*. From a stability point of view, underdrainage is more satisfactory since there is less chance of erosion at the downstream face and the slope can therefore be steeper but, on the other hand, seepage loss is smaller in dams resting on impermeable bases.

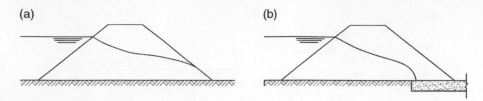

Fig. 2.21   Types of seepage through an earth dam. (a) Impermeable base. (b) Base permeable at down-stream end.

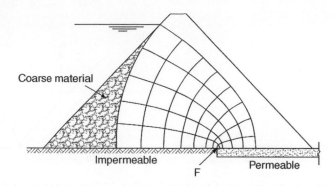

Fig. 2.22   Flow net for a theoretical earth dam.

2.14.3   Parabolic solutions for seepage through an earth dam

In Fig. 2.22 is shown the cross-section of a theoretical earth dam, the flow net of which consists of two sets of parabolas. The flow lines all have the same focus, F, as do the equipotential lines. Apart from the upstream end, actual dams do not differ substantially from this imaginary example, so that the flow net for the middle and downstream portions of the dam are similar to the theoretical parabolas (a parabola is a curve, such that any point along it is equidistant from both a fixed point, called the focus, and a fixed straight line, called the directrix). In Fig. 2.23, FC = CB.

The graphical method for determining the phreatic surface in an earth dam was evolved by Casagrande (1937) and involves the drawing of an actual parabola and then the correction of the upstream end. Casagrande showed that this parabola should start at the point C of Fig. 2.24 (which depicts a cross-section of a typical earth dam) where AC $\approx$ 0.3AB (the focus, F, is the upstream edge of the filter). To determine the directrix, draw, with compasses, the arc of the circle as shown, using centre C and radius CF; the vertical tangent to this arc is the directrix, DE. The parabola passing through C, with focus F and directrix DE, can now be constructed. Two points that are easy to establish are G and H, as FG = GD and FH = FD; other points can quickly be obtained using compasses. Having completed the parabola, a correction is made as shown to its upstream end so that the flow line actually starts from A.

This graphical solution is only applicable to a dam resting on a permeable material. When the dam is sitting on impermeable soil, the phreatic surface cuts the downstream slope at a distance (a) up the slope from the toe (Fig. 2.20a). The focus, F, is the toe of the dam, and the procedure is now to establish point C as before and draw the theoretical parabola (Fig. 2.25a). This theoretical parabola will actually cut the downstream face at a distance Δa above the actual phreatic surface; Casagrande established a relationship between a and Δa in terms of α, the angle of the downstream slope (Fig. 2.25b). In Fig. 2.25, the point J can thus be established, and the corrected flow line sketched in as shown.

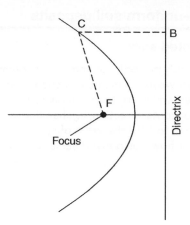

Fig. 2.23 The parabola.

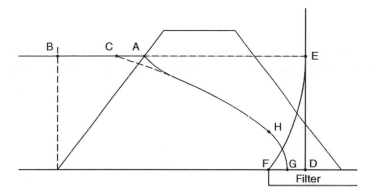

Fig. 2.24 Determination of upper flow line.

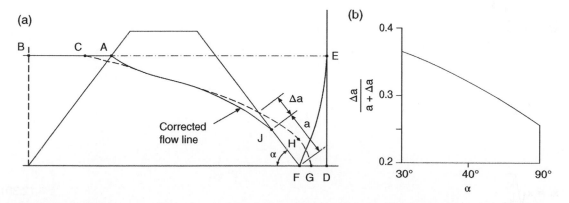

Fig. 2.25 Dam resting on an impermeable soil. (a) Construction for upper flow line. (b) Relationship between a and Δa (after Casagrande).

2.15   Seepage through non-uniform soil deposits

2.15.1   Stratification in compacted soils

Most loosely tipped deposits are probably isotropic, i.e. the value of permeability in the horizontal direction is the same as in the vertical direction. However, in the construction of embankments, spoil heaps, and dams, soil is placed and spread in loose layers which are then compacted. This construction technique results in a greater value of permeability in the horizontal direction, k_x, than that in the vertical direction (the anisotropic condition). The value of k_z is usually 1/5 to 1/10 the value of k_x.

The general differential equation for flow was derived earlier in this chapter (Equation 2.16):

$$k_x \frac{\partial^2 h}{\partial x^2} + k_y \frac{\partial^2 h}{\partial y^2} + k_z \frac{\partial^2 h}{\partial z^2} = 0$$

For the two-dimensional, i.e. anisotropic case, the equation becomes:

$$k_x \frac{\partial^2 h}{\partial x^2} + k_z \frac{\partial^2 h}{\partial z^2} = 0 \qquad (2.28)$$

Unless k_x is equal to k_z the equation is not a true Laplacian and cannot therefore be solved by a flow net. To obtain a graphical solution, the equation must be written in the form (i.e. divide through by k_z):

$$\frac{k_x}{k_z} \frac{\partial^2 h}{\partial x^2} + \frac{\partial^2 h}{\partial z^2} = 0$$

or

$$\frac{\partial^2 h}{\partial x_t^2} + \frac{\partial^2 h}{\partial z^2} = 0 \qquad (2.29)$$

where

$$\frac{1}{x_t^2} = \frac{k_x}{k_z} \frac{1}{x^2}$$

or

$$x_t^2 = x^2 \frac{k_z}{k_x}$$

i.e.

$$x_t = x\sqrt{\frac{k_z}{k_x}} \qquad (2.30)$$

This equation is Laplacian and involves the two coordinate variables x_t and z. It can be solved by a flow net provided that the net is drawn to a vertical scale of z and a horizontal scale of:

$$x_t = z\sqrt{\frac{k_z}{k_x}} \qquad\qquad (2.31)$$

2.15.2   Calculation of seepage quantities in an anisotropic soil

This is exactly as before:

$$q = kh\frac{N_f}{N_d} \qquad\qquad (2.32)$$

and the only problem is what value to use for k.

Using the transformed scale, a square flow net is drawn, and N_f and N_d are obtained. If we consider a 'square' in the transformed flow net, it will appear as shown in Fig. 2.26a. The same figure, drawn to natural scales (i.e. scale x = scale z), will appear as shown in Fig. 2.26b.

Let k' be the effective permeability for the anisotropic condition. Then k' is the operative permeability in Fig. 2.26a.

Hence, in Fig. 2.26a:

$$\text{Flow} = ak'\frac{\Delta h}{a} = k'\Delta h$$

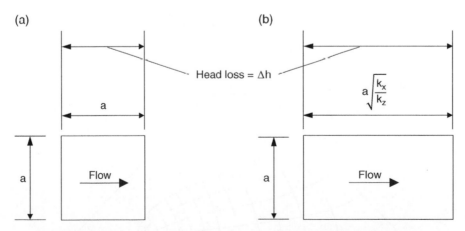

Fig. 2.26   Transformed and natural 'squares'. (a) Transformed. (b) Natural.

and, in Fig. 2.26b:

$$\text{Flow} = ak_x\frac{\Delta h}{a\sqrt{\frac{k_x}{k_z}}} = \sqrt{k_x k_z}\Delta h$$

i.e. the effective permeability, $k' = \sqrt{k_x k_z}$ (2.33)

Example 2.8: Seepage loss through dam (i)

The cross-section of an earth dam is shown in Fig. 2.27a. Assuming that the water level remains constant at 35 m, determine the seepage loss through the dam. The width of the dam is 300 m, and the soil is isotropic with $k = 5.8 \times 10^{-7}$ m/s.

Solution:

The flow net is shown in Fig. 2.27b. From it, we have $N_f = 4.0$ and $N_d = 14$.

$$q/\text{metre width of dam} = 5.8 \times 35 \times \frac{4.0}{14} \times 60 \times 60 \times 24 \times 10^{-7}$$
$$= 5.0 \times 10^{-1} \text{m}^3/\text{day}$$

Total seepage loss per day $= 300 \times 5.0 \times 10^{-1} = 150 \text{ m}^3/\text{day}$

(a)

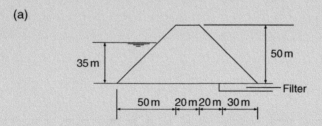

(b)

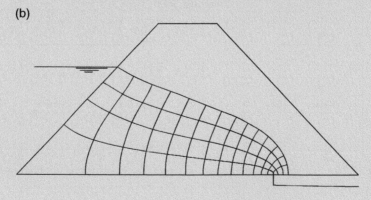

Fig. 2.27   Example 2.8. (a) The problem. (b) Flow net.

Example 2.9: Seepage loss through dam (ii)

A dam has the same details as in Example 2.8 except that the soil is anisotropic with $k_x = 5.8 \times 10^{-7}$ m/s and $k_z = 2.3 \times 10^{-7}$ m/s.
    Determine the seepage loss through the dam.

Solution:

Transformed scale for x direction, $x_t = x\sqrt{\frac{k_z}{k_x}} = x\sqrt{\frac{2.3}{5.8}} = 0.63x$

    This means that, if the vertical scale is 1 : 500, then the horizontal scale is 0.63 : 500 or 1 : 794.
    The flow net is shown in Fig. 2.28. From the flow net, $N_f = 5.0$ and $N_d = 14$.

$$k' = \sqrt{k_x k_z} = \sqrt{5.8 \times 2.3} \times 10^{-7} = 3.65 \times 10^{-7}\, \text{m/s}$$

$$\text{Total seepage loss} = 300 \times 3.65 \times 35 \times \frac{5.0}{14} \times 60 \times 60 \times 24 \times 10^{-7}$$

$$= 118\, \text{m}^3/\text{day}$$

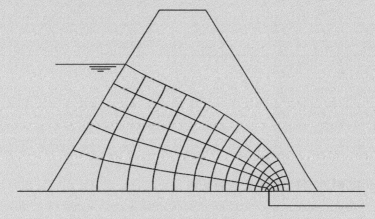

Fig. 2.28   Example 2.9.

Example 2.10: Seepage loss through dam (iii)

A dam has the same details as in Example 2.8, except that there is no filter drain at the toe. Determine the seepage loss through the dam.

Solution:

The flow net is shown in Fig. 2.29 and from it we see $N_f = 4.0$ and $N_d = 18$ (average). From the flow net, it is also seen that $a + \Delta a = 22.4$ m. Now $\alpha = 45°$, and hence:

$$\frac{\Delta a}{a + \Delta a} = 0.34 \text{ (refer to Fig. 2.25b).}$$

Hence $\Delta a = 7.6\,\text{m}$

$$\text{Total seepage loss} = 300 \times 5.8 \times \frac{4}{18} \times 35 \times 60 \times 60 \times 24 \times 10^{-7}$$
$$= 117\,\text{m}^3/\text{day}$$

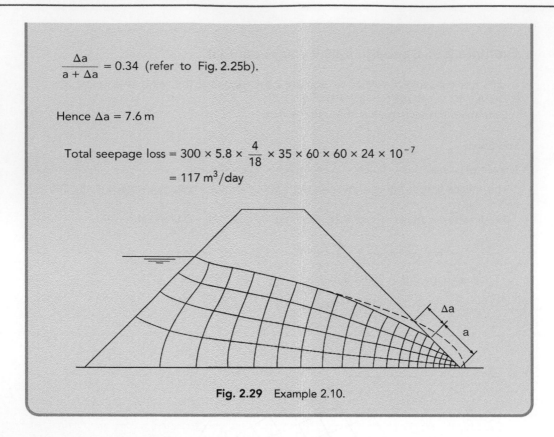

Fig. 2.29   Example 2.10.

2.15.3   Permeability of sedimentary deposits

A sedimentary deposit may consist of several different soils and it is often necessary to determine the average values of permeability in two directions, one parallel to the bedding planes and the other at right angles to them.

Let there be n layers of thicknesses H_1, H_2, H_3, ... H_n.

Let the total thickness of the layers be H.

Let k_1, k_2, k_3, ... k_n be the respective coefficients of permeability for each individual layer.

Let the average permeability for the whole deposit be k_x for flow parallel to the bedding planes, and k_z for flow perpendicular to this direction.

Consider flow parallel to the bedding planes:

$$\text{Total flow} = q = Ak_x i$$

where A = total area and i = hydraulic gradient.

This total flow must equal the sum of the flow through each layer, therefore:

$$Ak_x i = A_1 k_1 i + A_2 k_2 i + A_3 k_3 i + \cdots + A_n k_n i$$

Considering unit width of soil:

$$Hk_x i = i(H_1 k_1 + H_2 k_2 + H_3 k_3 + \cdots + H_n k_n)$$

hence

$$k_x = \frac{H_1 k_1 + H_2 k_2 + H_3 k_3 \cdots + H_n k_n}{H} \tag{2.34}$$

Considering flow perpendicular to the bedding planes:

$$\text{Total flow} = q = A k_z i = A k_1 i_1 = A k_2 i_2 = A k_3 i_3 = A k_n i_n$$

Considering unit area:

$$q = k_z i = k_1 i_1 = k_2 i_2 = k_3 i_3 = k_n i_n$$

Now

$$k_z i = k_z \frac{(h_1 + h_2 + h_3 + \cdots + h_n)}{H}$$

where h_1, h_2, h_3, etc., are the respective head losses across each layer.
Now

$$\frac{k_1 h_1}{H_1} = q; \quad \frac{k_2 h_2}{H_2} = q; \quad \frac{k_3 h_3}{H_3} = q; \quad \frac{k_n h_n}{H_n} = q$$

$$\Rightarrow h_1 = \frac{q H_1}{k_1}; \quad h_2 = \frac{q H_2}{k_2}; \quad h_3 = \frac{q H_3}{k_3}; \quad h_n = \frac{q H_n}{k_n}$$

$$\Rightarrow \frac{k_z \left(\dfrac{q H_1}{k_1} + \dfrac{q H_2}{k_2} + \dfrac{q H_3}{k_3} + \cdots + \dfrac{q H_n}{k_n} \right)}{H} = q$$

hence

$$k_z = \frac{H}{\dfrac{H_1}{k_1} + \dfrac{H_2}{k_2} + \dfrac{H_3}{k_3} + \cdots + \dfrac{H_n}{k_n}} \tag{2.35}$$

Example 2.11: Quantity of flow

A three-layered soil system consisting of fine sand, coarse silt, and fine silt in horizontal layers is shown in Fig. 2.30.

Beneath the fine silt layer, there is a stratum of water-bearing gravel with a water pressure of 155 kPa. The surface of the sand is flooded with water to a depth of 1 m.

Determine the quantity of flow per unit area in mm^3/s, and the excess hydraulic heads at the sand/coarse silt and the coarse silt/fine silt interfaces.

Solution:

$$k_z = \frac{12}{\frac{4}{2.0 \times 10^{-4}} + \frac{4}{4.0 \times 10^{-5}} + \frac{4}{2.0 \times 10^{-5}}} = 3.75 \times 10^{-5} \text{mm/s}$$

Taking the top of the gravel as datum:
Head of water due to artesian pressure = 15.5 m
Head of water due to groundwater = $3 \times 4 + 1 = 13$ m
Therefore, excess head causing flow = $15.5 - 13 = 2.5$ m.

$$\text{Flow} = q = Aki = 3.75 \times \frac{2.5}{12} \times 10^{-5} = 7.8 \times 10^{-6} \text{mm}^3/\text{s}$$

This quantity of flow is the same through each layer.
Excess head loss through fine silt:

$$\text{Flow} = 7.8 \times 10^{-6} = 2.0 \times 10^{-5} \times \frac{h}{4}$$

Therefore,

$$h = \frac{31.2 \times 10^{-6}}{2 \times 10^{-5}} = 1.56 \text{ m}$$

Excess head loss through coarse silt:

$$h = \frac{7.8 \times 10^{-6} \times 4}{4 \times 10^{-5}} = 0.78 \text{ m}$$

Excess head loss through fine sand:

$$h = \frac{7.8 \times 10^{-6} \times 4}{2 \times 10^{-4}} = 0.16 \text{ m}$$

Excess head at interface between fine and coarse silt

$$= 2.5 - 1.56 = 0.94 \text{ m}$$

Excess head at interface between fine sand and coarse silt

$$= 0.94 - 0.78 = 0.16 \text{ m}$$

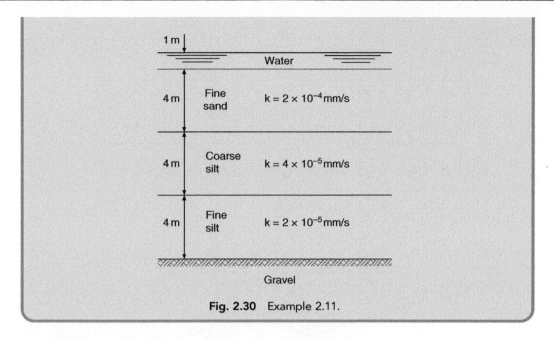

Fig. 2.30   Example 2.11.

2.15.4   Seepage through soils of different permeabilities

When water seeps from a soil of permeability k_1 into a soil of permeability k_2, the principle of the square flow net is no longer valid. If we consider a flow net in which the head drop across each figure, Δh, is a constant then, as has been shown, the flow through each figure is given by the expression:

$$\Delta q = k\Delta h \frac{b}{l} \tag{2.36}$$

If Δq is to remain the same when k is varied, then b/l must also vary. As an illustration of this effect, consider the case of two soils with $k_1 = k_2/3$.
   Then

$$\Delta q = k_1 \Delta h \frac{b_1}{l_1}$$

and

$$\Delta q = k_2 \Delta h \frac{b_2}{l_2} = 3k_1 \Delta h \frac{b_2}{l_2} \tag{2.37}$$

i.e.

$$\frac{b_1}{l_1} = 3\frac{b_2}{l_2}$$

If the portion of the flow net in the soil of permeability k_1 is square, then:

$$\frac{b_2}{l_2} = \frac{1}{3} \quad \text{or} \quad \frac{b_2}{l_2} = \frac{k_1}{k_2}$$

The effect on a flow net is illustrated in Fig. 2.31.

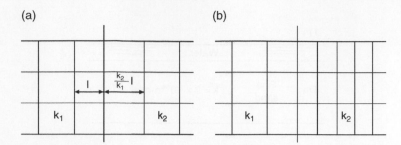

Fig. 2.31 Effect of variation of permeability on a flow net. (a) $k_2 > k_1$. (b) $k_2 < k_1$.

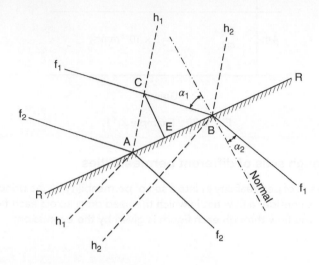

Fig. 2.32 Flow across an interface when the flow lines are at an angle to it.

2.15.5 Refraction of flow lines at interfaces

An interface is the surface or boundary between two soils. If the flow lines across an interface are normal to it, then there will be no refraction and the flow net appears as shown in Fig. 2.31. When the flow lines meet the interface at some acute angle to the normal, then the lines are bent as they pass into the second soil.

In Fig. 2.32, let RR be the interface of two soils of permeabilities, k_1 and k_2. Consider two flow lines, f_1 and f_2, making angles to the normal of α_1 and α_2 in soils 1 and 2, respectively.

Let f_1 cut RR in B and f_2 cut RR in A.

Let h_1 and h_2 be the equipotentials passing through A and B, respectively, and let the head drop between them be Δh.

With uniform flow conditions, the flow into the interface will equal the flow out. Consider flow normal to the interface.

In soil (1):

$$\text{Normal component of hydraulic gradient} = \frac{\text{Head drop along CE}}{\text{CE}}$$

$$\text{Head drop from A to E} = \Delta h \frac{AE}{CE} = \text{head drop from C to E}$$

$$\Rightarrow q_1 = ABk_1 \frac{\Delta h}{CE} \frac{AE}{AB} = k_1 \Delta h \frac{AE}{CE} = \frac{k_1 \Delta h}{\tan \alpha_1}$$

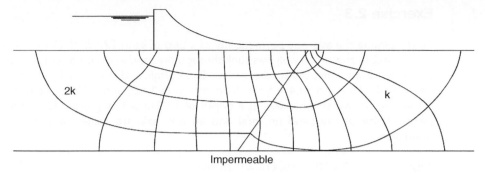

Fig. 2.33   Flow net for seepage through two soils of different permeabilities.

Similarly, it can be shown that, in soil (2):

$$q_2 = \frac{k_2 \Delta h}{\tan \alpha_2}$$

Now $q_1 = q_2$,

$$\Rightarrow \frac{k_1}{k_2} = \frac{\tan \alpha_1}{\tan \alpha_2} \tag{2.38}$$

A flow net which illustrates the effect is shown in Fig. 2.33.

Exercises

Exercise 2.1

In a falling head permeameter test on a fine sand, the sample had a diameter of 76 mm and a length of 152 mm with a standpipe of 12.7 mm diameter. A stopwatch was started when h was 508 mm and read 19.6 s when h was 254 mm. The test was repeated for a drop from 254 to 127 mm and the time was 19.4 s.
  Determine an average value for k in m/s.

Answer 1.5×10^{-4} m/s

Exercise 2.2

A sample of coarse sand 150 mm high and 55 mm in diameter was tested in a constant head permeameter. Water percolated through the soil under a head of 400 mm for 6.0 s and the discharge water had a mass of 400 g.
  Determine k in m/s.

Answer 1.05×10^{-2} m/s

Exercise 2.3

To determine the average permeability of a bed of sand 12.5 m thick overlying an impermeable stratum, a well was sunk through the sand and a pumping test carried out. After some time, the discharge was 850 kg/min and the drawdowns in observation wells 15.2 and 30.4 m from the pump were 1.625 and 1.360 m, respectively. If the original water table was at a depth of 1.95 m below ground level, find the permeability of the sand (in m/s) and an approximate value for the effective grain size.

Answer $k = 6.7 \times 10^{-4}$ m/s, $D_{10} \approx 0.26$ mm

Exercise 2.4

A cylinder of cross-sectional area 2500 mm^2 is filled with sand of permeability 5.0 mm/s. Water is caused to flow through sand under a constant head using the arrangement shown in Fig. 2.34.

Determine the quantity of water discharged in 10 minutes.

Answer 9×10^6 mm^3

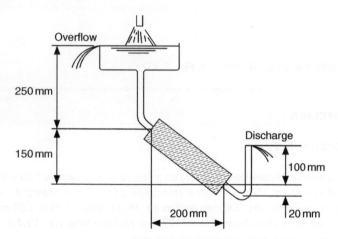

Fig. 2.34 Exercise 2.4.

Exercise 2.5

The specific gravity of particles of a sand is 2.54 and their porosity is 45% in the loose state and 37% in the dense state. What are the critical hydraulic gradients for these two states?

Answer 0.85, 0.97

Exercise 2.6

A large open excavation was made into a stratum of clay with a saturated unit weight of 17.6 kN/m^3. When the depth of the excavation reached 7.63 m the bottom rose, gradually cracked, and was flooded from below with a mixture of sand and water. Subsequent borings showed that the clay was underlain by a bed of sand with its surface at a depth of 11.3 m.

Determine the elevation to which water would have risen from the sand into a drill hole before excavation was started.

Answer   6.45 m above top of sand

Exercise 2.7

A soil deposit consists of three horizontal layers of soil: an upper stratum A (1 m thick), a middle stratum B (2 m thick), and a lower stratum C (3 m thick). Permeability tests gave the following values:

Soil A 3 × 10^{-1} mm/s
Soil B 2 × 10^{-1} mm/s
Soil C 1 × 10^{-1} mm/s

Determine the ratio of the average permeabilities in the horizontal and vertical directions.

Answer   1.22

Beneath the deposit there is a gravel layer subjected to artesian pressure, the surface of the deposit coinciding with the groundwater level. Standpipes show that the fall in head across soil A is 150 mm. Determine the value of the water pressure in the gravel.

Answer   80 kPa

Chapter 3

Stresses in the Ground

Learning objectives:

By the end of this chapter, you will have been introduced to:

- the stress conditions acting on an element of soil beneath the ground surface;
- the key principles of elastic material behaviour;
- the important concept of effective stress;
- methods for determining the vertical total stress, the pore water pressure, and the vertical effective stress for a range of situations;
- drained and undrained conditions;
- methods for determining the increase in vertical stress in soil, caused by applied loading.

3.1 State of stress in a soil mass

3.1.1 Stress–strain relationships

Before commencing a study of the material in this chapter, it is best to become familiar with the main terms used to describe the stress–strain relationships of a material. It is useful to begin by examining a typical stress–strain plot obtained for a metal (Fig. 3.1a).

Results such as those indicated in the figure would normally be obtained by subjecting a specimen of the metal to a tensile test and plotting the nominal values of tensile stress against the values of tensile strain, as the stress–strain relationship obtained is equally applicable in either tension or compression in the case of a metal.

Note: Nominal stress = actual load/original cross-sectional area of specimen, i.e. no allowance is made for reduction in area, due to necking, as the load is increased.

From the plot, it is seen that in the early stages of loading, up to point B, the stress is proportional to the strain. Unloading tests can also demonstrate that, up to the point A, the metal is elastic, in that it will return to its original dimensions if the load is removed. The limiting stress at which the elasticity effects are not quite complete, is known as the *elastic limit*, represented by point A. The limiting stress at which the linearity between stress and strain ceases is known as the *limit of proportionality*, point B.

Smith's Elements of Soil Mechanics, 10th Edition. Ian Smith.
© 2021 John Wiley & Sons Ltd. Published 2021 by John Wiley & Sons Ltd.
Companion website: www.wiley.com/go/smith/soilmechanics10e

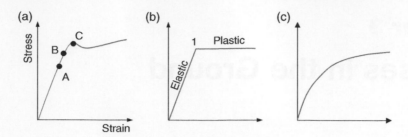

Fig. 3.1   Stress–strain relationships.

In most metals, points A and B occur so close together that they are generally assumed to coincide, i.e. elastic limit is assumed equal to the limit of proportionality.

Point C in Fig. 3.1a represents the yield point, i.e. the stress value at which there is a sudden drop of load, as illustrated, or the stress value at which there is a continuing extension with no further significant increase in load.

Figure 3.1a can be approximated to Fig. 3.1b which represents the ideal elastic–plastic material. In this diagram, point 1 represents the limit of elasticity and proportionality, and the point at which plastic behaviour occurs. The form of the compressive stress–strain relationships typical for all types of soil up to their peak values is as shown in Fig. 3.1c.

It is seen that the stress–strain relationship of a soil is never linear and, in order to obtain solutions, the designer is forced either to assume the idealised conditions of Fig. 3.1b or to solve a particular problem directly from the results of tests that subject samples of the soil to conditions that closely resemble those that are expected to apply *in situ*.

In most soil problems the induced stresses are either low enough to be well below the yield stress of the soil and it can be assumed that the soil will behave elastically (e.g. immediate settlement problems), or they are high enough for the soil to fail by plastic yield (bearing capacity and earth pressure problems), where it can be assumed that the soil will behave as a plastic material.

With soils, even further assumptions must be made if one is to obtain a solution. Generally, it is assumed that the soil is both homogeneous and isotropic. As with the assumption of perfect elasticity, these theoretical relationships do not apply in practice but can lead to realistic results when sensibly applied.

Deeper coverage of this subject and an introduction to constitutive modelling are given in Chapter 16.

3.1.2   Stresses within a soil mass

A major problem in geotechnical analysis is the estimation of the state of stress at a point at a particular depth in a soil mass.

Load acting on a soil mass, whether internal, due to its self-weight, or external, due to load applied at the boundary, creates stresses within the soil. If we consider an elemental cube of soil at the point considered, then a solution by elastic theory is possible. Each plane of the cube is subjected to a stress, σ, acting normal to the plane, together with a shear stress, τ, acting parallel to the plane. There are, therefore, a total of six stress components acting on the cube (see Fig. 3.2a). Once the values of these components are determined, then they can be compounded to give the magnitudes and directions of the principal stresses acting at the point considered.

Many geotechnical structures operate in a state of *plane strain*, i.e. one dimension of the structure is large enough for end effects to be ignored, and the problem can be regarded as one of two dimensions. The two-dimensional stress state is illustrated in Fig. 3.2b.

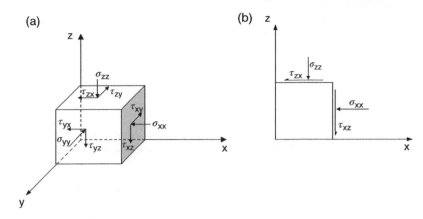

Fig. 3.2 Three- and two-dimensional stress states. (a) 3-D stress system. (b) 2-D stress system.

3.1.3 Normal stress and strain

Let us consider the cube of soil in Fig. 3.2b to have side length, L and to be subjected to a normal force, P acting on one face. This allows us to work in terms of unit area, A onto which P acts, thus creating an applied normal stress, σ. The magnitude of the normal stress is simply the force divided by area:

$$\sigma = \frac{P}{A} \tag{3.1}$$

Considering strain now, the normal strain, ε is defined as change in length, δL divided by original length, L (for relatively small values of δL):

$$\varepsilon = \frac{\delta L}{L} \tag{3.2}$$

For an elastic material, and as seen by the straight-line portion in Fig. 3.1a, stress and strain are proportional to each other. The constant of proportionality is called *Young's Modulus* or the *Modulus of Elasticity*, E and is defined as follows:

$$E = \frac{\sigma}{\varepsilon} \tag{3.3}$$

The relationship between stress and the strain that it causes, is termed the *stiffness* of the material and is instrumental in influencing settlement problems in geotechnical engineering.

Now, consider the force P to be acting in the z-direction. As the element of soil is compressed in the vertical direction (z) under P, the soil expands in both horizontal directions (x and y) as shown in Fig. 3.3.

For an *isotropic linear elastic material*, the vertical and horizontal strains are proportional and related by the following equation:

$$\varepsilon_x = \varepsilon_y = -\nu \varepsilon_z \tag{3.4}$$

The constant ν (nu) is known as *Poisson's ratio*.
Combining earlier expressions, we now have:

$$\varepsilon_x = \varepsilon_y = -\nu \varepsilon_z = -\nu \frac{\sigma_z}{E} \tag{3.5}$$

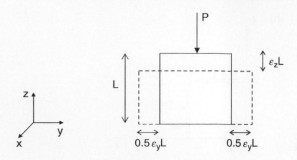

Fig. 3.3 Strains in the vertical and horizontal directions: isotropic linear elastic material.

Equally, had P been applied in either the x- or y-direction similar expressions would be derived for ε_x and ε_y.

Thus, we can define the following expressions (known as the *fundamental elastic equations*) to help us describe the response of an isotropic linear elastic material under a set of loading conditions, where the element is loaded in the x-, y-, and z-directions:

$$\left. \begin{aligned} \varepsilon_x &= \frac{1}{E}\left[\sigma_x - \nu\sigma_y - \nu\sigma_z\right] \\ \varepsilon_y &= \frac{1}{E}\left[\sigma_y - \nu\sigma_z - \nu\sigma_x\right] \\ \varepsilon_z &= \frac{1}{E}\left[\sigma_z - \nu\sigma_x - \nu\sigma_y\right] \end{aligned} \right\} \tag{3.6}$$

From these expressions, it is clear that if we know the values of E and ν for a material, we can determine the deformations that a particular set of loading conditions will cause.

As we know, soils are not isotropic linear elastic materials, so to analyse geotechnical problems of stress and strain, we need to consider the two-phase nature of soil (particles and water/air). If we consider the behaviour of the soil skeleton only, and drawing on the principles of effective stress described in Section 3.4, we can redefine the fundamental equations, as follows:

$$\left. \begin{aligned} \varepsilon_x &= \frac{1}{E'}\left[\sigma'_x - \nu'\sigma'_y - \nu'\sigma'_z\right] \\ \varepsilon_y &= \frac{1}{E'}\left[\sigma'_y - \nu'\sigma'_z - \nu'\sigma'_x\right] \\ \varepsilon_z &= \frac{1}{E'}\left[\sigma'_z - \nu'\sigma'_x - \nu'\sigma'_y\right] \end{aligned} \right\} \tag{3.7}$$

where the prime mark indicates that we are dealing with effective stress conditions.

3.1.4 Shear stress and strain

As well as normal forces, it is possible to apply a shear force to the surface of a material. If a tangential force, T is applied over an area A, the induced shear stress, τ is given by:

$$\tau = \frac{T}{A} \tag{3.8}$$

When a shear stress exists in an element of soil, the element will distort as represented by the dashed parallelogram in Fig. 3.4a. The stress causes a shear strain, shown by the angle γ.

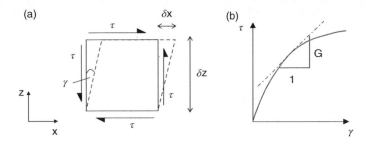

Fig. 3.4   Shear stress and strain.

From the figure it is seen that $\tan \gamma = \frac{\delta x}{\delta z}$ but, since the displacements are very small, $\tan \gamma$ is taken to equal γ.

The slope of the shear stress/shear strain plot is termed the *Shear Modulus*, G as shown in Fig. 3.4b. Over the elastic range (which is small in soils), G is constant, but at higher values of strain, G is defined as the slope at a certain point on the plot. This is quantified as the rate of change in shear stress divided by the rate of change in corresponding shear strain, and is defined as:

$$G = \frac{\delta \tau}{\delta \gamma} \tag{3.9}$$

Again, considering effective stresses, this becomes:

$$G' = \frac{\delta \tau'}{\delta \gamma} \tag{3.10}$$

It can be shown that in isotropic elastic material, G is related to E and ν by the following relationship:

$$G' = \frac{E'}{2(1 + \nu')} \tag{3.11}$$

3.1.5   Volumetric strain and bulk modulus

Consider again the cubic element of soil of Fig. 3.2a. If the element was to be subjected to three normal stresses (one acting on each of the x, y, and z planes) then the element would experience a volume reduction, which would be equal to the sum of the individual distortions in each of the three planes. Volumetric strain, ε_v, therefore is defined as follows:

$$\varepsilon_v = \varepsilon_x + \varepsilon_y + \varepsilon_z \tag{3.12}$$

If we combine this expression with the fundamental effective strain expressions from earlier, we have:

$$\varepsilon_v = \frac{1 - 2\nu'}{E'} \left[\sigma'_x + \sigma'_y + \sigma'_z \right] \tag{3.13}$$

Introducing the mean normal effective stress, p':

$$p' = \frac{\sigma'_x + \sigma'_y + \sigma'_z}{3} \tag{3.14}$$

we have:

$$\varepsilon_v = \frac{3(1-2\nu')p'}{E'} \tag{3.15}$$

and now introducing $K' = \dfrac{E'}{3(1-2\nu')}$, we have:

$$\varepsilon_v = \frac{p'}{K'} \tag{3.16}$$

K' is known as the effective *Bulk Modulus* or effective *Volumetric Modulus*.

In geotechnical analyses, the elastic parameters G' and K' are often used rather than E and ν as they enable us to consider the behaviour during shearing separately from compression. This is particularly relevant during constitutive modelling of soil behaviour, as demonstrated in Chapter 16.

3.1.6 Plane strain conditions

Under very long structures, such as a retaining wall or a strip foundation, or within a wide slope, the strain in the longitudinal direction will be zero (except at each end). Strain, therefore, only takes place in the cross-section plane and we refer to this condition as *plane strain*. Most geotechnical design work for walls, strip foundations, and slope stability consider the problem to be one of plane strain.

Under plane strain conditions, it can be shown that:

$$K'_{ps} = \frac{E'}{2(1-2\nu')(1+\nu')} \tag{3.17}$$

3.1.7 One-dimensional compression

Another loading condition experienced by soils in particular situations is where lateral restraint of the soil exists in all horizontal directions. Here, the soil can only experience compression in the vertical direction, and thus this situation is known as one-dimensional compression.

Such situations exist in the field beneath wide surface surcharges, such as raft foundations or wide embankments, and in the laboratory oedometer test (see Chapter 12) used to determine the consolidation properties.

In this loading condition, the lateral restraint provided by the adjacent soil (beneath the wide surface surcharge) or by the steel ring (oedometer test) means that all strains in the horizontal directions equal 0. Thus, distortion (compression) of the soil can only exist in the vertical direction, and therefore, we must consider Young's Modulus to be one-dimensional (E'_0).

It can be shown that:

$$E'_0 = \frac{E'(1-\nu')}{(1-2\nu')(1+\nu')} \tag{3.18}$$

For a saturated soil, with drainage prevented (i.e. undrained loading) $\nu' = 0.5$. In this case, E'_0 tends to infinity. i.e. no volume change can occur if the water cannot drain from the sample.

3.2   Total stress

The total vertical stress acting at a point in the soil (e.g. stress σ_{zz} in Fig. 3.2b) is due to the weight of everything that lies above that point including soil, water, and any load applied to the soil surface. Stresses induced by the weight of the soil subject the elemental cube to vertical stress only and they cannot create shear stresses under a level surface.

Total stress increases with depth and with unit weight, and the total vertical stress at depth z in the soil due to the weight of the soil acting above, as depicted in Fig. 3.5, is defined,

$$\sigma_z = \gamma z \tag{3.19}$$

where γ = unit weight of the soil

If the soil is multi-layered, the total vertical stress is determined by summing the stresses induced by each layer of soil (see Examples 3.1 and 3.2).

3.3   Pore water pressure

The term pore water was introduced in Section 2.2.1. Pore water experiences pressure known as the pore pressure or *pore water pressure*, u. The magnitude of the pore water pressure at a point in the soil depends on the depth below the water table and the flow conditions. In the case of a horizontal ground water table, we may be able to assume that no flow is taking place and the pore water pressure at a point beneath the ground water table can be established from the hydrostatic pressure acting. The magnitude of the pore water pressure at the water table is zero.

In Fig. 3.6, the pore water pressure is given by the hydrostatic pressure:

$$u = \gamma_w z_w \tag{3.20}$$

where

   z_w = the depth below the water table.

In situations where seepage is taking place, the pore water pressures can be established from a flow net.

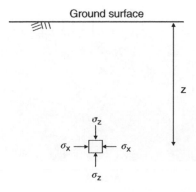

Fig. 3.5   Total stress in a homogeneous soil mass.

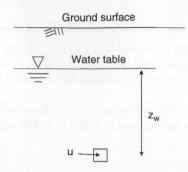

Fig. 3.6 Hydrostatic pore water pressure.

It is clear that pore water pressures are positive below the water table. Above the water table, however, the soil is saturated with capillary water in a state of suction, and here the pore water pressures will be negative.

$$u = -\gamma_w h_c$$

where

h_c = height of capillary rise.

3.4 Effective stress

The stress that controls changes in the volume and strength of a soil is known as the *effective stress*. In Chapter 1, it was seen that a soil mass consists of a collection of mineral particles with voids between them. These voids are filled with water, air and water, or air only (see Fig. 1.11).

For the moment, let us consider saturated soils only. When a load is applied to such a soil, it will be carried by the water in the soil voids (causing an increase in the pore water pressure) or by the soil skeleton (in the form of grain to grain contact stresses), or else it will be shared between the water and the soil skeleton as illustrated in Fig. 3.7. The portion of the total stress carried by the soil particles is known as the effective stress,

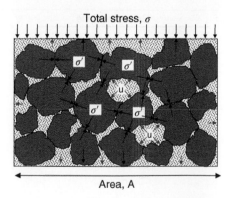

Fig. 3.7 Load carried by soil particles and pore water.

σ'. The load carried by the water gives rise to an increase in the pore water pressure, u, which, depending on the permeability, leads to water flowing under pressure out of the soil mass. This is called drainage and leads to soils possessing different strength characteristics before, during, and at the end of the drainage period. This, in turn, necessitates the need for us to understand the behaviour of the soil both immediately at the point of loading (i.e. when the soil is in an *undrained* state) and at a point in time long after the load has been applied (i.e. when the soil is in a *drained* state).

Terzaghi first presented the concept of effective stress in 1925 and again, in 1936, at the First International Conference on Soil Mechanics and Foundation Engineering, at Harvard University. He showed, from the results of many soil tests, that when an undrained saturated soil is subjected to an increase in applied normal stress, $\Delta\sigma$, the pore water pressure within the soil increases by Δu, and the value of Δu is equal to the value of $\Delta\sigma$. This increase in u caused no measurable changes in either the volumes or the strengths of the soils tested, and Terzaghi therefore used the term *neutral stress* to describe u, instead of the now more popular term pore water pressure.

Terzaghi concluded that only part of an applied stress system controls measurable changes in soil behaviour and this is the balance between the applied stresses and the neutral stress. He called these balancing stresses the effective stresses. He further explained that if a saturated soil fails by shear, the normal stress on the plane of failure, σ, also consists of the neutral stress, u, and an effective stress, σ' which led to the equation known to all geotechnical engineers:

effective stress = total stress − pore water pressure

$$\sigma' = \sigma - u \tag{3.21}$$

where the prime represents 'effective stress'.

This equation is applicable to all saturated soils.

Example 3.1: Total and effective stress

A 3 m layer of sand, of saturated unit weight 18 kN/m³, overlies a 4 m layer of clay, of saturated unit weight 20 kN/m³. If the groundwater level occurs within the sand at 2 m below the ground surface, determine the total and effective vertical stresses acting at the centre of the clay layer. The sand above groundwater level may be assumed to be saturated.

Solution:

For this sort of problem, it is usually best to draw a diagram to represent the soil conditions (see Fig. 3.8).

Total vertical stress at centre of clay = Stress due to total weight of soil above

$$\sigma_v = 2\,m\ saturated\ clay + 3\,m\ saturated\ sand$$
$$= (2 \times 20) + (3 \times 18) = 94\,kPa$$

Effective stress = Total stress − pore water pressure
$$\sigma' = 94 - 9.81(2 + 1) = 64.6\,kPa$$

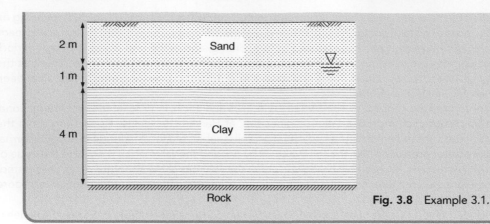

Fig. 3.8　Example 3.1.

Example 3.2:　Distributions of total stress, pwp, and effective stress

A 5 m deep deposit of silty sand lies above a 4 m deep deposit of gravel. The gravel is underlain by a deep layer of stiff clay. The ground water table is found 2 m below the ground surface. The soil properties are:

ρ_b sand above GWT = 1.7 Mg/m^3
ρ_{sat} sand below GWT = 1.95 Mg/m^3
ρ_{sat} gravel = 2.05 Mg/m^3

Draw the distributions of vertical total stress, pore water pressure, and vertical effective stress with depth down to the clay layer.

Solution:

The values of total stress, pore water pressure, and effective stress are calculated at the salient points through the soil profile. These points are where changes in conditions occur, such as the horizon between two soils.

Depth = 0 m :　$\sigma_z = 0$;　$u = 0$;　$\sigma' = 0$

Depth = 2 m :　$\sigma_z = 1.70 \times 9.81 \times 2 = 33.4$ kPa

　　　　　　　$u = 0$

　　　　　　　$\sigma'_z = 33.4 - 0 = 33.4$ kPa

Depth = 5 m :　$\sigma_z = 33.4 + (1.95 \times 9.81 \times 3) = 90.8$ kPa

　　　　　　　$u = 9.81 \times 3 = 29.4$ kPa

　　　　　　　$\sigma'_z = 90.8 - 29.4 = 61.4$ kPa

Depth = 9 m :　$\sigma_z = 90.8 + (2.05 \times 9.81 \times 4) = 171.2$ kPa

　　　　　　　$u = 9.81 \times 7 = 68.7$ kPa

　　　　　　　$\sigma'_z = 171.2 - 68.7 = 102.5$ kPa

The distributions are then plotted (see Fig. 3.9). Note that the values of total stress, pore water pressure, and effective stress all increase linearly between successive depths in the profile.

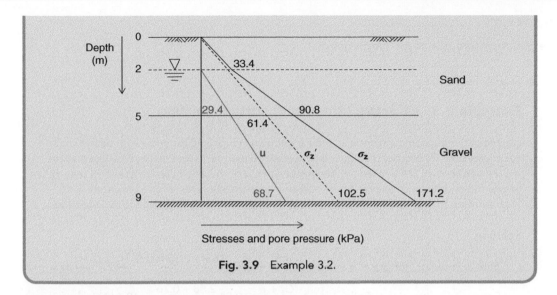

Fig. 3.9   Example 3.2.

3.5   Undrained and drained conditions in a soil

In all geotechnical analyses and design situations, the expressions *drained* and *undrained* are constantly used in reference to the state of soil or of loading conditions. Drained does not mean the soil is dry of water.

As we saw in Section 3.4, when a normal stress is applied to a saturated soil, the pore water pressure increases leading to the flow of water out of the voids. This increase in pore water pressure is known as the *excess pore water pressure*. At this stage, the soil is *undrained*. As the slightly pressurised water flows, the remaining water begins to experience a reduction in the excess pore water pressure. Eventually, the pore water pressure returns to its initial value (i.e. the excess pore water pressure is zero). The rate at which this water flow occurs varies depending on the type of soil. Granular soils (e.g. sands) have much higher permeability than cohesive soils (e.g. clays) so the water drains much more quickly in granular soils. When the excess pore water pressure equals zero, the soil is said to be *drained*.

It should be fairly obvious that granular soils can become *drained* very quickly. (Think of a bucket of water poured onto the sand on a beach – the water flows into the sand almost instantaneously.) Thus, granular soils are said to operate in a *drained state*. Consequently, when doing geotechnical analysis and design for granular soils, we will almost always do a *drained analysis*, or consider *drained conditions*.

Clays, however, have significantly lower permeabilities than granular soils and so drainage in clay takes much longer (years, as opposed to minutes/hours). Because of this, clays remain in an *undrained state* for a long time, as the excess pore water pressure very slowly dissipates and heads towards zero. When doing geotechnical analysis and design for cohesive soils, therefore, we will consider *undrained conditions* and will do an *undrained analysis*. We will carry out a drained analysis too since the soil will after all eventually reach a drained state.

In *undrained* analyses, we consider the effect of the *total stresses*, whereas in *drained* analyses we work with *effective stresses*. Thus, the terms *total stress analysis* and *effective stress analysis* are routinely used to describe *undrained* and *drained* analyses, respectively.

3.6   Stresses induced by applied loads

3.6.1   Stresses induced by uniform surface surcharge

In the case of a uniform surcharge spread over a large area, it can be assumed that the increase in vertical stress resulting from the surcharge is constant throughout the soil. Here, the vertical total stress at depth z, is given by

$\sigma_z = \gamma z + q$

where q is the magnitude of the surcharge (kPa).

Example 3.3: Effective stress with surface loading

Details of the subsoil conditions at a site are shown in Fig. 3.10 together with details of the soil properties. The ground surface is subjected to a uniform surcharge of 60 kPa and the groundwater level is 1.2 m below the upper surface of the silt. It can be assumed that the gravel has a degree of saturation of 50% and that the silt layer is fully saturated.

Determine the vertical effective stress acting at a point 1 m above the silt/rock interface.

Solution:

Bulk unit weight of gravel, $\gamma = \gamma_w \dfrac{G_s + eS_r}{1 + e} = 9.81 \dfrac{2.65 + 0.65 \times 0.5}{1 + 0.65} = 17.7 \text{ kN/m}^3$

Saturated unit weight of silt, $\gamma_{sat} = \gamma_w \dfrac{G_s + e}{1 + e} = 9.81 \dfrac{2.58 + 0.76}{1 + 0.76} = 18.6 \text{ kN/m}^3$

Effective vertical stress at 1 m above silt/rock interface

$= 60 + (1.8 \times 17.7 + 4.2 \times 18.6) - (3 \times 9.81)$

$= 140.6 \text{ kPa}$

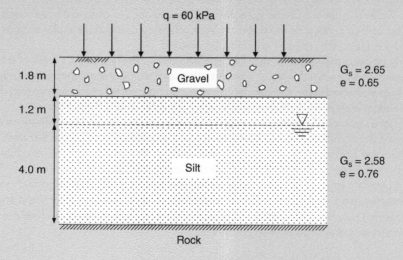

Fig. 3.10   Example 3.3.

Example 3.4: Applied wide loading

A shopping mall with underground parking is to be constructed at an urban site. The basement parking facility requires a very large raft foundation to be cast with its base at a depth of 2 m beneath the ground surface. During a site investigation, the soil profile was established and is summarised as follows:

Depth (m)	Soil type	Bulk density (Mg/m³)
0–4.0	Dense SAND	1.84
4.0	Ground water level	–
4.0–6.0	Dense gravely SAND	1.90
6.0–12.0	Glacial CLAY	1.95

The site is situated above an underground railway tunnel, whose crown is at a depth of 7 m beneath the ground surface.

Structural engineers have confirmed that the completed raft foundation will apply a uniform pressure of 200 kPa onto the soil, acting over a wide area.

Determine the magnitude of the initial, minimum, and maximum vertical total stress values that will act on the crown of the tunnel throughout this construction activity.

Solution:

Using the site investigation data, the soil profile, and problem can be presented visually as shown in Fig. 3.11.

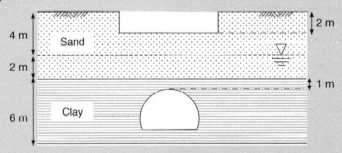

Fig. 3.11 Example 3.4.

At 7 m depth,
Initial stress:

$$\sigma = (1.84 \times 9.81 \times 4) + (1.90 \times 9.81 \times 2) + (1.95 \times 9.81 \times 1) = 128.6 \, \text{kPa}$$

Following excavation, the stress reduces as 2 m of the dense sand is removed:

$$\sigma = (1.84 \times 9.81 \times 2) + (1.90 \times 9.81 \times 2) + (1.95 \times 9.81 \times 1) = 92.5\,\text{kPa}$$

Following completion, a pressure of 200 kPa applies:

$$\sigma = 200 + (1.84 \times 9.81 \times 2) + (1.90 \times 9.81 \times 2) + (1.95 \times 9.81 \times 1) = 292.5\,\text{kPa}$$

Example 3.5: Changes in effective stress through time

A highway embankment is to be constructed as shown in Fig. 3.12. The compacted bulk density of the fill material will be 2.10 Mg/m^3.
The unit weights of the soils are:

Clay : $\gamma = 20\,\text{kN/m}^3$
Sand : $\gamma = 19.5\,\text{kN/m}^3$

Determine the vertical effective stress at points A and B for the following conditions:

(i) initially, before placement of fill,
(ii) immediately after placement of fill,
(iii) many years after placement of fill.

Solution:

(a). *Before construction*

Point A

Vertical total stress, $\sigma = 20.0 \times 2.0 = 40.0\,\text{kPa}$
Pore water pressure, $u = 9.81 \times 2.0 = 19.6\,\text{kPa}$
Vertical effective stress, $\sigma' = \sigma - u = 20.4\,\text{kPa}$

Point B

Vertical total stress, $\sigma = 20.0 \times 4.0 + 19.5 \times 1 = 99.5\,\text{kPa}$
Pore water pressure, $u = 9.81 \times 5.0 = 49.1\,\text{kPa}$
Vertical effective stress, $\sigma' = \sigma - u = 50.4\,\text{kPa}$

(a). *Immediately after construction*

Following construction, the embankment applies a wide surface surcharge:

$q = 2.10 \times 9.81 \times 5.0 = 103.0$ kPa.

The sand is drained and so water flows from this soil under the load (either horizontally or vertically into the gravel) and thus, there is no increase in pore water pressure. The clay, however, is undrained and the pore water pressure increases by 103.0 kPa.

Point A

$\sigma = 20.0 \times 2.0 + 103.0 = 143.0$ kPa
$u = 9.81 \times 2.0 + 103.0 = 122.6$ kPa
$\sigma' = \sigma - u = 20.4$ kPa

Point B

$\sigma = 20.0 \times 4.0 + 19.5 \times 1 + 103 = 202.5$ kPa
$u = 9.81 \times 5.0 = 49.1$ kPa
$\sigma' = \sigma - u = 153.4$ kPa

(b). *Many years after construction*

After many years, the excess pore water pressures in the clay will have dissipated. The pore water pressures will now be the same as they were initially.

Point A

$\sigma = 20.0 \times 2.0 + 103.0 = 143.0$ kPa
$u = 9.81 \times 2.0 = 19.6$ kPa
$\sigma' = \sigma - u = 123.4$ kPa

Point B

$\sigma = 20.0 \times 4.0 + 19.5 \times 1 + 103 = 202.5$ kPa
$u = 9.81 \times 5.0 = 49.1$ kPa
$\sigma' = \sigma - u = 153.4$ kPa

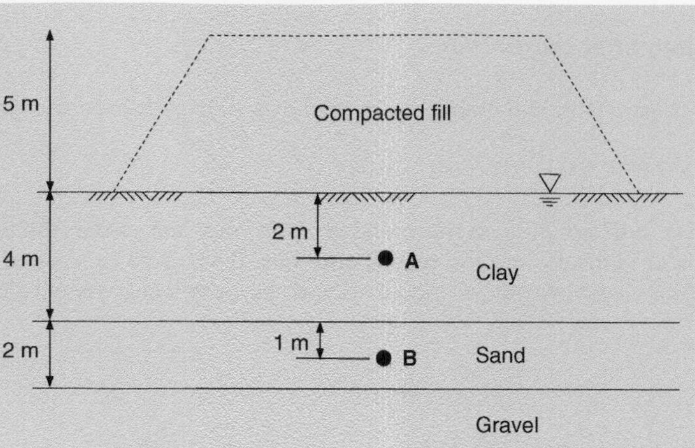

Fig. 3.12   Example 3.5.

3.6.2   Stresses induced by point load

The simplest case of applied loading has been illustrated in Examples 3.4 and 3.5. However, most loads are applied to soil through foundations of finite area so that the stresses induced within the soil directly below a particular foundation are different from those induced within the soil at the same depth but at some radial distance away from the centre of the foundation.

The determination of the stress distributions is challenging and the basic assumption used in most analyses is that the soil mass acts as a continuous, homogeneous, and elastic medium. The assumption of elasticity obviously introduces errors, but it leads to stress values that are of the right order and are suitable for most routine design work.

In most foundation problems, however, it is only necessary to be acquainted with the *increase* in vertical stresses (for settlement analysis) and the *increase* in shear stresses (for shear strength analysis).

Boussinesq (1885) evolved equations that can be used to determine the six stress components that act at a point in a semi-infinite elastic medium due to the action of a vertical point load applied on the horizontal surface of the medium.

His expression for the increase in vertical stress is:

$$\sigma_z = \frac{3Pz^3}{2\pi(r^2 + z^2)^{5/2}} \tag{3.22}$$

where

P = magnitude of concentrated load
$r = \sqrt{x^2 + y^2}$ (see Fig. 3.13, inset).

The expression has been simplified to:

$$\sigma_z = K\frac{P}{z^2}$$

where K is an influence factor.

Values of K against values of r/z are shown in Fig. 3.13.

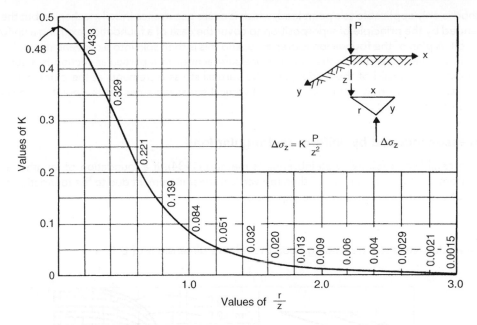

Fig. 3.13 Influence coefficients for vertical stress from a concentrated load. Based on Boussinesq (1885).

Example 3.6: Vertical stress increments beneath a point load

A concentrated load of 400 kN acts on the surface of a soil.

Determine the vertical stress increments at points directly beneath the load to a depth of 10 m.

Solution:

For points below the load r = 0 and at all depths r/z = 0, whilst from Fig. 3.13 it is seen that K = 0.48.

The results are presented below.

z (m)	z^2	$\dfrac{P}{z^2}$	$\Delta\sigma_z = K\dfrac{P}{z^2}$ (kPa)
0.5	0.25	1600.0	768.0
1.0	1.00	400.0	192.0
2.5	6.25	64.0	30.7
5.0	25.00	16.0	7.7
7.5	56.25	7.1	3.4
10.0	100.00	4.0	1.9

This method is only applicable to a point load, which is a rare occurrence in soil mechanics, but the method can be extended by the principle of superposition to cover the case of a foundation exerting a uniform pressure on the soil. A plan of the foundation is prepared, and this is then split into a convenient number of geometrical sections. The force due to the uniform pressure acting on a particular section is assumed to be concentrated at the centroid of the section, and the vertical stress increments at the point to be analysed due to all the sections are now obtained. The total vertical stress increment at the point is the summation of these increments.

3.6.3   Stresses induced by uniform rectangular load

These can be established following Steinbrenner's method (1934). If a foundation of length L and width B exerts a uniform pressure, p, on the soil then the vertical stress increment due to the foundation at a depth z below one of the corners is given by the expression:

$$\sigma_z = pI_\sigma$$

where I_σ is an influence factor depending upon the relative dimensions of L, B, and z.

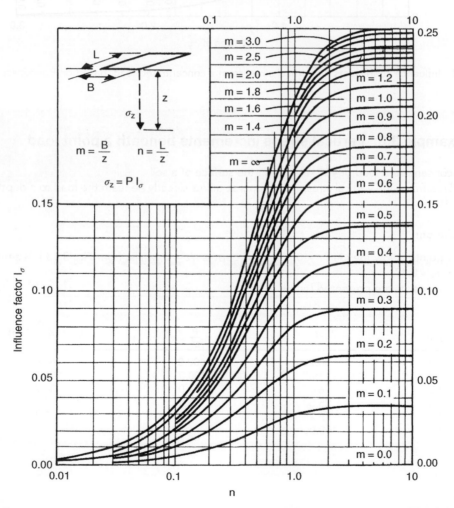

Fig. 3.14   Influence factors for the vertical stress beneath the corner of a rectangular foundation. Modified from Fadum (1948).

I_σ can be evaluated by the Boussinesq theory and values of this factor (which depend upon the two coefficients m = B/z and n = L/z) were prepared by Fadum in 1948 (Fig. 3.14).

With the use of this influence factor the determination of the vertical stress increment at a point under a foundation is very much simplified, provided that the foundation can be split into a set of rectangles or squares with corners that meet over the point considered.

To use Fig. 3.14, read vertically from x-axis (n value) to where it crosses the curve (m value) and read horizontally to the left vertical axis to read I_σ.

Note m and n are interchangeable. It doesn't matter which dimension is taken as L and which is taken as B.

Example 3.7: Vertical stress increments beneath a foundation

A 4.5 m square foundation exerts a uniform pressure of 200 kPa on a soil.

Determine the vertical stress increments due to the foundation load to a depth of 10 m below:

(i) its centre;
(ii) at a point 3 m below the foundation and 4 m from its centre along one of the axes of symmetry.

Solution:

(i) The square foundation can be divided into four squares whose corners meet at the centre O (Fig. 3.15a). Thus, L = B = 2.25 m. Establish the stress increment at regular depths, say every 2.5 m, to the total depth of 10 m.

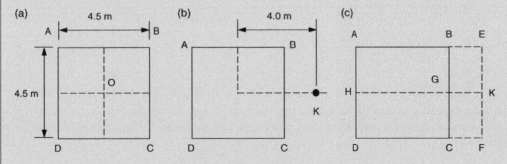

Fig. 3.15 Example 3.7.

z (m)	$m = \frac{B}{z}$	$n = \frac{L}{z}$	I_σ	$4\,I_\sigma$	σ_z (kPa)
2.5	0.90	0.90	0.16	0.64	128
5.0	0.45	0.45	0.07	0.28	56
7.5	0.30	0.30	0.04	0.16	32
10.0	0.23	0.23	0.02	0.08	16

(ii) b). The foundation is assumed to extend to the point K (Fig. 3.15c) and is now split into two rectangles, AEKH and HKFD.

For both rectangles:

$$m = \frac{B}{z} = \frac{2.25}{3} = 0.75; n = \frac{L}{z} = \frac{6.25}{3} = 2.08$$

From Fig. 3.14, $I_\sigma = 0.18$, therefore $\sigma_z = 0.18 \times 2 \times 200 = 70.4$ kPa. The effect of rectangles BEGK and KGCF must now be subtracted. For both rectangles:

$$m = \frac{2.25}{3} = 0.75; n = \frac{1.75}{3} = 0.58$$

From Fig. 3.14, $I_\sigma = 0.12$ (strictly speaking m is 0.58 and n is 0.75, but m and n are interchangeable in Fig. 3.14). Hence:

$$\sigma_z = 0.12 \times 2 \times 200 = 48.8 \text{ kPa}$$

Therefore, the vertical stress increment due to the foundation

$$= 70.4 - 48.8 = 21.6 \text{ kPa}$$

Circular foundations can also be solved by this method. The stress effects from such a foundation may be found approximately by assuming that they are the same as for a square foundation of the same area.

Example 3.8: Vertical stress increments beneath circular foundation

An oil storage tank, of diameter 20 m exerts a uniform pressure on the soil of 115 kPa. Determine the vertical stress increments for depths of 5, 10, 15, and 20 m beneath its centre.

Solution:

$$\text{Area of foundation} = \frac{\pi D^2}{4} = \frac{\pi \times 20^2}{4} = 314.2\ m^2$$

Length of side of square foundation of same area = $\sqrt{314.2}$ = 17.7 m

Now divide this 'imaginary' square into four squares whose corners meet at the centre of the foundation:

Length of side of each square = 17.7/2 = 8.85 m

z (m)	$m = n = \frac{L}{z}$	I_σ	$4\,I_\sigma$	σ_z (kPa)
5	1.77	0.23	0.92	105.8
10	0.88	0.16	0.64	73.6
15	0.59	0.10	0.40	46.0
20	0.44	0.07	0.28	32.2

3.6.4 Irregularly shaped foundations

It may not be possible to employ Fadum's method for irregularly shaped foundations, and a numerical solution is then only possible by the use of Boussinesq's coefficients, K, and the principle of superposition.

Computer software for calculating the vertical stress increments beneath irregularly shaped foundations is widely available and nowadays such software is routinely used to determine the stress values. Historically, however, the Newmark chart (Newmark, 1942) was used to determine the values and any reader interested in the use of the Newmark chart is guided to the seventh, or earlier, editions of this book for a full and detailed explanation.

3.6.5 Bulbs of pressure

If points of equal vertical pressure are plotted on a cross-section through the foundation, diagrams of the form shown in Fig. 3.16a and b are obtained.

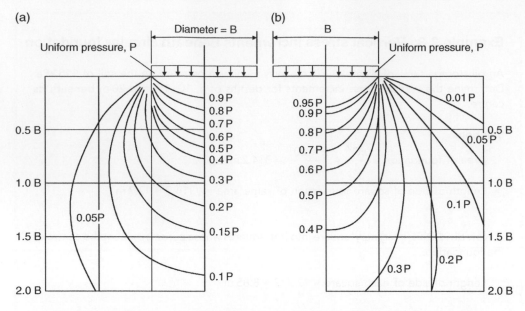

Fig. 3.16   Bulbs of pressure for vertical stress. (a) Circular footing. (b) Strip footing.

These diagrams are known as bulbs of pressure and constitute another method of determining vertical stresses at points below a foundation, or footing, that is of regular shape. The bulb of pressure for a square footing can be obtained approximately by assuming that it has the same effect on the soil as a circular footing of the same area.

In the case of a rectangular footing, the bulb of pressure will vary at cross-sections taken along the length of the foundation, but the vertical stress at points below the centre of such a foundation can still be obtained from the charts in Fig. 3.16 by either (i) assuming that the foundation is a strip footing or (ii) determining σ_z values for both the strip footing case and the square footing case and combining them by proportioning the length of the two foundations.

From a bulb of pressure, we can gain an idea of the depth of soil affected by a foundation. Significant stress values go down roughly to 2.0 times the width of the foundation, and Fig. 3.17 illustrates how the results from a plate loading test (see Chapter 7) may give quite misleading results if the proposed foundation is much larger. In the figure, the soft layer of soil is unaffected by the plate loading test but would be considerably stressed by the foundation.

Boreholes in a site investigation should therefore be taken down to a depth significantly greater than the width of the proposed foundation or until rock is encountered. Guidance on depth of boreholes is given in Section 7.2.3.

Small foundations will act together as one large foundation (Fig. 3.18) unless the foundations are at a greater distance apart (c/c) than five times their width, which is not usual. Boreholes for a building site investigation should therefore be taken down to a depth of significantly greater than the width of the proposed building – see Section 7.2.3.

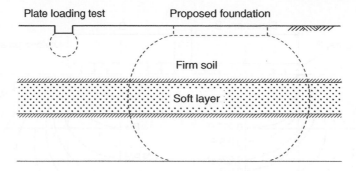

Fig. 3.17   Plate loading test may give misleading results.

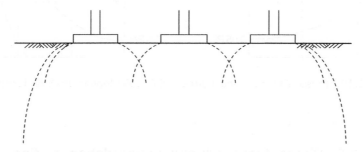

Fig. 3.18   Overlapping of pressure bulbs.

3.6.6   Shear stresses

In normal foundation design procedure, it is essential to check that the shear strength of the soil (see Chapter 4) will not be exceeded. The shear stress developed by loads from foundations of various shapes can be calculated and solutions for the case of a circular footing and for the case of a strip footing are shown in Fig. 3.19. From the figure, it is seen that, in the case of a strip footing, the maximum stress induced in the soil is p/π, this value occurring at points lying on a semicircle of diameter equal to the foundation width B. Hence, the maximum shear stress under the centre of a strip foundation occurs at a depth of 0.5B beneath the centre.

Shear stresses under a rectangular foundation

It is sometimes necessary to evaluate the shear stresses beneath a foundation in order to determine a picture of the likely overstressing in the soil. Often foundations are neither circular nor square but are rectangular, but Fig. 3.19a and b can be used to give a rough estimate of shear stress under the centre of a rectangular footing.

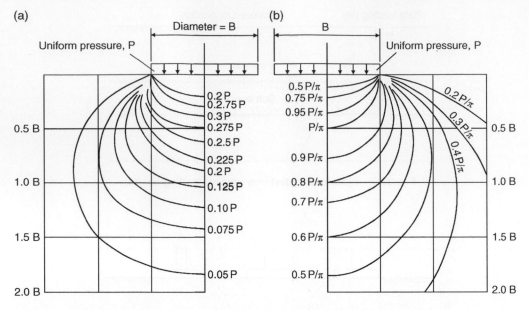

Fig. 3.19 Pressure bulbs of shear stress. (a) Circular footing. (b) Strip footing.

Example 3.9: Shear stress induced by foundation loading

A rectangular foundation has the dimensions 15 m × 5 m and exerts a uniform pressure on the soil of 600 kPa. Determine the shear stress induced by the foundation beneath the centre at a depth of 5 m.

Solution:

Strip footing, at a depth of 5 m:

$$\frac{z}{B} = \frac{5}{5} = 1.0$$

From Fig. 3.19b, τ at z = 1.0B:

$$\tau = \frac{0.8 \times 600}{\pi} = 153 \text{ kPa}$$

For a square footing:
Area = 5 × 5 = 25 m^2
Diameter of circle of same area:

$$\sqrt{\frac{25 \times 4}{\pi}} = 5.64 \text{ m}$$

Hence, the shear stress under a 5 m square foundation can be obtained from the bulb of pressure of shear stress for a circular foundation of diameter 5.64 m.

$$\frac{z}{B} = \frac{5}{5.64} = 0.89$$

From Fig. 3.19a, τ at $z = 0.89B$:

$$\tau = 0.2 \times 600 = 120 \text{ kPa}$$

These values can be combined if we proportion them to the respective areas (or lengths):

$$\tau = 120 + (153 - 120)\frac{15}{15 + 5} = 145 \text{ kPa}$$

The method is approximate, but it does give an indication of the shear stress values.

3.6.7 Contact pressure

Contact pressure is the actual pressure transmitted from the foundation to the soil. Throughout Section 3.6 it has been assumed that this contact pressure value, p, is uniform over the whole base of the foundation, but a uniformly loaded foundation will not necessarily transmit a uniform contact pressure to the soil. This is only possible if the foundation is perfectly flexible. The contact pressure distribution of a rigid foundation depends upon the type of soil beneath it. Fig. 3.20a and b shows the form of contact pressure distribution induced in a cohesive soil (a) and in a cohesionless soil (b) by a rigid, uniformly loaded, foundation.

On the assumption that the vertical settlement of the foundation is uniform, it is found from the elastic theory that the stress intensity at the edges of a foundation on cohesive soils is infinite. Obviously, local yielding of the soil will occur until the resultant distribution approximates to Fig. 3.20a.

For a rigid surface footing, sitting on sand the stress at the edges is zero as there is no overburden to give the sand shear strength, whilst the pressure distribution is roughly parabolic (Fig. 3.20b). The more the foundation is below the surface of the sand, the more shear strength there is developed at the edges of the foundation, with the result that the pressure distribution tends to be more uniform.

In the case of cohesive soil, which is at failure when the whole of the soil is at its yield stress, the distribution of the contact pressure again tends to uniformity.

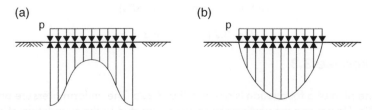

Fig. 3.20 Contact pressure distribution under a rigid foundation loaded with a uniform pressure, p. (a) Cohesive soil. (b) Cohesionless soil.

A reinforced concrete foundation is neither perfectly flexible nor perfectly rigid, the contact pressure distribution depending upon the degree of rigidity. This pressure distribution should be considered when designing for the moments and shears in the foundation, but in order to evaluate shear and vertical stresses below the foundation, the assumption of a uniform load inducing a uniform pressure is sufficiently accurate.

Exercises

Exercise 3.1

A raft foundation subjects its supporting soil to a uniform pressure of 300 kPa. The dimensions of the raft are 6.1 m by 15.25 m. Determine the vertical stress increments due to the raft at a depth of 4.58 m below it (i) at the centre of the raft and (ii) at the central points of the long edges.

Answer   (i) 192 kPa, (ii) 132 kPa

Exercise 3.2

A concentrated load of 85 kN acts on the horizontal surface of a soil. Plot the variation of vertical stress increments due to the load on horizontal planes at depths of 1, 2 and 3 m directly beneath it.

Answer   Figure 3.21.

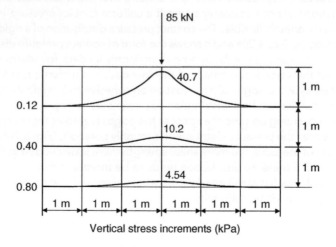

Vertical stress increments (kPa)

Fig. 3.21   Exercise 3.2.

Exercise 3.3

The plan of a foundation is given in Fig. 3.22a. The uniform pressure on the soil is 40 kPa. Determine the vertical stress increment due to the foundation at a depth of 5 m below the point X, using Fig. 3.14.

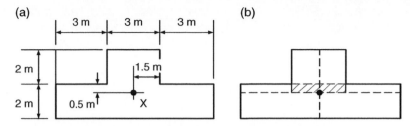

Fig. 3.22 Exercise 3.3. (a) The problem. (b) Selection of rectangles.

Note: In order to obtain a set of rectangles whose corners meet at a point, a section of the foundation area is sometimes included twice and a correction made. For this particular problem, the foundation area must be divided into six rectangles (Fig. 3.22b); the effect of the shaded portion will be included twice and must therefore be subtracted once.

Answer   11.1 kPa

Exercise 3.4

A load of 500 kN is uniformly distributed through a pad foundation of dimensions 1.0 m × 1.5 m. Determine the increase in vertical stress at a depth of 2.0 m below one corner of the foundation.

Answer   36 kPa

Chapter 4
Shear Strength of Soils

Learning objectives:

By the end of this chapter, you will have been introduced to:
- the theory behind the shear strength of soil;
- states of stress on failure planes and the Mohr circle diagram;
- internal friction, shearing resistance, and cohesion;
- undrained and drained shear strength parameters;
- laboratory shear strength testing: shear box and the various triaxial tests.

The property that enables a material to remain in equilibrium when its surface is not level is known as its *shear strength*. Soils in liquid form have virtually no shear strength and even when solid, have shear strengths of relatively small magnitudes compared with those exhibited by steel or concrete.

The shear strength of a soil in any direction is the maximum shear stress that can be applied to the soil in that direction. When this maximum has been reached, the soil is regarded as having failed. The shear strength of a soil depends on various factors including soil type, density, and loading conditions, and can be quantified through particle interlock *friction* and particle-to-particle cementation, known as *cohesion*.

4.1 Shear strength of soil

4.1.1 Frictional resistance to movement

Consider a block of weight W resting on a horizontal plane (Fig. 4.1a). The vertical reaction, R, equals W, and there is consequently no tendency for the block to move. If a small horizontal force, H, is now applied to the block and the magnitude of H is such that the block still does not move, then the reaction R will no longer act vertically but becomes inclined at some angle, α, to the vertical (Fig. 4.1b).

The angle α is called the angle of obliquity and is the angle that the reaction on the plane of sliding makes with the normal to that plane. If H is slowly increased in magnitude, a stage will be reached at which sliding is imminent; as H is increased the value of α will also increase until, when sliding is imminent, α has reached a limiting value, ϕ. If H is now increased still further the angle of obliquity, ϕ will not become greater and the block, having achieved its maximum resistance to horizontal movement, will move. ϕ is known as the *angle of internal friction*, or the *angle of shearing resistance*.

Smith's Elements of Soil Mechanics, 10th Edition. Ian Smith.
© 2021 John Wiley & Sons Ltd. Published 2021 by John Wiley & Sons Ltd.
Companion website: www.wiley.com/go/smith/soilmechanics10e

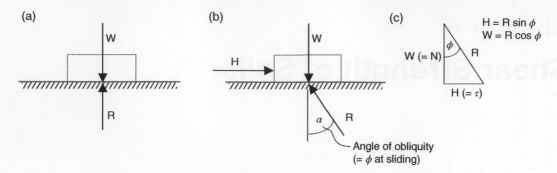

Fig. 4.1   Friction on horizontal plane.

By considering the equilibrium of forces, first in the horizontal direction and then in the vertical direction, it is seen that (Fig. 4.1c):

Horizontal component of $R = H = R \sin \phi$

Vertical component of $R = W = R \cos \phi$

The frictional resistance to sliding is the horizontal component of R and, as can be seen from the triangle of forces in Fig. 4.1c, equals $N \tan \phi$ where N equals the normal force on the surface of sliding (in this case $N = W$). As α only achieves the value ϕ when sliding occurs, it is seen that the frictional resistance is not constant and varies with the applied load until movement occurs. The term $\tan \phi$ is known as the *coefficient of friction*.

If we consider W acting on unit area, we are able to work in terms of stress rather than in terms of force:

$$\tau = \sigma_n \tan \phi \tag{4.1}$$

where

τ = shear stress (kPa)
σ_n = normal stress acting on plane (=force/area) (kPa)

Since the angle of obliquity is maximum when equal to ϕ, it follows that the value of τ in the above expression is the maximum value of shear stress for the particular value of σ_n. In this case, τ is referred to as the *shear strength* of the soil.

From the above relationship, we can see that $\tau \propto \sigma_n$. Therefore, we can plot a straight-line graph of τ against σ_n (Fig. 4.2).

The plane on which the block rests is analogous to a failure surface within a soil mass as represented in Fig. 4.3.

So, if we can obtain ϕ for a granular soil, we can establish the shear strength mobilised for a given normal stress. If, at any point in a soil mass, the shear stress becomes equal to the shear strength, failure of the soil will occur at that point.

4.1.2   Complex stress

When a body is acted upon by external forces, then any plane within the body will be subjected to a stress that is generally inclined to the normal to the plane. Such a stress has both a normal and a tangential component and is known as a *compound*, or *complex, stress* (Fig. 4.4).

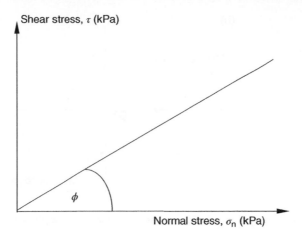

Fig. 4.2 Shear stress – normal stress relationship.

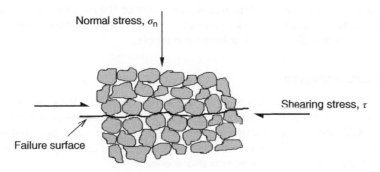

Fig. 4.3 Failure surface in a soil mass.

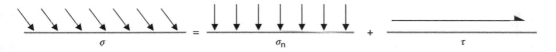

Fig. 4.4 Complex stress.

Principal plane

A plane that is acted upon by a normal stress only is known as a *principal plane*; there is no tangential, or shear, stress present. As is seen in the next section dealing with principal stress, only three principal planes can exist in a stressed mass.

Principal stress

The normal stress acting on a principal plane is referred to as a *principal stress*. At every point in a soil mass, the applied stress system that exists can be resolved into three principal stresses that are mutually orthogonal. The principal planes corresponding to these principal stresses are called the major, intermediate, and minor principal planes, and are so named from a consideration of the principal stresses that act upon

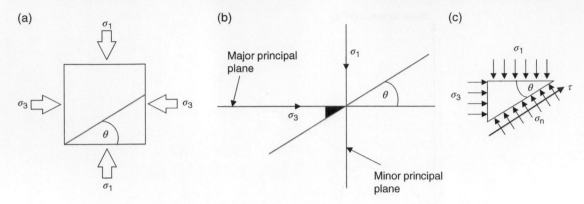

Fig. 4.5 Stress induced by two principal stresses, σ_1 and σ_3, on a plane inclined at θ to σ_3.

them. The largest principal stress, σ_1, is known as the *major principal stress* and acts on the *major principal plane*. Similarly, the intermediate principal stress, σ_2, acts on the intermediate principal plane whilst the smallest principal stress, σ_3, called the *minor principal stress*, acts on the *minor principal plane*. Critical stress values and obliquities generally occur on the two planes normal to the intermediate plane so that the effects of σ_2 can be ignored and a two-dimensional solution is possible.

4.1.3 The Mohr circle diagram

Figure 4.5a shows a small element of soil acted upon by a major principal stress, σ_1, and a minor principal stress, σ_3. A plane within the element, inclined at an angle θ to the direction of the major principal plane, is also shown. Figure 4.5b shows the situation within the element: the major principal plane, acted upon by the major principal stress, σ_1, and the minor principal plane, acted upon by the minor principal stress, σ_3. Again, the inclined plane within the element is also shown.

By considering the equilibrium of the element, it can be shown that on the plane inclined at the angle θ, there is a shear stress, τ, and a normal stress, σ_n (Fig. 4.5c). The magnitudes of these stresses are:

$$\tau = \frac{\sigma_1 - \sigma_3}{2} \sin 2\theta \tag{4.2}$$

$$\sigma_n = \sigma_3 + (\sigma_1 - \sigma_3) \cos^2\theta \tag{4.3}$$

These formulae lend themselves to graphical representation, and it can be shown that the locus of stress conditions for all planes through a point is a circle, generally called a Mohr circle. To draw a Mohr circle diagram a specific convention must be followed: all normal stresses (including principal stresses) being plotted along the axis OX whilst shear stresses are plotted along the axis OY. For most cases, the axis OX is horizontal and OY is vertical, but the diagram is sometimes rotated to give correct orientation. The convention also assumes that the direction of the major principal stress is parallel to axis OY, i.e. the direction of the major principal plane is parallel to axis OX.

A Mohr circle diagram is shown in Fig. 4.6. To draw the diagram, first lay down the axes OX and OY, then set off OA and OB along the OX axis to represent the magnitudes of the minor and major principal stresses, respectively, and finally construct the circle with diameter AB. This circle is the plot of stress conditions for all planes passing through the point A, i.e. a plane passing through A and inclined to the major principal plane at angle θ cuts the circle at D. The coordinates of the point D are the normal and shear stresses on that plane. Point A is the pole of all planes within the element of soil.

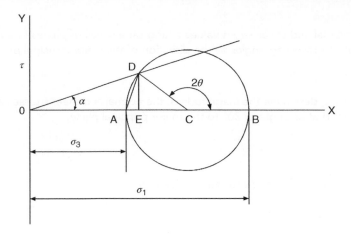

Fig. 4.6 Mohr's circle.

AD intercepts the x-axis at angle θ. So, $AD = AB \cos \theta$.
Now we can see the derivations of Equations (4.2) and (4.3):

$$\text{Normal stress} = \sigma_n = OE = OA + AE = \sigma_3 + AD \cos \theta$$
$$= \sigma_3 + AB \cos^2 \theta \qquad (4.4)$$
$$= \sigma_3 + (\sigma_1 - \sigma_3) \cos^2 \theta$$

$$\text{Shear stress} = \tau = DE = DC \sin(180° - 2\theta)$$
$$= DC \sin 2\theta \qquad (4.5)$$
$$= \frac{\sigma_1 - \sigma_3}{2} \sin 2\theta$$

It is also worth noting that,

- the abscissa of the centre of the Mohr circle (point C) is $\frac{1}{2}(\sigma_1 + \sigma_3)$.
- the radius of the Mohr circle = $\frac{1}{2}(\sigma_1 - \sigma_3)$.

In Fig. 4.6, OE and DE represent the normal and shear stress components of the complex stress acting on plane AD. From triangle ODE, it can be seen that this complex stress is represented in the diagram by the line OD, whilst the angle DOB represents the angle of obliquity, α of the resultant stress on plane AD.

Example 4.1: Mohr's circle

An element of soil experiences an applied major principal stress, $\sigma_1 = 300$ kPa and a minor principal stress, $\sigma_3 = 100$ kPa.

(a) Plot the normal and shear stress values acting on a series of planes in the element, inclined at the following angles to the direction of the major principal plane:

$\theta = 0, 22.5°, 45°, 67.5°, 90°$

(b) From the plot, determine the magnitude of the normal and shear stresses acting on a plane inclined at an angle of 30° to the major principal plane.

Solution:

(a) Determine values of σ_n and τ for each value of θ.

$$\sigma_n = \sigma_3 + (\sigma_1 - \sigma_3)\cos^2\theta$$
$$\tau = \frac{\sigma_1 - \sigma_3}{2}\sin 2\theta$$

θ	σ_n	τ
0	300.0	0.0
22.5	270.7	70.7
45	200.0	100.0
67.5	129.3	70.7
90	100.0	0.0

The plot of τ against σ_n values is shown in Fig. 4.7. As is seen, these values fall on the Mohr circle for this element of soil, illustrating the relationships between θ, σ_n, and τ.

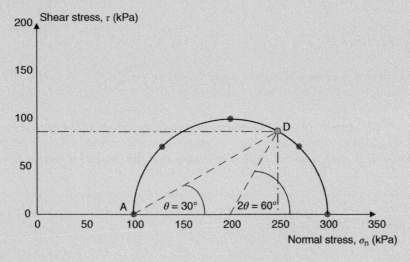

Fig. 4.7   Example 4.1.

(b) For a plane inclined at an angle $\theta = 30°$, we can establish the stress conditions on the Mohr circle either by drawing a straight line inclined at 30° from the pole, A or by drawing a line inclined at 60° (i.e. $2 \times \theta$) from the centre of the circle. This gives us Point D. At this point, $\sigma_n = 250$ kPa and $\tau = 87$ kPa.

This example illustrates how the Mohr circle can be used to rapidly establish the magnitude of σ_n and τ acting on any plane within the element of soil.

Limit conditions

We saw earlier that the maximum shearing resistance is developed when the angle of obliquity equals its limiting value, ϕ. For this condition, the line OD becomes a tangent to the stress circle, inclined at angle ϕ to axis OX (Fig. 4.8).

An interesting point that arises from Fig. 4.8 is that the failure plane is not the plane subjected to the maximum value of shear stress. The criterion of failure is maximum obliquity, not maximum shear stress. Hence, although the plane AE in Fig. 4.8 is subjected to a greater shear stress than the plane AD, it is also subjected to a larger normal stress and therefore the angle of obliquity is less than on AD, which is the plane of failure.

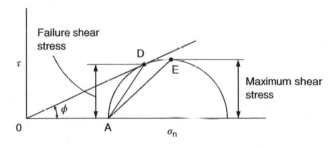

Fig. 4.8 Mohr circle diagram for limit shear resistance.

Example 4.2: Angle of shearing resistance and angle of failure plane

On a failure plane in a purely frictional mass of dry sand, the total stresses at failure were: shear = 3.5 kPa; normal = 10.0 kPa.

Determine (a) by calculation and (b) graphically, the resultant stress on the plane of failure, the angle of shearing resistance of the soil, and the angle of inclination of the failure plane to the major principal plane.

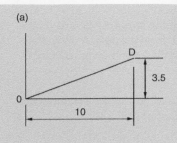
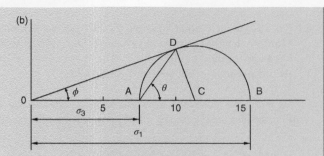

Fig. 4.9   Example 4.2.

Solution:

(a) By calculation:

The soil is frictional therefore the strength envelope must go through the origin. The failure point is represented by point D in Fig. 4.9a with coordinates (10, 3.5).

Resultant stress = OD = $\sqrt{3.5^2 + 10^2}$ = 10.6 kPa

$\tan\phi = \dfrac{3.5}{10} = 0.35$

$\Rightarrow \phi = 19.3°$

$\theta = \dfrac{\phi}{2} + 45° = 54.6°$

(b) Graphically:

The procedure (Fig. 4.9b) is first to draw the horizontal (normal stress) and vertical (shear stress) axes and then, to a suitable scale, set off point D with coordinates (10, 3.5). Join OD (this line represents the strength envelope). The stress circle is tangential to OD at the point D; draw line DC perpendicular to OD to cut the normal stress axis at C, C being the centre of the circle.

With centre C and radius CD draw the circle establishing the points A and B on the x-axis.

By scaling, OD = resultant stress = 10.6 kPa. With protractor, ϕ = 19°; θ = 55°.

Note: From the diagram, we see that OA = σ_3 = 7.6 kPa and OB = σ_1 = 15 kPa.

Strength envelopes

If ϕ is assumed constant for a certain material, then the shear strength of the material can be represented by a pair of lines passing through the origin, O, at angles $+\phi$ and $-\phi$ to the axis OX (Fig. 4.10). These lines comprise the Mohr strength envelope for the material.

In Fig. 4.10, a state of stress represented by circle A is quite stable as the circle lies completely within the strength envelope. Circle B is tangential to the strength envelope and represents the condition of incipient failure, since a slight increase in stress values will push the circle over the strength envelope and failure will occur. Circle C cannot exist as it is beyond the strength envelope.

Relationship between ϕ and θ

In Fig. 4.11, $\angle DCO = 180° - 2\theta$.

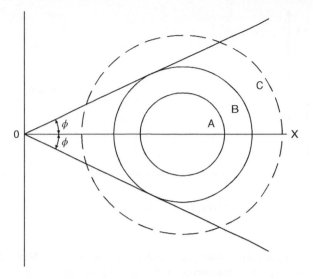

Fig. 4.10   Mohr strength envelope.

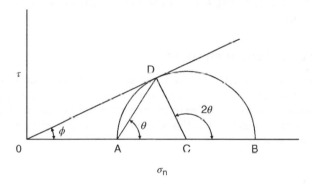

Fig. 4.11   Relationship between ϕ and θ.

In triangle ODC: $\angle DOC = \phi$, $\angle ODC = 90°$, $\angle OCD = 180° - 2\theta$. These angles summate to 180°, i.e.

$$\phi + 90° + 180° - 2\theta = 180°$$

hence

$$\theta = \frac{\phi}{2} + 45° \qquad\qquad (4.6)$$

4.1.4   Cohesion

It is possible to make a vertical cut in silts and clays and for this cut to remain standing, unsupported, for some time. This cannot be done with a dry sand which, on removal of the cutting implement, will slump until its slope is equal to an angle known as the *angle of repose*. In silts and clays, therefore, some other factor must contribute to shear strength. This factor is called cohesion and results from the state of drainage within the soil mass: water in a cohesive soil cannot drain quickly and so the soil is claimed to be in an undrained state,

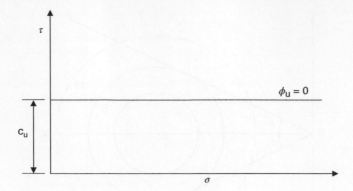

Fig. 4.12 A cohesive soil subjected to undrained conditions and zero total normal stress will still exhibit a shear stress, c_u.

and the undrained cohesion is providing strength to the soil. In terms of the Mohr diagram, this means that the strength envelope for the soil, for undrained conditions, no longer goes through the origin but intercepts the shear stress axis (see Fig. 4.12). The value of the intercept, to the same scale as σ_n, is the undrained cohesion and is given the symbol c_u. In Fig. 4.12, as the soil is in an *undrained* state, the angle of shearing resistance, ϕ_u is equal to zero.

4.1.5 Coulomb's law of soil shear strength

It can be seen that the shear resistance offered by a particular soil is made up of the two components of friction and cohesion. Frictional resistance does not have a constant value but varies with the value of normal stress acting on the shear plane, whereas cohesive resistance has a constant value which is independent of the value of σ_n. In 1776, Coulomb suggested that the equation of the strength envelope of a soil could be expressed by the straight-line equation,

$$\tau_f = c + \sigma \tan \phi \tag{4.7}$$

where

τ_f = shear stress at failure, i.e. the shear strength;
c = unit cohesion;
σ = total normal stress on failure plane; and
ϕ = angle of shearing resistance.

The equation gave satisfactory predictions for sands and gravels, for which it was originally intended, but it was not so successful when applied to silts and clays. The reasons for this are now well known and are that the drainage conditions under which the soil is operating together with the rate of the applied loading have a considerable effect on the amount of shearing resistance the soil will exhibit. None of this was appreciated in the 18th century, and this lack of understanding continued more or less until 1925 when Terzaghi published his theory of *effective stress*.

Note: It should be noted that there are other factors that affect the value of the angle of shearing resistance of a particular soil. They include the effects of such properties as the amount of friction between the soil particles, the shape of the particles and the degree of interlock between them, the density of the soil and its previous stress history.

Effective stress, σ'

If a soil mass is subjected to the action of compressive forces applied at its boundaries then the stresses induced within the soil at any point can be estimated by the theory of elasticity, described in Section 3.1. For most soil problems, estimations of the values of the principal stresses σ_1, σ_2, and σ_3 acting at a particular point are required. Once these values have been obtained, the values of the normal and shear stresses acting on any plane through the point can be determined.

At any point in a saturated soil, each of the three principal stresses consists of two parts:

(1) u, the neutral pressure acting in both the water *and* in the solid skeleton in every direction with equal intensity;
(2) the balancing pressures $(\sigma_1 - u)$, $(\sigma_2 - u)$ and $(\sigma_3 - u)$.

As explained in Section 3.4, Terzaghi's theory is that only the balancing pressures, i.e. the effective principal stresses, influence volume, and strength changes in saturated soils:

Principal effective stress = Principal normal stress − Pore water pressure

i.e.

$\sigma_1' = \sigma_1 - u$, etc.

4.1.6 Modified Coulomb's law

Shear strength depends upon effective stress and not total stress. Coulomb's equation must therefore be modified in terms of effective stress and becomes:

$$\tau_f = c' + \sigma' \tan \phi' \tag{4.8}$$

where

c' = unit cohesion, with respect to effective stresses;
σ' = effective normal stress acting on failure plane; and
ϕ' = angle of shearing resistance, with respect to effective stresses.

It is seen that, dependent upon the loading and drainage conditions, it is possible for a clay soil to exhibit purely frictional shear strength (i.e. to act as a $c' = 0$ or ϕ' soil), when it is loaded under drained conditions, or to exhibit only cohesive strength (i.e. to act as a '$\phi = 0$' or 'c_u' soil), when it is loaded under undrained conditions. It follows then that at an interim stage, the clay can exhibit both cohesion and frictional resistance (i.e. to act as a $c' - \phi'$ soil).

4.1.7 The Mohr–Coulomb yield theory

Over the years, various yield theories have been proposed for soils. The best-known ones are: the Tresca theory, the von Mises theory, the Mohr–Coulomb theory, and the Critical State theory.

The Mohr–Coulomb theory does not consider the effect of strains or volume changes that a soil experiences on its way to failure; nor does it consider the effect of the intermediate principal stress, σ_2. Nevertheless, satisfactory predictions of soil strength are obtained and, as it is simple to apply, the Mohr–Coulomb theory is widely used in the analysis of most practical problems which involve soil strength.

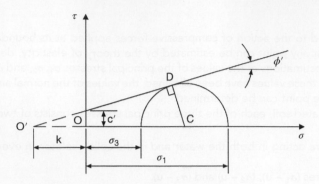

Fig. 4.13   Mohr circle diagram.

The Mohr strength theory is really an extension of the Tresca theory, which in turn was probably based on Coulomb's work – hence the title. The theory assumes that the difference between the major and minor principal stresses is a function of their sum, i.e. $(\sigma_1 - \sigma_3) = f(\sigma_1 + \sigma_3)$. Any effect due to σ_2 is ignored.

The Mohr circle has been discussed earlier and a typical example of a Mohr circle diagram is shown in Fig. 4.13. The intercept on the shear stress axis of the strength envelope is the intrinsic pressure, i.e. the strength of the material when under zero normal stress. As we know, this intercept is called cohesion in soil mechanics.

In Fig. 4.13:

$$\sin \phi = \frac{DC}{O'C} = \frac{\frac{1}{2}(\sigma_1 - \sigma_3)}{k + \frac{1}{2}(\sigma_1 + \sigma_3)} = \frac{\sigma_1 - \sigma_3}{2k + \sigma_1 + \sigma_3}$$

Hence

$$\sigma_1 - \sigma_3 = 2k \sin \phi + (\sigma_1 + \sigma_3) \sin \phi$$

Now

$$k = c \cot \phi$$
$$\Rightarrow (\sigma_1 - \sigma_3) = 2c \cos \phi + (\sigma_1 + \sigma_3) \sin \phi \tag{4.9}$$

which is the general form of the Mohr–Coulomb theory.

The equation can be expressed in terms of either total stress (as shown) or effective stress:

$$\sigma_1' - \sigma_3' = 2c' \cos \phi' + (\sigma_1' + \sigma_3') \sin \phi' \tag{4.10}$$

It is seen therefore that when $c' = 0$,

$$\sigma_1' - \sigma_3' = (\sigma_1' + \sigma_3') \sin \phi' \tag{4.11}$$

4.2   Determination of the shear strength parameters

The shear strength of a soil is controlled by the effective stress that acts upon it and it is therefore obvious that a geotechnical analysis involving the operative strength of a soil should be carried out in terms of the effective stress parameters ϕ' and c'. This is the general rule and, as you might expect, there is at least one exception.

The case of a fully saturated clay subjected to undrained loading is more appropriately analysed using total stress values and c_u than using an effective stress approach. As will be illustrated in later chapters, such a situation can arise in both slope stability and bearing capacity problems.

It is seen therefore, that both the values of the undrained parameter c_u, and of the drained parameters, ϕ' and c' are generally required. They are obtained from the results of laboratory tests carried out on representative samples of the soil with loading and drainage conditions approximating to those in the field where possible. The tests in general use are the direct shear or shear box test, the triaxial test, and the unconfined compression test – an adaptation of the triaxial test.

4.2.1   The shear box test

The apparatus consists of a brass box, split horizontally at the centre of the soil specimen. The soil is gripped by perforated metal grilles, behind which porous discs can be placed if required to allow the sample to drain (see Fig. 4.14). The usual plan size of the sample is 60 × 60 mm; but for testing granular materials, such as gravel or stony clay, it is necessary to use a larger box, generally 300 × 300 mm, although even greater dimensions are sometimes used. Shear box testing is carried out in accordance with procedures set out in BS EN ISO 17892-10 (BSI, 2018). Additional details on the testing procedures are given by Head and Epps (2011).

The shear box sits in a watertight carriage into which water may be poured for testing soils in saturated conditions. A vertical load is applied to the top of the sample by means of weights placed onto a hanger as part of the overall shear box frame assembly. As the shear plane is predetermined in the horizontal direction, the vertical load is also the normal load on the plane of failure. Having applied the required vertical load, a shearing force is gradually exerted on the box from an electrically driven jack. The shear force is measured by means of a load transducer connected to a computer.

By means of additional transducers (fixed to the shear box), it is possible to determine both the horizontal and the vertical strains of the test sample at any point during shear:

$$\text{Horizontal strain (\%)} = \frac{\text{Displacement of box}}{\text{Length of sample}}$$

The shearing force reading is taken at fixed horizontal displacements, and failure of the soil specimen is indicated by a sudden drop in the magnitude of the reading or a levelling off in successive readings. In most cases, the computer plots a graph of the shearing force against horizontal strain as the test progresses. Failure of the soil is visually apparent from a turning point in the graph (for dense granular soils or overconsolidated clays) or a levelling off of the graph (for loose granular soils and normally consolidated clays).

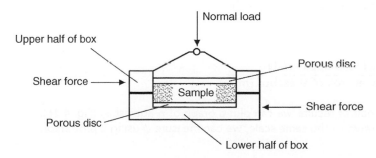

Fig. 4.14   Shear box assembly.

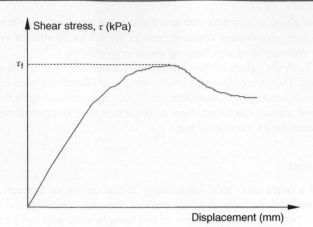

Fig. 4.15   Shear box: shear stress against displacement.

The apparatus is predominantly used for drained tests. Although undrained tests on silts and sands are not possible (because drainage will occur), the test procedure can be modified to maintain constant volume conditions during shear by adjusting the hanger weights. This procedure, in effect, gives an undrained state.

A sand can be tested either dry or saturated. If dry, there will be no pore water pressures and the intergranular pressure will equal the applied stress. If the sand is saturated, the pore water pressure will be zero due to the quick drainage, and the intergranular pressure will again equal the applied stress.

The shear box test is not suitable for undrained testing of clay soils as the apparatus cannot entirely stop drainage from occurring. Drained tests on saturated clays, however, are possible and the test procedure involves two stages: (i) the sample is consolidated under the application of the normal load at the required normal pressure; (ii) once full consolidation is achieved (confirmed through vertical transducer readings), the sample is sheared at a very low rate of strain to ensure no pore water pressures are developed.

From the results obtained, a graph is plotted of the shear stress against displacement of the lower half of the box (Fig. 4.15).

The maximum of the graph occurs at the point of failure. The shear stress at this stage (τ_f) is the shear stress at failure, for the particular normal stress applied.

With a series of tests, with different applied normal loads, we can establish ϕ.

To calculate τ_f:

$$\tau_f = \frac{\text{maximum shear force at failure (kN)}}{\text{cross-sectional area of shear box (m}^2)}$$

To calculate σ_n:

$$\sigma_n = \frac{\text{normal load applied (kN)}}{\text{cross-sectional area of shear box (m}^2)}$$

Then, using each pair of results, we can plot a graph of τ_f against σ_n (Fig. 4.16).

If both axes are drawn to the same scale, we can measure ϕ using a protractor.

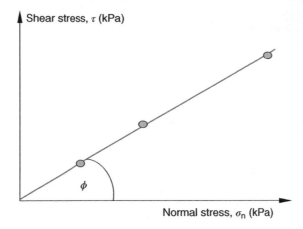

Fig. 4.16   Shear box: shear stress against normal stress.

Example 4.3: Shear box test (i)

Drained shear box tests were carried out on a series of soil samples with the following results:

Test no.	Normal stress (kPa)	Shear stress at failure (kPa)
1	100	48
2	200	89
3	300	130
4	400	172

Each sample was fully consolidated before shearing at a low rate of strain.
Determine the cohesion and the angle of shearing resistance of the soil, with respect to effective stress.

Solution:

In this case, both the normal and the shear stresses at failure are known, so there is no need to draw stress circles and the four failure points may simply be plotted. These points must lie on the strength envelope and the best straight line through the points will establish it (Fig. 4.17).
From the plot, $c' = 7$ kPa and $\phi' = 22°$.

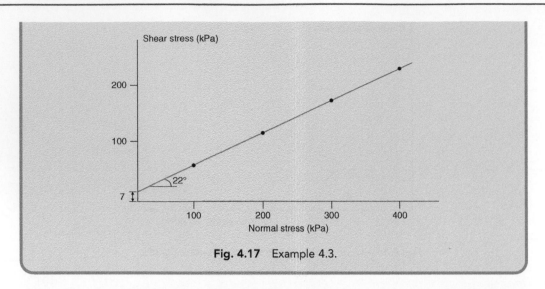

Fig. 4.17   Example 4.3.

Example 4.4: Shear box test (ii)

The following results were obtained from drained shear box tests carried out on a set of undisturbed soil samples:

Normal load (kN)	0.2	0.4	0.8
Strain (%)	Shearing force (N)		
0	0	0	0
1	21	33	45
2	46	72	101
3	70	110	158
4	89	139	203
5	107	164	248
6	121	180	276
7	131	192	304
8	136	201	330
9	138	210	351
10	138	217	370
11	137	224	391
12	136	230	402
13		234	410
14		237	414
15		236	416
16			417
17			417
18			415

The cross-sectional area of the box was 3600 mm^2 and the test was carried out in a fully instrumented shear box apparatus.

Determine the strength parameters of the soil in terms of effective stress.

Solution:

The plot of load transducer readings against strain is shown in Fig. 4.18a.

From this plot, the maximum readings for normal loads of 0.2, 0.4, and 0.8 kN were 138, 237, and 417 N.

(a)

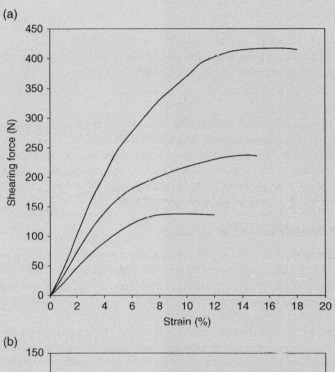

(b)

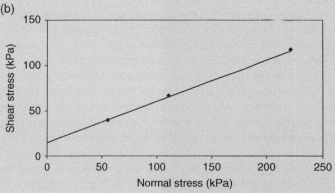

Fig. 4.18 Example 4.4.

For this particular case, the maximum readings could obviously have been obtained directly from the tabulated results, but viewing the plots is sometimes useful to demonstrate whether one of the sets of readings differs from the other two.

The shear stress at each maximum load reading is calculated.

Normal load (kN)	Normal stress (kPa)	Shear force (N)	Shear stress (kPa)
0.2	$\dfrac{0.2 \times 10^6}{3600} = 56$	138	$\dfrac{0.138 \times 10^6}{3600} = 38$
0.4	111	237	66
0.8	222	417	116

The plot of shear stress to normal stress is given in Fig. 4.18b.

The effective stress envelope is obtained by drawing a straight line through the three points. The strength parameters are $\phi' = 25°$ and $c' = 13\,kPa$.

As mentioned earlier, the shear box test subjects the soil to direct shear failure, where movement along the failure surface can only occur in the one direction. The results from the test therefore are only really applicable to soils in the field experiencing direct shear conditions. An example where soils experience movement along a single plane, where movement in the perpendicular planes is prevented (known as *plane strain* conditions as well as direct shear), is beneath a long foundation, or in a long slope. Where the strength properties of a soil in such a situation are required, the shear box test is appropriate.

4.2.2   The effect of density on shear strength

The initial density of a soil tested in a shear box determines the way in which the soil will behave during shearing. If two samples of the same soil are tested in the shear box, one sample placed in a loose state and the other compacted to a higher density, the plots of both measured shear stress and vertical displacement to horizontal displacement will be of the forms shown in Fig. 4.19a and c. The strength envelopes (shear stress against normal stress) are shown in Fig. 4.19b.

If the movement of pore water is restricted, the shear strength of the sand will be affected: the dense sand will have negative pore pressures induced in it, causing an increase in shear strength, while a loose sand will have positive pore pressures induced with a corresponding reduction in strength (Fig. 4.19a and b). A practical application of this effect occurs when a pile is driven into sand; the load on the sand being applied so suddenly that, for a moment, the water it contains has no time to drain away.

It is seen from Fig. 4.19c that as the shear box test progresses, the initially dense soil expands in volume (dilates) – indicted by the increase in vertical displacement of the sample – until a steady volume is reached. In contrast, the initially loose soil contracts until some steady state of volume is reached. This steady state of volume (referred to as the *critical volume*) is found to be the same value for both samples and, as shearing continues, the volume remains constant. The corresponding density of the soil at this volume is known as the *critical density*.

The critical volume is not evident in Fig. 4.19c, but if we consider the changes in void ratio that are happening during shearing, we can identify the critical volume (Fig. 4.19d). Here the loose sample starts with a high void ratio, and the dense sample starts with a lower one. As the test progresses, the void ratios converge thus indicating the presence of the critical volume, and the existence of conditions which create a state known as the *critical state*.

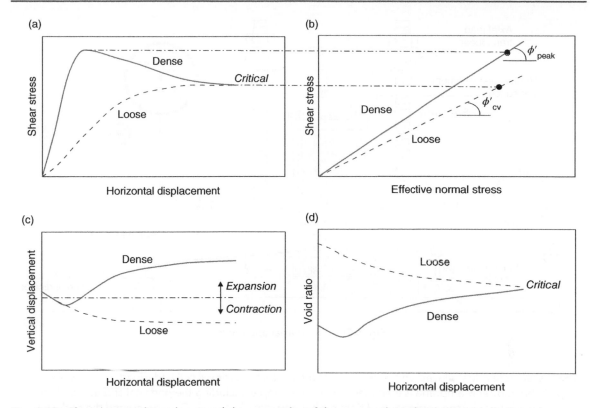

Fig. 4.19 Shear box results on loose and dense samples of the same soil: (a) shear stress v's horizontal displacement; (b) shear stress v's normal stress; (c) vertical displacement v's horizontal displacement; (d) void ratio v's horizontal displacement.

From Fig. 4.19, we see that both the ultimate shear strength and the ultimate void ratio are the same for both samples. Thus, it is seen that soils will reach the critical state when undergoing continuous shearing at constant stress and constant volume. The critical state, and *critical state theory* in general, is described in Chapter 5.

4.2.3   Triaxial testing

The triaxial test subjects the soil specimen to three compressive stresses at right angles to each other, one of the three stresses being increased until the sample fails in shear. Its great advantage is that the plane of shear failure is not predetermined as in the shear box test. Triaxial testing is performed in accordance with procedures set out in BS EN ISO 17892-8 and BS EN ISO 17892-9 (BSI, 2018) and additional details on the testing procedures are given by Head and Epps (2011, 2013) and Lade (2016). The test apparatus is shown in Fig. 4.20.

Details of the different tests carried out using the triaxial apparatus are given in Sections 4.2.4–4.2.9. The sample preparation and the test setup procedure, however, are similar for all tests and are described here.

The soil sample tested is cylindrical with a height equal to twice its diameter. The usual sizes are 76 mm high × 38 mm diameter and 200 mm high × 100 mm diameter. Although triaxial specimens can be remoulded (i.e. prepared from disturbed material, formed into a cylinder of the required dimensions inside a metal former), most triaxial testing is carried out on undisturbed samples. (See Chapter 7 for details on soil sampling.)

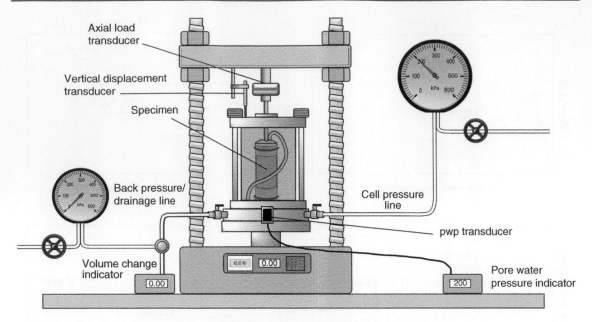

Fig. 4.20   The triaxial apparatus.

The test sample is first placed on the pedestal of the base of the triaxial cell and a loading cap is placed on its top. A thin rubber membrane is then placed over the sample, including the pedestal and the loading cap, and made watertight by the application of tight rubber ring seals, known as 'O' rings, around the pedestal and the loading cap.

The upper part of the cell, which is cylindrical and generally made of Perspex, is next fixed to the base and the assembled cell is filled with de-aired water. The water is then subjected to a predetermined value of pressure, known as the *cell pressure*, which is kept constant throughout the length of the test. It is this water pressure that subjects the sample to an all-round pressure.

The additional axial stress is created by an axial load applied through a load transducer, in a similar way to that in which the horizontal shear force is applied in the shear box apparatus. By the action of an electric motor, the axial load is gradually increased at a constant rate of strain and as the axial load is applied, the sample suffers continuous compressive deformation. The amount of this vertical deformation is measured by a displacement transducer. Readings from the displacement transducer and corresponding readings of axial load are taken throughout the test until the sample fails. With this data, software is used to plot the variation of axial load on the sample against vertical strain.

Note: In the past, soil laboratories made use of dial gauges to measure displacement and proving rings to measure applied loads. Some laboratories still use such equipment, and any reader interested in an explanation and examples of their use is guided to the sixth, or earlier, editions of this book.

Determination of the additional axial stress

From the load transducer, it is possible at any time during the test to determine the additional axial load that is being applied to the sample.

During the application of this load, the sample experiences shortening in the vertical direction with a corresponding expansion in the horizontal direction. This means that the cross-sectional area of the sample varies, and it has been found that very little error is introduced if the cross-sectional area is evaluated on the assumption that the volume of the sample remains unchanged during the test. Thus, the cross-sectional area at failure is found from:

$$\text{Cross sectional area} = \frac{\text{Volume of sample}}{\text{Original length} - \text{Vertical deformation}}$$

Principal stresses

The intermediate principal stress, σ_2, and the minor principal stress, σ_3, are equal and are the radial stresses caused by the cell pressure, p_c. The major principal stress, σ_1, consists of two parts: the cell water pressure acting on the ends of the sample and the additional axial stress from the load transducer, q. To ensure that the cell pressure acts over the whole area of the end cap, the bottom of the plunger is drilled so that the pressure can act on the ball seating.

From this we see that the triaxial test can be considered as happening in two stages (Fig. 4.21): the first being the application of the cell water pressure (p_c, i.e. σ_3), while the second is the application of a deviator stress (q, i.e. $[\sigma_1 - \sigma_3]$).

A set of at least three samples is tested. The deviator stress is plotted against vertical strain and the point of failure of each sample is obtained. The Mohr circles for each sample are then drawn and the best common tangent to the circles is taken as the strength envelope (Fig. 4.22). A small curvature occurs in the strength envelope of most soils, but this effect is slight and for all practical work the envelope can be taken as a straight line.

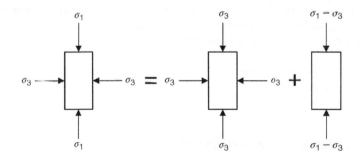

Fig. 4.21   Stresses in the triaxial test.

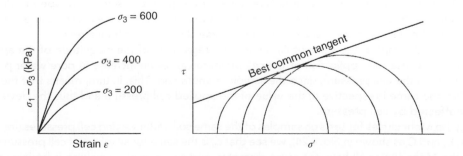

Fig. 4.22   Typical triaxial test results.

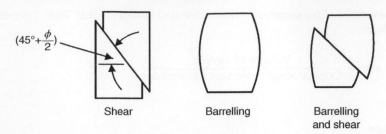

$(45° + \dfrac{\phi}{2})$

Shear   Barrelling   Barrelling and shear

Fig. 4.23 Types of failure in the triaxial test.

Types of failure

Not all soil samples will fail in pure shear; there are generally some barrelling effects as well (see Fig. 4.23). In a sample that fails completely by barrelling, there is no definite failure point; the deviator stress simply increases slightly with strain. In this case, an arbitrary value of the failure stress is taken as the stress value at 20% strain.

The triaxial tests

The triaxial test equipment is an extremely versatile piece of kit and can be used to determine both the undrained and the drained strength parameters of the soil. The main benefit of the triaxial test is that specimens are tested under conditions that reasonably replicate the conditions of the soil in the field. This is achieved through the application of the all-round confining pressure imparted onto the sample through the pressurised water in the cell. Further, since no failure plane is predetermined in the sample, the sample fails in a manner akin to that likely to be experienced by the soil *in situ*.

The triaxial apparatus is used to carry out three different tests:

(1) The unconsolidated undrained (UU) test
(2) The consolidated undrained (CU) test
(3) The consolidated drained (CD) test

4.2.4 The unconsolidated undrained (UU) test

This test is also known as the *quick undrained test* and is carried out to determine the total stress parameter c_u – the undrained cohesion. In the test, the sample, at its natural water content, is prevented from draining during shear. It is sheared therefore, immediately after the application of the cell pressure. A sample can be tested in 15 minutes or less, so that there is no time for any pore pressures developed to dissipate or to distribute themselves evenly throughout the sample. Measurements of pore water pressure are therefore not possible, and the results of the test can only be expressed in terms of total stress.

For fully saturated samples, the value of c_u is constant, regardless of the magnitude of the applied cell pressure. This is because the application of the cell pressure causes an increase in pore water pressure of the same amount as the cell pressure, since the sample is undrained. This, in turn, means that the effective stresses remain the same irrespective of the value of the applied cell pressure and thus the deviator stress $(\sigma_1 - \sigma_3)$ is unaffected by cell pressure.

By plotting the Mohr circles for tests on samples of the same soil under varying cell pressures (for example, three tests A, B, and C as shown in Fig. 4.24), we see that c_u is the same for all values of cell pressure. Further, we see that the Mohr circles all have the same diameter, and so the Mohr envelope is horizontal, i.e. the undrained angle of shearing resistance, $\phi_u = 0$.

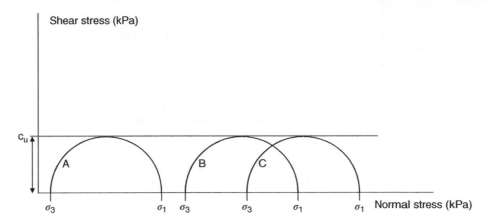

Fig. 4.24 Mohr circles for saturated clay.

A variation of the single-stage UU test, where a single sample is tested at one cell pressure, is to carry out a multi-stage test, where the sample is tested at three different confining pressures in the one test. In the multi-stage test, the cell pressure is increased (typically doubled) as the sample begins to fail during shearing. As the sample fails, the cell pressure is swiftly increased and readings of axial force against strain continue to be taken. The test is stopped when failure occurs under the third value of cell pressure or when 20% axial strain is reached.

Multi-stage testing can be used to observe the deviator stress-axial strain response under the three cell pressures. Significant increases in deviator stress for increased cell pressures reveals that the sample is not fully saturated. Mohr circles from a multi-stage test will reveal three different values of c_u and the correct single c_u value must be carefully interpreted from these Mohr circles. Common practice is to select the conservative, minimum value.

Example 4.5: Quick undrained triaxial test

A sample of clay was subjected to an undrained triaxial test with a cell pressure of 100 kPa and the additional axial stress necessary to cause failure was found to be 188 kPa.

(a) Determine the undrained cohesion, c_u;
(b) Determine the value of additional axial stress that would be required to cause failure of a further sample of the soil, if it was tested undrained with a cell pressure of 200 kPa.

Solution:

(a) The Mohr circle representing the conditions of the first test is drawn:

$\sigma_3 = 100$ kPa and $\sigma_1 = 188 + 100 = 288$ kPa.

The circle is shown on the left in Fig. 4.25 and the strength envelope representing the condition that $\phi_u = 0°$ is now drawn as a horizontal line tangential to the stress circle. It can be seen from the figure that $c_u = 94$ kPa.

This value could also have been obtained numerically from: $c_u = \dfrac{\sigma_1 - \sigma_3}{2}$

(b) Draw the Mohr circle with $\sigma_3 = 200$ kPa and tangential to the strength envelope. Where this circle cuts the normal stress axis it gives the value of σ_1, which is seen to be 388 kPa.

The additional axial stress required for failure therefore $= (\sigma_1 - \sigma_3)$
$$= (388 - 200) = 188 \text{ kPa}$$

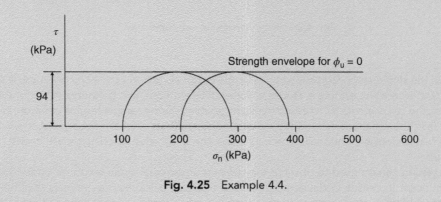

Fig. 4.25 Example 4.4.

4.2.5 The consolidated drained (CD) test

This test is used to determine the effective stress parameters c' and ϕ'.

Test set up and saturation

A porous disc is placed on the pedestal before the test sample is placed in position so that water can drain out from the soil. The triaxial cell is then assembled, filled with de-aired water, and pressurised. If the sample is not already saturated – confirmed through the B value (see Section 4.3) – it can be saturated using a process of pressure increments between cell pressure acting externally on the sample and a back pressure applying pressure into the sample. This process has the effect of bringing air out of the voids and into solution, thus increasing the degree of saturation of the sample. The cell pressure is increased by an amount $\Delta\sigma_3$ and the back pressure is then increased by a slightly smaller amount. The change in pore water pressure, Δu in the soil is determined through readings from the pore water pressure transducer (Fig. 4.20). The process is repeated in increments until $\Delta u / \Delta\sigma_3 > 0.95$. At that point, the soil is considered adequately saturated and the consolidation stage can commence.

Consolidation stage

The drainage tap is opened. The cell pressure creates a pore water pressure within the soil sample and the apparatus is left until the sample has consolidated, i.e. until this excess pore water pressure has been dissipated by water seeping out through the drainage line and into a volume measuring device (Fig. 4.20). This process usually takes about a day but is quicker if a porous disc is installed beneath the loading cap and joined to the pedestal disc by connecting strips of vertical filter paper placed on the outside of the sample but within the rubber membrane. The pore pressure is measured so that the point when full consolidation has been reached can be identified. Full consolidation occurs when the excess pore water pressure has reduced to zero and the volume change has reduced to less than 0.1% of the sample volume.

This type of consolidation (where the sample is consolidated due to an all-round confining pressure subjecting the sample to equal compressive stresses in all directions) is known as *isotropic consolidation*. Alternatively, consolidation can be carried out by applying an axial stress too so that stresses in the vertical direction are different than those in the horizontal. Such consolidation is known as *anisotropic consolidation*. These terms give rise to the CD test being sometimes referred to as the *isotropically consolidated drained* (CID) test or the *anisotropically consolidated drained* (CAD) test, depending on the method of consolidation adopted.

Shearing stage

When consolidation is complete, the sample is sheared by applying a deviator stress at such a low rate of strain that any pore water pressures induced in the sample have time to dissipate through the porous discs and into the drainage line. In this test, the pore water pressure is therefore always zero and the effective stresses are consequently equal to the applied stresses.

The main drawback of the drained test is the length of time it takes, with the attendant risk of testing errors. An average test time for a clay sample is about three days but with some soils a test may last as long as two weeks.

As with all triaxial testing, failure of the sample is observed when the readings of axial force or deviator stress begin to drop off as the shearing progresses. By testing three samples of the same soils at different cell pressures, a set of effective stress Mohr circles can be drawn and the failure envelope, and hence c' and ϕ', are established.

Example 4.6: Consolidated drained triaxial test

A series of CD triaxial tests were performed on a soil. Each test was continued until failure and the data obtained during the test were:

Test no.	Cell pressure (kPa)	Deviator stress (kPa)
1	200	370
2	300	575
3	400	762

Plot the relevant Mohr stress circles and hence determine the strength envelope of the soil with respect to effective stress.

Solution:

σ'_3 (kPa)	Deviator stress $(\sigma'_1 - \sigma'_3)$ (kPa)	σ'_1 (kPa)
200	370	570
300	575	875
400	762	1162

The Mohr circle diagram is shown in Fig. 4.26. The circles are drawn first and then, by constructing the best common tangent to these circles, the strength envelope is obtained. In this case, it is seen that the soil is cohesionless as there is no cohesive intercept. By measurement, $\phi' = 29°$.

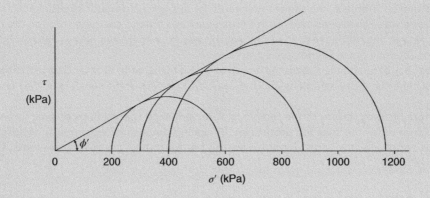

Fig. 4.26   Example 4.5.

4.2.6   The consolidated undrained (CU) test

This test is used to determine the effective stress parameters c' and ϕ'.

The CU test is far more common than the CD test, mainly because it has the advantage that the shearing part of the test can be carried out in only two to three hours. Shorter tests are less expensive than longer CD ones and clearly the results are available more quickly, so these are the reasons why this test is much more popular than the CD test. In addition, from the total stress values measured in the test, the undrained shear strength c_u can be established for any of the cell pressures applied – see Example 4.8.

The sample is saturated (if required) and consolidated exactly as for the drained test, but after consolidation, the drainage connection is shut off and the sample is sheared under undrained conditions. The application of the deviator stress induces pore water pressures (which are measured), and the effective deviator stress is then simply the total deviator stress less the pore water pressure.

Although the sample is sheared undrained, the rate of shear must be slow enough to allow the induced pore water pressures to distribute themselves evenly throughout the sample. For most soils, a strain rate of 0.05 mm/min is satisfactory, which means that the majority of samples can be sheared in under three hours.

As with the consolidated drained test, depending on the method of consolidation adopted, the test may be referred to as an *isotropically consolidated undrained* (CIU) test or as an *anisotropically consolidated undrained* (CAU) test. The CAU test is more appropriate for samples taken from significant depth, where a reasonable amount of stress release through sampling occurs. The anisotropic consolidation process can be used to better simulate the *in situ* conditions where the sample will have experienced significant one-dimensional consolidation prior to sampling (see Chapter 12 for explanation of one-dimensional consolidation).

Example 4.7: Consolidated undrained triaxial test (i)

The following results were obtained from a series of consolidated undrained triaxial tests carried out on undisturbed samples of a compacted soil:

Cell pressure (kPa)	Additional axial load at failure (N)	Pore water pressure at failure (kPa)
200	148	150
400	236	280
600	398	360

Each sample, originally 76 mm long and 38 mm in diameter, experienced a vertical deformation of 5.1 mm.

Draw the strength envelope and determine the drained shear strength properties of the soil.

Solution:

$$\text{Volume of sample} = \frac{\pi \times 38^2}{4} \times 76 = 86\,193\,\text{mm}^3$$

$$\text{Therefore, cross-sectional area at failure} = \frac{86193}{76-5.1} = 1216\,\text{mm}^2.$$

Effective cell pressure, σ'_3 (kPa)	Deviator stress, $(\sigma'_1 - \sigma'_3)$ (kPa)	Major principal stress, σ'_1 (kPa)
$200 - 150 = 50$	$\dfrac{0.148 \times 10^6}{1216} = 122$	172
$400 - 280 = 120$	$\dfrac{0.236 \times 10^6}{1216} = 194$	314
$600 - 360 = 240$	$\dfrac{0.398 \times 10^6}{1216} = 327$	567

The Mohr circles and the strength envelope are shown in Fig. 4.27. From the diagram $\phi' = 21°$ and $c' = 23$ kPa.

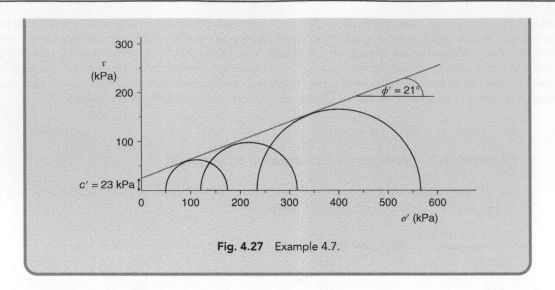

Fig. 4.27   Example 4.7.

4.2.7   Testing with back pressures

It should be noted that, with some soils, the reduction of the pore water pressure to atmospheric during the consolidation stage of a triaxial test on a saturated soil sample can cause air dissolved in the water to come out of solution. If this happens, the sample is no longer fully saturated, and this can affect the results obtained during the shearing part of the test.

To maintain a state of occlusion in the pore water, i.e. the state where air can no longer exist in a free state but only in the form of bubbles, its pressure can be increased by applying a pressure (known as a *back pressure*) to the water in the drainage line (the soil water can still drain from the sample). The back pressure ensures that air does not come out of solution and, by applying the same increase in pressure to the value of the cell pressure, the effective stress situation is unaltered.

The technique can also be used to create full saturation during the consolidation and shearing of partially saturated natural or remoulded soils for both the drained and consolidated undrained triaxial tests. In these cases, back pressure values often as high as 650 kPa are necessary to achieve full saturation.

Example 4.8: Consolidated undrained triaxial test (ii)

A series of undisturbed samples from a normally consolidated clay was subjected to consolidated undrained tests. The samples were taken from the same depth in the clay deposit.
The results were:

Cell pressure, σ_3 (kPa)	Deviator stress at failure (kPa)	Pore water pressure at failure (kPa)
200	118	104
400	240	220
600	352	320

(a) Plot the Mohr circles in terms of both total and effective stresses.
(b) Determine the effective shear strength parameters.
(c) If the lateral stress acting on the samples in the field was estimated to be 400 kPa, determine the likely *in situ* undrained shear strength.

Solution:

Total stresses
(1) $\sigma_1 = 200 + 118 = 318$ kPa
(2) $\sigma_1 = 400 + 240 = 640$ kPa
(3) $\sigma_1 = 600 + 352 = 952$ kPa

Effective stresses
(1) $\sigma_3' = 200 - 104 = 96$ kPa; $\sigma_1' = 118 + 96 = 214$ kPa
(2) $\sigma_3' = 400 - 220 = 180$ kPa; $\sigma_1' = 240 + 180 = 420$ kPa
(3) $\sigma_3' = 600 - 320 = 280$ kPa; $\sigma_1' = 352 + 280 = 632$ kPa

(a) The two Mohr circle diagrams are shown in Fig. 4.28.
(b) The values of c' and ϕ' are obtained from by direct measurement: c' = 0; ϕ' = 23°.
(c) The undrained shear strength is c_u and can be established from the appropriate total stress Mohr circle, i.e. where the lateral pressure is equal to the *in situ* value (σ_3 = 400 kPa): c_u = radius of Mohr circle = 120 kPa.

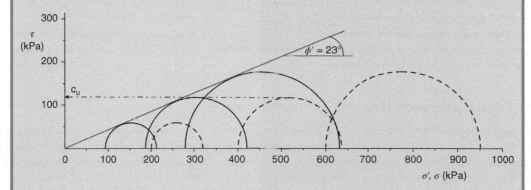

Fig. 4.28   Example 4.8.

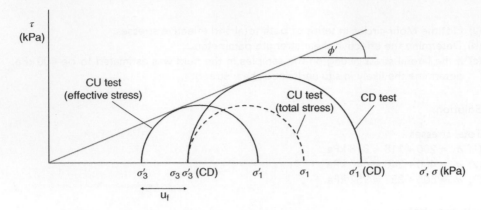

Fig. 4.29 Mohr circles for undrained and drained tests at same cell pressure, σ_3.

4.2.8   Comparison of strength parameters obtained from different triaxial tests

Example 4.8 illustrates that two different Mohr–Coulomb failure envelopes are defined during the same undrained test: (i) through plotting in terms of total stress; (ii) through plotting in terms of effective stress. It is seen then, that for a single value of cell pressure in a consolidated undrained test, two Mohr circles can be drawn – one in terms of total stress and one in terms of effective stress. A consolidated drained test on the same soil at the same cell pressure would generate a third and larger Mohr circle as shown in Fig. 4.29. The increased deviator stress in the drained test is a result of the volume reduction during the test causing the soil to become stiffer and therefore stronger. The strength parameters c' and ϕ' are the same of course, regardless of which test is used to determine them.

As we have seen, the effective stress strength parameters c' and ϕ' can be established from either the consolidated drained or the consolidated undrained test. Since the CU test is both quicker and cheaper, it is routinely used in preference to the CD test.

4.2.9   Triaxial extension test

In most geotechnical situations, the vertical stress acting on the soil exceeds the horizontal stress and to this end, adoption of the triaxial compression test to establish the strength parameters is appropriate since this same vertical to horizontal stress relationship applies in the test. Occasionally however, the horizontal stress acting on an *in situ* soil exceeds the vertical stress, and using the triaxial compression test to establish the strength parameters would yield unrelated values to the situation the soil experiences in the field. Thus, where a soil is likely to experience a horizontal stress greater than the vertical stress (e.g. in the lower reaches of a slip circle: either beside a shallow foundation (see Chapter 10), or in a cohesive slope (see Chapter 14), the shear strength parameters are more appropriately determined through the triaxial extension test.

In the normal triaxial test, the soil sample is subjected to an all-round water pressure and fails under an increasing axial load. This is known as a compression test in which $\sigma_1 > \sigma_2 = \sigma_3$.

From Equation (4.11), we see that when the cohesive intercept, c' is equal to zero, as is the case for granular soils, silts, and normally consolidated clays, then the relevant form of the Mohr–Coulomb equation is:

$$\sigma_1' - \sigma_3' = \sigma_1'\sin\phi' + \sigma_3'\sin\phi'$$

i.e.

$$\sigma'_{1f}(\text{max}) = \sigma'_{3f}\frac{1 + \sin \phi'}{1 - \sin \phi'}$$

where σ'_{1f} and σ'_{3f} are the respective stresses at failure.

It is possible to fail the sample in axial tension by first subjecting it to equal pressures σ'_1 and σ'_3 and then gradually reducing σ'_1 below the value of σ'_3 until failure occurs. This test is known as an extension test and the Mohr–Coulomb expression becomes:

$$\sigma'_{1f}(\text{min}) = \sigma'_{3f}\frac{1 - \sin \phi'}{1 + \sin \phi'} \text{ where } \sigma'_1 < \sigma'_2 = \sigma'_3$$

The Mohr circle diagram showing the maximum and minimum values of σ'_1 for a fixed value of σ'_3 is shown in Fig. 4.30. In the triaxial compression test, the stress state is $\sigma'_1 > \sigma'_2 = \sigma'_3$, and in the triaxial extension test, the stress state is $\sigma'_1 < \sigma'_2 = \sigma'_3$.

The symbols used in Fig. 4.30 might be confusing to a casual observer. Strictly speaking, for the extension test, σ'_{1f} (min) should really be given the symbol σ'_{3f} and its accompanying σ'_{3f} given the symbol σ'_{1f}. To avoid this sort of confusion between major and minor principal stresses, it has become standard practice to designate the axial effective stress as σ'_a, and the radial effective stress as σ'_r.

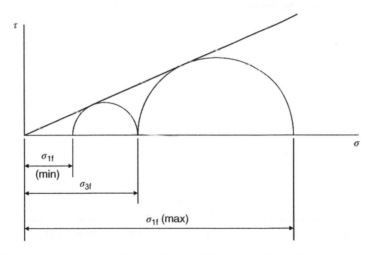

Fig. 4.30 Mohr circle diagram for triaxial compression and tension tests.

4.2.10 The unconfined compression test

In this test (Fig. 4.31), no all-round pressure is applied to the soil specimen and the results obtained give a measure of the *unconfined* compressive strength, q_u of the soil. The test is only applicable to cohesive soils and, although not as popular as the triaxial test, it is used where a rapid result is required. An electric motor within the base unit drives the platen supporting the specimen upwards and the load carried by the soil is recorded by the load transducer. The vertical strain is recorded by a displacement transducer and the load–displacement curve is plotted on a computer connected to the system. The load and strain readings at failure are used to give a direct measure of the unconfined compressive strength of the soil. The apparatus

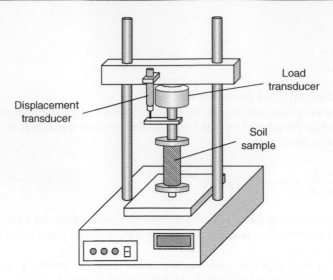

Fig. 4.31   The unconfined compression test.

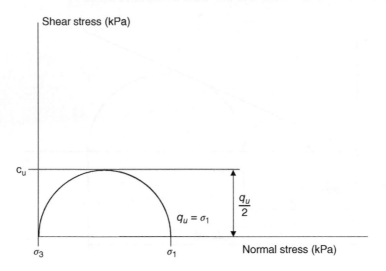

Fig. 4.32   The unconfined compression test Mohr circle.

is only capable of carrying out an undrained test on a clay sample with no radial pressure applied and the test takes about a minute to carry out.

The unconfined compressive strength is the maximum applied normal stress that the sample withstands before failure. This value is σ_1 on the Mohr circle. As we have seen, σ_3 on the Mohr circle is the all-round lateral pressure on the sample, which in this test is equal to zero since the sample is unconfined. The Mohr circle for the unconfined compression test therefore is as shown in Fig. 4.32. It is seen from the figure that the undrained cohesion, c_u, the intercept of the failure envelope with the shear stress axis, is equal to half q_u; i.e. the shear strength of an unconfined sample of soil is half the compressive strength. The undrained cohesion, c_u therefore, is often referred to as the *undrained shear strength*. They are the same: since, in an undrained state, all of the strength of a soil comes from its cohesion.

4.3   The pore pressure coefficients A and B

These coefficients were proposed by Skempton in 1954 and are now almost universally accepted. The relevant theory is set out below.

$$\text{Volumetric strain} = \frac{\text{Change in volume}}{\text{Original volume}} = \frac{-\Delta V}{V}$$

(ΔV is negative when dealing with compressive stresses as is the general case in soil mechanics.)

Consider an elemental cube of unit dimensions and acted upon by compressive principal stresses σ_1, σ_2, and σ_3 (Fig. 4.33).

On horizontal plane (2,3):

$$\text{Compressive strain} = \frac{\sigma_1}{E}$$

Lateral strain from stresses σ_2 and $\sigma_3 = -\left(\frac{\nu\sigma_2}{E} + \frac{\nu\sigma_3}{E}\right)$ where $\nu = $ Poisson's ratio.

i.e. total strain on this plane $= \dfrac{\sigma_1}{E} - \dfrac{\nu}{E}(\sigma_2 + \sigma_3)$

Similarly, strains on the other two planes are:

$$\frac{\sigma_2}{E} - \frac{\nu}{E}(\sigma_3 + \sigma_1)$$
$$\frac{\sigma_3}{E} - \frac{\nu}{E}(\sigma_1 + \sigma_2)$$

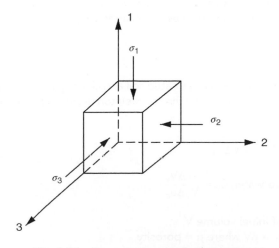

Fig. 4.33   Compressive principal stresses.

Now it can be shown that, no matter what the stresses on the faces of the cube, the volumetric strain is equal to the sum of the strains on each face.

$$-\frac{\Delta V}{V} = \frac{(\sigma_1 + \sigma_2 + \sigma_3)}{E} - \frac{2\nu}{E}(\sigma_1 + \sigma_2 + \sigma_3)$$

i.e.

$$-\frac{\Delta V}{V} = \frac{1-2\nu}{E}(\sigma_1 + \sigma_2 + \sigma_3)$$

Compressibility of a material is the volumetric strain per unit pressure, i.e. for a soil skeleton,

$$C_c = \frac{-\Delta V}{V} \text{ per unit pressure increase}$$

Average pressure increase = $\frac{1}{3}(\sigma_1 + \sigma_2 + \sigma_3)$. Therefore, for a perfectly elastic soil:

$$C_c = \frac{3(1-2\nu)}{E}\frac{(\sigma_1 + \sigma_2 + \sigma_3)}{(\sigma_1 + \sigma_2 + \sigma_3)} = \frac{3(1-2\nu)}{E}$$

Consider a sample of saturated soil subjected to an undrained triaxial test. The applied stress system for this test has already been discussed and is shown in Fig. 4.21. The pore water pressure, u, produced during the test will be made up of two parts corresponding to the application of the cell pressure and the deviator stress.

Let

u_a = pore pressure due to σ_3
u_d = pore pressure due to $(\sigma_1 - \sigma_3)$.

If we consider the effects of small total pressure increments $\Delta\sigma_3$ and $\Delta\sigma_1$ then $\Delta\sigma_3$ will cause a pore pressure change Δu_a and $\Delta\sigma_1 - \Delta\sigma_3$ will cause a pore pressure change Δu_d.

Effect of $\Delta\sigma_3$

When an all-around pressure is applied to a saturated soil and drainage is prevented, the proportions of the applied stress carried by the pore water and by the soil skeleton depend upon their relative compressibilities:

compressibility of the soil, $C_c = \dfrac{-\Delta V}{V\Delta\sigma_3}$

compressibility of the pore water, $C_v = \dfrac{-\Delta V_v}{V_v\Delta u_a}$

Consider a saturated soil of initial volume V.
Then volume of pore water = nV where n = porosity.
Assume a change in total ambient stress = $\Delta\sigma_3$.

Assume that the change in effective stress caused by this total stress increment is $\Delta\sigma'_3$ and that the corresponding change in pore water pressure is Δu_a. Then,

decrease in volume of soil skeleton $= C_c V \sigma'_3$

and

decrease in volume of pore water $= C_v n V \Delta u_a$

With no drainage, these changes must be equal:

i.e.

$$C_c V \Delta\sigma'_3 = C_v n V \Delta u_a$$

$$\Delta\sigma'_3 = \frac{nC_v}{C_c} \Delta u_a$$

Now

$$\Delta\sigma'_3 = \Delta\sigma_3 - \Delta u_a$$

$$\frac{nC_v}{C_c} \Delta u_a = \Delta\sigma_3 - \Delta u_a$$

or

$$\Delta u_a \left(1 + \frac{nC_v}{C_c} \right) = \Delta\sigma_3$$

$$\Rightarrow \Delta u_a = \frac{\Delta\sigma_3}{1 + \dfrac{nC_v}{C_c}}$$

i.e.

$$\Delta u_a = B\Delta\sigma_3$$

$$\text{or, } B = \frac{\Delta u_a}{\Delta\sigma_3} \tag{4.12}$$

where $B = \dfrac{1}{1 + \dfrac{nC_v}{C_c}}$

The compressibility of water is of the order of 1.63×10^{-7} kPa.

Typical results from soil tests are given in Table 4.1 and show that, for all saturated soils, B can be taken as equal to 1.0 for practical purposes.

Table 4.1 Compression of saturated soils.

Soil type	Soft clay	Stiff clay	Compact silt	Loose sand	Dense sand
n (%)	60	37	35	46	43
C_c (m^2/kN)	4.79×10^{-4}	3.35×10^{-5}	9.58×10^{-5}	2.87×10^{-5}	1.44×10^{-5}
B	0.9998	0.9982	0.9994	0.9973	0.9951

Effect of $\Delta\sigma_1 - \Delta\sigma_3$

Increase in effective stresses:

$$\Delta\sigma_1' = (\Delta\sigma_1 - \Delta\sigma_3) - \Delta u_d$$
$$\Delta\sigma_2' = \Delta\sigma_3' = -\Delta u_d$$

Change in volume of soil skeleton,

$$\Delta V_c = -C_c V\left(\Delta\sigma_1' + 2\Delta\sigma_3'\right)$$

i.e.

$$\Delta V_c = -V\frac{C_c}{3}\left[(\Delta\sigma_1 - \Delta\sigma_3) - 3\Delta u_d\right]$$

Now

$$\Delta V_v = -C_v n \Delta u_d V \text{ and } \Delta V_c \text{ must equal } \Delta V_v$$

$$\Rightarrow \frac{1}{3}C_c(\Delta\sigma_1 - \Delta\sigma_3) - C_c\Delta u_d = C_v n \Delta u_d$$

or

$$\Delta u_d(C_c + nC_v) = C_c(\Delta\sigma_1 - \Delta\sigma_3)$$

$$\Rightarrow \Delta u_d = \frac{1}{1 + \dfrac{nC_v}{C_c}}\frac{1}{3}(\Delta\sigma_1 - \Delta\sigma_3)$$

$$= B \times \frac{1}{3}(\Delta\sigma_1 - \Delta\sigma_3)$$

Now

$$\Delta u = \Delta u_a + \Delta u_d$$

$$\Rightarrow \Delta u = B\left[\Delta\sigma_3 + \frac{1}{3}(\Delta\sigma_1 - \Delta\sigma_3)\right]$$

Generally, a soil is not perfectly elastic, and the above expression must be written in the form:

$$\Delta u = B[\Delta\sigma_3 + A(\Delta\sigma_1 - \Delta\sigma_3)] \tag{4.13}$$

where A is a coefficient determined experimentally.

The expression is often written in the form:

$$\Delta u = B\Delta\sigma_3 + \overline{A}(\Delta\sigma_1 - \Delta\sigma_3) \text{ where } \overline{A} = AB \tag{4.14}$$

$\overline{A}$ and B can be obtained directly from the undrained triaxial test. As has been shown, for a saturated soil B = 1.0 and the above expression must be:

$$\Delta u = \Delta\sigma_3 + A(\Delta\sigma_1 - \Delta\sigma_3) \tag{4.15}$$

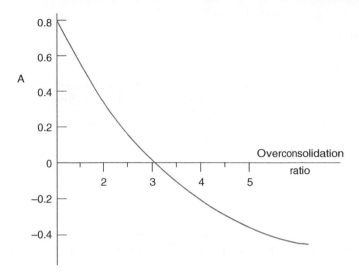

Fig. 4.34   Effects of overconsolidation on the pore pressure coefficient A.

Values of A

For a given soil, A varies with both the stress value and the rate of strain, due mainly to the variation of Δu_d with the deviator stress. The value of Δu_d under a particular stress system depends upon such factors as the degree of saturation and whether the soil is normally consolidated or overconsolidated. The value of A must be quoted for some specific point, e.g. at maximum deviator stress or at maximum effective stress ratio (σ'_1/σ'_3). At maximum deviator stress it can vary from 1.5 (for a highly sensitive clay) to –0.5 (for a heavily overconsolidated clay).

Variation of A

An important effect of overconsolidation is its effect on the pore pressure coefficient A. With a normally consolidated clay, the value of A at maximum deviator stress, A_f, is virtually the same in a consolidated undrained test no matter what cell pressure is used, but with an overconsolidated clay, the value of A_f falls off rapidly with increasing overconsolidation ratio (Fig. 4.34).

   Overconsolidation ratio is the ratio of preconsolidation pressure divided by the cell pressure used in the test. When the overconsolidation ratio is 1.0, the soil is normally consolidated. See Section 4.4 for overconsolidation definitions.

Example 4.9: Pore pressure coefficient A

A series of consolidated undrained triaxial tests were carried out on undisturbed samples of an overconsolidated clay.

Results were:

Cell pressure (kPa)	Deviator stress at failure (kPa)	Pore water pressure at failure (kPa)
100	410	−65
200	520	−10
400	720	80
600	940	180

(i) Plot the strength envelope for the soil with respect to effective stresses.
(ii) If the preconsolidation pressure to which the clay had been subjected was 800 kPa, plot the variation of the pore pressure coefficient A_f with the overconsolidation ratio.

Solution:

The Mohr circle diagrams are shown in Fig. 4.35a. When a pore pressure is negative, the principle of effective stress still applies, i.e. $\sigma' = \sigma - u$; for a cell pressure of 100 kPa, $\sigma_1 = 510$ kPa, and $u = -65$ kPa, so that

$$\sigma'_3 = 100 - (-65) = 165\ \text{kPa and } \sigma'_1 = 510 - (-65) = 575\ \text{kPa}$$

After consolidation in a consolidated undrained test (i.e. when shear commences) the soil is saturated, $B = 1$, and hence the pore pressure coefficient $\overline{A} = A$.

σ_3 (kPa)	o/c ratio	$A = \dfrac{\Delta u_d}{\Delta\sigma_1 - \Delta\sigma_3}$
100	8	−65/410 = −0.16
200	4	−0.02
400	2	0.111
600	1.33	0.19

The results are shown plotted in Fig. 4.35b.

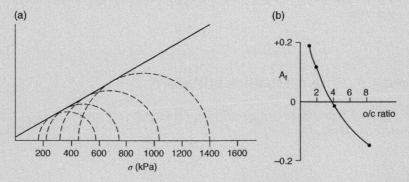

Fig. 4.35   Example 4.9.

Example 4.10: Pore pressure coefficients

The following results were obtained from an undrained triaxial test on a compacted soil sample using a cell pressure of 300 kPa. Before the application of the cell pressure, the pore water pressure within the sample was zero.

Strain (%)	σ_1 (kPa)	u (kPa)
0.0	300	120
2.5	500	150
5.0	720	150
7.5	920	120
10.0	1050	80
15.0	1200	10
20.0	1250	–60

(i) Determine the value of the pore pressure coefficient B and state whether or not the soil was saturated.

(ii) Plot the variation of deviator stress with strain.

(iii) Plot the variation of the pore pressure coefficient A with strain.

Solution:

$$B = \frac{\Delta u_a}{\Delta \sigma_3} = \frac{120}{300} = 0.4$$

The soil was partially saturated as B was less than 1.0.

Strain (%)	Δu_d	$(\Delta \sigma_1 - \Delta \sigma_3)$	$\bar{A}$	$A\left(=\dfrac{\bar{A}}{B}\right)$
2.5	30	200	$\dfrac{30}{200} = 0.15$	0.375
5.0	30	420	0.071	0.178
7.5	0	620	0	0
10.0	–40	750	–0.053	–0.133
15.0	–110	900	–0.122	–0.305
20.0	–180	950	–0.188	–0.473

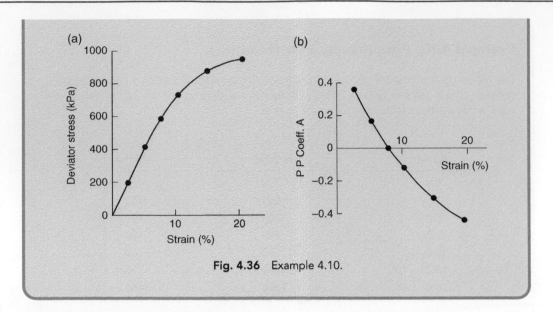

Fig. 4.36   Example 4.10.

4.4   Behaviour of soil during shearing

We saw in Section 4.2.2 that the behaviour of granular soil under shear depends on the initial density of the soil. Similarly, the shear strength of cohesive soil also depends on the initial density, which in turn is influenced by the stress history of the soil. Before continuing the subject, it is useful to introduce the following definitions.

- *Overburden pressure*: The overburden pressure at a point in a soil mass is simply the vertical pressure due to the weight of the material above it. The *effective overburden* is the pressure from this material, less the pore water pressure due to the depth of water extending from the water table to the point of interest.
- *Normally consolidated clay*: Clay which, at no time in its history, has been subjected to pressures greater than its existing overburden pressure.
- *Overconsolidated clay*: Clay which, during its history, has been subjected to pressures greater than its existing overburden pressure. One cause of overconsolidation is the erosion of material that once existed above the clay layer. Boulder clays are overconsolidated, as the many tons of pressure exerted by the mass of ice above them has been removed.
- *Preconsolidation pressure*: The maximum value of pressure exerted on an overconsolidated clay before the pressure was relieved.
- *Overconsolidation ratio (OCR)*: The ratio of the value of the effective preconsolidation pressure to the value of the presently existing effective overburden pressure. A normally consolidated clay has an OCR = 1.0 whilst an overconsolidated clay has an OCR > 1.0.

4.4.1   Undrained shear

The shear strength of a soil, if expressed in terms of total stress, corresponds to Coulomb's Law, i.e.

$$\tau_f = c_u + \sigma \tan \phi_u$$

where

c_u = unit cohesion of the soil, with respect to total stress;
ϕ_u = angle of shearing resistance of soil, with respect to total stress (=0); and
σ = total normal stress on plane of failure.

For saturated cohesive soils tested in undrained shear, it is found that τ_f has a constant value being independent of the value of the cell pressure σ_3, as indicated in Fig. 4.37. The main exception to this is a fissured clay.

Hence, and as we saw in Sections 4.2.4 and 4.2.10, $\phi_u = 0$ when a saturated cohesive soil is subjected to undrained shear. Hence:

$$\tau = c_u = \frac{1}{2}(\sigma_1 - \sigma_3) \tag{4.16}$$

Because of this, and as we have seen earlier, the term c_u is referred to as the *undrained shear strength* of the soil. The undrained shear strength is used in geotechnical design for analyses to assess stability during or soon after construction. Such analyses, referred to as *short-term* analyses, consider the soil in an undrained state and use total stresses in the calculations.

If the results of an undrained test are to be quantified in terms of effective stress, the nature of the test must be considered. In the standard unconsolidated undrained triaxial test, the soil sample is placed in the triaxial cell, the drainage connection is removed, the cell pressure is applied, and the sample is immediately sheared by increasing the axial stress. Any pore water pressures generated throughout the test are not allowed to dissipate.

If, for a particular undrained shear test carried out at a cell pressure p_c, the pore water pressure generated at failure is u, then the effective stresses at failure are:

$$\sigma'_1 = \sigma_1 - u; \sigma'_3 = \sigma_3 - u = p_c - u$$

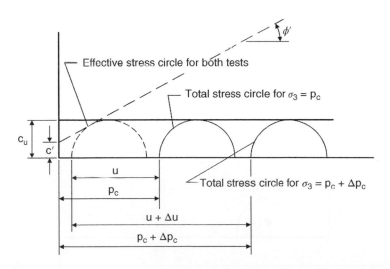

Fig. 4.37   Strength envelope for a saturated cohesive soil subjected to an undrained shear test.

Recall that in a saturated soil the pore pressure parameter B = 1.0. From this, and as already demonstrated in Section 4.2.4, it is seen that if the test is repeated using a cell pressure of $p_c + \Delta p_c$, the value of the undrained strength of the soil will be the same as that obtained from the first test. This is because the increase in the cell pressure, Δp_c will induce an increase in pore water pressure, Δu of the same magnitude (i.e. $\Delta u = \Delta p_c$). The effective stress circle at failure will therefore be the same as for the first test (Fig. 4.37), the soil acting as if it were purely cohesive. It is therefore seen that there can only be one effective stress circle at failure, independent of the cell pressure value, in an undrained shear test on a saturated soil.

4.4.2   Drained and consolidated undrained shear

The triaxial forms of these shear tests have already been described in Sections 4.2.5 and 4.2.6. It is generally accepted that, for all practical purposes, the values obtained for the drained parameters, c' and ϕ', from either test are virtually the same.

The c' value for normally consolidated clays is negligible and can be taken as zero in virtually every situation. A normally consolidated clay therefore has an effective stress strength envelope similar to that shown in Fig. 4.38 and, under drained conditions, will behave as if it were a frictional material.

The effective stress envelope for an overconsolidated clay is shown in Fig. 4.39. Unless unusually high cell pressures are used in the triaxial test, the soil will be sheared with a cell pressure less than its preconsolidation pressure value. The resulting strength envelope is slightly curved with a cohesive intercept c'. As the curvature is very slight, it is approximated to a straight line inclined at ϕ' to the normal stress axis.

In Fig. 4.39, the point A represents the value of cell pressure that is equal to the preconsolidation pressure. At cell pressures higher than this, the strength envelope is the same as for a normally consolidated clay, the value of ϕ' being increased slightly. If this line is projected backwards, it will pass through the origin.

Due to the removal of stresses during sampling, even normally consolidated clays will have a slight degree of overconsolidation and may give a small c' value, usually so small that it is difficult to measure and has little importance.

The shearing characteristics of silts are similar to those of normally consolidated clays.

The behaviour of saturated normally consolidated and overconsolidated clays in undrained shear is illustrated in Fig. 4.40 which shows the variations of both deviator stress and pore water pressure during shear.

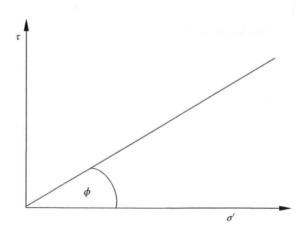

Fig. 4.38   Strength envelope for a normally consolidated clay subjected to a drained shear test.

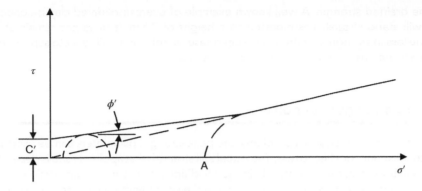

Fig. 4.39 Strength envelope for an overconsolidated soil subjected to a drained shear test.

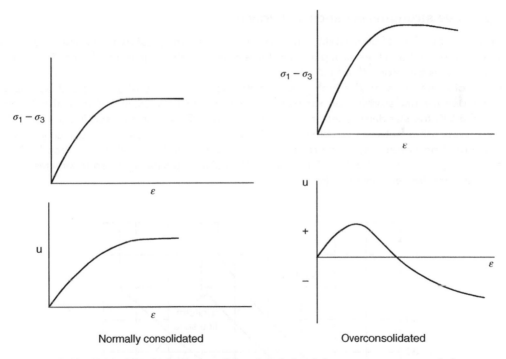

Normally consolidated   Overconsolidated

Fig. 4.40 Typical results from consolidated undrained shear tests on saturated clays.

An overconsolidated clay is considerably stronger at a given pressure than it would be if normally consolidated, and also tends to dilate during shear whereas a normally consolidated clay will consolidate. Hence, when an overconsolidated clay is sheared under undrained conditions, negative pore water pressures are induced, the effective stress is increased, and the undrained strength is much higher than the drained strength – the exact opposite to a normally consolidated clay.

If an excavation is made through overconsolidated clay, the negative pressures set up give an extremely high undrained strength, but these pore pressures gradually dissipate, and the strength falls by as much as

60 or 80% to the drained strength. A well-known example of overconsolidated clay is London clay, which when first cut, will stand virtually unsupported to a height of 7.5 m. It does not remain stable for long, and so great is the loss in strength that there have been cases of retaining walls built to support the soil, failing under the gradually increasing applied pressures from the soil.

4.5 Operative strengths of soils

For the solution of most soil mechanics problems, the peak strength parameters can be used, i.e. the values corresponding to maximum deviator stress. The actual soil strength that applies *in situ* is dependent upon the type of soil, its previous stress history, the drainage conditions, the form of construction, and the form of loading. Obviously, the shear tests chosen to determine the soil strength parameters to be used in a design should reflect the conditions that will actually prevail during and after the construction period.

The variations of strength properties of different soils are described below.

4.5.1 Operative strengths of sands and gravels

These soils have high values of permeability, and any excess pore water pressures generated within them are immediately dissipated. For all practical purposes, these soils operate in the drained state. The appropriate strength parameter is therefore ϕ', with $c' = 0$.

In granular soils, the value of ϕ' is highly dependent upon the density of the soil and, as it is difficult to obtain inexpensive undisturbed soil samples, its value is generally estimated from the results of *in situ* tests. In the UK, the standard penetration test (see Chapter 7) is the most used *in situ* test and a very approximate relationship between the blow count N and the angle of shearing resistance ϕ' is shown in Fig. 4.41. It should be noted that the corrected value for N (i.e. the $(N_1)_{60}$ value described in Chapter 7) can be used in conjunction with Fig. 4.41. The value of ϕ' obtained from the figure approximates to the peak angle obtained from drained triaxial tests.

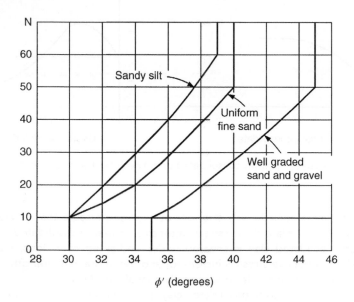

Fig. 4.41 Relationship between N and ϕ'.

Other factors, besides the value of N, such as the type of minerals, the effective size, the grading, and the shape of the particles are acknowledged to have an effect on the value of ϕ', but in view of the rough-and-ready method used to determine the value of N, any attempt at refinement seems fairly pointless.

4.5.2   Operative strengths of silts

These soils rarely occur in a pure form in the UK and are generally mixed with either sand or clay. It is therefore usually possible to classify silty soils as being either granular or clayey. When there is a reasonable amount of clay material within the soil, there should be little difficulty in obtaining undisturbed samples for strength evaluation. With sandy silts, estimated values for ϕ' can be obtained from the results of the standard penetration test.

4.5.3   Operative strengths of clays

The field consistency behaviour of clays can be used as an indicator to the range of undrained shear strength likely to be experienced for that clay, as summarised in Table 4.2.

Due to the low permeability of these soils, any excess pore water pressures generated within them will not dissipate immediately. The first step in any design work is to determine whether the clay is *normally consolidated* or *overconsolidated*.

Soft or normally consolidated clay

A clay with an undrained shear strength, c_u of not more than 40 kPa is classified as a soft clay and will be normally consolidated (or lightly overconsolidated). Such clays, when subjected to undrained shear, tend to develop positive pore water pressures (as illustrated in Fig. 4.40), so that during and immediately after construction, the strength of the soil is at its minimum value.

After completion of the construction, over a period of time, the soil will achieve its drained condition and will then be at its greatest strength.

Overconsolidated clay

With these soils, any pore water pressures generated during shear will be negative. This means simply that the clay is at its strongest during and immediately after construction. The weakest strength value will occur once the soil achieves its fully drained state, the operative strength parameters then being c' and ϕ'.

Table 4.2   Undrained shear strength of clays.

Consistency	c_u (kPa)	Field behaviour
Hard	> 300	Brittle
Very stiff	300–150	Brittle or very tough
Stiff	150–75	Cannot be moulded in fingers
Firm	75–40	Can just be moulded in fingers
Soft	40–20	Easily moulded in fingers
Very soft	<20	Exudes between fingers if squeezed

4.6   Sensitivity of clays

If the strength of an undisturbed sample of clay is measured and it is then re-tested at an identical water content, but after it has been *remoulded* to the same dry density, a reduction in strength is often observed.

The *sensitivity* of clays is defined as the ratio of the undisturbed undrained strength to the remoulded undrained strength:

$$\text{Sensitivity, } S_t = \frac{\text{Undisturbed undrained strength}}{\text{Remoulded undrained strength}}$$

Normally, consolidated clays tend to have sensitivity values varying from 5 to 10 but certain clays in Canada and Scandinavia have sensitivities as high as 100 and are referred to as *quick clays*. Video footage of two classic quick clay flows in Norway (Rissa in 1978 and Alta in 2020) is freely available online.

Sensitivity can vary slightly depending upon the water content of the clay. Generally, overconsolidated clays have negligible sensitivity, but some quick clays have been found to be overconsolidated. A classification of sensitivity is given in Table 4.3.

Thixotropy

Some clays, if kept at a constant water content, regain a portion of their original strength after remoulding with time. This property is known as *thixotropy*.

Liquidity index I_L

The definition of this index has already been given in Chapter 1 (Equation 1.6):

$$I_L = \frac{w - w_P}{I_p}$$

where w is the natural water content.

This index probably more usefully reflects the properties of plastic soil than the generally used consistency limits w_P and w_L. Liquid and plastic limit tests are carried out on remoulded soil in the laboratory, but the same soil, in its *in situ* state (i.e. undisturbed), may exhibit a different consistency at the same water content as the laboratory specimen, due to sensitivity effects. It does not necessarily mean, therefore, that a soil found to have a liquid limit of 50% will be in the liquid state if its *in situ* water content w is also 50%.

Table 4.3   Sensitivity classification.

S_t	Classification
1	Insensitive
1–2	Low
2–4	Medium
4–8	Sensitive
8–16	Extra sensitive
>16	Quick (can be up to 150)

If w is greater than the test value of w_L then I_L is >1.0 and it is fairly obvious that if the soil was remoulded it would be transformed into a slurry. In such a case, the soil is probably an unconsolidated sediment with an undrained shear strength, c_u, in the order of 15–50 kPa.

Most cohesive soil deposits have I_L values within the range 1.0–0.0. The lower the value of w, the greater the amount of compression that must have taken place and the nearer I_L will be to zero.

If w is less than the test value of the plastic limit, then I_L < 0.0 and the soil cannot be remoulded (as it is outside the plastic range). In this case, the soil is most likely a compressed sediment. Soil in this state will have a c_u value varying from 50 to 250 kPa.

4.7   Residual strength of soil

In an investigation concerning the stability of a clay slope, the normal procedure is to take representative samples, conduct shear tests, establish the strength parameters c′ and ϕ′ from the peak values of the tests, and conduct an effective stress analysis. For this analysis, the shear strength of the soil, as we have already seen, can be expressed by the equation:

$$\tau_f = c' + \sigma' \tan \phi'$$

There have been many cases of slips in clay slopes which have afforded a means of checking this procedure. Knowing the mass of material involved and the location of the slip surface, it is possible to deduce the value of the average shear stress on the slip surface, τ, at the time failure occurred. It has often been found that τ is considerably less than τ_f especially with slopes that have been in existence for some years.

Figure 4.42a shows a typical stress-strain relationship obtained in a drained shear test on a clay. Normal practice is to stop the test as soon as the peak strength has been reached, but if the test is continued it is found that as the strain increases the shear strength decreases and finally levels out. This constant stress value is termed the *critical*, or *constant volume*, strength, τ_{cv}, of the clay (similar to granular soils as explained in Section 4.2.2). If the strain increases significantly, the clay will reach a state of lowest strength known as the *residual strength*. The strength envelopes from the three sets of strength values are shown in Fig. 4.42b.

Residual strength tests can be carried out in the ring shear apparatus, developed by Bishop et al. (1971). A thin annular soil specimen is sheared by clamping it between two metal discs which are then rotated in opposite directions. The test is not common in commercial soils laboratories mainly because it is complicated and non-standardised. It also takes a long time to carry out a test. The apparatus, however, is occasionally still used by researchers and is considered a reliable means of determining residual strengths of cohesive soils (Fig. 4.43).

4.7.1   Residual strength of clays

The reduction from peak to residual strength in clays is considered to result primarily from the formation of extremely thin layers of fine particles orientated in the direction of shear. These particles would originally have been in a random state of orientation and must therefore have had a greater resistance to shear than when they became parallel to each other in the shear direction.

The development of residual strength in a soil is a continuous process. If at a particular point the soil is stressed beyond its peak strength, its strength will decrease, and additional stress will be transmitted to other points in the soil. These points become overstressed and decrease in strength, and the failure process continues once it has started (unless the slope slips) until the strength at every point along the potential slip surface has been reduced to residual strength.

Clays, especially overconsolidated deposits, contain fissures, such as those in London clay which occur some 150–200 mm apart. These fissures are already established points of weakness, the strength between

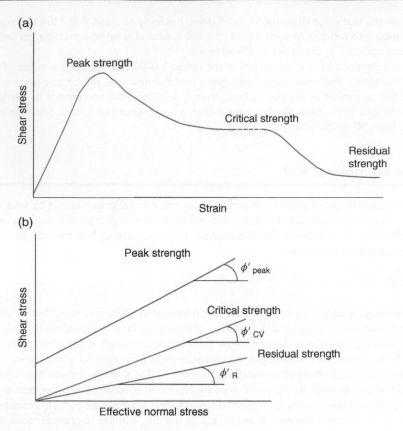

(a)

Peak strength

Critical strength

Residual strength

Shear stress

Strain

(b)

Peak strength

ϕ'_{peak}

Critical strength

ϕ'_{CV}

Residual strength

ϕ'_R

Shear stress

Effective normal stress

Fig. 4.42 The peak, critical, and residual strengths of clays.

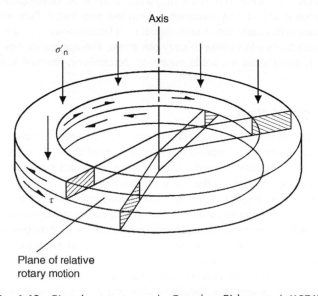

Axis

σ'_n

τ

Plane of relative
rotary motion

Fig. 4.43 Ring shear test sample. Based on Bishop et al. (1971).

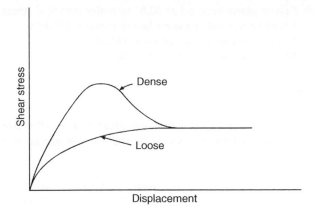

Fig. 4.44   Stress–displacement characteristics of sands.

their contact surfaces probably being about residual. An important feature of fissures is that they can tend to act as stress concentrators at their edges, leading to overstressing beyond the peak strength and hence to a progressive strength decrease.

Tests carried out by Skempton indicate that the residual strength of clay under a particular effective stress is the same, whether the clay was normally or overconsolidated. Hence, in any clay layer, provided the particles are the same, the value of ϕ'_r will be constant.

4.7.2   Residual strength of silts and silty clays

From a study of case records, Skempton showed that the value of ϕ'_r decreases with increasing clay percentage. Sand-sized particles, being roughly spherical in shape, cannot orientate themselves in the same way as flaky clay particles and when they are present in silts or clays, the residual strength becomes greater as the percentage of sand increases.

4.7.3   Residual strength of sands

Shear tests on sand indicate that the stress–displacement curve for the loose and dense states are as shown in Fig. 4.44. This was explained in Section 4.2.2. The residual strength is seen to correspond to the peak strength of the loose density and is usually reached fairly quickly in one travel of the shear box, succeeding reversals having little effect.

Exercises

Exercise 4.1

A soil sample is tested to failure in a drained triaxial test using a cell pressure of 200 kPa. The effective stress parameters of the soil are known to be $\phi' = 29°$ and c' = 0.

Determine the inclination of the plane of failure, with respect to the direction of the major principal stress, and the magnitudes of the stresses that will act on this plane. What is the maximum value of shear stress that will be induced in the soil?

Answer   Failure plane inclined at 30.5° to major principal stress.
Effective normal stress on failure plane = 295 kPa.
Shear stress on failure plane = 161 kPa.
Maximum shear stress = 184 kPa.

Exercise 4.2

A soil has an effective angle of shearing resistance, $\phi' = 20°$ and an effective cohesion, c' of 20 kPa. Determine the value of the vertical stress at failure if the soil is subjected to:

(a) a drained triaxial extension test with a cell pressure of 250 kPa.
(b) a drained triaxial compression test with the same cell pressure?

Answer   (a) 95 kPa   (b) 567 kPa

Exercise 4.3

The readings below were taken during two shear box tests carried out on samples of the same sand. In both cases, the normal stress was 200 kPa.

Horizontal displacement (mm)	Shear stress (kPa)	
	Test 1	Test 2
0	0	0
0.5	59	73
1.0	78	118
1.5	91	143
2.0	99	150
2.5	106	149
3.0	111	139
3.5	113	133
4.0	114	126
4.5	116	122
5.0	116	120

Draw the shear stress/displacement curves for the two tests, and determine the peak and constant volume values of angle of shearing resistance.

Answer $\phi' = 37°$, $\phi' = 30°$

Exercise 4.4

The following results were obtained from a drained triaxial test on a soil:

Cell pressure (kPa)	Additional effective axial stress at failure (kPa)
200	200
400	370
600	540

Determine the cohesion and angle of shearing resistance of the soil with respect to effective stresses.

Answer $\phi' = 17°$, c′ = 11 kPa

Exercise 4.5

Undisturbed samples were taken from a compacted fill material and subjected to consolidated undrained triaxial tests. Results were:

Cell pressure (kPa)	Additional axial stress at failure (kPa)	Pore water pressure at failure (kPa)
200	140	50
400	256	128
600	380	200

Determine the shear strength parameters of the soil.

Answer   c′ = 0, $\phi' = 19°$

Exercise 4.6

An undrained triaxial test carried out on a compacted soil gave the following results:

Strain (%)	Deviator stress (kPa)	Pore water pressure (kPa)
0	0	240
1	240	285
2	460	300
3	640	270
4	840	200
5	950	160

7.5	1100	110
10.0	1150	75
12.5	1170	55
15.0	1150	50

The cell pressure was 400 kPa, and before its application the pore water pressure in the sample was zero.

(i) Determine the value of the pore pressure coefficient B.
(ii) Plot deviator stress (total) against strain.
(iii) Plot pore water pressure against strain.
(iv) Plot the variation of the pore pressure coefficient A with strain.

Answer (i) 0.6

Exercise 4.7

The following results were obtained from a shear box test carried out on a set of soil samples. The apparatus made use of a proving ring to measure the shear forces.

Normal load (kN)	0.2	0.4	0.6
Strain (%)	Proving ring dial gauge readings (no. of divisions)		
1	8.5	16.5	28.0
2	16.0	27.0	39.0
3	22.5	34.9	46.8
4	27.5	39.9	52.3
5	31.3	45.0	56.6
6	33.4	46.0	59.7
7	33.4	47.6	61.7
8	33.4	47.6	62.7
9		47.6	62.7
10			62.7

The cross-sectional area of the box was 3600 mm^2 and one division of the proving ring dial gauge equalled 0.01 mm. The calibration of the proving ring was 0.01 mm deflection equalled 8.4 N.
   Determine the strength parameters of the soil.

Answer $\phi' = 32°$ and $c' = 43$ kPa

Chapter 5

Stress Paths and the Critical State

Learning objectives:

By the end of this chapter, you will have been introduced to:
- stress paths in both two- and three-dimensional spaces;
- the theory behind isotropic consolidation;
- drained and undrained stress paths in the triaxial test;
- a simplified overview of critical state soil mechanics;
- the critical state line, drained and undrained planes and the Roscoe and Hvorslev state boundary surfaces.

5.1   Stress paths in two-dimensional space

As we saw in Chapter 4, the state of stress in a soil sample can be shown graphically by a Mohr circle diagram. In a triaxial compressive test, the axial strain of the test specimen increases up to failure and the various states of stress that the sample experiences from the start of the test until failure could obviously be represented by a series of Mohr circles. The same stress states can be represented in a much simpler form by expressing each successive stress state as a point. The line joining these successive points is known as a *stress path*.

Stress paths can be of several forms and we have already seen the stress–strain relationships plotted in $\tau - \sigma$ space to show triaxial test results. Further ahead, Fig. 12.10 shows the $e - \log p$ compression curves for normally consolidated clay. These two cases are examples of two-dimensional stress paths.

If a Mohr circle diagram of total stress is examined (Fig. 5.1), the point of maximum shear has the coordinates p and q where:

$$p = \frac{\sigma_1 + \sigma_3}{2} \quad \text{and} \quad q = \frac{\sigma_1 - \sigma_3}{2}$$

where σ_1 and σ_3 are the principal total stresses.

Smith's Elements of Soil Mechanics, 10th Edition. Ian Smith.
© 2021 John Wiley & Sons Ltd. Published 2021 by John Wiley & Sons Ltd.
Companion website: www.wiley.com/go/smith/soilmechanics10e

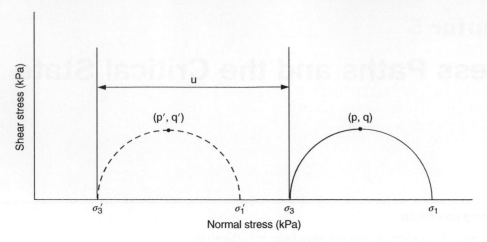

Fig. 5.1 Points of maximum shear stress.

In terms of effective stresses, σ'_1 and σ'_3, the point of maximum shear has the coordinates p' and q' where:

$$p' = \frac{\sigma'_1 + \sigma'_3}{2} \text{ and } q' = \frac{\sigma'_1 - \sigma'_3}{2}$$

If a soil is subjected to a range of values of σ'_1 and σ'_3, the point of maximum shear stress can be obtained for each stress circle; the line joining these points, in the order that they occurred, is termed the *stress path* or *stress vector* of maximum shear. Any other point instead of maximum shear can be used to determine a stress path, e.g. the point of maximum obliquity, but the stress paths of maximum shear are not only simple to use but are also more applicable to consolidation work (see Chapter 12).

Typical effective stress paths obtained from a series of consolidated undrained triaxial tests on samples of normally consolidated clay together with the effective stress Mohr circles at failure are shown in Fig. 5.2. The effective stress paths are denoted by the dashed lines.

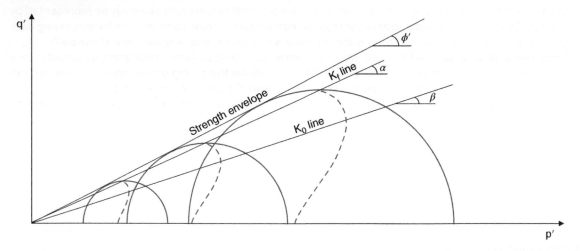

Fig. 5.2 Typical effective stress paths obtained from consolidated undrained triaxial tests on a normally consolidated clay.

Also shown in Fig. 5.2 is the strength envelope – the line tangential to the stress circles, inclined at ϕ' to the normal stress axis.

The K_f line

If each Mohr circle is considered, it is seen that the ratio σ'_3/σ'_1 is a constant, to which the symbol K_f is applied. If the points of maximum shear for each effective stress circle p'_f and q'_f are joined together (i.e. the apexes of the Mohr circles), the stress path of maximum shear stress at failure is obtained. This line is called the K_f *line* or the *instability line* and is inclined at angle α to the normal stress axis. Using trigonometry, we see that tan $\alpha = \sin \phi'$.

Example 5.1: K_f line

Using the data from the consolidated drained test of Example 4.6, determine the slope angle, α of the K_f line for the soil and the angle of shearing resistance ϕ'.

Solution:

Determine the maximum points of shear, p' and q' for each cell pressure and plot the results.

σ'_3	σ'_1	$p' = \dfrac{\sigma'_1 + \sigma'_3}{2}$	$q' = \dfrac{\sigma'_1 - \sigma'_3}{2}$
200	570	385	185
300	875	588	288
400	1162	781	381

The values of p' and q' fall on the K_f line as shown in Fig. 5.3.
The angle α is determined as 26°.

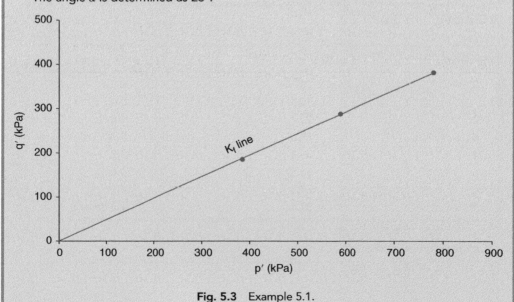

Fig. 5.3 Example 5.1.

Since $\tan \alpha = \sin \phi'$, ϕ' is easily determined:

$$\phi' = \sin^{-1}(\tan 26°) \Rightarrow \phi' = 29°$$

This is the same value (as would be expected) determined in Example 4.6.

The K_0 line

For a soil undergoing one-dimensional consolidation (see Chapter 12), the ratio σ_3'/σ_1' is again constant and its value is given the symbol K_0. Plotting the maximum shear stress points of these stress circles enables the stress path for one-dimensional consolidation, the K_0 line, to be determined; this line is inclined at angle β to the normal stress axis.

K_0 is known as the *coefficient of earth pressure at rest*. For consolidation work, K_0 may be defined for a soil with a history of one-dimensional strain as the ratio:

$$K_0 = \frac{\text{Lateral effective stress}}{\text{Vertical effective stress}}$$

Lateral earth pressure is covered in Chapter 8.

Example 5.2: Effective stress paths: normally consolidated clay

Consolidated undrained triaxial tests carried out on representative undisturbed samples of a normally consolidated clay gave the following results:
Cell pressure = 35 kPa

Strain (%)	Deviator stress (kPa)	Pore water pressure (kPa)
0	0	0
1	10.4	0.4
2	20.7	4.8
3	29.0	9.7
4	33.2	13.2
5	35.8	16.6
6	37.3	18.1
6.8	37.8	19.3 (failure)

Cell pressure = 70 kPa

Strain (%)	Deviator stress (kPa)	Pore water pressure (kPa)
0	0	0
1	20.7	4.1
2	42.7	12.8
3	54.4	22.1
4	63.4	30.4
5	66.1	34.0
6	71.7	37.9
7	75.8	40.7 (failure)

Draw the effective stress paths for undrained shear obtained from the tests and establish the value for the coefficient of earth pressure, K_0.

Solution:

The first step is to plot the two effective stress paths. The calculations are best set out in tabular form:

Cell pressure = 35 kPa

Strain	$\sigma_1 - \sigma_3$	u	$q = \frac{\sigma_1' - \sigma_3'}{2}$	$p' = \frac{\sigma_1' + \sigma_3'}{2}$
0	0	0	0	35
1	10.4	0.4	5.2	39.8
2	20.7	4.8	10.4	40.6
3	29.0	9.7	14.5	39.8
4	33.2	13.2	16.6	38.4
5	35.8	16.6	17.9	36.3
6	37.3	18.1	18.7	35.6
6.8	37.8	19.3	18.9	34.6

Cell pressure = 70 kPa

0	0	0	0	70
1	20.7	4.1	10.4	76.3
2	42.7	12.8	21.4	78.6
3	54.4	22.1	27.2	75.1
4	63.4	30.4	31.7	71.3
5	66.1	34.0	33.1	69.1
6	71.7	37.9	35.9	68.0
7	75.8	40.7	37.9	67.2

The stress paths are shown in Fig. 5.4 and the K_f line is drawn. The K_f line passes through the points of failure of each test and, since the clay is normally consolidated, the origin too. On the diagram, the strain contours (lines joining equal strain values on the two test paths) are also indicated. These will be used in Example 12.9.

From the K_f line, α is established as 29°. tan α (=sin ϕ') = tan 29° = 0.554.

$\Rightarrow K_0 = 1 - 0.554 = 0.446$

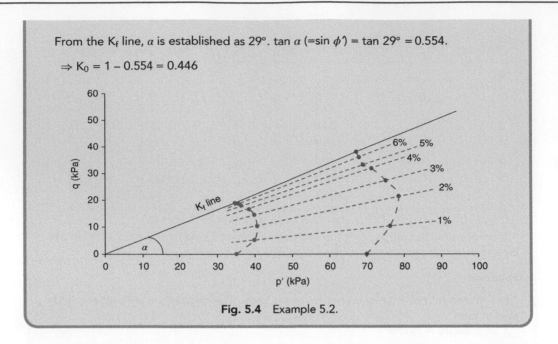

Fig. 5.4   Example 5.2.

Example 5.3: Effective stress paths: overconsolidated clay

CIU triaxial tests, carried out on U100 samples of overconsolidated clay at low rates of strain, gave the following results:

Cell pressure = 100 kPa

Strain (%)	Deviator stress (kPa)	Pore water pressure (kPa)
0	0.0	0.0
1	11.9	13.2
2	25.1	19.7
3	33.3	23.4
4	38.2	25.2
5	48.1	26.2
6	62.9	26.0
7	74.8	25.5
8	86.8	25.2
9	96.8	25.0
10	107.7	25.0
11	118.6	24.9
12	122.9	24.9
13	127.3	24.9 (failure)
14	125.8	24.8

Cell pressure = 200 kPa

Strain (%)	Deviator stress (kPa)	Pore water pressure (kPa)
0	0	0.0
1	19.9	15.1
2	41.9	25.4
3	55.5	29.8
4	63.7	32.7
5	80.1	36.0
6	104.8	39.1
7	124.7	41.0
8	144.6	42.0
9	161.3	42.5
10	179.5	41.8
11	197.6	41.6
12	204.8	41.4
13	212.2	41.2
14	217.3	41.1 (failure)

Draw the effective stress paths for undrained shear obtained from the tests and establish the equation of the K_f line.

Solution:

Similar to Example 5.2, the calculations are best set out in tabular form:
Cell pressure = 100 kPa

Strain (%)	Deviator stress (kPa)	Pore water pressure (kPa)	$q = \frac{\sigma'_1 - \sigma'_3}{2}$	$p' = \frac{\sigma'_1 + \sigma'_3}{2}$
0	0.0	0.0	100.0	0.0
1	11.9	13.2	92.8	6.0
2	25.1	19.7	92.9	12.6
3	33.3	23.4	93.3	16.7
4	38.2	25.2	93.9	19.1
5	48.1	26.2	97.8	24.0
6	62.9	26.0	105.4	31.4
7	74.8	25.5	111.9	37.4
8	86.8	25.2	118.2	43.4
9	96.8	25.0	123.4	48.4
10	107.7	25.0	128.9	53.9
11	118.6	24.9	134.4	59.3
12	122.9	24.9	136.5	61.4
13	127.3	24.9	138.8	63.7
14	125.8	24.8	138.1	62.9

Cell pressure = 200 kPa

Strain (%)	Deviator stress (kPa)	Pore water pressure (kPa)	$q = \dfrac{\sigma'_1 - \sigma'_3}{2}$	$p' = \dfrac{\sigma'_1 + \sigma'_3}{2}$
0	0	0.0	200.0	0.0
1	19.9	15.1	194.9	10.0
2	41.9	25.4	195.6	21.0
3	55.5	29.8	198.0	27.8
4	63.7	32.7	199.2	31.9
5	80.1	36.0	204.1	40.1
6	104.8	39.1	213.3	52.4
7	124.7	41.0	221.4	62.4
8	144.6	42.0	230.3	72.3
9	161.3	42.5	238.2	80.7
10	179.5	41.8	248.0	89.8
11	197.6	41.6	257.2	98.8
12	204.8	41.4	261.0	102.4
13	212.2	41.2	264.9	106.1
14	217.3	41.1	267.6	108.7

The stress paths are shown in Fig. 5.5 and the K_f line is drawn. The K_f line passes through the points of failure of each test.

From the K_f line, α is established as 23°. tan α = tan 23° = 0.42. The q' axis intercept is 14 kPa. The equation of the K_f line therefore is

$$K_f = 0.42\, p' + 14$$

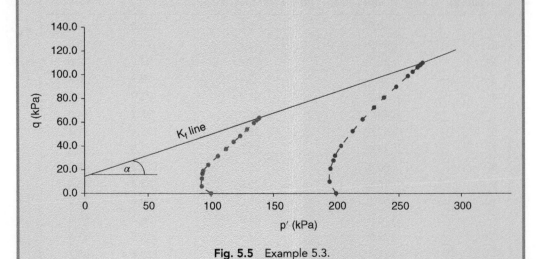

Fig. 5.5 Example 5.3.

5.2   Stress paths in three-dimensional space

Having considered two-dimensional stress paths, it is now appropriate to examine the form of these paths if they were plotted in three-dimensional space. To describe the behaviours during shear and compression, we introduce three parameters:

the mean normal stress, p
the deviatoric stress, q
the specific volume, v

With reference to Fig. 3.2a, and representing the vertical stress by σ_1, and the two horizontal stresses as σ_2 and σ_3 (where $\sigma_2 = \sigma_3$), it is seen that the mean normal stress, p, acting on the element of soil may be given by:

$$p = \frac{1}{3}(\sigma_1 + 2\sigma_3) \tag{5.1}$$

and as we saw in Chapter 4, the deviator stress, q is defined as:

$$q = (\sigma_1 - \sigma_3) \tag{5.2}$$

Similar expressions apply for effective stress:

$$p' = \frac{1}{3}(\sigma'_1 + 2\sigma'_3) \tag{5.3}$$

$$q' = (\sigma'_1 - \sigma'_3) \tag{5.4}$$

In Chapter 1, we saw that the void ratio, e, is the volume of voids, V_v, divided by the volume of solids, V_s. The specific volume, v, is the total volume of soil that contains a unit volume of solids:

$$v = \frac{V}{V_s} = \frac{V_S + V_V}{V_S} = 1 + e \tag{5.5}$$

The advantage of the p' and q' parameters is their association with the strains that they cause. Changes in p' are associated with volumetric strains and changes in q' with shear strains.

For the general three-dimensional state, such as a natural soil in the field p' and q' are defined as:

$$p' = \frac{1}{3}(\sigma'_1 + \sigma'_2 + \sigma'_3) \tag{5.6}$$

$$q' = \frac{1}{\sqrt{2}}\sqrt{\left[\left(\sigma'_1 - \sigma'_2\right)^2 + \left(\sigma'_2 - \sigma'_3\right)^2 + \left(\sigma'_3 - \sigma'_1\right)^2\right]} \tag{5.7}$$

and in the triaxial test, where $\sigma'_2 = \sigma'_3$, we see that Equations (5.6) and (5.7) reduce to Equations (5.3) and (5.4).

5.3   Isotropic consolidation

As we have seen, most soil samples tested in the triaxial apparatus are isotropically consolidated, i.e. consolidated under an all-round water pressure, before the commencement of the shearing stage of the test. Other forms of consolidation exist, such as one-dimensional consolidation, and these are discussed in Chapter 12.

5.3.1   Isotropically consolidated clay

The form of the compression curve for an isotropically consolidated clay is shown in Fig. 5.6a. The curve is indicated by the line ABC. It should be noted that the plot is in the form of a v : p' plot, the vertical axis being 0 : v and the horizontal axis 0 : p'. The v : ln p' plot is shown in Fig. 5.6b and from this diagram we see that, if we are prepared to ignore the slight differences between the expansion and the recompression curves, the semi-log plot of the isotropic consolidation curve for most clays can be assumed to be made up from a set of straight lines and to have the idealised form of Fig. 5.6c.

Any point on the line ABC represents normal consolidation whereas a point on the line BD, or indeed any point below ABC, represents overconsolidation. As line DB represents the idealised condition that the expansion and recompression curves coincide, it is probably best to give it a new name, and it is therefore usually called the *swelling line*. The line ABC in Fig. 5.6a is known as the *normal consolidation line* (NCL).

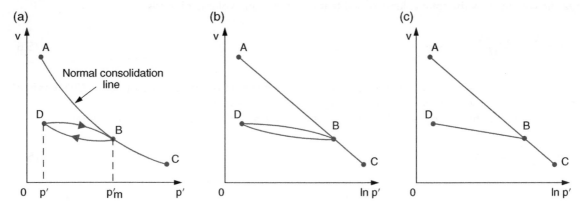

Fig. 5.6   Typical shape of the isotropic normal consolidation curves of a saturated cohesive soil.

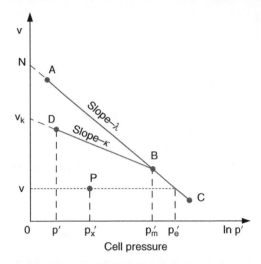

Fig. 5.7   Idealised form of v : ln p' plot.

If the maximum previous pressure on a swelling line is p'_m and the pressure at D, a point on the swelling line, is p' then we can say that the degree of overconsolidation represented by point D is $R_p = p'_m/p'$. (Note the use of the subscript '$_p$' in R_p to indicate isotropic consolidation.)

Figure 5.7 is a close-up of Fig. 5.6c. In the diagram, let the slope of AC, the normal isotropic consolidation line, be $-\lambda$, and the slope of the swelling line, DB, be $-\kappa$.

N (capital nu) is the specific volume of a soil normally consolidated at p' value of 1.0 kPa. This gives ln $p' = 0$. Then the equation of line AC is

$$v = N - \lambda \ln p'$$

A swelling line, such as BD, can lie anywhere beneath the line AC as its position is dependent upon the value of the maximum pressure on the line, p_m, which determines the position of B.

Let v_κ be the specific volume of an overconsolidated soil at $p' = $ unity (i.e. 1.0 kPa). Then the equation of line DB is

$$v = v_\kappa - \kappa \ln p'$$

λ, N, and κ are measured values and must be found from appropriate tests.

Note: The normal consolidation line AC is often referred to as the λ *line* (i.e. the *lambda line*), and the swelling line BD is often called the κ *line* (i.e. the *kappa line*).

5.3.2   Equivalent isotropic consolidation pressure, p'_e

Consider a particular specific volume, v. Then the value of consolidation pressure which corresponds to v on the normal isotropic consolidation curve is known as the *equivalent consolidation pressure* and is given the

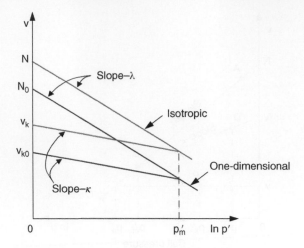

Fig. 5.8 Isotropic and one-dimensional consolidation.

symbol p'_e. In Fig. 5.7, the point P represents a soil with a specific volume, v, and an existing effective consolidation pressure p'_x. The procedure for determining p'_e is illustrated in the diagram. Note that as P is below AB, it represents a state of overconsolidation.

For a normally consolidated clay, subjected to an undrained triaxial test, $p'_e = \sigma'_3$ but with drained tests, p'_e will vary.

5.3.3 Comparison between isotropic and one-dimensional consolidation

Note: The subject of one-dimensional consolidation is covered in Chapter 12 and some of the text in this short section refers to subject matter discussed in that chapter.

If a sample of clay is subjected to one-dimensional consolidation in an oedometer and another sample of the clay is subjected to isotropic consolidation in a triaxial cell, then the idealised forms of the v : ln p′ plots for the tests will be more or less as illustrated in Fig. 5.8.

The values of the slopes of the two normal consolidation lines are very close and, for all practical purposes, can both be assumed to be equal to $-\lambda$. Similarly, the slopes of the swelling lines can both be taken as equal to $-\kappa$.

Note that the values of ln p′ for the one-dimensional test are taken as equal to ln σ′, where σ′ = the normal stress acting on the oedometer sample.

As the compression index C_c is expressed in terms of common logarithms, we see that:

$$\lambda = \frac{C_c}{2.3}$$

5.4 Stress paths in the triaxial apparatus

The triaxial apparatus enables many different loading conditions to be used in sample testing. The simplest condition is that of triaxial compression. In this test, the cell pressure remains constant, while the axial stress is slowly increased to shear the sample.

5.4.1 Drained compression test

This test was described in Section 4.2.5. Refer to Fig. 5.9 which shows the test paths from a drained compression test and draws on Equations (5.3) and (5.4). At the beginning of the test, the cell pressure is set at a kPa. This pressure is applied to all faces of the cylindrical sample ($\sigma'_1 = \sigma'_3$) so that p' = a. Since the sample is drained, u = 0. The state of the sample is represented by point A and since u = 0, A represents both total and effective stresses.

During the compression stage, the axial stress σ'_1 increases and thus so do p' and q' and the stress path travels from A to B where point B represents a stress point later on in the test (Fig. 5.9a). The change of q' ($\delta q'$) is equal to $\delta \sigma'_1$ (since the cell pressure, σ'_3 remains constant, i.e. $\delta \sigma'_3 = 0$):

$$\delta q' = \delta q = \delta \sigma'_1$$

From Equation (5.3), it follows that the change of p' ($\delta p'$) is equal to one-third $\delta \sigma'_1$ (again, because $\delta \sigma'_3 = 0$). This is illustrated in Fig. 5.9b. Hence,

$$\delta p' = \delta p = \delta \sigma'_1 / 3$$

and therefore, the slope of the total and effective stress paths is $\delta q / \delta p = 3$.

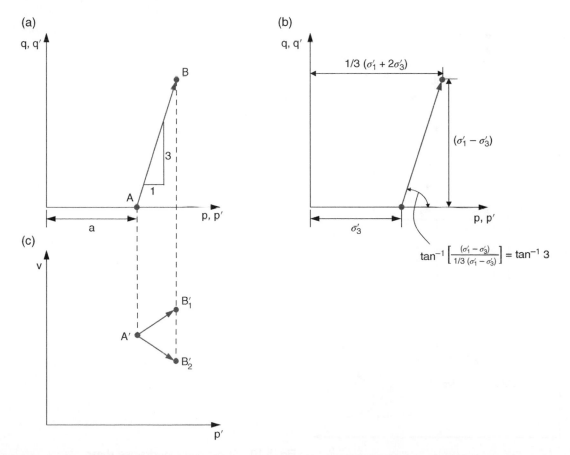

Fig. 5.9 Test paths for a drained compression test in the triaxial apparatus. Source: Based on Atkinson and Bransby (1978).

The path can also be plotted in v : p' space (Fig. 5.9c). It is not possible to define the direction of the path A'B' unless changes in specific volume are measured in the test. The change in specific volume that takes place during the test may be positive (e.g. path A'B$_1'$) or negative (e.g. path A'B$_2'$). So without measurement of v, all that can be deduced is that the 'B' points (B$_1'$, B$_2'$ etc.) lie on the projection at constant p' from the point B in q' : p' space.

It is seen that the stress path is fixed in q' : p' space, but not fixed in v : p' space. From an initial state (point A), all samples will follow a slope of 3 in q' : p' space until the sample fails. Without measurement of the magnitude and sign of the volume change during shearing, it is not possible to know the exact path in v : p' space.

If we plot the effective stress paths for several drained tests, however, we obtain a set of paths similar to those shown in Fig. 5.10. For the q' − p' plane (Fig. 5.10a), the plot consists of straight lines which are, as we have seen, inclined to the horizontal at $\tan^{-1}(3)$. The points B$_1$, B$_2$, and B$_3$ are the stress points at failure and it is seen that these fall on a straight line known as the *critical state line* (CSL) – see Section 5.5.1.

The stress paths in the v − p' plane are illustrated in Fig. 5.10b. The points C$_1$, C$_2$, and C$_3$ represent the failure points after drained shear, so the void ratio values, and therefore the specific volume values, at these points are less than those at the corresponding A points. It is seen that the failure points C$_1$, C$_2$, and C$_3$ lie on a curved line similar to the normal consolidation line. This clearly indicates that the CSL can be identified in both q' : p' space and v : p' space.

(a)

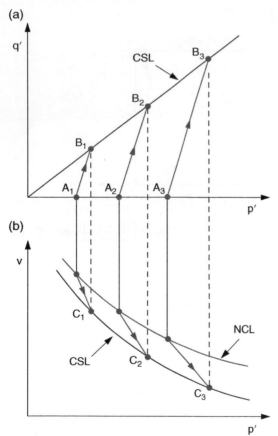

(b)

Fig. 5.10   Stress paths for drained shear.

5.4.2   Undrained compression test

This test was described in Section 4.2.6. In the test, the drainage tap is closed so that no drainage can occur during compression. The cell pressure is kept constant, but this time because drainage cannot occur, changes in pore water pressure occur during the test and the sample responds to the changes in total stress at constant volume (since no water drains out).

Typical stress paths for this test are shown in Fig. 5.11. There are three plots this time: total stress $(q : p)$, effective stress $(q' : p')$, and specific volume $(v : p')$. It is seen that the total stress path (TSP) is identical to that experienced in the drained test (Fig. 5.9a).

If we measure the pore water pressure, u throughout the shearing stage (i.e. as q' increases), we can establish the effective stress path (ESP) A′B′ (top right plot) by subtracting u from p at regular intervals, i.e. $p' = p - u$. The dashed line A′X, which mimics the TSP, is only drawn to enable the ESP to be easily established.

The lower plot shows the $v : p'$ space. In the undrained test, because no water can drain out, the volume of the sample remains constant throughout the test. Thus, the test path here is simply a line of constant v from A″ to B″, which project from points A′ and B′ in $q' : p'$ space.

If we consider the $q' - p'$ plane, then we can plot the effective stress paths for undrained shear in a manner similar to the previous two-dimensional stress paths.

The resulting diagram is shown in Fig. 5.12a. The points A_1, A_2, and A_3 lie on the isotropic normal consolidation line and their respective stress paths reach the failure boundary at points B_1, B_2, and B_3. As the tests are undrained, the values of void ratio at points B_1, B_2, and B_3 are the same as they were when the soil was at the stress states A_1, A_2, and A_3, respectively. Knowing the e values, we can determine the values of specific volume and prepare the corresponding plot on the $v - p'$ plane (Fig. 5.12b).

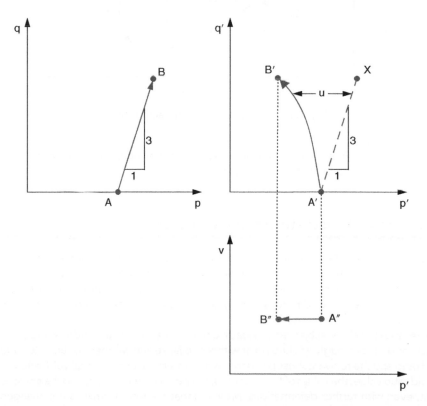

Fig. 5.11   Test paths for an undrained compression test in the triaxial apparatus. Based on Atkinson and Bransby (1978).

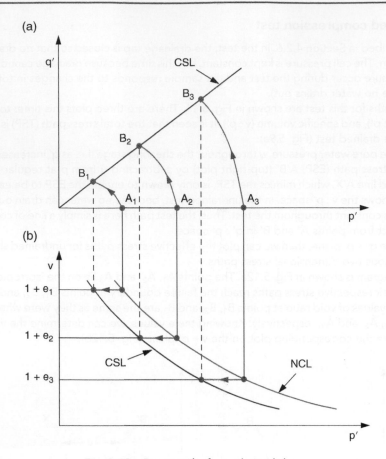

Fig. 5.12   Stress paths for undrained shear.

It is seen that the failure points B_1, B_2, and B_3 lie on a straight line in the $q' - p'$ plane and on a curve, similar to the normal consolidation curve, in the $v - p'$ plane.

5.5   Introduction to critical state soil mechanics

In Section 4.2.2, we saw that during a drained test, the void ratio of a soil changes during shear. If several samples of the same soil are tested at different initial densities it is found that, if the rate of shearing is constant, the samples all fail at the same void ratio (Fig. 4.19d). If the deformation is allowed to continue, the sample will remain at the same void ratio and only deform by shear distortion. This condition is referred to as the *critical state*.

If a saturated, remoulded clay is subjected to a loading that creates a constant and low rate of increasing strain, the clay will reach, and pass through, a failure point without collapse and will then continue to suffer deformation as both the void ratio and the relevant stress paths follow a yield surface until a critical void ratio value is achieved. At this critical void ratio value, the values of the void ratio, the pore water pressure, and the stresses within the soil remain constant, even with further deformations, provided that the rate of strain is not changed.

This important concept led to the theory of *critical state*, a soil model that brings together the relationships between its shear strength and its void ratio, and which can be applied to any type of soil. The theory has been established as a research tool for several years and is regularly used in geotechnical limit state design.

Critical state soil mechanics is a specialised topic and if a deep understanding of the subject is to be gained, reference to specialised texts is required. This section of the book merely offers a simplistic and short introduction to the topic of the critical state. Readers interested in developing a thorough knowledge of the subject are referred to the texts by Muir Wood (1991) and Atkinson (2007).

5.5.1   The critical state line

As is seen in Figs 5.10 and 5.12, the failure points, regardless of whether tested undrained or drained, all lie on a single line. This line is called the *critical state line (CSL)*.

A set of drained and undrained triaxial compression test results on samples of Weald clay (Parry, 1960) are shown in Fig. 5.13. The (p', q) points obtained from each of the test results are plotted in Fig. 5.13a and the (p', v) points are plotted in Fig. 5.13b.

We can further confirm from these diagrams that there is a single line of failure points within the q' : p' : v space which projects as a straight line on to the q' − p' plane and projects as a curved line, close to the normal consolidation line, on to the v − p' plane. The position of this line, the *critical state line*, is illustrated in Fig. 5.14.

5.5.2   The equation of the critical state line

The line's projection on to the q' − p' plane is a straight line with the equation q = Mp', where M is the slope of the line.

The projection of the critical state line on to the v − p' plane is curved but if we consider the projection on to the v − ln p' plane, we obtain a straight line with a slope that can be assumed to be equal to the slope of the normal consolidation line. Figure 5.15 shows the v : ln p'_f plot for Parry's results.

If we use the symbol Γ to represent the value of v which corresponds to ln p' = 0 (i.e. a p' value of 1.0 kPa), then the equation of the straight line projection is

$$v = \Gamma - \lambda \ln p' \tag{5.8}$$

which can be written as:

$$\frac{\Gamma - v}{\lambda} = \ln p'$$
$$p' = \exp\left(\frac{\Gamma - v}{\lambda}\right)$$

Hence, the critical state line is that line which satisfies the two equations:

$$q = Mp' \tag{5.9}$$

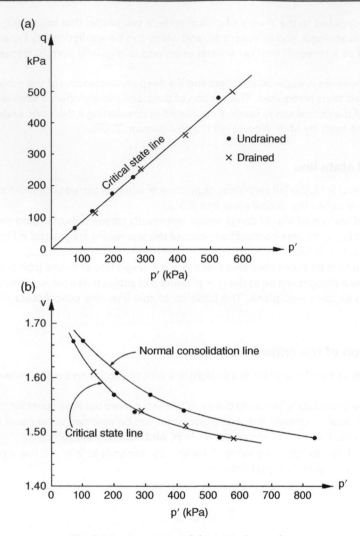

Fig. 5.13 Projection of the critical state line.

$$p' = \exp\left(\frac{\Gamma - v}{\lambda}\right) \tag{5.10}$$

Referring back to Section 5.3, it is seen that λ is the slope of the normal consolidation line and N is the specific volume at $\ln p' = 0$. The values of M, N, Γ, and λ vary with the type of soil. From Figs 5.13 and 5.15, it can be determined that the values for remoulded Weald clay are approximately M = 0.85; N = 2.13; Γ = 2.09, and λ = 0.10.

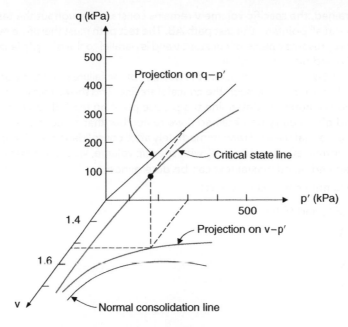

Fig. 5.14   Position of the critical state line.

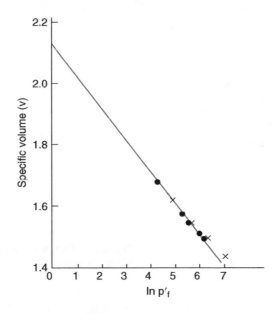

Fig. 5.15 $v : \ln p'_f$ plot.

5.6   Undrained and drained planes

5.6.1   Undrained planes

Refer to Fig. 5.16. A sample in an undrained triaxial test will be initially isotropically compressed to point A and then slowly sheared until it fails at point B on the critical state line. The test path AB can be projected into q' : p' space, as shown by path A_1B_1.

Since the test is undrained, the specific volume v remains constant throughout the test and therefore v at point A = v at point B = v at all points on the test path AB. The test path must therefore remain on the shaded rectangle ACDE, which represents a plane of constant v and is parallel to the q' – p' plane. The plane ACDE is considered as the *undrained plane* through A.

Hence, knowing the position of A, we can obtain the position of the failure point B, by drawing the undrained plane through A and noting where it intersects the critical state line. This shows that the initial state of the sample and the test conditions completely determine the precise point on the CSL at which the sample will fail.

Expressions for p' and q' can easily be obtained if we remember that, since the sample is undrained, the void ratio at failure e_f must equal the void ratio immediately after consolidation, e_0. If a sample is isotropically compressed to a mean normal effective stress, p_0' with specific volume, v_0, the stresses q_f' and p_f' and specific volume v_f at failure in an undrained triaxial test can be determined:

$v_f = v_0$ (since v does not change during test)

Developing Equations (5.9) and (5.10):

$$p_f' = \exp\left(\frac{\Gamma - v_0}{\lambda}\right)$$

$$q_f' = M \exp\left(\frac{\Gamma - v_0}{\lambda}\right)$$

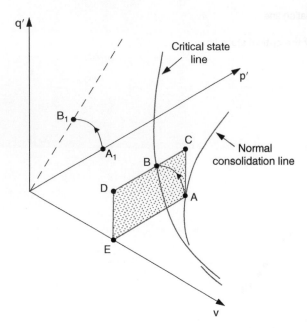

Fig. 5.16 Path followed by an undrained test in q' : p' : v space.

Example 5.4: Failure conditions in an undrained test

A sample of Weald clay is consolidated in a triaxial cell under a cell pressure of 200 kPa and is then sheared undrained. Using the following properties, determine the values of q', p' and v' at failure.

$N = 2.13, \lambda = 0.10, \Gamma = 2.09, M = 0.85$

Solution:

Since the sample is isotropically consolidated,

$$v_0 = N - \lambda \ln p_0' = 2.13 - 0.1 \ln (200) = 1.60$$
$$\Rightarrow v_f = 1.60$$

$$p_f' = \exp \left(\frac{\Gamma - v_0}{\lambda} \right) = \exp \left(\frac{2.09 - 1.60}{0.1} \right) = 134 \text{ kPa}$$

$$q_f' = Mp_f' = 0.85 \times 134 = 114 \text{ kPa}$$

5.6.2 Drained planes

Refer to Fig. 5.17. A sample in a drained triaxial test will be initially isotropically compressed to point A and then slowly sheared until it fails at point B on the critical state line. The test path AB can be projected into $q' : p'$ space, as shown by path A_1B_1. Note: as we saw in Section 5.4.1, the projection of the drained stress path on to the $q' - p'$ plane is a straight line inclined at an angle $\tan^{-1} 3$ to the horizontal.

Since the test is drained, the specific volume v changes through the test and therefore the drained plane must be parallel to the v-axis (i.e. Line AA_1 is parallel to v-axis). The test path must therefore remain in the shaded rectangle ACB_1A_1, during the test. The plane ACB_1A_1 is considered as the *drained plane* through A.

It is seen from Fig. 5.17, that if the initial conditions of the test are specified as those at point A, then the point of failure B will occur at the intersection of the drained plane and the critical state line. Further, if we now look at the path on the $q' : p'$ space (Fig. 5.18), we can establish the location of B via geometry.

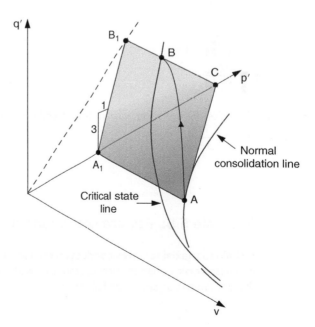

Fig. 5.17 Path followed by a drained test in $q' : p' : v$ space.

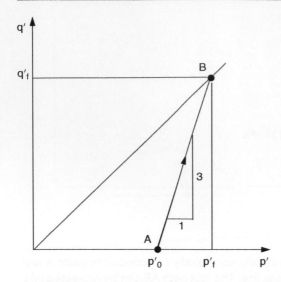

Fig. 5.18 Path followed by a drained test in q′ : p′ space.

From Fig. 5.18,

$$q'_f = 3(p'_f - p'_0)$$

and since $q'_f = Mp'_f$ we see that:

$$q'_f = Mp'_f = 3p'_f - 3p'_0$$

i.e. $M = 3 - \dfrac{3p'_0}{p'_f}$

$$\Rightarrow p'_f = \dfrac{3p'_0}{3 - M}$$

$$\Rightarrow q'_f = \dfrac{3Mp'_0}{3 - M}$$

and from Equation (5.8) we can obtain the specific volume at failure, v_f:

$$v_f = \Gamma - \lambda \ln\left(\dfrac{3p'_0}{3 - M}\right)$$

Example 5.5: Failure conditions in a drained test

A drained triaxial test was conducted on a sample of the same clay from Example 5.4. The initial isotropic consolidation pressure p'_0 was 200 kPa. Calculate the values of q′, p′, v and the volumetric strain, ε_v at failure.

Solution:

From before, $v_0 = 1.60$

$$q_f' = \frac{3M\, p_0'}{3-M} = \frac{3 \times 0.85 \times 200}{3-0.85} = 237\, \text{kPa}$$

$$p_f' = q_f'/M = 237/0.85 = 279\, \text{kPa}$$

$$v_f = \Gamma - \lambda \ln p_f' = 2.09 - 0.10 \ln(279) = 1.53$$

$$\varepsilon_v = \delta v/v_0 = (1.6 - 1.53)/1.6 = 4.4\%$$

Example 5.6: Drained and undrained failure conditions

An undisturbed laboratory sample of clay has the following properties:
  $M = 0.78$, $\lambda = 0.19$, $N = 2.80$, and $\Gamma = 2.72$.
  Samples of the clay were isotropically consolidated in a triaxial cell under cell pressures of 100, 200, and 400 kPa.

(a). For each cell pressure, determine the mean principal stress at failure, the deviatoric compression stress at failure, the major principal stress at failure, and the ultimate void ratio, if the shearing stages were carried out under drained conditions.
(b). If the same soil was tested under undrained conditions, calculate the deviatoric compression stress and the major principal stress at failure.

Solution:

(a). *Drained shearing*
  Cell pressure, $p_0' = 100\, \text{kPa}$

$$p_f' = \frac{3p_0'}{(3-M)} = \frac{3 \times 100}{(3-0.78)} = 135.2\, \text{kPa}$$

$$q_f' = Mp_f' = 0.78 \times 135.2 = 105.4\, \text{kPa}$$

$$\sigma_1' = q_f' + \sigma_3' = 105.4 + 100 = 205.4\, \text{kPa}$$

$$v_f = \Gamma - \lambda \ln p_f' = 2.72 - 0.19 \ln(135.2) = 1.788$$

The ultimate void ratio, e_f occurs at failure:

$$e_f = v_f - 1 = 1.788 - 1 = 0.788$$

Similar calculations for cell pressures of 200 and 400 kPa yield the following results:

p_0' (kPa)	p_f' (kPa)	q_f' (kPa)	σ_1' (kPa)	e_f
200	270.3	210.8	410.8	0.656
400	540.5	421.6	821.6	0.524

(b). *Undrained shearing*

$$v_0 = N - \lambda \ln p_0' = 2.80 - 0.19 \ln (100) = 1.925$$
$$\Rightarrow v_f = 1.925$$
$$p_f' = \exp \left(\frac{\Gamma - v_0}{\lambda} \right) = \exp \left(\frac{2.72 - 1.925}{0.19} \right) = 65.6 \text{ kPa}$$
$$q_f' = M p_f' = 0.78 \times 65.6 = 51.2 \text{ kPa}$$
$$\sigma_1' = q_f' + \sigma_3' = 51.2 + 100 = 151.2 \text{ kPa}$$

Similar calculations for cell pressures of 200 and 400 kPa yield the following results:

p_0' (kPa)	v_f	p_f' (kPa)	q_f' (kPa)	σ_1' (kPa)
200	1.793	131.3	102.4	302.4
400	1.662	262.5	204.8	604.8

5.7   State boundaries

5.7.1   The Roscoe surface

For any value of the consolidation pressure p_0', there will be a corresponding position for A and hence an infinite number of possible planes, drained or undrained, on which stress paths travelling from A to B may lie. A number of planes with their stress paths are shown in Figs 5.19 (undrained) and 5.20 (drained).

If we place the two sets of stress paths together, we see that they appear to lie on a three-dimensional surface bounded by the critical state line at the top and by the normal consolidation line at the bottom. It can be shown that both sets of stress paths do lie on this surface by the technique of normalisation. If we take the results of a set of undrained and drained compression tests and divide the test q' and p' values by their corresponding p_0' values, the resulting plots tend to lie on a single unique line of the form illustrated in Fig. 5.21. Undrained and drained stress paths plotted in $q' : p' : v$ space, therefore, lie on the same three-dimensional surface. This surface is called the *Roscoe surface*.

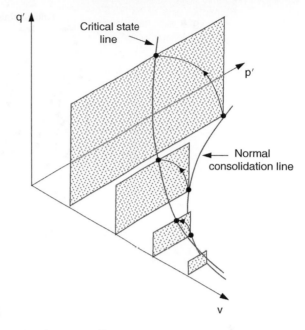

Fig. 5.19 Four undrained paths in q' : p' : v space.

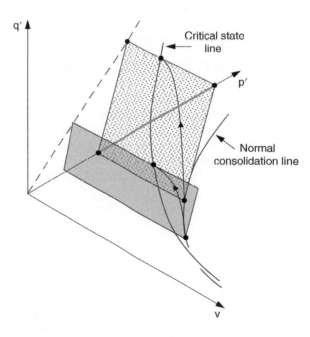

Fig. 5.20 Two drained paths in q' : p' : v space.

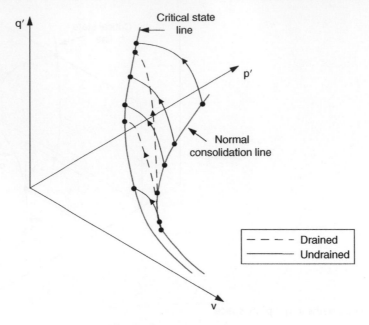

Fig. 5.21   The Roscoe surface.

Example 5.7: Roscoe surface

A sample of saturated clay had an initial volume of 86.2 ml and was isotropically consolidated at a cell pressure of 300 kPa, which ensured normal consolidation. During consolidation, 6.2 ml of water was measured as expelled into the drainage line and the void ratio of the sample at this stage was estimated to be 0.893. The sample was then subjected to drained shear, and readings of deviator stress and volume change were taken at increments of axial strain with the following results:

ε_a (%)	$\sigma'_1 - \sigma'_3$ (kPa)	ΔV (ml)
5	210	2.47
10	330	5.12
15	415	6.72
20	478	7.76
22 (failure)	507	8.08

If N = 2.92 and λ = 0.18 for the soil, plot the stress path normalised to p'_e.

Solution:

Volume after consolidation, $V_0 = 86.2 - 6.2 = 80.0\,\text{ml}$

Volumetric strain, $\varepsilon_v = \frac{\Delta V}{V_0}$

Let the specific volume at a particular value of volumetric strain, ε_v, be

$$v = (1 + e)$$

then:

$$\varepsilon_v = \frac{\Delta V}{V_0} = \frac{(1 + e) - (1 + e_0)}{1 + e_0} = \frac{V - V_0}{V_0}$$

$$\Rightarrow v = v_0(1 - \varepsilon_v)$$

Remembering that:

$$p'_e = \exp\left(\frac{N - v}{\lambda}\right)$$

we can now determine the values tabulated below:

ε_a (%)	$\sigma'_1 - \sigma'_3$ (kPa)	ΔV (ml)	ε_v	v	p'_e (kPa)	p' (kPa)	$\frac{p'}{p'_e}$	$\frac{q}{p'_e}$
0	0	0	0	1.893	300.0	300.0	1.0	0
5	210	2.47	0.031	1.834	417.1	370.0	0.89	0.503
10	330	5.12	0.064	1.772	588.6	410.0	0.70	0.56
15	415	6.72	0.084	1.734	727.0	438.3	0.60	0.57
20	478	7.76	0.097	1.709	835.3	459.3	0.55	0.572
22	507	8.08	0.101	1.702	868.4	469.0	0.54	0.584

The normalised plot is shown in Fig. 5.22.

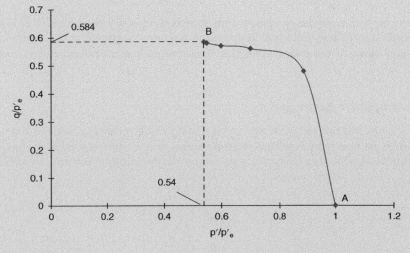

Fig. 5.22   Example 5.7.

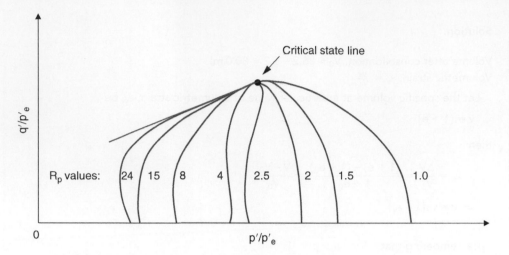

Fig. 5.23 Undrained stress paths of overconsolidated clay.

5.7.2 Overconsolidated clays – Hvorslev surface

We have established that the stress paths of normally consolidated clays lie on the Roscoe surface and, in order to complete the picture, we must determine whether the stress paths of overconsolidated clays also lie on this surface or whether they have a unique surface of their own.

Figure 5.23 shows a series of normalised stress paths of undrained shear obtained from tests carried out on overconsolidated clays. The values of *isotropic overconsolidation ratio* R_p are indicated for each stress path. As expected, with $R_p = 1.0$ the stress path lies on the Roscoe surface (as the soil is normally consolidated). For lightly overconsolidated clays, i.e. for R_p values up to about 2.5, the stress paths rise upwards, more or less vertically in the initial loading stage, towards the Roscoe surface but, before reaching it, they bend slightly and gradually make their ways to the critical state line where failure occurs.

The stress paths for the more heavily overconsolidated clays initially rise up more or less vertically and then incline inwards during the final loading stages to become tangential to a common straight line as they make their way towards the critical state line. This straight line is a boundary known as the *Hvorslev surface*.

5.7.3 The overall state boundary

The Roscoe surface has the property that any stress state outside it cannot exist in a soil. It is a boundary between possible stress states and impossible stress states and is therefore usually referred to as a state boundary. The Hvorslev surface is a similar state boundary and links up with the Roscoe surface at the critical

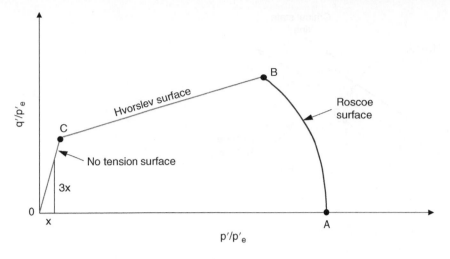

Fig. 5.24 The overall state boundary surface.

state line (point B in Fig. 5.24). The Hvorslev surface cannot extend to the $\frac{q'}{p_e'}$ axis because of the line of no tension, OC, which rises from the origin at a slope of 1 : 3.

Refer again to Fig. 5.24. If we assume that soil cannot carry tension then tension failure must occur if ever σ_a' is less than σ_r'. Now the lowest possible value of q is 0 which means that the tension failure boundary must pass through the origin. The highest possible value for q will occur when $\sigma_r' = 0$ and q therefore equals σ_a'. At this stage, $p' = \sigma_a'/3$ which means that $\frac{q'}{p'} = 3$.

Hence, if we select some value for $\frac{p'}{p_e'}$ say x, then $p_e' = \frac{p'}{x}$ and $\frac{q'}{p_e'} = 3x$. The unified plot of the complete state boundary, in $\frac{q'}{p_e'} : \frac{p'}{p_e'}$ space is shown in Fig. 5.24 and in $q' : p' : v$ space in Fig. 5.25.

5.7.4 Equation of the Hvorslev surface

We have established that the Hvorslev surface is a straight line when projected in $\frac{q'}{p_e'} : \frac{p'}{p_e'}$ space, as shown in Fig. 5.26.

We can therefore write the equation for the surface in the form $y = mx + c$, i.e.

$$\frac{q}{p_e'} = c + \frac{mp'}{p_e'}$$

i.e.

$$q = cp_e' + mp'$$

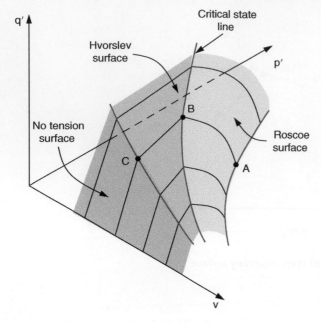

Fig. 5.25   The overall state boundary surface.

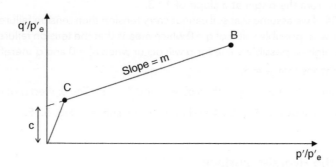

Fig. 5.26   The Hvorslev surface.

Now

$$p'_e = \exp\left(\frac{N-v}{\lambda}\right)$$

i.e.

$$q = c \exp\left(\frac{N-v}{\lambda}\right) + mp'$$

But, as the Hvorslev surface cuts the critical state line at point B, it must satisfy the critical state equations:

$$q' = Mp' \quad \text{and} \quad v = \Gamma - \lambda \ln p'$$

Substituting for q' and v gives

$$Mp' = c \exp\left(\frac{N-\Gamma+\lambda\ln p'}{\lambda}\right) + mp'$$

$$= c\left[\exp\left(\frac{N-\Gamma}{\lambda}\right)\right][\exp(\ln p')] + mp'$$

$$= cp' \exp\left(\frac{N-\Gamma}{\lambda}\right) + mp'$$

i.e.

$$c = \frac{M-m}{\exp\left(\frac{N-\Gamma}{\lambda}\right)} = (M-m)\exp\left(\frac{\Gamma-N}{\lambda}\right)$$

$$\Rightarrow q = (M-m)\left[\exp\left(\frac{\Gamma-N}{\lambda}\right)\right]\left[\exp\left(\frac{N-v}{\lambda}\right)\right] + mp'$$

$$= (M-m)\exp\left(\frac{\Gamma-v}{\lambda}\right) + mp'$$

5.8   Residual and critical strength states

The stress conditions that apply at the critical state line represent the ultimate strength of the soil (i.e. its critical state strength), and this is the lowest strength that the soil will reach provided that the strains within it are reasonably uniform and not excessive in magnitude. The residual strength of a soil operates, in the case of clays, only after the soil has been subjected to considerable strains with layers of soil sliding over other layers.

It is important that the difference between these two strengths is appreciated. It has been shown that the residual angle of shearing resistance of London clay, ϕ_r, can be less than 10°, whereas the same soil at critical state conditions has an angle of shearing resistance ϕ_{cv} of more than 20°.

Exercises

Exercise 5.1

During a CD triaxial test on a normally consolidated clay with cell pressure, σ_3 set at 50 kPa, readings of deviator stress at various values of axial strain were as follows:

Axial strain (%)	$\sigma_1' - \sigma_3'$ (kPa)
1	10
2	25
5	45
10	70
15	90

Draw the Mohr circle for each state of stress in the sample throughout the test and thereafter draw the total stress path for the test. What is the slope of the TSP to the horizontal?

Answer   Mohr circle diagram – see companion website.
   Slope of TSP = 45°

Exercise 5.2

An undrained triaxial test was conducted on a sample of clay with the following constants: N = 3.15, λ = 0.18, Γ = 3.10, M = 0.87. The initial isotropic consolidation cell pressure was 200 kPa. Calculate the values of v, p', and q' at failure.

Answer

v_f = 2.196; p_f' = 151 kPa; q_f' = 131 kPa

Exercise 5.3

A sample of clay was subjected to isotropic normal consolidation at a pressure of 350 kPa. The sample was then sheared in a drained state.
   Determine the values of q', p', and v at failure if the properties of the clay were

M = 0.89; N = 2.87; Γ = 2.76 and λ = 0.16.

Answer

q_f' = 443 kPa; p_f' = 498 kPa; v_f = 1.77

Exercise 5.4

A sample of clay was subjected to isotropic normal consolidation during a laboratory triaxial test, while the cell pressure was maintained at 300 kPa. The sample was then sheared in a drained state. Determine the values of deviator stress q', the mean stress p' and the specific volume, v at failure for the following properties of the clay:

M = 0.87;   N = 2.82; Γ = 2.67; λ = 0.18

Answer

q_f' = 367 kPa; p_f' = 423 kPa; v_f = 1.58

Exercise 5.5

An undisturbed laboratory sample of clay has the following properties:

M = 0.8, λ = 0.15, N = 2.46, and Γ = 2.32.

Samples of the clay were isotropically consolidated in a triaxial cell under cell pressures of 50, 100, and 200 kPa then sheared to failure in a drained test.

For each cell pressure, determine the mean principal stress at failure, the deviatoric compression stress at failure, the major principal stress at failure, and the ultimate void ratio.

Answer

Cell pressure = 50 kPa : p'_f = 68.2 kPa; q'_f = 54.5 kPa; σ'_1 = 104.5 kPa; e_f = 0.687

Cell pressure = 100 kPa : p'_f = 136.4 kPa; q'_f = 109.1 kPa; σ'_1 = 209.1 kPa; e_f = 0.583

Cell pressure = 200 kPa : p'_f = 272.7 kPa; q'_f = 218.2 kPa; σ'_1 = 418.2 kPa; e_f = 0.479

Exercise 5.6

If the sample of soil from Exercise 5.4 was sheared in an undrained test, determine values at failure of: the specific volume, the mean principal stress, the deviatoric compression stress at failure, and the major principal stress.

Answer

Cell pressure = 50 kPa : v_f = 1.873; p'_f = 19.7 kPa; q'_f = 15.7 kPa; σ'_1 = 65.7 kPa

Cell pressure = 100 kPa : v_f = 1.769; p'_f = 39.3 kPa; q'_f = 31.5 kPa; σ'_1 = 131.5 kPa

Cell pressure = 200 kPa : v_f = 1.665; p'_f = 78.6 kPa; q'_f = 62.9 kPa; σ'_1 = 262.9 kPa

Part II

Geotechnical Codes and Standards and Site Investigation

Chapter 6

Eurocode 7

Learning objectives:

By the end of this chapter, you will have been introduced to

- the first and second generations of the European geotechnical design code, Eurocode 7;
- the procedures to be followed in carrying out geotechnical design to both generations of the code;
- geotechnical design by calculation and the use of partial factors to quantify uncertainties in limit state design;
- several examples illustrating how to use both generations of the code to carry out limit state geotechnical design.

6.1 Preface to Chapter 6

The Eurocodes were first published in the lead up to 2010 and that set of documents has become known as the *first generation* of Eurocodes. In the period 2014–2021, the codes were reviewed, and amendments (including complete new Parts) will be published from 2021 onwards – referred to as the *second generation* of codes.

This means that currently the first generation of Eurocodes are still in use, but from the mid-2020s, the second generation will become available. It is unclear when use of the second generation will become mandatory for public sector works, but eventually structural and geotechnical design will be carried out in accordance with the second generation of the codes. Accordingly, this chapter covers both generations of Eurocode 7, and in subsequent chapters where the design processes are explained, both generations of the code will be covered.

After a common introduction (Section 6.2), this chapter is split into two sections: *Section A* covering the first generation of Eurocode 7 (EN 1997-1:2004 (BSI, 2004) and EN 1997-2:2007 (BSI, 2007a)) and *Section B* covering the second generation of the code (EN 1997-1:202x, EN 1997-2:202x and EN 1997-3:202x)*.

202x indicates that the specific year of publication of each part is unknown at time of writing.

Smith's Elements of Soil Mechanics, 10th Edition. Ian Smith.
© 2021 John Wiley & Sons Ltd. Published 2021 by John Wiley & Sons Ltd.
Companion website: www.wiley.com/go/smith/soilmechanics10e

6.2   Introduction to the Eurocodes

6.2.1   The Eurocode Programme

The Eurocode Programme was initiated to establish a set of harmonised technical rules for the design of building and civil engineering works across Europe. The rules are known collectively as the Structural Eurocodes which comprise a series of 10 European Standards, EN 1990–EN 1999, providing a common approach for the design of buildings and other civil engineering works and construction products. Eurocode 7 (EN 1997) is the document that concerns geotechnical design and we will look at the use of this Eurocode throughout the following chapters of this book.

It is the European Commission's intention that the Eurocodes become the recommended means for the structural design of works throughout the European Union (EU) and the European Free Trade Association (EFTA). They establish principles and requirements for achieving safety, serviceability and durability of structures and their adoption is leading to more common practice in structural and geotechnical design across Europe.

The Eurocodes are published by the *Comité Européen de Normalisation* (CEN) – the European Committee for Standardisation – and have been under development by CEN since 1989. The Eurocode programme actually started in 1975 following a decision of the then Commission of the European Community, but it was only when the work was passed to CEN that significant progress began. Draft versions of the codes (known as *Euronorm Vornorms* (ENV)) were produced throughout the 1990s and publication of the final versions (*Euronorms* (EN)) commenced in 2002. By 2010, all the Eurocodes had been published and these are now used in all member states and certain other nations.

As mentioned in Section 6.1, we are now in the period of the second generation of Eurocodes. It is anticipated that these versions of the codes will become available between 2021 and 2025.

The Eurocodes timeline is depicted in Fig. 6.1.

6.2.2   Scope of the Eurocodes

Since publication, the Eurocodes have become the reference design codes throughout the European member states and the structural designs of all public-sector works must now be Eurocode-compliant.

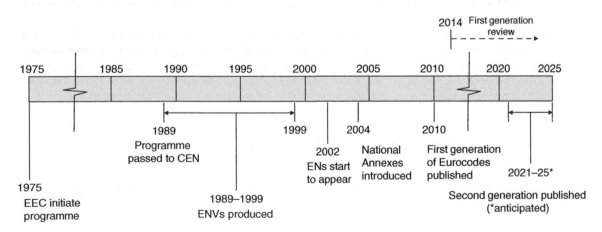

Fig. 6.1   Eurocodes timeline.

The Eurocodes cover the basis of structural design, actions on structures, design with each of the main structural materials, geotechnical design and design of structures for earthquake resistance. The ten Eurocodes are:

EN 1990   Eurocode: Basis of structural design (named *Basis of structural and geotechnical design* in second generation)
EN 1991   Eurocode 1: Actions on structures
EN 1992   Eurocode 2: Design of concrete structures
EN 1993   Eurocode 3: Design of steel structures
EN 1994   Eurocode 4: Design of composite steel and concrete structures
EN 1995   Eurocode 5: Design of timber structures
EN 1996   Eurocode 6: Design of masonry structures
EN 1997   Eurocode 7: Geotechnical design
EN 1998   Eurocode 8: Design of structures for earthquake resistance
EN 1999   Eurocode 9: Design of aluminium structures

Three new Eurocodes, covering (i) Design of structural glass, (ii) Design of fibre-polymer composite structures and (iii) Design of membrane structures, are under development.

6.2.3   Eurocode Parts and National Annexes

With the exception of EN 1990, each Eurocode consists of a number of Parts that cover particular technical aspects. EN 1997 (2004/2007) comprises two parts and EN 1997 (202x) comprises three parts as described later in this chapter.

To ensure that the safety of a design remains a national and not a European responsibility, each country has published a *National Annex* for each Part of the 10 Eurocodes. The National Annex (NA) is a document containing country specific information, rules and parameters and must be used alongside the main Eurocode document. The country specific parameters declared in a National Annex are referred to as *nationally determined parameters* (NDPs).

6.2.4   Design philosophy

The design philosophy adopted in Eurocode 7 is the same as that adopted in all the Eurocodes and advocates the use of limit state design to ensure that the serviceability limit states are not exceeded. *Serviceability limit states* are those states that, if exceeded, render the structure unsafe even though no collapse situation is reached, such as excessive deflection, settlement or rotation. In contrast to the traditional method of the use of lumped factors of safety, the standard promotes the use of *partial factors of safety* and thus reflected a significant shift from traditional UK geotechnical design practice. Where once the design methods treated material properties and loads in an unmodified state and applied a Factor of Safety at the end of the design process to allow for the uncertainty in the unmodified values, Eurocode 7 guides the designer to modify each parameter early in the design by use of the partial factor of safety. This approach sees the *representative* or *characteristic* value of the parameters (e.g. loads, soil strength parameters, etc.) converted to the design value by combining it with the particular partial factor of safety for that parameter. Worked examples in the following chapters will help the student to follow this approach to design.

Section A: Eurocode 7 – first generation (EN 1997:2004 and 2007)

6.3   Eurocode 7 – first generation

Eurocode 7 – Geotechnical design was published in two parts in the UK by the British Standards Institution as:

- BS EN 1997-1:2004 Part 1: General rules
- BS EN 1997-2: 2007 Part 2: Ground investigation and testing

6.3.1   Contents of Eurocode 7

The contents of both parts of Eurocode 7 are shown in Fig. 6.2. At first glance it appears that Part 1 covers more subject area and, whilst it arguably does, it's important to appreciate that Part 2 is a significantly longer document.

As the titles of the two documents indicate, Part 1 covers the general rules for design whilst Part 2 covers ground investigation practice. The two documents rely on each other for use and it would be very rare that one part is used in isolation from the other. The use of Part 2 is described in Chapter 7.

It is important to realise that by itself Eurocode 7 is not the only European standard that is used on a geotechnical project. Other Eurocodes (most notably EN 1990, EN 1991 and EN 1998) are involved as are several ISO testing and execution standards.

6.3.2   Using Eurocode 7: basis of geotechnical design

The clauses throughout Eurocode 7 are considered as either *Principles* (identified by the letter P immediately preceding the clause) or *Application Rules*. Principles are unique statements or definitions that must be adopted. Application Rules offer examples of how to ensure that the Principles are adhered to and thus offer guidance to the designer in following the Principles.

Part 1 – General rules

    Foreword
1. General
2. Basis of Geotechnical design
3. Geotechnical data
4. Supervision of construction, monitoring and maintenance
5. Fill, dewatering, ground improvement and reinforcement
6. Spread foundations
7. Pile foundations
8. Anchorages
9. Retaining structures
10. Hydraulic failure
11. Overall stability
12. Embankments
        Annexes A–J

167 pages

Part 2 – Ground investigation and testing

    Foreword
1. General
2. Planning of ground investigation
3. Soil and rock sampling and groundwater measurements
4. Field tests in soil and rock
5. Laboratory tests on soil and rock
6. Ground investigation report
        Annexes A–X

196 pages

Fig. 6.2   Contents of Eurocode 7 (EN 1997) Parts 1 and 2.

Section 2 of EN 1997-1 describes the basis of geotechnical design and the code states that the limit states should be verified by one of four means: by (i) calculation, (ii) prescriptive measures, (iii) experimental models and load tests, or (iv) an observational method. In this book we shall concentrate solely on geotechnical design by calculation (see Section 6.4) although a prescriptive measure for the determination of presumed allowable bearing values is touched upon in Chapter 10.

To facilitate an appropriate design, projects are considered as falling into one of three *Geotechnical Categories* based on the complexity of the geotechnical design together with the associated risks. Category 1 is for small projects with negligible risk, Category 2 is for conventional structures (e.g. foundations, retaining walls, embankments) and Category 3 is for structures not covered by Categories 1 and 2. It is obvious that most routine geotechnical design work will fall into Geotechnical Category 2.

6.4   Geotechnical design by calculation

To enable the limit states to be checked, the design values of the geotechnical parameters, the ground resistance and the actions (e.g. forces or loads), must be determined. Thereafter, a geotechnical analysis is employed to show that the particular limit state being checked will not be exceeded. A typical sequence of the processes involved in the design calculations is shown in Fig. 6.3.

The design values of actions, F_d are derived by multiplying the *representative* values, F_{rep} by the appropriate partial factor of safety, γ_F. The design values of geotechnical parameters, X_d are derived by dividing the *characteristic* values, X_k by the appropriate partial factor of safety, γ_M. The resistance, R is derived from the design values of actions and ground parameters. The design resistance, R_d can either be taken as equal to R or as equal to a reduced value of R, which is derived by dividing by an additional partial factor of safety, γ_R. The choice of which partial factor of safety to use is governed by the nature of the action and by the *design approach* (see Section 6.7) being used. Actions are classified as permanent (G) (either *favourable* or *unfavourable*), variable (Q), accidental (A) or seismic (AE). The 'effects of actions' are also considered in the design.

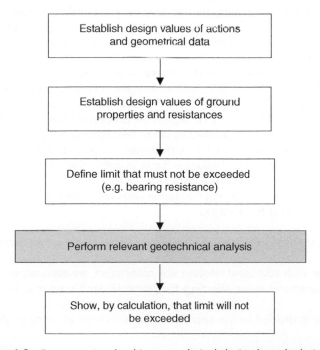

Fig. 6.3 Processes involved in geotechnical design by calculation.

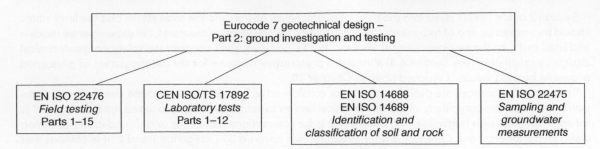

Fig. 6.4   ISO International and Technical Standards.

6.4.1   Characteristic values of geotechnical parameters

A number of codes and standards are used when designing a geotechnical project. Although the bulk of design work will involve following the rules of Eurocode 7 Part 1, the designer must also use Part 2 and other documents, as appropriate.

The Scope of Eurocode 7, Part 2 states that Part 2 is intended to be used in conjunction with Part 1 and provides rules supplementary to Part 1 related to site investigation and the testing required to establish geotechnical parameters. Part 2 does not cover the standardisation of laboratory and field tests themselves. Hence several International Standards Organisation (ISO) International and Technical Standards also play a part in the design process as indicated in Fig. 6.4.

EN 1990:2002 (BSI, 2002a) provides a statistical procedure for establishing a characteristic material property from a set of the material's property values. Such a procedure works well for man-made materials but is not applicable for use on soils.

Defining the characteristic values of geotechnical parameters is therefore challenging and EN 1997-1 states that these must be based on results from laboratory and field tests, complemented by well-established experience. Further, addressing the fact that the statistical procedure is not readily applicable to soil (e.g. since only a small number of test results will likely exist for a single soil property), the code states that the characteristic value should be taken as a *cautious estimate* of the value affecting the occurrence of the limit state.

It is appropriate to look at the issues affecting the selection of a *cautious estimate* of a geotechnical parameter. Fig. 6.5 (Fig. 1.1 in EN 1997-2:2007 (BSI, 2007a)) indicates the general framework for the selection of derived values of geotechnical properties from which the cautious estimate can be made.

The top part of Fig. 6.5 (above the dashed line) depicts the processes linked to Eurocode 7 Part 2 where the values of the geotechnical parameter are derived by lab or field testing; either directly or through some sort of correlation process. The lower part of the figure illustrates the processes described in Eurocode 7 Part 1 where the cautious estimate of the derived values is made to define the characteristic value, which is subsequently used to derive the design value of the parameter.

Thus, the procedure involved in determining the design values of geotechnical parameters from test results may be considered as comprising three stages:

1. field and laboratory test measurements are interpreted (using any required correlation) to derive a test result;
2. all test results, together with additional relevant site information, are assessed so that a *cautious estimate* of the geotechnical parameter value affecting the particular limit state may be made (the characteristic value);
3. the characteristic value is divided by the appropriate partial factor of safety to yield the design value.

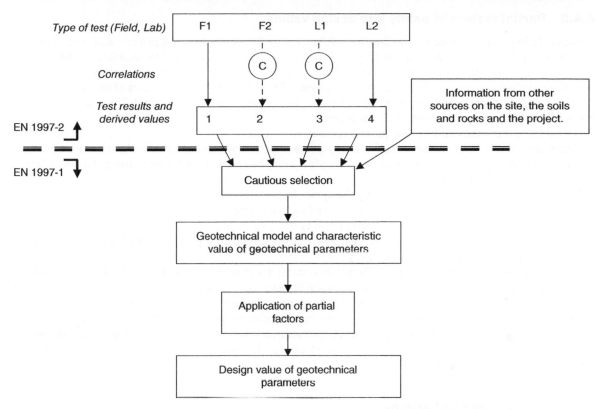

Fig. 6.5 General framework for the selection of derived values of geotechnical properties. Based on EN1997-2:2007 (BSI, 2007a).

Bond and Harris (2008) give guidance on establishing a cautious estimate of a geotechnical parameter whilst illustrating the variation in values that can result. Addressing the same issue, Simpson (2011) adds interpretation to the Eurocode text to aid users of the code to understand the processes involved. It can be argued that only a skilled and experienced geotechnical engineer can interpret test results and other factors rigorously enough to establish a reliable cautious estimate. To help address this, Bond (2011) has developed a procedure for determining the characteristic value based on simple statistical methods which goes some way in helping designers overcome the challenges in establishing the cautious estimate. Furthermore, Schneider and Schneider (2013) present a simplified statistical approach based on the mean and standard deviation or coefficient of variance of a soil property, combined with consideration of the vertical extent of the influencing failure mechanism, to determine the characteristic value.

Also helpful is an illustration by Hicks (2013) on the potential use of the random finite element method in determining characteristic values by using it to quantify the combined effects of spatial averaging soil properties along a failure plane with the fact that failure planes tend to follow the path of least resistance.

In offering alternatives to the cautious estimate approach, Eurocode 7 Part 1 states that both statistical methods and standard tables of characteristic values can be used if sufficient geotechnical measurements/results exist. However, as stated above, the likelihood of large enough data sets of geotechnical test results existing is small.

6.4.2 Partial factors of safety and design values

The calculation method prescribed in Eurocode 7 Part 1 is the limit state design approach used in conjunction with a partial factor method. The use of partial factors of safety ensure that the reliability of the various components of the design are assessed individually, rather than assessing the overall safety of the system as is the practice with a global factor of safety approach. This means that partial factors of safety are applied to all actions, material properties and resistances for each limit state being checked. There are thus many partial factors of safety that have to be considered.

Partial factors are denoted by the general symbol, γ.

(Note: this symbol is, of course, already used by geotechnical engineers to represent 'unit weight' (referred to as weight density in Eurocode 7) and thus a bit of care may be needed initially when using this Greek letter gamma to represent the partial factor.)

Partial factors on actions are denoted by the symbol, γ_F.

Partial factors on material properties are denoted by the symbol, γ_M.

Partial factors on resistances are denoted by the symbol, γ_R.

Partial factors for specific parameters are identified by the subscript; e.g. partial factor of safety on coefficient of shearing resistance, $\tan \phi$ is denoted by γ_ϕ.

The partial factors to be used for the different limit states are provided in Annex A of EN 1997-1:2004 (BSI, 2004). The National Annex can offer national choice for each partial factor.

The verification of any limit state involves an assessment of the effect of the *design* actions against the magnitude of the *design* resistance being offered by the structure or the ground. These design values are obtained by combining the characteristic values with appropriate partial factors of safety. Once the design values have been established, the geotechnical analysis is performed to check that the effects of the design actions do not exceed the design resistance.

6.4.3 Design values of actions

The sequence of taking a characteristic value of an action through to the design effect of the action involves multiplying the characteristic value, F_k by a correlation factor, ψ (in accordance with EN 1990:2002 (BSI, 2002a)) then multiplying the resulting representative value, F_{rep} by a partial factor of safety, γ_F to yield the design value of the action, F_d. The design effects of the action then depend on the limit state under consideration but could be, for example, a sliding force or a moment.

The sequence is thus:

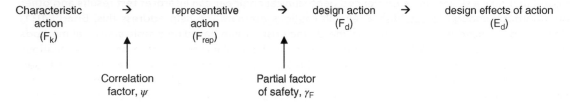

i.e.

$$F_{rep} = F_k \times \psi \ (\psi \le 1.0; \psi = 1.0 \text{ for persistent actions})$$
$$F_d = F_{rep} \times \gamma_F$$

Note 1: Characteristic self-weights are calculated from characteristic weight density values (e.g. see Example 6.3).

Note 2: In geotechnical design work, an assumption may be made that any structural action has already been combined with the correlation factor, ψ. This makes sense as a structural action coming onto a geotechnical structure will almost certainly be the result of a structural design process.

Example 6.1: Design value of action

A representative action has magnitude 200 kN. Considering a partial factor of safety, $\gamma_F = 1.35$, determine the design value of the action.

Solution:

$$F_d = F_{rep} \times \gamma_F = 200 \times 1.35 = 270 \text{ kN}$$

6.4.4 Design values of geotechnical parameters

Defining characteristic values of geotechnical parameters has already been discussed in Section 6.4.1. To reach the design value of the geotechnical parameter, the characteristic value is divided by the appropriate partial factor of safety.

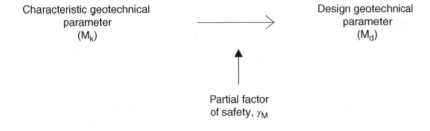

i.e.

$$M_d = \frac{M_k}{\gamma_M}$$

Example 6.2: Design value of geotechnical parameters

Determine the design values of the following characteristic soil strength properties, using the partial factors of safety provided:

$$c_{u;k} = 40 \text{ kPa}; \; c'_k = 5 \text{ kPa}; \; \phi'_k = 27°$$

$$\gamma_{cu} = 1.4; \gamma_{c'} = 1.25; \gamma_{\phi'} = 1.25$$

Solution:

$$c_{u;d} = \frac{40}{\gamma_{cu}} = \frac{40}{1.4} = 28.5 \text{ kPa}$$

$$c'_d = \frac{5}{\gamma_{c'}} = \frac{5}{1.25} = 4 \text{ kPa}$$

$$\phi'_d = \tan^{-1}\left(\frac{\tan\phi'}{\gamma_{\phi'}}\right) = \tan^{-1}\left(\frac{\tan 27°}{1.25}\right) = 22.2°$$

6.4.5 Design values of geometrical data

The action and material partial factors include an allowance for minor variations in geometrical data and, in cases where minor variation in the geometrical data will not affect the structural reliability, no further safety margin on the geometrical data should be required. However, in cases where deviations in the geometrical data might have a significant effect on the reliability of a structure (e.g. in the case of 'unplanned future excavations' in front of a retaining structure, see Examples 9.5 and 9.6) the design geometrical value can be adjusted by a nominal amount, following guidance provided in EN 1997-1:2004 (BSI, 2004).

6.4.6 Design effects of actions

When assessing the stability or strength resistance of the structure, it is the 'effects of actions' that are considered. These effects include the internal forces, moments, stresses and strains within the structural members plus any deflection or rotation of the structure as a whole.

(i) During the verification of geotechnical strength (where the GEO limit state (see Section 6.7) is used) some effects of the actions will depend on the strength of the ground in addition to the magnitude of the applied action and the dimensions of the structure. Thus, the effect, E of an action in the GEO limit state is a function of the action, the material properties and the geometrical dimensions.

  i.e.

$$E_d = E\{F_d; X_d; a_d\}$$

where

E_d is the design effect of the action, and
F_d is the design action; X_d is the design material property; a_d is the design dimension,
and where

$E\{\ldots\}$ indicates that the effect, E is a function of the terms in the parenthesis.

  The above expression relates to situations where the effect responds linearly to the design action, i.e. where the magnitude of the effect is directly proportional to the magnitude of the design action. For situations where the ratio of effect to action is non-linear, E_d should be established by considering the representative action and then applying a partial factor γ_E directly to the effect:

$$E_d = \gamma_E E\{F_{rep}; X_d; a_d\}$$

(ii) Similarly, during the verification of static equilibrium (where the EQU limit state (see Section 6.6) is used) some effects of the actions (both destabilising and stabilising) will depend on the strength of the ground in addition to the magnitude of the applied action and the dimensions of the structure. Thus, the effect of an action in the EQU limit state, whether it is a stabilising or a destabilising action, is a function of the action, the material properties and the geometrical dimensions.

i.e.

$$E_{dst;d} = E\{F_d; X_d; a_d\}_{dst}$$

where

$E_{dst;d}$ is the design effect of the destabilising action, and

$$E_{stb;d} = E\{F_d; X_d; a_d\}_{stb}$$

where

$E_{stb;d}$ is the design effect of the stabilising action.

6.4.7 Design resistances

Equation 6.6 in EN 1990:2002 (BSI, 2002a) indicates that the design resistance depends on material properties and the structural dimension. However, in geotechnical design, many resistances depend on the magnitude of the actions and so EN 1997-1 redefines Equation 6.6 from EN 1990 to include the contribution made by the design action. The clause actually offers three methods of establishing the design resistance, R_d:

$$R_d = R\{F_d; X_d; a_d\} \text{ or } R_d = \frac{R\{F_d; X_k; a_d\}}{\gamma_R} \text{ or } R_d = \frac{R\{F_d; X_d; a_d\}}{\gamma_R}$$

Annex B of Eurocode 7 Part 1 offers guidance on which of the three formulae above to use for each *design approach* (see Section 6.7).

6.5 Ultimate limit states

Eurocode 7 lists five limit states to be considered in the design process:

EQU: the loss of equilibrium of the structure or the supporting ground when considered as a rigid body and where the internal strengths of the structure and the ground do not provide resistance (e.g. Fig. 6.6a). In effect, this is where the wall is constructed on a rigid foundation such as rock, or hard soil. This limit state is satisfied if the sum of the design values of the effects of destabilising actions, $E_{dst;d}$ is less than or equal to the sum of the design values of the effects of the stabilising actions, $E_{stb;d}$ together with any contribution through the resistance of the ground around the structure, T_d, i.e. $E_{dst;d} \leq E_{stb;d} + T_d$. In most cases, the contribution to stability from the resistance of the ground around the structure will be minimal so T_d will be taken as zero.

GEO: failure or excessive deformation of the ground, where the soil or rock is significant in providing resistance (e.g. Fig 6.6b–e). This limit state is satisfied if the design effect of the actions, E_d is less than or equal to the design resistance, R_d, i.e. $E_d \leq R_d$. This is the limit state for which most routine geotechnical structures are checked.

STR: failure or excessive deformation of the structure, where the strength of the structural material is significant in providing resistance (e.g. Fig. 6.6f). As with the GEO limit state, the STR is satisfied if the design effect of the actions, E_d is less than or equal to the design resistance, R_d, i.e. $E_d \leq R_d$.

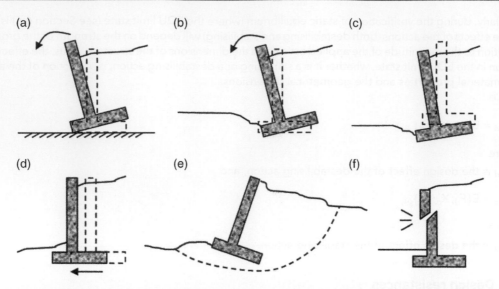

Fig. 6.6   Limit states for earth retaining structures. (a) Overturning – rigid base (EQU limit state). (b) Overturning (GEO limit state). (c) Bearing failure (GEO limit state). (d) Forward sliding (GEO limit state). (e) Ground failure (GEO limit state). (f) Structural failure (STR limit state).

UPL: the loss of equilibrium of the structure or the supporting ground by vertical uplift due to water pressures (buoyancy) or other actions (e.g. Fig. 6.7a). This limit state is verified by checking that the sum of the design permanent and variable destabilising vertical actions, $V_{dst;\,d}$ is less than or equal to the sum of the design stabilising permanent vertical action, $G_{stb;\,d}$ and any additional resistance to uplift, R_d such as the friction force T_d shown in Fig. 6.7a, i.e. $V_{dst;\,d} \leq G_{stb;\,d} + R_d$.

HYD: hydraulic heave, internal erosion and piping in the ground as might be experienced, for example, at the base of a braced excavation. This limit state is verified by checking that the design total pore water pressure, $u_{dst;\,d}$ or seepage force, $S_{dst;\,d}$ at the base of the soil column under investigation is less than or equal to the total vertical stress, $\sigma_{stb;\,d}$ at the bottom of the column, or the submerged unit weight, $G'_{stb;\,d}$ of the same column, i.e. $u_{dst;\,d} \leq \sigma_{stb;\,d}$ or $S_{dst;d} \leq G'_{stb;d}$ (e.g. Fig. 6.7b and Example 2.6).

The EQU, GEO and STR limit states are the most likely ones to be considered for routine design. Furthermore, in the design of retaining walls and foundations it is likely that limit state GEO will be the prevalent state for determining the size of the structural elements.

6.6   The EQU limit state

To check this limit state, the equilibrium of the structure when considered as a rigid body is assessed. The procedure is shown in Fig. 6.8. There are two sides of the analysis to consider represented by the large shaded areas: *destabilising* actions and effects and *stabilising* actions and effects.

For both the destabilising and the stabilising aspects, the representative actions are combined with the appropriate partial factors of safety to yield the design values. The analysis (typically a moment equilibrium analysis for the EQU state) is then performed and the magnitudes of the effects of the actions are compared to assess stability.

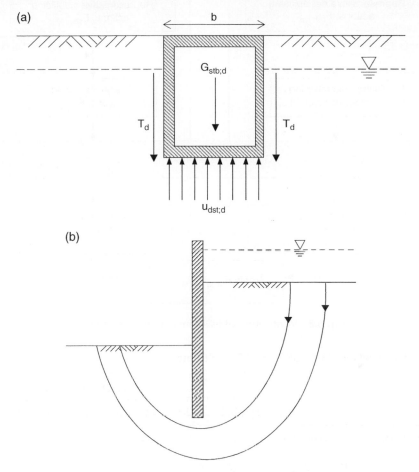

Fig. 6.7   (a) Uplift of a buried hollow structure. (b) Example where heave may occur. Based on EN1997-1:2004 (BSI, 2004).

An example of this limit state is the overturning of a gravity retaining wall resting on a rigid foundation, such as rock or hard soil. In the case of such a wall (e.g. Fig. 6.6a), the *destabilising actions* are the forces tending to push the wall over (e.g. the active thrust behind wall) and the *stabilising actions* are the forces resisting the overturning (i.e. the self-weight of the wall). The *effects* of the actions (both the stabilising and the destabilising) are the moments created by the actions. Thus, verification that the limit state requirement against overturning is satisfied requires that the overturning moment, $E_{dst;\,d}$ is less than or equal to the restoring moment, $E_{stb;\,d}$.

i.e. $E_{dst;\,d} \leq E_{stb;\,d}$

The partial factors for use in the EQU limit state are listed in Eurocode 7 Part 1, Annex A and are reproduced in Table 6.1. It is important to remember that the National Annex can provide alternative values to those published in Annex A and indeed the UK National Annex publishes different material partial factors for the EQU limit state (NA to BS EN 1997-1:2004 (BSI, 2007b)).

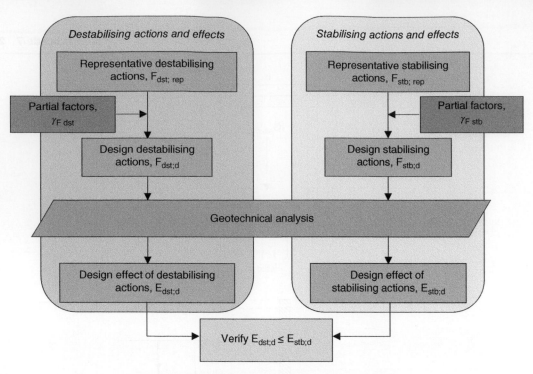

Fig. 6.8 Verification of EQU limit state for stability.

Table 6.1 Partial factor sets for EQU, GEO and STR limit states.

Parameter		Symbol	EQU	GEO/STR – partial factor sets						
				A1	A2	M1	M2	R1	R2	R3
Permanent action (G)	Unfavourable	$\gamma_{G;\,dst}/$ $\gamma_{G;\,unfav}$	1.1	1.35	1.0					
	Favourable	$\gamma_{G;\,stb}/$ $\gamma_{G;\,fav}$	0.9	1.0	1.0					
Variable action (Q)	Unfavourable	$\gamma_{Q;\,dst}/$ γ_{Q}	1.5	1.5	1.3					
	Favourable	–	–	–	–	–				
Accidental action (A)	Unfavourable	$\gamma_{A;\,dst}$	1.0	1.0	1.0					
	Favourable	–	–	–	–	–				
Coefficient of shearing resistance (tan ϕ')		$\gamma_{\phi'}$	1.25			1.0	1.25			
Effective cohesion (c')		$\gamma_{c'}$	1.25			1.0	1.25			
Undrained shear strength (c$_u$)		γ_{cu}	1.4			1.0	1.4			
Unconfined compressive strength (q$_u$)		γ_{qu}	1.4			1.0	1.4			
Weight density (γ)		γ_{γ}	1.0			1.0	1.0			
Bearing resistance (R$_v$)		γ_{Rv}						1.0	1.4	1.0
Sliding resistance (R$_h$)		γ_{Rh}						1.0	1.1	1.0
Earth resistance (R$_e$)		γ_{Re}						1.0	1.4	1.0

Note: weight density $\equiv$ unit weight

Example 6.3: EQU limit state

Consider a simple reinforced concrete gravity retaining wall ($\gamma_c = 25$ kN/m^3) of width 2 m retaining a homogeneous granular fill to a height of 4 m as shown in Fig. 6.9. The resultant active thrust due to the retained soil is equal to 66.5 kN and the lateral thrust from the surcharge is equal to 15.1 kN. Check the safety of the wall against the EQU limit state of Eurocode 7. Assume that the wall rests on a stiff layer.

Solution:

The first step is to consider which partial factors of safety we need from Table 6.1. In this example we require factors for (i) the destabilising actions (both permanent and variable) and (ii) the stabilising actions, i.e. $\gamma_{G; dst}$, γ_Q and $\gamma_{G; stb}$.

The actions acting are:

P_a, P_q – destabilising actions
Weight of wall, W – stabilising action

Design actions:

$P_{a;d} = 66.5 \times \gamma_{G; dst} = 66.5 \times 1.1 = 73.2$ kN
$P_{q;d} = 15.1 \times \gamma_Q = 15.1 \times 1.5 = 22.7$ kN
$W_d = 2 \times 4 \times 25 \times \gamma_{G; stb} = 200 \times 0.9 = 180$ kN

Design effect of actions:

Destabilising moment, $M_{dst; d}$
$= (73.2 \times \frac{4}{3}) + (22.7 \times 2)$
$= 143$ kNm
Stabilising moment, $M_{stb; d}$
$= 180 \times 1.0$
$= 180$ kNm

Since $M_{stb; d} \geq M_{dst; d}$ the EQU limit state requirement is satisfied.

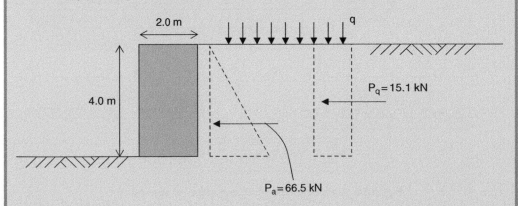

Fig. 6.9 Example 6.3.

6.7 The GEO limit state and design approaches

To check this limit state, the strength of the ground is assessed to ensure ground failure will not occur. The procedure is shown in Fig. 6.10.

There are two sides of the analysis to consider represented by the large shaded areas: *actions and effects* and *material properties and resistance*. The representative actions are combined with the appropriate partial factors of safety to yield the design values. The material properties are then combined with their partial factors of safety to yield the design material properties. The analysis is then performed, and the design effect of the actions is compared to the design resistance to assess safety.

An example of this limit state is the forward sliding of a gravity retaining wall. In the case of such a wall (e.g. Fig. 6.6d), the *design effects of the actions* (i.e. the forward sliding caused by the active thrust behind wall) and the *design resistance* (i.e. the force resisting sliding along the base of the wall) are established. Verification that the limit state requirement against sliding is satisfied requires that the effect of the actions, E_d is less than or equal to the ground resistance, R_d.

i.e.

$$E_d \leq R_d$$

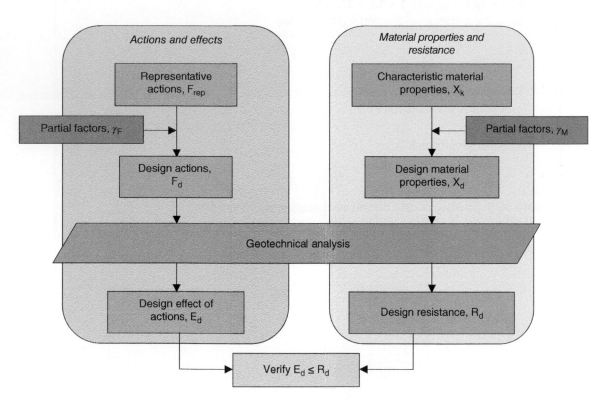

Fig. 6.10 Verification of GEO limit state for strength.

6.7.1   Design approaches

When checking the GEO and STR limit state requirements, one of three design approaches is used: *Design Approach 1*, *Design Approach 2* or *Design Approach 3*. This choice of three approaches reflects the Europe-wide adoption of the Standard and offers designers in different nations an approach most relevant to their needs. The UK National Annex to EN 1997-1 states that Design Approach 1 is to be used in the UK and worked examples in the following chapters illustrate the use of this method.

As mentioned earlier, the choice of partial factors to be used is dependent on the design approach being followed (for the GEO and STR limit states). For each design approach, a different combination of partial factor sets is used to verify the limit state. For Design Approach 1 (for retaining walls and shallow footings), two combinations are available, and the designer would normally check the limit state using each combination, except on occasions where it is obvious that one combination will govern the design. (The combination of partial factor sets for Design Approach 1 is different for pile foundations – see Chapter 11.)

Design Approach 1 :Combination 1 : A1 + M1 + R1
Combination 2 : A2 + M2 + R1
Design Approach 2 :A1 + M1 + R2
Design Approach 3 :A* + M2 + R3

(*Note.* A∗: use set A1 on structural actions, set A2 on geotechnical actions.)

The sets for actions (denoted by A), material properties (denoted by M) and ground resistance (denoted by R) for each design approach are given in Table 6.1. Also given in the table are the partial factors for the EQU limit state.

Example 6.4: Design approaches; design actions

A concrete foundation is to be cast into a soil deposit as shown in Fig. 6.11. The foundation has a representative self-weight, W of 50 kN.

During a check for bearing resistance (see Chapter 10), the vertical representative actions $V_{G;\,k}$ and $V_{Q;\,k}$ are considered as *unfavourable*. Determine the design values of each action, for each design approach.

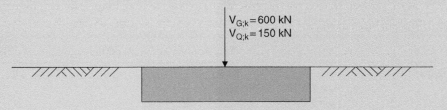

$V_{G;k} = 600$ kN
$V_{Q;k} = 150$ kN

Fig. 6.11   Example 6.4.

Solution:

The design values of the actions are achieved by multiplying the representative actions by the appropriate partial factors of safety from Table 6.1.

e.g.   Design Approach 1, Combination 1 (DA 1-1)

$$G_d = G_k \times \gamma_G = 600 \times 1.35 = 810 \, kN$$

		$V_{G;d}$ (kN)	$V_{Q;d}$ (kN)	$W_{G;d}$ (kN)
DA 1–1	$\gamma_G = 1.35$	810	225	67.5
	$\gamma_Q = 1.5$			
DA 1–2	$\gamma_G = 1.0$	600	195	50
	$\gamma_Q = 1.3$			
DA 2	$\gamma_G = 1.35$	810	225	67.5
	$\gamma_Q = 1.5$			
DA 3	$\gamma_G = 1.0$	600	195	50
	$\gamma_Q = 1.3$			

Example 6.5: Design approaches; design geotechnical parameters

The ground beneath the foundation shown in Fig. 6.11 has the following characteristic values: $c_{u;k} = 40 \, kPa$; $c'_k = 5 \, kPa$; $\phi'_k = 25°$.

Determine the design values of each property, for each design approach.

Solution:

The design values of the geotechnical properties are achieved by dividing the characteristic values by the appropriate partial factors of safety from Table 6.1.

e.g.   Design Approach 1, Combination 2 (DA 1-2)

$$\phi'_d = \tan^{-1}\left(\frac{\tan \phi'}{\gamma_{\phi'}}\right) = \tan^{-1}\left(\frac{\tan 25°}{1.0}\right) = 20.5°$$

		$C_{u;d}$ (kPa)	c'_d (kPa)	ϕ'_d (°)
DA 1–1	$\gamma_{Cu} = 1.0$	40	5	25
	$\gamma_{c'} = 1.0$			
	$\gamma_{\phi'} = 1.0$			
DA 1–2	$\gamma_{Cu} = 1.4$	28.6	4	20.5
	$\gamma_{c'} = 1.25$			
	$\gamma_{\phi'} = 1.25$			
DA 2	$\gamma_{Cu} = 1.0$	40	5	25
	$\gamma_{c'} = 1.0$			
	$\gamma_{\phi'} = 1.0$			
DA 3	$\gamma_{Cu} = 1.4$	28.6	4	20.5
	$\gamma_{c'} = 1.25$			
	$\gamma_{\phi'} = 1.25$			

6.7.2   The over-design factor and the degree of utilisation

When checking the ultimate limit state (for any of the five ultimate limit states) it may be helpful to represent the degree of safety of the system by either the *over-design factor* or the *degree of utilisation*. These expressions are quite simply the ratio of the design resistance to the effects of the actions (ODF) and its reciprocal (DoU). i.e.

Over-design factor, $\Gamma = \dfrac{R_d}{E_d}$

Degree of utilisation, $\Delta = \dfrac{E_d}{R_d}$

Example 6.6: GEO limit state: forward sliding

Return to the retaining wall of Example 6.3. Assume now that the wall is founded upon a clay of characteristic undrained strength 75 kPa. Check the safety of the wall against forward sliding by checking the GEO limit state of Eurocode 7 for all three design approaches.

Solution:

1. Design Approach 1
   Combination 1: (A1 + M1 + R1)
   Design Material Properties:

   $$c_{u;d} = \frac{75}{\gamma_{cu}} = \frac{75}{1} = 75 \text{ kPa}$$

   Design Actions and effect of actions:
   The active and surcharge thrusts are unfavourable.

   Active thrust: $P_{a;d} = 66.5 \times \gamma_G = 66.5 \times 1.35 = 89.8 \text{ kN/m}$
   Surcharge thrust: $P_{q;d} = 15.1 \times \gamma_Q = 15.1 \times 1.5 = 22.7 \text{ kN/m}$
   Total sliding force, $E_d = 89.8 + 22.7 = 112.5 \text{ kN/m}$

   Design Resistance:

   $R_d = c_{u;d} \times B = 75 \times 2 = 150 \text{ kN/m}$ (since $\gamma_{Rh} = 1.0$, see Table 6.1)

   From the results it is seen that the limit state is satisfied since the design resistance (150 kN/m) is greater than the sliding force (112.5 kN/m).
   This result may be presented by the *over-design factor*, Γ, or the *degree of utilisation*, Δ:

   $$\Gamma = \frac{150}{112.5} = 1.33$$
   $$\Delta = 1/1.33 = 75\%$$

Combination 2: (A2 + M2 + R1)

$$c_{u;d} = \frac{75}{\gamma_{cu}} = \frac{75}{1.4} = 53.6 \, \text{kPa}$$

$P_{a;d} = 66.5 \times \gamma_G = 66.5 \times 1.0 = 66.5 \, \text{kN/m}$

$P_{q:d} = 15.1 \times \gamma_Q = 15.1 \times 1.3 = 19.6 \, \text{kN/m}$

Total sliding force, $E_d = 66.5 + 19.6 = 86.1 \, \text{kN/m}$

$R_d = c_{u;d} \times B = 53.6 \times 2 = 107.2 \, \text{kN/m}$ (again $\gamma_{Rh} = 1.0$)

$$\Gamma = \frac{107.2}{86.1} = 1.25$$

$\Delta = 1/_{1.25} = 80\%$

In conclusion, the GEO limit state requirement is satisfied since $R_d \geq E_d$ in both combinations. Combination 2 is more critical and thus 'governs' the design.

2. Design Approach 2 (A1 + M1 + R2)

$$c_{u;d} = \frac{75}{\gamma_{cu}} = \frac{75}{1} = 75 \, \text{kPa}$$

$P_{a;d} = 66.5 \times \gamma_G = 66.5 \times 1.35 = 89.8 \, \text{kN/m}$

$P_{q:d} = 15.1 \times \gamma_Q = 15.1 \times 1.5 = 22.7 \, \text{kN/m}$

$E_d = 89.8 + 22.7 = 112.5 \, \text{kN/m}$

$$R_d = \frac{c_{u;d} \times B}{\gamma_{Rh}} = \frac{75 \times 2}{1.1} = 136.4 \, \text{kN/m}$$

$$\Gamma = \frac{136.4}{112.5} = 1.21$$

$\Delta = 1/_{1.21} = 83\%$

i.e. limit state requirement ok

3. Design Approach 3 (A* + M2 + R3)

   *use A1 on structural actions, A2 on geotechnical actions

$$c_{u;d} = \frac{75}{\gamma_{cu}} = \frac{75}{1.4} = 53.6 \, \text{kPa}$$

$P_{a;d} = 66.5 \times \gamma_G = 66.5 \times 1.0 = 66.5 \, \text{kN/m}$

$P_{q:d} = 15.1 \times \gamma_Q = 15.1 \times 1.3 = 19.6 \, \text{kN/m}$

Total sliding force, $E_d = 66.5 + 19.6 = 86.1 \, \text{kN/m}$

$R_d = c_{u;d} \times B = 53.6 \times 2 = 107.2 \, \text{kN/m}$ (since $\gamma_{Rh} = 1.0$)

$$\Gamma = \frac{107.2}{86.1} = 1.25$$

$\Delta = 1/_{1.25} = 80\%$

i.e. limit state requirement ok.

6.8   Serviceability limit states

Serviceability limit states are those that result in excessive settlement, heave or ground vibration and, whilst the structure at such a state is unlikely to collapse (i.e. reach the ultimate limit state), the structure will nonetheless be considered unsafe.

Only brief guidance is given in Eurocode 7 Part 1 on the checking of serviceability limit states. It is stated that verification for serviceability limit states requires that the effects of the actions, E_d is less than or equal to the limiting values of the effects, C_d.

i.e. $E_d \leq C_d$

The effects of the actions, E_d include deformations, settlements, ground heave and vibrations etc.

The values of partial factors for serviceability limit states should normally be taken equal to 1.0 though the National Annex can set different values. The limiting value for a particular serviceability deformation such as settlement must be agreed during the design of the supported structure and EN 1997-1 Annex H provides brief guidance on limiting values of structural deformation and foundation movement.

6.9   Geotechnical design report

At the end of the design process, all the calculations, drawings and ground investigation data are compiled together into the geotechnical design report. Guidance on the contents of this document is given in Section 7.7.

Section A of this chapter has explained the design processes and techniques used in Eurocode 7, *first generation* (EN 1997-1:2004 (BSI, 2004) and EN 1997-2:2007 (BSI, 2007a)). Additional explanation on the use of the code is given by Bond and Harris (2008) and Simpson (2011) and a review of how the first generation of Eurocode 7 affected geotechnical design was given by Orr (2012).

Section B now describes the procedures involved in using the *second generation* of the code (EN 1997:202x). At the time of writing, this code is unpublished, and the descriptions given are based on the contents of the draft code: prEN 1990:2020 (CEN, 2020), prEN 1997-1:2018 (CEN, 2018a), prEN 1997-2:2018 (CEN, 2018b) and prEN 1997-3:2018 (CEN, 2018c). Beneficially though, major changes are not anticipated between the draft and the final published versions.

Note: Readers are recommended to consult the companion website for details of any changes in the final versions of the codes which affect the procedures or numerical values described in Section B, and in later chapters.

Section B: Eurocode 7 – second generation (EN 1997: 202x)

6.10   Eurocode 7 – second generation

The main objectives behind the development of the second generation of the Eurocodes were to improve the ease of use of the standards, to implement recent developments in design, and to improve the harmonisation of practice between nations – specifically by reducing the number of NDPs and national practice alternatives. The work involved in developing the new Eurocode 7 was detailed and has been described by Bond *et al.* (2019a), van Seters and Franzen (2019), Estaire *et al.* (2019), Franzen *et al.* (2019), Norbury *et al.* (2019) and Bond *et al.* (2019b).

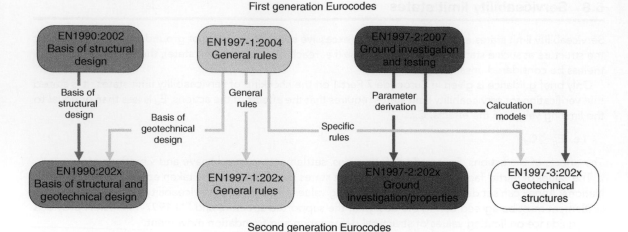

Fig. 6.12 Redistribution of content from first to second generations. Bond *et al.* (2019a).

The headline changes between the first and the second generations of EN 1997 are:

- whereas the first generation EN 1997 has 2 parts; the second generation has 3 parts:

  EN 1997-1: 202x *Geotechnical design – General rules*
  EN 1997-2: 202x *Geotechnical design – Ground investigation*
  EN 1997-3: 202x *Geotechnical design – Geotechnical structures*
- various sections of EN 1997-1:2004 (BSI, 2004) and EN 1997-2:2007 (BSI, 2007a) have been redistributed, as illustrated in Fig. 6.12;
- different approaches now exist for checking the limit states requirements, and the three letter abbreviations, EQU, GEO, STR etc. have been removed;
- new sections dealing with reinforced soil, ground improvement, rock engineering and dynamic loading have been added;
- there is improved specification for dealing with water pressures;
- there is consideration of the inclusion of geological determined parameters (GDPs) – a variant of NDPs, to give nations scope to include parameters specific to the nation's geological conditions.

As seen from Fig. 6.12, some of the original EN 1997-1 has been moved into the new EN 1990. Accordingly, EN 1990:202x has been renamed as *Basis of Structural and Geotechnical Design*. This also means that in considering the use of EN 1997:202x, we now must also rely significantly on EN 1990:202x.

As with the first generation, the verification of a limit state in the second generation of Eurocode 7 can be performed by calculation using partial factors, by prescriptive methods, by testing or by using the observational method. The following sections consider only the calculation using partial factors method and describe how the new version of the code is used to validate geotechnical design propositions.

We begin by considering the new EN 1990, before looking at the new EN 1997.

6.11 Basis of structural and geotechnical design – EN 1990:202x

6.11.1 Introduction

In the first generation of Eurocode 7, assessment of the reliability of a design was made through the use of partial factors together with (for the GEO and STR limit states) the adoption of one of three design approaches (see Section 6.7). Since different nations were free to specify which design approach had to be adopted, this complex approach led to significant difficulty in nation to nation assessments of designs, particularly on multi-national projects. To address this, the design approaches have been removed from the second generation of the code, and the general aspects of geotechnical design have been added to EN 1990. With the inclusion of geotechnical design in EN 1990, definitions of some limit states requirements have had to change, specifically where combined effects of different limit states occur.

6.11.2   Design actions and the design effects of actions

In the same way as the first generation of Eurocode 7 handled uncertainties in magnitudes of actions, effects of actions, material properties and resistances through the application of partial factors of safety, the second generation of the code does likewise.

As with EN 1997-1:2004 (BSI, 2004) actions in the second generation of the code are classified as permanent (G) (either *favourable* or *unfavourable*), variable (Q) always *unfavourable*, accidental (A) or seismic (A_E). In static equilibrium and uplift limit states verifications, permanent actions are considered as either *stabilising* or *destabilising*.

For actions, characteristic values of the action (denoted by the subscript k) are multiplied by the partial factor to determine the design value.

e.g.

$$G_d = G_k \times \gamma_G$$

$$Q_d = Q_k \times \gamma_Q$$

Actions cause an effect, e.g. a foundation transmits the applied load to the soil, which causes vertical movement (settlement). In the first generation of the head Eurocode (EN 1990:2002 (BSI, 2002a)) the *design effect* was established by considering the design action and the design geometry, with no consideration of the material properties. The second generation of EN 1990 addresses this by introducing *coupling* between action-effects and material properties. In effect, this is the inclusion of the formulae for the design effects of actions from EN 1997-1:2004 (BSI, 2004), as listed in Section 6.4.6.

i.e.

$E_d = E\{\Sigma(\gamma_F F_{rep}); a_d; X_{Rd}\}$ – for cases where the effect responds linearly to the action;
$E_d = \gamma_E E\{F_{rep}; X_d; a_d\}$ – for cases where the action causes a disproportionate effect (i.e. non-linear response).

where the symbols have the same meanings as defined in Section 6.4.6 and X_{Rd} is the material property used in the assessment of R_d.

The partial factors to be used on actions (γ_G, γ_Q), and on the effects of actions (γ_E), are provided in EN 1990:202x (see Section 6.13.1).

6.11.3   Design material properties and design resistance

As with the approach used in the first generation of Eurocode 7, design material properties are established by dividing the *representative value* of the property by the appropriate partial factor.

e.g.

$$c'_d = \frac{c'_{rep}}{\gamma_{c'}}$$

$$\tan \phi'_d = \frac{\tan \phi'_{rep}}{\gamma_{\tan \phi}}$$

The *representative value* of a material property, X_{rep} is obtained by multiplying the *characteristic value* by a conversion factor, η which brings in allowance for the effects of scale, moisture, temperature, ageing of materials, anisotropy, stress path or strain level:

$$X_{rep} = \eta X_k \tag{6.1}$$

where X_k is the nominal (characteristic) value of the ground property.

The *nominal value*, X_k is the *cautious estimate* of the value affecting the occurrence of the limit state – obtained in the same manner as described in Section 6.4.1.

Typically, η may be taken as 1.0, so invariably X_{rep} will equal X_k.

The coupling effect mentioned in Section 6.11.2 also applies to the determination of the resistance. The resistance of a soil to an applied action is a function of the soil properties, the geotechnical structure's geometry and any other applied actions. To this end, the uncertainties in actions, material and geometry must be considered in the determination of the *design resistance*, R_d.

R_d may be established by applying a *material partial factor* to the material property, or by applying a *resistance partial factor* to the representative resistance, R. These two approaches give rise to two methods of determining the design resistance: a *material factor approach (MFA)* or a *resistance factor approach (RFA)*.

i.e.

$$R_d = R\left\{ \frac{X_{rep}}{\gamma_M}; F_d; a_d \right\} \text{(material factor approach, MFA)} \tag{6.2a}$$

or

$$R_d = \frac{R\{X_{rep}; F_d; a_d\}}{\gamma_R} \text{(resistance factor approach, RFA)} \tag{6.2b}$$

The partial factors to be used on material properties (γ_M) are provided in EN 1997-1:202x and those on resistances (γ_R) are provided in EN 1997-3:202x (see Section 6.13.2).

6.11.4   Consequence classes

In order to satisfy the design requirements of any structure, several clauses in EN 1990:202x must be followed. One of the key clauses is that concerning the consequences of failure. By classifying the consequence of failure of a structure or structural element in terms of potential loss of life and economical/social/environmental loss, a contribution to the procedures for specifying the minimum requirements for reliability management (see Section 6.12.1) is made. Table 6.2 shows the consequence class (CC) categories (0–4) as defined in EN 1990:202x, together with values of consequence factor, K_F which are used in the ULS checks (see Section 6.13.1).

Note that consequence classes 0 and 4 are beyond the scope of the Eurocodes, and so a Eurocode design will only consider CC's 1–3. Note also that the CC is chosen based on the more severe of the third and fourth columns of Table 6.2.

Table 6.2   Definition of consequence classes.

Consequence class	Description of consequence	Loss of human life	Economic, social, environmental consequences	Consequence factor, K_F
CC4	Highest	Extreme	Huge	–
CC3	Higher	High	Very great	1.1
CC2	Normal	Medium	Considerable	1.0
CC1	Lower	Low	Small	0.9
CC0	Lowest	Very low	Insignificant	–

Based on prEN 1990:2020 (CEN, 2020).

6.11.5   Design cases

A combination of actions for which a set of values of partial factors is applicable, is known as a *design case*. EN 1990:202x defines four design cases (DC 1–4) and the applicability of each to geotechnical ultimate limit states is shown in Table 6.3, which is based on tables in EN 1990:202x and EN 1997-1:202x. Also shown are the minimum and fixed partial factor values for both permanent (γ_G) and variable (γ_Q) actions, and for the effects of actions (γ_E).

The partial factors to use both on actions and on the effects of actions in the design of particular geotechnical structures depend on the identified design case. The specific partial factors values are provided in EN 1990:202x and are called upon for specific design situations in EN 1997-3:202x (see Section 6.13.1 and Table 6.6).

Table 6.3 is helpful in noting which design cases are appropriate for a particular design situation and limit state check. However, specific guidance on the verification methods to be considered for each different type of geotechnical structure (e.g. slope, retaining wall, shallow foundation etc.) is given in Part 3 of Eurocode 7. This Part of the code is structured such that the design rules and procedures for each geotechnical structure

Table 6.3   Applicability of design cases in geotechnical design.

	Design cases (and minimum, or fixed, values of partial factors)					
	DC1	DC2		DC3	DC4
		DC2a	DC2b		
	Structural resistance	Static equilibrium and uplift		Geotechnical design	
Ultimate limit state	$\gamma_Q > \gamma_G > 1.0$	$\gamma_Q > \gamma_G > 1.0$	$\gamma_G = 1.0$ $\gamma_Q > 1.0$	$\gamma_G = 1.0$ $\gamma_Q > 1.0$	$\gamma_E > 1.0$ $\gamma_Q > 1.0$
Failure by rupture or excessive deformation	✓			✓	✓
Loss of rotational equilibrium or loss of vertical equilibrium due to uplift forces		✓	✓		
Time-dependent effects and liquefaction	✓			✓	✓

are provided in separate sections. Within each section, subclause 6 gives the specific guidance on the ultimate limit state verification methods which may be used. This guidance includes the combination of design case and, ground property (γ_M) or resistance (γ_R) partial factors values, to be used when either an MFA or RFA is being followed. Further explanation is given in Section 6.13.

6.12   Design of a geotechnical structure – EN 1997: Parts 1, 2 and 3 (202x)

As we have now just seen, EN 1990:202x contains guidance and instruction on some principles concerning geotechnical design. Hence, during the design of a geotechnical structure, EN 1990:202x must be used alongside all three parts of EN 1997:202x. There is a four-stage process employed in the design of any geotechnical structure, which can be represented by the tasks and outputs shown in Fig. 6.13.

With reference to the figure, the following stages are identified:

1. reliability management – pre-assessment to determine the *geotechnical category*
2. site investigation – which includes the development of the *ground model*
3. verification of the design – which includes the development of the *geotechnical design model*
4. construction (execution)

and each of these is described in the following sections.

6.12.1   Reliability management

Reliability management encapsulates the preliminary stages of the design where information about the site, the proposed structure, the consequence class (CC) and other relevant information is all brought together to aid the designer in assessing the aspects to be included in the design verification. Using all of the information above, the *Geotechnical Category* (GC) is determined for the proposed structure. The GC is established by linking the CC with a further class – the *Geotechnical Complexity Class* (GCC). The GCC is established using Table 6.4 and, thereafter, Table 6.5 is used to establish the GC. The content in both tables are NDPs and therefore may be adjusted in a National Annex.

Once the geotechnical category is determined, the minimum requirements for a number of design aspects (e.g. designer qualification and experience level, amount of ground investigation, design check level, validation of calculation models) are set. This procedure ensures appropriate rigour and robustness in all aspects of the geotechnical design.

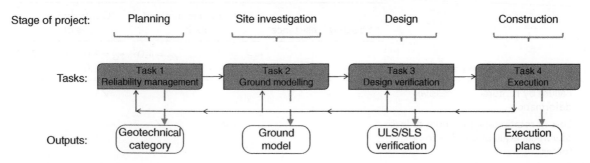

Fig. 6.13   Key stages in the design and construction of a geotechnical structure. Based on Estaire *et al.* (2019).

Table 6.4   Geotechnical complexity class.

Geotechnical complexity class	Complexity	General features causing uncertainty
GCC3	Higher	Any of the following apply: • considerable uncertainty regarding ground conditions, • highly variable or difficult ground conditions, • significant sensitivity to groundwater conditions • significant complexity of the ground-structure interaction
GCC2	Normal	Covers everything not contained in the features of GCC 1 and GCC3
GCC1	Lower	All the following conditions apply: • negligible uncertainty regarding the ground conditions • uniform ground conditions • low sensitivity to groundwater conditions, • low complexity of the ground-structure-interaction

Data from prEN 1997-1:2018 (CEN, 2018a).

Table 6.5   Relationships between geotechnical categories, consequences classes, and geotechnical complexity classes.

Consequence class (CC)	Geotechnical complexity class (GCC)		
	Lower (GCC1)	Normal (GCC2)	Higher (GCC3)
Higher (CC3)	GC2	GC3	GC3
Normal (CC2)	GC2	GC2	GC3
Lower (CC1)	GC1	GC2	GC2

Data from prEN 1997-1:2018 (CEN, 2018a).

6.12.2   Development of the ground model

The *ground model* is the term used to describe the package of all information pertaining to the site which helps the designer understand the site environment, the ground and groundwater conditions and all other relevant information. The ground model constantly evolves throughout the site investigation as increasing amounts of information are sourced and discovered. This includes information obtained at all stages of the site investigation from the desk study and walkover survey through to the *ground investigation*. The ground investigation is the stage where investigation pits and boreholes are made, samples taken, and field and lab testing is performed to fully establish the soil, rock and groundwater properties. The ground investigation must be planned based on the information retrieved from the desk study and walkover surveys carried out. This ensures an effective and efficient ground investigation.

The sources of information and the procedures used to obtain the information at all stages of a site investigation are described in detail in Chapter 7. By referring to this chapter, it is clearly seen that the ground model is most significantly informed by the content of the *Ground Investigation Report (GIR)* – see Section 7.7.

One critical aspect of the ground model is the assessment of ground properties. Ground properties to be used in any calculation model (i.e. the analysis method used to assess the safety of the design against the occurrence of a particular limit state) can be established directly (e.g. a test result) or indirectly (e.g. a mean value from a set of results). Depending on the method used to determine the ground property, the value of the property is referred to as:

- a *derived value* – where the value is derived by calculation or correlation from laboratory or field test results;
- a *nominal value* – a cautious estimate of the value of the property that affects the occurrence of a particular limit state. This is, in effect, the method used in the first generation of Eurocode 7 (and described in Section 6.4.1);
- a *characteristic value* – a statistically determined value of the property that affects the occurrence of a particular limit state, quantified by the probability of the value not being attained. This might be, for example, a mean value from a set of test results, or any other fractile of the probability distribution of the set.
- a *representative value* – this is either the nominal or the characteristic value, when combined with a specific conversion factor. The conversion factor accounts for various external factors which might affect the value of the property and which are not considered in the determination of the value. Such factors include temperature, effects of scale and any ageing of the soil.

6.12.3   Development of the geotechnical design model

The ground model is developed into the geotechnical design model (GDM) within the design verification stage (refer to Fig. 6.13). This is the process wherein the ground model combines with the specific requirements for geotechnical structures. The GDM can be considered as *the physical, mathematical, or numerical representation of the geotechnical system used for the purposes of analysis, design, and verification containing ground information for engineering design purposes developed for a particular design situation and limit state* – EN 1997-1:202x.

Both the ground model and the GDM continue to evolve during the design verification stage as more information feeds into the evolving design. These processes are therefore dynamic, and the design methodology should be such that the data used in all calculation models should be regularly reviewed, validated and updated or supplemented as additional facts come to light and/or as new observations are made which affect the design. The actual process of verifying the design is introduced in the following section.

6.12.4   Design verification

As mentioned earlier, the verification of a limit state can be performed by calculation using partial factors, by prescriptive methods, by testing or by using the observational method. Verification by the partial factor method is established as the most common approach used and the techniques and procedures involved are explained in detail in Section 6.13.

6.12.5   Water actions

In the first generation of Eurocode 7, the situations where groundwater actions should be factored in order to establish design values were not particularly clear. The second generation of the code helps to address this by specifying representative water actions as permanent, variable or accidental (depending on their variation with time) to which partial factors are applied accordingly, depending on the particular design situation.

A particular challenge in determining actions from groundwater pressures is that the pressures vary throughout the seasons as the groundwater level fluctuates. This gives uncertainty in establishing the correct action from the water pressure. EN 1997-1:202x deals with this uncertainty by stating that the representative value of the permanent water action, $G_{w,rep}$ be established through consideration of the annual probability of exceedance of a particular groundwater level.

Specifically, representative values of groundwater actions, $G_{w,rep}$ can be evaluated from either:

- a single permanent value, equal to the characteristic upper, $G_{wk;sup}$ or characteristic lower, $G_{wk;inf}$ value of groundwater pressure, whichever is more adverse – typically considered over a time period of $T = 50$ years, or
- the combination of a permanent value, G_{wk} (equal to the mean value of the groundwater pressure) and a variable value, $Q_{w,rep}$ (the representative value of the variation in groundwater pressure). $Q_{w,rep}$ is the amplitude of the groundwater pressure time series, T.

It is clear therefore, that to establish $G_{w,rep}$, a record of groundwater pressure variation through time is required. In order to acquire knowledge of the groundwater level changes through time (and therefore the groundwater pressure in a particular situation, as covered in Chapter 2), groundwater measurements and data from historical groundwater records must be recorded during the site investigation and incorporated into the ground model. Groundwater measuring is described in Chapter 7.

If there is insufficient data to statistically determine the characteristic values G_{wk}, the values are taken as cautious estimates of the most adverse values likely to occur during the particular design situation.

6.13   Verification by the partial factor method

In general, when checking an ultimate limit state for a particular situation, calculations should verify that the design effect of the actions, E_d (see Section 6.11.2) is less than or equal to the design resistance, R_d (see Section 6.11.3).

i.e.

$$E_d \leq R_d \tag{6.3}$$

This is no different than the check used in the first generation of the code (see Section 6.7).

Related, when checking a *serviceability limit state*, and as seen in the first generation of the code (see Section 6.8), the effects of the actions, E_d should be less than or equal to the particular limiting design value, $C_{d,SLS}$ (e.g. maximum permitted settlement).

i.e.

$$E_d \leq C_{d,SLS} \tag{6.4}$$

In order to establish the design value of the effects of actions, E_d, partial factors of safety taken from EN 1990 are applied to the actions or the effects, depending on which design case is being considered (refer to Section 6.11.5).

Ultimate limit states

For ultimate limit states, in order to establish the design value of the resistance, R_d, a material factor approach or a resistance factor approach is followed which draws material partial factors (γ_M) from EN 1997-1, and resistance partial factors (γ_R) from EN 1997-3, as required and appropriate to the particular geotechnical structure under analysis.

The ultimate limit states checked in a geotechnical design include those situations described in Section 6.5 (though termed differently in the second generation of Eurocode 7):

- *Failure by rupture:* where failure in the ground or structural element occurs because the ultimate strength of the material is reached.

- *Failure of the ground due to excessive deformation:* where failure occurs in a ductile manner (without rupture) and the limit state is caused due to excessive deformation of the ground.
- *Failure by loss of static equilibrium of the structure or ground:* this may be due to:
  - *Loss of rotational equilibrium:* assessed using Design Case 2 by verifying that any destabilising design moments are less than or equal to the stabilising design moment about the point of rotation.
  - *Loss of vertical equilibrium due to uplift forces:* the loss of equilibrium of the structure or the supporting ground by vertical uplift due to water pressures (buoyancy) or other actions. Assessed by ensuring that the design stabilising forces exceed the design destabilising uplift forces.
- *Hydraulic failure caused by seepage*: In the presence of groundwater flow, checks against *hydraulic heave* and *internal erosion and piping* must be carried out.

Additional ultimate limit state checks should be carried out on failure due to time-dependent effects or cyclic effects acting on the ground.

Serviceability limit states

For serviceability limit state calculations to determine $C_{d,SLS}$, the partial factors for all soil properties are taken as 1.0 unless the National Annex provides a different value.

6.13.1   Determining design values of actions and effect of actions

As explained in Section 6.11.2, design actions are obtained from the characteristic value, multiplied by the appropriate partial factor of safety. The values of the partial factors depend on the design case being followed. Table 6.3 shows the ultimate limit states for which each of the four design cases may be used in the verification.

The following descriptions from EN 1997-1:202x capture the essence of each design case:

- *Design Case 1:* partial factors > 1.0 are applied to unfavourable actions. DC1 can be used with the Resistance Factor Approach (RFA) – with partial factors on actions – or with the Material Factor Approach (MFA).
- *Design Case 2*: a combined verification of strength and static equilibrium, partial factors are applied to actions in two different combinations, a and b. DC2 is static equilibrium with partial factors applied on actions.
- *Design Case 3*: partial factors = 1.0 are applied to most actions (except to variable actions). DC3 is used primarily with the MFA.
- *Design Case 4*: partial factors are applied to effects of actions (and not to actions). DC4 can be used with the RFA or the MFA.

The specific values of partial factors for each design case are given in Table 6.6.

The values in Table 6.6 come from EN 1990:202x but may be adjusted in a National Annex. The parameter K_F is the consequence factor whose magnitude depends on the consequence class (see Section 6.11.4). Partial factors for accidental actions (not shown) are obtained from EN 1991:202x.

Table 6.6 Partial factors on actions and effects for persistent and transient design situations.

Action or effect				Partial factors γ_F and γ_E for Design Cases 1–4				
					DC2			
				DC1	a	b	DC3	DC4
Type	Group	Symbol	Resulting effect	Structural resistance	Static equilibrium and uplift		Geotechnical design	
Permanent action (G_k)	All	γ_G	Unfavourable/ destabilising	1.35 K_F	1.35 K_F	1.0	1.0	G_k is not factored
	Water	$\gamma_{G,w}$		1.2 K_F	1.2 K_F	1.0	1.0	
	All	$\gamma_{G,stb}$	Stabilising	Not used	1.15	1.0	Not used	
	Water	$\gamma_{G,w,stb}$			1.0	1.0		
	All	$\gamma_{G,fav}$	Favourable	1.0	1.0	1.0	1.0	
Variable action (Q_k)	All	γ_Q	Unfavourable	1.5 K_F	1.5 K_F	1.5 K_F	1.3	$\gamma_{Q,1}\big/\gamma_{G,1}$
	Water	$\gamma_{Q,w}$		1.35 K_F	1.35 K_F	1.35 K_F	1.15	1.0
	All	$\gamma_{Q,fav}$	Favourable			0		
Effects of actions (E)		γ_E	Unfavourable	Effects are not factored				1.35 K_F
		$\gamma_{E,fav}$	Favourable					1.0

Note: $\gamma_{Q,1}\big/\gamma_{G,1}$ indicates γ_Q of DC1 divided by γ_G of DC1.

Example 6.7: Design actions and effects of actions

Consider again the foundation of Example 6.4, shown in Fig. 6.14. The foundation forms part of a warehouse storage facility where failure of the foundation would result in a low risk of loss of life, but considerable economic loss. Checks are to be made for resistance to failure of the foundation through bearing failure of the soil due to the vertical actions.

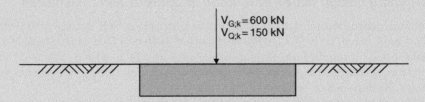

$V_{G;k} = 600$ kN
$V_{Q;k} = 150$ kN

Fig. 6.14 Example 6.7.

Solution:

Since the type of failure mechanism under consideration is geotechnical, we would use either an MFA or RFA for the design. This means that Design Cases 1, 3 and 4 should be considered and used to establish the design values of the actions and, thereafter, the design effect of the actions.

The consequence class is established using Table 6.2. The more severe factor of the consequences determines the consequence class: the consequence class therefore is CC2 and the consequence factor, $K_F = 1.0$.

In this example, the design effect is simply the sum of the vertical actions. It is clear that all the actions are *unfavourable* – since they all contribute to the total force causing the bearing pressure. The partial factors are obtained from Table 6.6.

Design Case 1:

$$V_{G;d} = V_{G;k} \times 1.35\,K_F = 600 \times 1.35 = 810\,kN$$
$$V_{W;d} = V_{W;k} \times 1.35\,K_F = 50 \times 1.35 = 67.5\,kN$$
$$V_{Q;d} = V_{Q;k} \times 1.5\,K_F = 150 \times 1.5 = 225\,kN$$
$$E_d = V_{G;d} + V_{W;d} + V_{Q;d} = 810 + 67.5 + 225 = 1102.5\,kN$$

Design Case 3:

$$V_{G;d} = V_{G;k} \times \gamma_G = 600 \times 1.0 = 600\,kN$$
$$V_{W;d} = V_{W;k} \times \gamma_G = 50 \times 1.0 = 50\,kN$$
$$V_{Q;d} = V_{Q;k} \times \gamma_Q = 150 \times 1.3 = 195\,kN$$
$$E_d = V_{G;d} + V_{W;d} + V_{Q;d} = 600 + 50 + 195 = 845\,kN$$

Design Case 4:

$$\gamma_Q = \gamma_{Q,1}/\gamma_{G,1} = \frac{1.5 \times K_F}{1.35 \times K_F} = 1.11$$
$$V_{Q;d} = V_{Q;k} \times \gamma_Q = 150 \times 1.11 = 166.5\,kN$$
$$V_{G;d} = V_{G;k} = 600\,kN$$
$$V_{W;d} = V_{W;k} = 50\,kN$$
$$E = V_{G;d} + V_{W;d} + V_{Q;d} = 600 + 50 + 166.5 = 816.5\,kN$$
$$E_d = E \times \gamma_E = 816.5 \times 1.35 \times K_F = 1102\,kN$$

6.13.2 Determining design values of material properties and resistances

As mentioned already, each geotechnical situation has its own unique limit states against which the design should be verified using Equation (6.3): $E_d \le R_d$. We have just seen how E_d is established. The design resistance, R_d is established using standard geotechnical analyses (particular to each geotechnical situation) which are described in later chapters. In all cases of determining ground resistance, and as Equations (6.2a) and (6.2b) demonstrate, R_d depends on soil properties.

The design values of soil properties are established by dividing the representative value of the property by the relevant partial factor, as explained in Section 6.11.3. The partial factors for the material properties are provided in Part 1 (EN 1997-1), grouped in two sets: M1 and M2 as shown in Table 6.7. The particular set to be used for specific geotechnical structures is specified in Part 3 (EN 1997-3).

K_M, similar to K_F (see Table 6.2), is a consequence factor whose magnitude depends on the consequence class (see Section 6.11.4). Values of K_M are given in Table 6.8. It is worth noting that consequence factors for resistance, K_R also exist in some situations, and where applicable these are provided in EN 1997-3.

As introduced in Section 6.11.3, ultimate limit states involving failure of the ground are verified using either a material factor approach (MFA) or a resistance factor approach (RFA). The set of material partial factors (M1 or M2), the design case and the partial factor value for resistance to be used for specific geotechnical structures are specified in EN 1997-3. Table 6.9 shows, for the case of a shallow foundation, the combinations of design case and the set of material partial factors to be used for both the MFA and RFA. The combinations in Table 6.9 also apply to sliding and bearing checks for gravity retaining walls.

The values in Tables 6.7–6.9 are NDPs and therefore may be adjusted in a National Annex. Similar tables exist for other geotechnical structures (e.g. slopes, retaining walls, pile foundations) and these are included in the relevant chapters later on in this book.

Table 6.7 Partial factors on ground properties for persistent and transient design situations.

Ground parameter	Symbol	Set M1	Set M2
Effective shear strength, τ'	$\gamma_{\tau s}$	1.0	1.25 K_M
Effective cohesion, c'	γ_c	1.0	1.25 K_M
Coefficient of internal friction, $\tan \phi'$	$\gamma_{\tan\phi}$	1.0	1.25 K_M
Coefficient of ground/structure interface friction, $\tan \delta$	$\gamma_{\tan\delta}$	1.0	1.0 K_M
Undrained shear strength, c_u	γ_{cu}	1.0	1.4 K_M
Unconfined compressive strength, q_u	γ_{qu}	1.0	1.4 K_M

Data from prEN 1997-1:2018 (CEN, 2018a).

Table 6.8 Consequence factors K_M for materials in geotechnical structures.

Consequence class	Description of consequence	Consequence factor, K_M
CC3	Higher	1.1
CC2	Normal	1.0
CC1	Lower	0.9

Table 6.9 Partial factors for the verification of ground resistance of spread foundations for persistent and transient design situations.

Verification check	Partial factor	Symbol	Material factor approach* (a)	Material factor approach* (b)	Resistance factor approach
Bearing and sliding resistance	Actions and effects of actions	γ_F and γ_E	DC1	DC3	DC1 or DC4
	Ground properties	γ_M	M1	M2	M1
	Bearing resistance	γ_{Rv}	Not factored		1.4
	Sliding resistance	γ_{Rh}	Not factored		1.1

*Both combinations should be verified: (a) structural resistance check, (b) geotechnical resistance check.
Data from prEN 1997-3:2018 (CEN, 2018c).

Example 6.8: Design material properties and resistances

The ground beneath the foundation of Example 6.7 is a clay and has the following representative values: $c_{u;\,rep} = 100$ kPa; $\gamma = 19.8$ kN/m^3.

If the foundation is square, with side length = 2 m, founded at a depth of 1 m below the ground surface, all actions act centrally, and the base of the foundation and the ground surface are horizontal, check the undrained bearing resistance limit state using both the material factor approach and the resistance factor approach.

Note: This example draws on undrained bearing resistance formulae derived in Chapter 10, which the reader may wish to familiarise themselves with before following this example.

Solution:

The weight density, γ is not usually factored so the unaltered value is used in the calculations.

Using bearing resistance formulae from Chapter 10, we can establish the various required factors and the undrained bearing resistance.

$$N_{cu} = 5.14, s_{cu} = 1.2, d_{cu} = 1.15$$

Material factor approach:
For the MFA, both combinations (see Table 6.9) must be verified:

- In the geotechnical check (DC3 and M2), the safety is applied to the material property, i.e. c_u,

$$c_{u;d} = \frac{c_{u;rep}}{\gamma_{cu}} = \frac{100}{1.4 \times K_M} = 71.4 \text{ kPa} \dots \text{ since } K_M = 1.0(CC2)$$

Design bearing resistance, $R_d = A'(c_{u;d}N_{cu}s_{cu}d_{cu} + q)$

$$= 4 \times (71.4 \times 5.14 \times 1.2 \times 1.15 + (19.8 \times 1))$$
$$= 2105 \text{ kN}$$

- In the structural resistance check (DC1 and M1), $\gamma_{cu} = 1.0$.

$$\Rightarrow c_{u;d} = \frac{100}{1.0} = 100 \text{ kPa}$$
$$R_d = 4 \times (100 \times 5.14 \times 1.2 \times 1.15 + (19.8 \times 1))$$
$$= 2916 \text{ kN}$$

Resistance factor approach:

In the RFA, the safety is applied to the resistance.

Undrained bearing resistance, $R_{Nu} = A'(c_u N_{cu} s_{cu} d_{cu} + q)$

$\quad = 4 \times (100 \times 5.14 \times 1.2 \times 1.15 + (19.8 \times 1))$

$\quad = 2916 \text{ kN}$

$R_d = \dfrac{R_{Nu}}{\gamma_{Rv}} = \dfrac{2916}{1.4} = 2083 \text{ kN}$

Design checks:

MFA:

- (DC3 & M2) : $R_d = 2105 \text{ kN}$; $E_d = 845 \text{ kN} \Rightarrow R_d > E_d$…Limit state check ok
- (DC1 & M1) : $R_d = 2916 \text{ kN}$; $E_d = 1102.5 \text{ kN} \Rightarrow R_d > E_d$…Limit state check ok

RFA:

- $R_d = 2083 \text{ kN}$; $E_d = 1102 \text{ kN} \Rightarrow R_d > E_d$…Limit state check ok

For convenience, all of the tables in Section B of this chapter are published side by side in a downloadable document on the companion website.

6.14   Execution, Monitoring and Reporting

6.14.1   Execution and monitoring

The geotechnical design is only part of the overall process which sees a project realised from inception to construction, as illustrated in Fig. 6.13. Following the design, construction (referred to as *execution* in the Eurocodes) can commence. The construction phase implements the design, and monitoring of movement, deflection, vibration and other recordable site measurements is made to inform control measures. Revisions to the design are sometimes necessary as a result of site observations during construction. Ground and structural monitoring can therefore be part of the design process, even though they take place following completion of the main design work. Unforeseen ground conditions, not picked up during the site investigation, might necessitate moderate or even significant changes to the original design to ensure limit state compliance.

To accompany the construction processes, many *execution standards* exist. These are documents which provide guidance and specification on particular geotechnical construction activities and are referred to before and throughout construction. EN 1997-3 refers to relevant execution standards in the appropriate sections of the code so that the design process for a particular geotechnical structure captures input from these standards. Some of the key execution standards are listed in Table 6.10.

Table 6.10   Execution standards referred to in EN 1997-3:202x.

Section in EN 1997-3	Geotechnical structure/process	Execution standard
4. Slopes, cuttings and embankments	Earthworks	EN 16907
5. Spread foundations	Concrete foundations	EN 13670
6. Piled foundations	Bored piles	EN 1536
	Displacement piles	EN 12699
	Micropiles	EN 14199
7. Retaining structures	Sheet pile walls	EN 12063
	Diaphragm walls	EN 1538
8. Anchors	Grouted anchors	EN 1537
9. Reinforced soil structures	Reinforced fill structures	EN 14475
	Soil nailing	EN 14490
10. Ground improvement	Grouting	EN 12715
	Jet grouting	EN 12716
	Deep mixing	EN 14679
	Deep vibration	EN 14731
	Vertical drainage	EN 15237

Bond *et al.* (2019b).

6.14.2   Reporting

All of the findings, calculations and decisions made through the design and execution phases are compiled into four separate reports:

- Ground investigation report (GIR)
- Geotechnical design report (GDR)
- Geotechnical construction record (GCR)
- Geotechnical test reports

Each report contains specific information and recommendations relevant to the particular stage or aspect of the project.

The amount and level of detail contained in the reports depends on the geotechnical category: GC3 structures, for example, will necessitate significantly more detailed reporting than a GC2 structure.

Ground investigation report (GIR)

The observations, results and findings of a site investigation (see Chapter 7) are compiled into a ground investigation report which is then incorporated into the geotechnical design model and referenced in the geotechnical design report.

Geotechnical design report (GDR)

Along with the GIR and GDM, the geotechnical design report contains the results from the calculations performed to verify the safety and serviceability during the geotechnical design. Further, in addition to the calculations, relevant drawings, BIM models and specific design recommendations are also included, as is a plan of supervision and monitoring for the site.

Geotechnical construction record (GCR)

The geotechnical construction record contains information on construction processes, supervision, monitoring and inspection of each construction phase as well as on the final structure. The documentation is a record of all the techniques used and the deliverables of the project management process and can help with future maintenance, design of additional works and decommissioning of the works.

Geotechnical test report

If specific geotechnical testing has taken place as part of the construction phase (e.g. pile load testing, anchorage pull-out tests etc.) the test description, ground conditions and test results are reported in the specific test report.

Chapter 7
Site Investigation

Learning objectives:

By the end of this chapter, you will have been introduced to:

- the purpose and the planning of site investigations and the sources of information used in the desk study phase;
- the various methods of ground exploration that are available;
- soil and rock sampling techniques and methods of groundwater measurement;
- some of the more common field tests employed to establish *in situ* properties;
- the development of the ground model and the compilation of the ground investigation report.

A site investigation is an essential part of the preliminary design work on any important geotechnical structure. Its purpose is to obtain information about the proposed site that can be used by the engineers to achieve a safe and economical design. Information retrieved during a site investigation can be very diverse and can include information about the past history of the site and ground information such as the sequences of strata and the depth of the groundwater level. During the *ground investigation* phase, samples of soil and rock can be taken for identification and laboratory testing, and *in situ* testing may be performed.

The list of primary objectives of a site investigation is given in BS 5930: *Code of practice for ground investigations* (BSI, 2015) and includes:

- to assess the general suitability of the site for the proposed works;
- to enable an adequate and economic design to be prepared;
- to foresee, and provide against, difficulties that may arise during construction due to ground and other local conditions;
- to predict any adverse effect of the proposed construction on neighbouring structures.

In addition, a site investigation is often necessary to assess the safety of an existing structure or to investigate a case where failure has occurred.

Smith's Elements of Soil Mechanics, 10th Edition. Ian Smith.
© 2021 John Wiley & Sons Ltd. Published 2021 by John Wiley & Sons Ltd.
Companion website: www.wiley.com/go/smith/soilmechanics10e

7.1   Eurocode 7 and execution standards

As we saw in Chapter 6, it is Part 2 of both generations of Eurocode 7 (EN 1997-2:2007 (BSI, 2007a) and EN 1997-2:202x) that deals with ground investigation and testing. Prior to the publication of EN 1997-2:2007, site investigation practice in the UK uniquely followed the guidelines offered in BS 5930. Although BS 5930 remains in current use, the procedures of EN 1997-2, together with the guidance provided in the many testing standards that are used in conjunction with it, are routinely followed. Some of these relevant testing standards, published by CEN and cited in Eurocode 7, are listed below:

EN ISO 14688:2018 (BSI, 2018a) *Geotechnical investigation and testing – identification and classification of soil* (2 parts)
EN ISO 14689:2018 (BSI, 2018b) *Geotechnical investigation and testing – identification, description and classification of rock* (1 part)
EN ISO 17892:2014–2019 (BSI, 2014–2019) *Geotechnical investigation and testing – laboratory testing of soil* (12 parts)
EN ISO 22282:2012 (BSI, 2012) *Geotechnical investigation and testing – geohydraulic testing* (6 parts)
EN ISO 22475:2006–2011 (BSI, 2006–2011) *Geotechnical investigation and testing – sampling of soil, rock and groundwater* (3 parts)
EN ISO 22476:2009–2020 (BSI, 2009–2020) *Geotechnical investigation and testing – field testing* (15 parts)
EN ISO 22477:2016–2018 (BSI, 2016–2018) *Geotechnical investigation and testing – testing of geotechnical structures* (4 parts)

In accordance with the UK National Annex to EN 1997-2:2007 (BSI, 2007a), in the UK, BS 1377:1990 (BSI, 1990) remains the standard for all laboratory testing of soils, with the exception of the fall cone test, which is covered instead by EN ISO 17892-6:2017. However, it is anticipated that EN ISO 17892 will become the standard for all soils testing in the UK in due course. See Chapters 1, 2, 4, 12 and 15 for details of some of the more routine laboratory tests.

The practices used in site investigations have been around for many years and whilst EN 1997-2 does offer a different strategy for the carrying out of ground investigation and testing work, the established equipment and procedures in use remain largely unchanged. BS 5930, similar to BS 1377, contains information referred to as *non-contradictory complementary information* (NCCI). Such information does not contradict any of the principles in the Eurocodes and therefore can be used to complement all other information used in the design. Thus, BS 5930 currently remains in use in the UK.

The procedures used in the development of both the *ground model* and the *geotechnical design model* have been outlined in Section 6.12. It is important to realise that in all Eurocode compliant design work, the principal purpose of site investigation is to enable the development of these two models. The rest of this chapter is arranged to align with the stages involved in those procedures.

7.2   Planning of ground investigations

The most significant (and the most expensive) part of a site investigation is the ground investigation (i.e. that stage where the ground profile and groundwater levels are established and where samples of soil and rock are taken for identification and testing). In order to maximise the value and relevance of the information and data gleaned during the ground investigation, it is critical that the investigation is well planned. Careful planning ensures that a cost-efficient investigation is achieved and that all the information required for the geotechnical design is obtained. This careful planning is achieved by performing several pre-ground investigation information searches, assessments and analyses.

7.2.1 Desk study

The desk study is generally the first stage in a site investigation. The size and extent of the study will vary according to the nature of the project and the anticipated ground conditions. It involves collecting and collating published information about the site under investigation and pulling it all together to build a conceptual model of the site. In terms of a project being designed to Eurocode 7, the conceptual model develops into the ground model which we referred to in Chapter 6. This model can then be used to guide the rest of the investigation, especially the ground investigation. Much of the information gathered at the desk study stage is contained in maps (online and hard copies), published reports, aerial photography (online and drone) and personal recollection.

Sources of information

The sources of information available to the engineer include geological maps, topographical maps (Ordnance Survey maps), soil survey maps, aerial photographs, mining records, groundwater information, existing site investigation reports, local history literature, meteorological records and river and coastal information. Details of a few of these are provided below but a thorough description of the sources of desk study information is given by SISG (2013).

Geological maps

Geological maps provide information on the extent of rock and soil deposits at a particular site. The significance of the geological information must be correctly interpreted by the engineer to assist in the further planning of the site investigation. Traditionally, hard copy geological maps were used in the desk study but nowadays much review is done online instead. The British Geological Survey's (BGS) *Geology of Britain viewer* and *GeoIndex* are online facilities for viewing geological information across the UK. Free and subscription services are available: the subscription service offering more detailed information for all locations.

Topographical maps

Ordnance Survey maps provide information on, for example, the relief of the land, site accessibility, and the landforms present. A study of the sequence of maps for the same location produced at different periods in time, can reveal features which are now concealed and identify features which are experiencing change. As with geological maps, print editions of topographical maps exist, though most research is done using digital, online versions.

Soil survey maps

A pedological soil survey involves the classification, mapping and description of the surface soils in the area and is generally of main interest to agriculturists. The soil studied is the top 1–1.5 m, which is the part of the profile that is significantly affected by vegetation and the elements. The maps produced give a good indication of the surface soil type and its drainage properties. The surface soil type can often be related to the parent soil lying beneath, and so soil types below 1.5 m can often be interpreted from the maps.

Aerial photographs and imagery

With careful interpretation of aerial photographs, it is possible to deduce information on landforms, topography, land use, historical land use, and geotechnical behaviour. The photographs allow a visual inspection of a site when access to the site is restricted. Freely accessible satellite imagery such as Google™ Earth is now a much-used source of aerial photography.

Existing site investigation reports

These can often be the most valuable source of geotechnical information. If a site investigation has been performed in the vicinity in the past, then information may already exist on the rock and soil types, drainage, access, etc. The report will likely also contain details of the properties of the soils and test results.

7.2.2   Site reconnaissance

A walk over the site can often help to give an idea of the work that will be required. Differences in vegetation often indicate changes in subsoil conditions, and any cutting, quarry or river on or near the site should be examined. Site access, overhead restrictions and signs of slope instability are further examples of aspects that can be observed during the walk over survey. The information observed during the survey is used to complement the desk study information so that the ground investigation can be well planned.

7.2.3   Planning field investigations and laboratory tests

In order to obtain quantitative data on the soil and rock types and properties, the ground investigation is performed. This phase involves the sampling of the ground using recognised sampling procedures and specialist equipment. The extent of the sampling and subsequent testing depends mainly on the size and nature of the proposed structure but is also influenced by the degree of variability of the soils on the site. Investigation points are locations on the site where profiling and sampling of the ground occurs. The ground is investigated and sampled using various methods, as described in Section 7.3.

Spacing of ground investigation points

Guidance on the *spacing* of investigation points is given in Part 2 of Eurocode 7 (both generations) and are summarised in Table 7.1 (EN 1997-2:2007) and Table 7.2 (EN 1997-2:202x). The values in Table 7.2 apply to *Geotechnical Category 2* structures (see Section 6.12.1 for definitions of Geotechnical Categories). As can be seen in both tables, the nature of the project influences significantly the recommended extent and number of investigation points.

Minimum depth of ground investigation points

Guidance on the *minimum depth* of investigation points is given in both generations of Eurocode 7 and are summarised in Table 7.3 (first generation) and Table 7.4 (second generation). The fundamental difference between the two generations of the code on this aspect is that guidance on the minimum depths is provided in Part 2 of the first generation (EN 1997-2:2007) whereas it is provided in Part 3 of the second generation (EN 1997-3:202x).

Table 7.1   Guidance values for spacing and pattern of investigation points.

Structure	Spacing	Layout
High-rise and industrial structures	15–40 m	Grid pattern
Large area	≤60 m	Grid pattern
Linear structures (e.g. roads, railways, walls etc.)	20–200 m	Linear
Special structures (e.g. bridges, stacks, machinery foundations)	2–6 investigation points per foundation	
Dams and weirs	25–75 m	Along relevant sections

Data from EN 1997-2:2007 (BSI, 2007a).

Table 7.2 Maximum spacing, X_{max} and minimum number, N_{min} of ground investigation points for *Geotechnical Category 2* structures.

Structures		X_{max} (m)	N_{min}
Low-rise structures		30	3
High-rise structures	4–10 storeys	25	4
	11–20 storeys	20	5
	>20 storeys	15	6
Estate roads, parking areas and pavements		40	2
Power lines, wind turbines		1 per pylon	
Wind turbines		2 per turbine	
Linear structures	<3 m high	100	–
	≥3 m high	50	–
Silos and tanks		15	3
Bridge piers		1 per pier/base	
Surface excavations		25	3

Data from EN1997-2:202x.

Table 7.3 Guidance values for minimum depth (z_a) of investigation points.

Foundation type	Depth criteria*
Shallow foundations for high-rise and civil engineering projects	*The greater of:* $z_a = 6$ m; $z_a = 3.0b_f$ (see Fig. 7.1a)
Raft foundation	$z_a \geq 1.5b_B$ (see Fig. 7.1b)
Linear structures: roads and airfields	$z_a \geq 2$ m
Linear structures: canals and pipelines	*The greater of:* $z_a = 2$ m; $z_a = 1.5b_{Ah}$ (see Fig. 7.1c)
Pile foundations	*The greater of:* $z_a = 5$ m; $z_a = 1.0b_g$ $z_a = 3.0D_F$ (see Fig. 7.1d)

*The depths are measured from the reference levels shown in Fig. 7.1.
Data from EN 1997-2:2007 (BSI, 2007a).

First generation

The values in Table 7.3 apply to *Geotechnical Category 2* structures (see Section 6.3.2 for definitions of Geotechnical Categories). As can be seen, the nature of the project influences significantly the recommended minimum depth of investigation points.

Table 7.4 Guidance values for minimum depths of investigation points.

Geotechnical structure	Depth criteria*
Embankments and cuttings	$z_a = 1.5\,h$ (see Fig. 7.1e)
Shallow foundations for low-rise structures in Geotechnical Category 1	$z_a = 2\,m$
Shallow foundations for low-rise structures in Geotechnical Category 2	The greater of: $z_a = 3\,m$; $z_a = 3.0b_f$ (see Fig. 7.1a)
Shallow foundations for high-rise structures	The greater of: $z_a = 6\,m$; $z_a = 3.0b_f$ (see Fig. 7.1a)
Raft foundation	$z_a \geq 1.5b_B$ (see Fig. 7.1b)
Pile foundations in soil or weak rock	The greater of: $d_{min} = 5\,m$; $d_{min} = 3D$; $d_{min} = p_{group}$ (max 25 m) (see Fig. 7.1d)
Pile foundations in strong rock	The greater of: $d_{min} = 2\,m$; $d_{min} = 3D$; (see Fig. 7.1d)

*The depths are measured from the reference levels shown in Fig. 7.1.
Data from EN1997-3:202x.

The depth of investigations can be reduced to $z_a = 2\,m$ where the foundations in Table 7.3 are constructed on competent strata with distinct geology.

Second generation

The depth of investigation points should be adequate to cover the zone of influence of the structure. This is the practice that has generally been adopted within the ground investigation industry for many years and guidance on the zone of influence for different geotechnical structures, together with recommended minimum depths of investigation, is given in EN 1997-3:202x. These are listed in Table 7.4.

7.3   Site exploration methods

7.3.1   Trial pits

A trial pit is simply a hole excavated in the ground that is large enough, if necessary, for an inspection cradle to be lowered, thus permitting a close examination of the exposed sides. The pit is created by removing successive layers of soil using a hydraulic excavator until the required depth is reached. Progression by cuts of depth about 400 mm is quite common. The excavated soil is usually placed beside the pit to enable easy

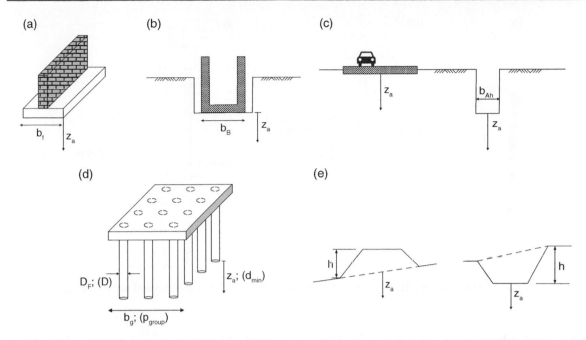

Fig. 7.1 Minimum depths of investigation points for various structures. Symbols in brackets in (d) refer to EN 1997-3:202x (Table 7.4). (a) Shallow foundation. (b) Raft foundation. (c) Roads and trenches. (d) Piles and pile groups. (e) Embankments and cuttings. Based on Bond *et al.* (2019b).

backfilling once the pit is ready to be closed up again. The sides of the trial pit are never assumed to be stable and, if personnel are to enter the pit to perform close inspection of the soil, to take samples or to perform *in situ* testing, the sides of the pit must be fully supported.

Groundwater conditions can be accurately established from a trial pit and undisturbed block soil samples are obtainable relatively easily. In addition, undisturbed samples can be obtained using cylindrical steel sampling tubes gently pushed into the soil by the excavator bucket.

Below a depth of about 4 m, the challenges of side support and the removal of excavated material become increasingly important and the cost of trial pits increases rapidly. In excavations below groundwater level the expense may be prohibitive. Trial pits should not be made at locations where pad foundations might be cast later in the project.

7.3.2   Hand excavated boreholes

A hand auger can be used in soft and loose soils for creating a borehole of up to about 6 m (using extension rods) and is useful for site exploration work in connection with roads. A choice of auger types exist, each of which is used for a specific type of soil. In clay soils a clay auger as shown in Fig. 7.2a is used, whereas in sands and gravels, the gravel auger (Fig. 7.2b) is used. The auger is connected to drill rods and to a rotary engine arrangement above ground surface which is held stable by two operators standing at each side holding stabilising handles. For shallow depth boreholes, the auger can be connected to short rods and to a cross bar at the top to enable the auger to be turned by a single operator and advanced into the soil.

Hand excavated boreholes are useful for cheap, rapid sampling and assessment of ground conditions where only one or two locations on a site are of interest. For larger scale investigations, the boreholes

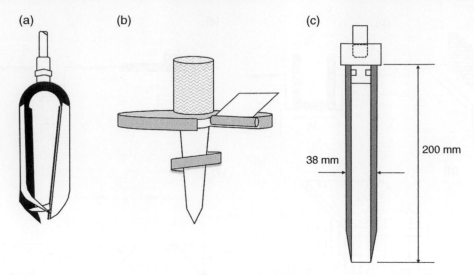

Fig. 7.2   Hand augers: (a) clay auger; (b) gravel auger; (c) 38 mm undisturbed sample tube.

will invariably be created using full scale drilling equipment. All samples of soil retrieved from hand excavated boreholes are classified as *disturbed*. However, 38 mm diameter *undisturbed* samples can be obtained from the undisturbed soil below the bottom of the borehole (see Section 7.4.1) using the sampling tube shown in Fig. 7.2c.

7.3.3   Boreholes

In most ground investigations, several boreholes are required and these are often taken down to many metres in depth. Thus, specialist drilling equipment is required to form these. The operation is carried out dry where possible. Where advancing the borehole is difficult because of ground conditions, water may be added into the borehole to ease the cutting at the base. Such a technique is called *wash boring*. If water is added, regardless of quantity, any soil samples retrieved will no longer be at their natural water content, and in some cases will possess a particle size distribution different from the natural state. This must be borne in mind when the addition of water to aid boring is being considered.

Two main methods of forming boreholes exist: *cable percussion boring* and *rotary drilling*.

Cable percussion boring

This method is sometimes referred to as the *shell and auger* method. The equipment is shown in Fig. 7.3.
The principle of operation is:

- the A-frame (which is transported to site in its collapsed state, towed by a four-wheel drive vehicle) is erected at the location of the borehole and stabilised;
- the winch, powered by the portable diesel generator, lifts the cutting tool (Fig. 7.4) towards the top of the A-frame. In clay soils, the *clay-cutter* is used; in more sandy and gravelly soils the *shell* is used;
- the winch brake is released, and the tool is allowed to fall freely into the soil;
- the cutting tool drives into the soil and the borehole soil is forced inside the tool;

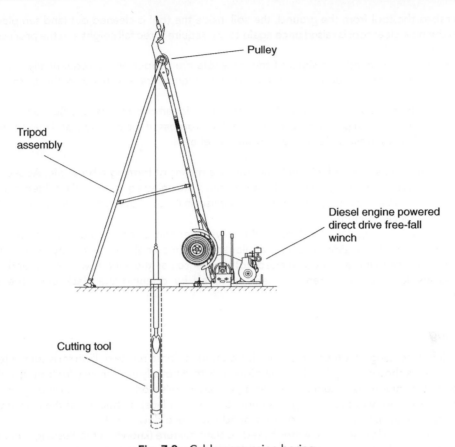

Fig. 7.3 Cable percussion boring.

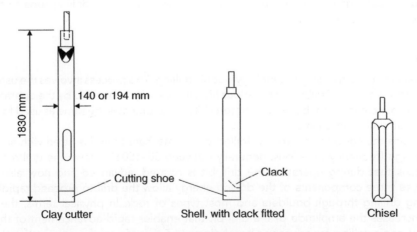

Fig. 7.4 Cable percussion cutting tools and samplers.

- the winch raises the tool from the ground, the soil inside the tool is cleaned out (and sampled if appropriate) and the now clear tool is raised once again to the required free fall height and the process repeated over and over;
- as the borehole is advanced, the sides of the borehole are supported by sequentially installing steel tube sections known as the borehole casing. These sections are advanced as the borehole itself is advanced;
- undisturbed samples of soils are retrieved in steel or plastic sampling tubes (see Section 7.4.1);
- the driller operating the equipment records a log of the progress of the borehole and makes any observations such as soil types, obstructions, groundwater level etc.

This is an extremely versatile and relatively inexpensive means of forming a borehole. As seen, it can be used in different soil types and this is essential for any borehole forming equipment. If boulders or cobbles are encountered, these can be broken down using a heavy chisel in place of the cutting tool until the obstruction is clear, then progress can continue.

In clay soils, the soil is simply wedged inside the clay-cutter and is removed by hand from inside using steel bars pushed through the side slots. In granular soils the material is retrieved by means of the shell. This cutting tool is fitted with a clack (a hinged lid) that closes as the shell is withdrawn and retains the loose particles inside. The soil is removed and sampled by opening the clack once the shell is at ground level.

Rotary drilling

Rotary drilling involves using a high-powered, truck mounted motor to rotate drilling rods connected to a drill bit into the ground as shown in Fig. 7.5. The technique was traditionally used mainly for boring and sampling rock, although the technique is becoming increasingly used in soils work too. The heavy-duty drill bit (interchangeable types exist for whether boring or sampling is taking place) is attached at the end of the drilling rods and rotates at high speeds to cut into the ground and move downward.

The drilling rods are hollow so that a water-based coolant mixture (known as the *flushing medium*) can be pumped down them and out through the holes in the drill bit into the surrounding space within the borehole. This fluid has several functions: it acts as both a coolant and as a lubricant to aid the cutting process, it provides pressure balance during drilling to resist inflow of groundwater to the borehole and it provides the means by which the cuttings of soil and rock are pumped up around the drilling rods to the surface for removal.

Sonic drilling

A fairly recent development in drilling technology is sonic drilling. This process involves the use of a sonic drill head at the ground surface that is vibrated at various high frequencies controlled by the operator, depending on the particular ground conditions being encountered. The equipment is the same as used for rotary drilling except for the sonic drill head addition.

The head contains the conventional rotary drilling/coring mechanism and is fitted with an oscillator. The oscillator sends high frequency vibrations, generally between 50–150 Hz, down the drill rods and sampler barrel. This means that during operation the drill bit is rotated, advanced and now also vibrated into the ground. These three components of the drilling energy allow the drill to proceed rapidly through the ground including drilling through boulders and most types of rock. In physical terms, the resonance of the vibrations increases the amplitude of the drill bit that enables rapid advancement of the borehole to be made. Indeed, sonic drilling can advance a borehole up to 5 times faster than conventional rotary drilling. Sonic drilling is also useful for retrieving continuous, relatively undisturbed, soil samples and rock cores.

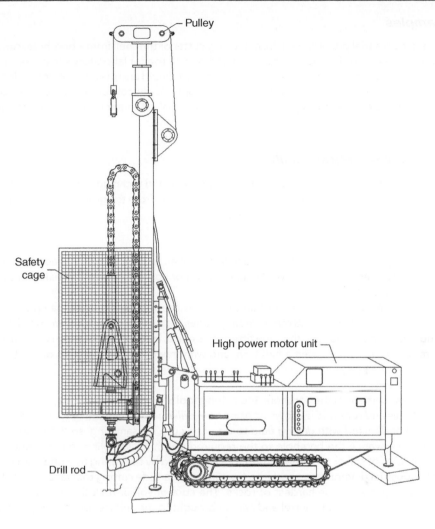

Fig. 7.5 Rotary drilling rig.

7.4 Soil and rock sampling

Soil and rock samples are taken regularly during the ground investigation so that specific ground properties required in the geotechnical design can be established.

7.4.1 Soil sampling

In general, two types of soil sample can be obtained: *disturbed sample* and *undisturbed sample*. Within this general division, five sampling categories (A–E) can be identified: each category yielding different quality classes of sample (see Section 7.4.3).

Disturbed samples

The soil excavated from a trial pit, or the soil from the clay cutter or the shell from a borehole, can be collected as disturbed samples. Such soil has been remoulded and is of no use for laboratory shear strength tests but is useful for identification, classification and chemical tests such as liquid and plastic limit determination, particle size distribution and sulphate testing. Disturbed samples are usually collected in plastic sampling bags or airtight tins or jars and are labelled to give the borehole (or trial pit) number, the depth and a description of the contents.

Undisturbed samples (cohesive soil)

Undisturbed samples can be achieved using different equipment and techniques in different situations. There will always be an element of disturbance to any sample of soil taken from the ground, but that disturbance can be minimal if care and appropriate methods are used.

  (i) *Trial pits*

  In a trial pit, samples can be cut out by hand if care is taken. Such a sample is placed in an airtight container and as a further measure to avoid change in water content, it may be sealed in paraffin wax.

 (ii) *Hand excavated boreholes*

  In a hand excavated borehole, the hand auger can be used to obtain useful samples for unconfined compression tests and employs 38 mm sampling tubes with a length of 200 mm (Fig. 7.2c). The auger is first removed from the rods and the tube fitted in its place, after which the tube is driven into the soil at the bottom of the borehole, given a half turn, and withdrawn. Finally, the ends of the tube are sealed with end caps.

(iii) *Rotary core drilling*

  During the advancement of the borehole, the cutting tool is used (Fig. 7.6a). This process can be applied to both hard soils and rock and is known as *open hole drilling* when applied to rock. The soil cuttings are too disturbed and mixed with *drilling fluid* to be of any use for sampling, so the method is really only used to rapidly advance the borehole to the required depth for sampling to commence.

  To take samples of soil or rock, the cutting tool is replaced by the coring tool attached to a core barrel (Fig. 7.6b). Industrial diamonds are cast into the tungsten carbide cutting tool face to enable the cutting shoe to cut through even the hardest of rocks. The core of soil or rock that is cut during this coring process is collected in the core barrel and can be brought to the surface for labelling and identification

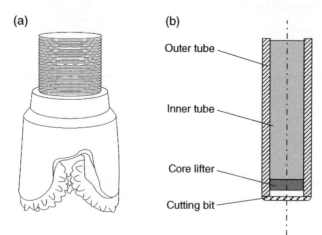

(a)

(b)

Outer tube

Inner tube

Core lifter

Cutting bit

Fig. 7.6   Rotary drilling bits: (a) cutting bit; (b) double-tube core barrel.

and transported to the geotechnical laboratory for testing. The core barrel can actually take one of three forms: single-tube, double-tube or triple-tube.

A single-tube barrel has the same diameter as the drilling rods and is connected directly to the cutting shoe. Samples retrieved in a single-tube core barrel experience a fair degree of disturbance, so double and triple-tube barrels are often used in preference.

As suggested by the name, a double-tube barrel comprises two tubes: an outer one which is attached to the coring bit and rotated by the drill rods, and a non-rotating inner one into which the core sample passes as the cutting bit is advanced as shown in Fig. 7.6b. An extension of the double-tube arrangement is to include a sample liner within the inner tube. This is known as a triple-tube core barrel.

(iv) *Cable percussion borehole – open tube sampler*

With the cable percussion boring rig, 100 mm diameter undisturbed samples, commonly referred to as *U100 samples*, are collected in a steel sampling tube fitted with a cutting shoe, which is driven into the soil under the percussive action of the falling weight assembly. Two approaches to retrieving the sample exist: a standard system using an open tube and cutting shoe as shown in Fig. 7.7a and a plastic liner system where the liner fits inside a larger steel tube and cutting shoe assembly (Fig. 7.7b). The degree of disturbance (see Section 7.4.2) is different between both systems.

During driving any entrapped water, air or slush can escape through a non-return valve fitted in the driving head at the top of the tube. After collection, the sample is sealed with end caps at both ends. If the sample is to be stored for a long time, a paraffin wax coating can be applied to each end of the soil in the tube to prevent long-term changes in water content.

Despite their popularity in the UK, with the implementation of EN ISO 22475-1:2019 (*International Standard for Geotechnical Investigation and Testing*), U100 samples are likely to become less used in the site investigation industry in the long-term. This standard recognises that U100 samples are not wholly appropriate for use for certain geotechnical tests (see Section 7.4.3) and shows that other methods of soil sampling retrieve better quality samples.

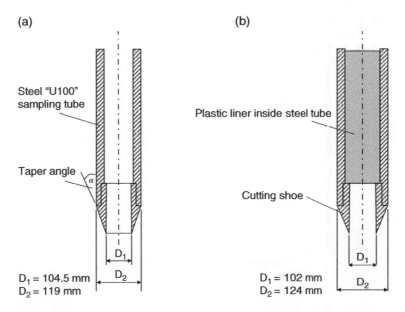

Fig. 7.7 Cable percussion equipment to obtain undisturbed samples: (a) standard system; (b) plastic liner system.

(v) *Thin-walled open tube sampler*

For soils such as soft clays and silts that are sensitive to disturbance, a thin-walled sampling tube can be used. Because of the softness of the soil to be collected, the tube is simply machined at its end to form a cutting edge and does not have a separate cutting shoe. The thin-walled sampler is similar in appearance to the sample tube shown in Fig. 7.7a but can have an internal diameter of up to about 200 mm.

These sampling techniques involve the removal of the boring rods from the hole, the replacement of the cutting edge with the sampler, the reinsertion of the rods, the collection of the sample, the removal of the rods, the replacement of the sampler with the cutting edge and, finally, the reinsertion of the rods so that boring may proceed. This is a most time-consuming operation and for deep bores, such as those that occur in site investigations for offshore structures, techniques have been developed to enable samplers to be inserted down through the drill rods so that soil samples can be collected much more quickly.

(vi) *Piston sampler*

A piston sampler is a specific thin-walled sampler for use in weak soils such as soft clays and slurry materials. A hydraulically powered piston sits neatly within the sampling tube and the assembly, with the piston locked in place at the cutting end of the tube, is carefully lowered to the bottom of the already formed borehole. The piston is connected by piston rods which pass through a sliding joint in the sampler head assembly so that the sample tube and the piston can move vertically independently of each other (Fig. 7.8).

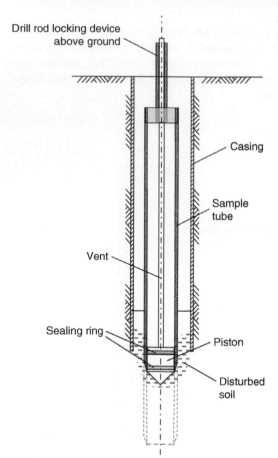

Drill rod locking device
above ground

Casing

Sample
tube

Vent

Sealing ring

Piston

Disturbed
soil

Fig. 7.8 Thin-walled stationary piston sampler.

The sample is taken from the undisturbed zone of soil beneath the borehole by releasing the lock between piston and tube and pushing the sampling tube into the ground. The piston remains still during this operation.

Once the tube is fully driven into the soil, the undisturbed sample fully occupies the sampling tube and the assembly is rotated slightly to shear the sample from the ground. The tube and piston are then locked together again and raised to the surface. A suction pressure is imparted within the sampling tube, which ensures that the sample is delicately held intact.

(vii) *Continuous sampler*

In some cases, particularly where the soil consists of layers of clay, separated by thin bands of sand and silt and even peat, it may be necessary to obtain a continuous core of the soil deposits for closer examination in the laboratory. Such sampling techniques are highly specialised and require the elimination of friction between the soil sample and the walls of the sampler. A sampler which reduces side friction by the use of thin strips of metal foil placed between the soil and the tube was developed by Kjellman *et al.* (1950) and is capable of collecting a core 68 mm in diameter and up to 25 m in length. The techniques of the thin foil sampler are still in use today though rarely are such lengths of sample taken.

(viii) *Window sampler*

Window sampler tubes, about 1 m in length, are driven into the ground using the percussive or jacking methods described earlier. The sampler possesses a slot, or window, cut on one side through which the soil can be inspected and sampled.

Undisturbed samples (sands)

If care is taken it may be possible to extract a sand sample by cutting from the bottom or sides of a trial pit. In a borehole, above groundwater level, sand is damp and there is enough temporary cohesion to allow samples to be collected in sampling tubes, but below groundwater level tube sampling is not possible. Various techniques employing chemicals or temporarily freezing the groundwater have been tried, but they are expensive and not very satisfactory.

Owing to the fact that sand is easily disturbed during sampling and transportation, any tests on the soil in the undisturbed state should be carried out on the site, the usual practice being to use the results of penetration tests (see Section 7.6) instead of sampling.

7.4.2 Degree of sample disturbance

No matter how careful the technique employed, there will inevitably be some disturbance of the soil during its collection as an 'undisturbed' sample, the least disturbance occurring in samples cut from the floor or sides of a trial pit. With sample tubes, jacking is preferable to hammering although if the blows are applied in a regular pattern there is little difference between the two.

There are various measures that can be used to assess the degree of sample disturbance based on the dimensions of the sampling tube (BS EN ISO 22475-1:2019) but the most commonly used is the area ratio, C_a:

$$C_a = \frac{D_2^2 - D_1^2}{D_1^2} \times 100 \qquad\qquad (7.1)$$

where D_2 and D_1 are the external and internal diameters of the cutting shoe respectively as shown in Fig. 7.7.

It is generally agreed that, for good undisturbed 100 mm diameter samples, the area ratio should not exceed 25%, but in fact most cutting heads have area ratios about 28%. For 38 mm samples the area ratio should not exceed 20%. Thin-walled sample tubes, of any diameter, have an area ratio of about 10%.

Example 7.1: Area ratio

Determine the area ratios for the two U100 sampling systems shown in Fig. 7.7.

Solution:

(a) standard system:

$$C_a = \frac{D_2^2 - D_1^2}{D_1^2} \times 100 = \frac{119^2 - 104.5^2}{104.5^2} = 29.7\%$$

(b) plastic liner system:

$$C_a = \frac{D_2^2 - D_1^2}{D_1^2} \times 100 = \frac{124^2 - 102^2}{102^2} = 47.8\%$$

7.4.3 Categories of sampling methods and quality classes of samples

A major consideration during sampling a particular stratum is to ensure that the type of sample taken is appropriate for the tests required and that the results obtained from the testing are then relevant to the actual conditions in the field. To ensure accurate and reliable test results, the quality of the sample must therefore be appropriate for that particular test. For example, a sample of soil taken for shear strength determination by triaxial testing must be undisturbed, whereas a sample for particle size distribution can be a disturbed sample.

The Part 2's of both generations of Eurocode 7 offer guidance which is used, alongside EN ISO 22475 (the *International Standard for Geotechnical Investigation and Testing – Sampling of soil, rock and groundwater*), to identify the type of laboratory tests that can be performed from different quality classes of sample, which in turn are achievable from the different sampling methods available. At the time of writing, EN ISO 22475 is under revision and anticipated to be published in 2021. The rest of this section is based on the content of the draft version, EN ISO 22475-1:2019.

Five quality classes of samples are considered (Classes 1–5), with Class 1 being the highest quality (least disturbed, most representative of actual *in situ* conditions and appropriate for use in shear strength and compressibility testing) and Class 5 being the lowest quality (i.e. completely disturbed and only of use in identifying the sequence of layers in the ground).

Five categories of sampling (Categories A, B, C, D and E) are considered. They are related to the best obtainable laboratory quality class of soil samples as shown in Table 7.5.

The aim of each category of sampling listed in Table 7.5 is given in ISO 22475-1:2019:

- Category A sampling: to obtain samples in which structure, texture, consistency and in-situ stresses are intact;
- Category B sampling: to obtain samples in which the structure, texture and consistency are intact;

Table 7.5 Quality classes of soil samples for laboratory testing and sampling categories to be used.

Soil properties:	Quality class of sample				
	1	2	3	4	5
Unchanged soil properties:					
Soil type	*	*	*	*	*
Particle size	*	*	*	*	
Water content	*	*	*		
Density, density index, permeability	*	*			
Compressibility, shear strength, stiffness	*				
Properties that can be determined:					
Sequence of layers	*	*	*	*	*
Boundaries of strata – broad	*	*	*	*	
Boundaries of strata – fine	*	*			
Atterberg limits, particle density, organic content	*	*	*	*	
water content	*	*	*		
Density, density index, porosity, permeability	*	*			
compressibility, shear strength, stiffness	*				
Sampling categories	A	B	C	D	E

Based on ISO 22475-1:2019.

- Category C sampling: to obtain samples in which the structure and texture are intact;
- Category D sampling: to obtain samples in which the structure is intact.

Category E sampling only obtains samples where all the *in situ* soil properties have changed due to the drilling process. From these samples, only a rough indication of the soil type and strata thickness can be determined.

Different sampling equipment obtain different quality classes of samples: whether sampling by drilling, or using samplers, the quality class of obtained sample differs. Thus, the testing required should really influence the sampling category to adopt which, in turn, influences the particular sampling equipment to use. However, often the types of soils on site are not established until drilling so adjustments to the expected sampling equipment may become necessary during drilling. The possible quality classes of sample achievable in different soils, for different sampling equipment, are given in Tables 7.6 and 7.7.

Table 7.6 Sampling by drilling – sampling categories for different soil types.

Sampling method	Flushing medium used?	Suitable soil types	Sampling category
Rotary core drilling:			
Single-tube corebarrel	N	Clay, silt, fine sand	D
	Y	Clay, cemented soils, boulders	D
Hollow stem auger with corebarrel	N	Clay, silt, sand, organic soils	B (A)
Double-tube corebarrel	Y	Clay, cemented soils, boulders	C (A–B)
Triple-tube corebarrel			A
Auger drilling	N	All soils above GWT, fine soils below GWT	D (C)
Hammer driving:			
Percussion: clay cutter: cutting edge inside	N	Fine soil	B (A)
		Coarse soil	C (B)
Percussion: clay cutter: cutting edge outside	N	Gravels* and finer soils	D
Rotary percussive drilling	Y	Clay, silt, fine sand	B (A)
		Sand	D (C)
Sonic drilling	N	Fine soil	D (C)
		Coarse soil	D
Cable percussion drilling	N	Clay, silt above GWT	D (C)
		Coarse soil	E (D)

Key: B (A), C (B) – the category in brackets is only achievable in favourable conditions, else the first category applies.
*Where particle size <1/3 internal diameter of sampler.

Table 7.7 Sampling using samplers – sampling categories for different soil types.

Type of sampler	Suitable for:	Sampling category
Thin-walled open tube sampler	Soft fine soils	A
	Medium sand	C (B)
	Stiff fine soils	B (A)
Thick-walled open tube sampler	Soft to stiff fine soils with some gravel	C (B)
Thin-walled piston sampler	Fine and sensitive soils	A
	Sand above GWT	C
Thick-walled piston sampler	Fine and sensitive soils	B (A)
Large cylinder sampler (250 mm diameter)	Clay, silt	A
SPT sampler	Sand, silt, clays	D
Window sampler	Silt, clay	D (C)

Key: B (A), C (B) – the category in brackets is only achievable in favourable conditions, else the first category applies.

7.4.4   Rock sampling

The most common form of taking rock samples is through rotary coring, as described in Section 7.4.1. As with soils, five categories of sampling methods (Categories A–E) are considered, depending on the quality class (1–5) of the rock sample required. Again, as with soils, there is a 1 : 1 relationship of sampling methods to sample quality classes (i.e. method A yields quality class 1 samples, method B yields quality class 2 etc.).

Quality class 1 samples are those which are more or less completely intact from which measurements of strength, deformation, density and permeability relate directly to the *in situ* condition. Quality class 2 samples include those that comprise broken segments of core where test results from the segments would not necessarily relate directly to the state in the undisturbed rock mass. Quality class 3 samples are those where the structure and constituents of the rock are so disturbed that only a limited set of identification and chemical tests can be performed on them.

Category E sampling methods may only be used to obtain quality class 5 samples which allow only the identification of the rock type, particle size and mineralogy to be made as well as establishing the bounds of any rock strata.

The quality of the rock sample can be assessed by determining three rock-core quantification parameters: total core recovery (TCR), rock quality designation (RQD) and solid core recovery (SCR), each of which is determined as a ratio of specific lengths of the different segments of the core sample to the total length of the sample. Details are given in EN ISO 22475-1.

7.5   Groundwater measurements

It is not possible to determine accurate groundwater conditions during the boring and sampling operations, except possibly in granular soils. Therefore, the determination of the groundwater conditions is made separately by installing *open* or *closed* groundwater measuring systems into the ground. Open systems are methods that measure the water head via an observation well and open standpipe. Closed systems measure the water pressure at a certain location via a measurement device inserted into the soil.

7.5.1   Open systems

In clays and silts it takes some time for water to fill in a borehole, and the normal procedure for obtaining the groundwater level is to insert an open-ended tube, usually 50 mm in diameter and perforated at its end, into the borehole (Fig. 7.9a). A filter to prevent the inflow of soil particles is placed around the perforated end. The

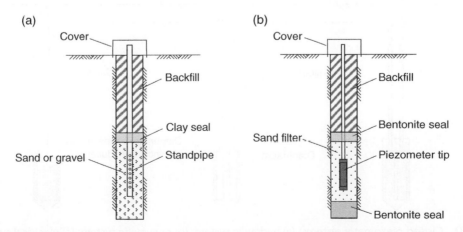

Fig. 7.9   Ground water level observation in a borehole: (a) standpipe; (b) Casagrande type standpipe piezometer.

tube is packed around with gravel and sealed in position with puddle clay and the borehole is then backfilled and sealed at the ground surface to prevent ingress of rainwater. The arrangement is simply known as a stand-pipe. Observations, using a dipmeter, should be taken for several weeks until equilibrium is achieved.

By inserting more than one tube, different strata can be cut off by puddled clay and the various water heads obtained separately. When a general water level is to be obtained, the gravel is usually extended to within a short distance of the top of the borehole and then sealed with the puddle clay.

Open-ended tubes tend to exhibit a slow response to rapid pore water pressure changes that can be caused, for example, by tidal variations or changes in foundation loadings. Casagrande type standpipe piezometers are more commonly used. They have a porous intake filter and are sealed into the soil above and below the intake end with either bentonite plugs or with a bentonite/cement grout seal (Fig. 7.9b).

7.5.2 Closed systems

For soils of medium to low permeability, open standpipes cannot be used because of the large time lag involved. When a faster response is necessary a closed piezometer system, such as a hydraulic, a pneumatic or an electrical system, is used instead of an open one. The systems are illustrated in Fig. 7.10.

Hydraulic system

In a hydraulic system, a ceramic filter unit is connected by twin, narrow nylon tubing to a pressure transducer housed in a pressure readout unit at the ground surface. The tubes and filter are filled with de-aired water. Changes in pore water pressure in the ground cause a change in the flow of water within the equipment which in turn is detected by the pressure transducer. The pressure is read directly from, and recorded by, the reader.

Pneumatic system

A pneumatic piezometer contains a flexible diaphragm housed within a protective casing and connected to a sensor at the ground surface via twin pneumatic tubes. The outer aspect of the diaphragm is in contact with the saturated soil and is pushed inwards within the housing as a result of the pore water pressure acting on it.

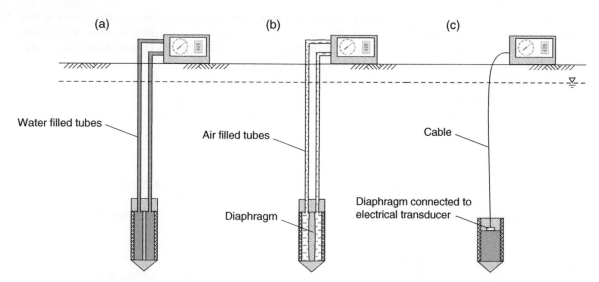

Fig. 7.10 Closed piezometer systems: (a) hydraulic system; (b) pneumatic system; (c) electrical system.

A flow of dry compressed air is passed from the sensor to the inner aspect of the diaphragm until the point that the diaphragm is forced back outwards. At this point the gas pressure is equal to the pore water pressure and this value is simply read from the sensor unit.

Electrical system

The electrical piezometer also incorporates a diaphragm within a protective housing operated within a borehole. The diaphragm is connected to a transducer and the water pressure acting on the diaphragm causes a measurable response. The signal from the transducer is transmitted to a readout device at the surface and then converted to display the pore water pressure.

The advantages of the electrical system are that (i) pressure is measured at the tip so that piezometric levels below the gauge house level can be recorded, (ii) the ancillary equipment is compact, and (iii) the time response of these instruments to pore pressure changes is fairly rapid. Disadvantages include the fact that the readings from an electric tip depend upon an initial calibration that cannot be checked once the tip has been installed, and the risk of calibration drift (especially if the tip is to be in operation for some time).

Instrumentation in geotechnical engineering is dealt with in detail by Dunnicliff (1993) and a review of piezometers within boreholes has been described by Mikkelsen and Green (2003).

7.6   Field tests in soil and rock

During a ground investigation, field tests can be conducted to provide additional ground stratification information and to obtain geotechnical parameters for the design. The tests are arranged such that the data they reveal complement the soil and rock sampling so that all the information retrieved from the ground investigation is linked.

A range of tests exist, each of which is used to gain specific information, and the following tests are recognised in EN 1997-2:2007 and ISO 22476 (the *International Standard for Geotechnical Investigation and Testing – Field testing*):

- cone penetration test;
- pressuremeter and dilatometer tests;
- standard penetration test;
- dynamic probing;
- weight sounding test;
- field vane test;
- flat dilatometer test;
- plate loading test.

Some of these tests are more commonly used than others. Some of the most common in the UK are described below.

7.6.1   Cone penetration test (CPT)

This test, sometimes referred to as the Dutch cone penetrometer as it was originally developed in The Netherlands, is described fully in EN ISO 22476-1:2012 and involves a cone penetrometer at the end of a series of stiff cylindrical rods being pushed vertically into the ground at a constant rate of penetration. A record of the resistance to the movement of the cone against depth is taken so that changes in soil strata and other soil

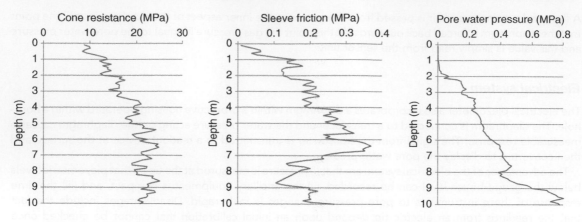

Fig. 7.11   Plots of cone resistance, sleeve friction and pore pressure with depth from CPTU test.

strength considerations can be identified. The movement of the cone is resisted by both the ground ahead of the cone as it advances through the soil, plus the friction acting on the side of the cone as it is pushed into the ground.

The cone has an apex angle of 60° and diameter of 35.7 mm, giving an end area of 1000 mm². The cone is forced downwards at a steady rate (15–25 mm/s) through the soil by means of a load from a vehicle-mounted hydraulic jack transmitted to rigid push rods.

The penetrometer is a sophisticated device and contains sensors for measuring the force resisting the cone, the side friction placed on the unit (known as sleeve friction) and, where required, the pore pressure acting on the cone penetrometer. Where pore water pressures are recorded, the test is known as piezocone penetration (CPTU) testing. To ensure the instrument is travelling vertically into the ground, an internal inclinometer is also included. This device records the angle of the instrument from the vertical.

The results obtained from the CPT/CPTU tests are recorded as plots of the measured value (cone resistance, sleeve friction, pore pressure) versus depth. These plots can then be used to visually and rapidly assess the strength of the ground profile, and also to provide data for foundation settlement analysis. Example plots are shown in Fig. 7.11.

A full description of cone penetration testing and its application in geotechnical and geo-environmental engineering is given by Lunne *et al.* (1997).

7.6.2   Standard penetration test (SPT)

This test is generally used to determine the bearing capacity of sands or gravels and is conducted with a split spoon sampler (a sample tube which can be split open longitudinally after sampling) with internal and external diameters of 35 and 50 mm respectively (Fig. 7.12). A full guide on the methods and use of the SPT is given by Clayton (1995) and the test specification is given in EN ISO 22476-3:2011.

The sampler, connected to a sequence of drive rods, is lowered down the borehole until it rests on the layer of cohesionless soil to be tested. It is then driven into the soil for a length of 450 mm by means of a 63.5 kg hammer free-falling 760 mm onto an *anvil* fixed at the top of the rods, for each blow. The number of blows required to drive the last 300 mm is recorded and this figure is designated as the N-value or the penetration resistance of the soil layer. The first 150 mm of driving is ignored because of possible loose soil in the bottom of the borehole from the boring operations. After the tube has been removed from the borehole it can be opened and its contents examined.

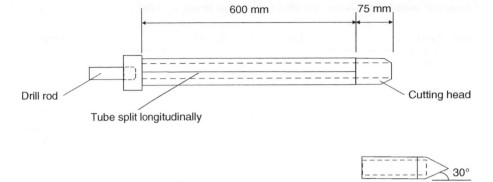

Fig. 7.12 Standard penetration test sampler.

In gravelly sand, damage can occur to the cutting head of the sampler and a 60° solid cone can be fitted in its place. In such a case the test is recorded as SPT(C). The N-value derived from such soils appears to be of the same order as that obtained when the cutting head is used in finer soils.

Correction factors to the measured N-value

The N-value observed from the test is affected by different features of the testing procedure and ground conditions. To take these into consideration, a number of correction factors can be applied to achieve a more appropriate N-value. The most significant factors address energy losses delivered by the hammer assembly and the effect of overburden pressure acting on the soil under test. EN ISO 22476-3:2011, Annex A offers the following corrections:

(i) *Energy delivered to the drive rods*

An energy ratio, E_r, measuring the ratio of the energy applied to the driving rods to the theoretical energy available from the hammer, can be used to assess the energy loss in the hammer system. The N-value is adjusted to a reference energy ratio of 60% through the following expression:

$$N_{60} = \frac{E_r}{60} N \tag{7.2}$$

where N_{60} denotes the N-value corrected for energy losses for a system operating with an energy ratio E_r.

(ii) *Effect of overburden pressure in sands*

An important feature of the standard penetration test is the influence of the effective overburden pressure on the N-value. Sand can exhibit different N-values at different depths even though its density index is constant. The effect of the overburden pressure can be taken into account by combining the N-value by the relevant Correction factor, C_N listed in Table 7.8.

The N_{60} value can now be corrected against a normalised effective vertical stress $\sigma'_v = 100 \, \text{kPa}$:

$$(N_1)_{60} = \frac{E_r \times N \times C_N}{60} \tag{7.3}$$

Table 7.8   Correction factors for overburden effective vertical stress, σ_v' (kPa).

Type of consolidation	Density index, I_D (%)	Correction factor, C_N
Normally consolidated	40–60	$\dfrac{200}{100 + \sigma_v'}$
	60–80	$\dfrac{300}{200 + \sigma_v'}$
Over consolidated	–	$\dfrac{170}{70 + \sigma_v'}$

Table 7.9   Correction factors for rod length in sands.

Rod length (m)	Correction factor λ
>10	1.0
6–10	0.95
4–6	0.85
3–4	0.75

where $(N_1)_{60}$ denotes the N-value corrected for energy losses and normalised for effective vertical overburden stress.

(iii) *Energy losses due to the length of the rods*

Where rods of length less than 10 m are used, a correction can be applied to the blow count for sands to allow for energy losses. The correction factors are given Table 7.9.

The $(N_1)_{60}$ expression can now be extended to allow for the energy losses in the rods:

$$(N_1)_{60} = \frac{E_r \times N \times C_N \times \lambda}{60} \tag{7.4}$$

Example 7.2: Standard Penetration Test

A silty sand of density index, $I_D = 45\%$ was subjected to a standard penetration test at a depth of 3 m. Groundwater level occurred at a depth of 1.5 m below the surface of the soil which was saturated throughout and had a unit weight of 19.3 kN/m³. The average N count was 15 and the anvil was fixed to the rods at 1 m above the ground surface.

During calibration of the test equipment, the energy applied to the top of the driving rods was measured as 346 J.

Determine the $(N_1)_{60}$ value for the soil.

Solution:

Theoretical energy of hammer $= m \times g \times h = 63.5 \times 9.81 \times 0.76 = 473\,J$

Energy ratio, $E_r = \dfrac{346}{473} = 73\%$

Effective overburden pressure $= 3 \times 19.3 - 1.5 \times 9.81 = 43\,\text{kPa}$

From Table 7.8, correction factor, $C_N = \dfrac{200}{100 + \sigma'_v} = \dfrac{200}{100 + 43} = 1.40$

Length of rods $= 3 + 1 = 4\,\text{m}$; from Table 7.9, $\lambda = 0.75$

$$(N_1)_{60} = \frac{E_r \times N \times C_N \times \lambda}{60} = \frac{73 \times 15 \times 1.40 \times 0.75}{60} = 19$$

Table 7.10   Correlation between Normalised blow count $(N_1)_{60}$ and density index I_D.

State of density	$(N_1)_{60}$	Density Index, I_D (%)
Very loose	0–3	0–15
Loose	3–8	15–35
Medium	8–25	35–65
Dense	25–42	65–85
Very dense	42–58	85–100

Correlations between blow count and density index

Terzaghi and Peck (1948) evolved a qualitative relationship between the density index of normally consolidated sand and the N-value and, later, Gibbs and Holtz (1957) put figures to this relationship. More recent work has adjusted the figures to the normalised blow count $(N_1)_{60}$ and these are published in EN 1997-2:2007 (BSI, 2007a), Annex F and reproduced in Table 7.10.

7.6.3   Dynamic probing test

In this test, a cone of steel of apex angle 90° and 35 mm diameter is advanced into the ground from a free-falling hammer, in a rig assembly, striking extension rods at a driving rate of 15–30 blows per minute. The height of free-fall is 500 mm and the hammer can have mass 10, 30, 50 or 63.5 kg depending on whether a *light, medium, heavy* or *super heavy* test is being undertaken. The number of blows required to advance the cone for multiples of 100 mm is recorded. This is referred to as the N_{10} *value* for that particular 100 mm depth range. For the super heavy test, the number of blows required to advance 200 mm is recorded instead – the N_{20} *value*. If required, after each 1.0 m of penetration, the rods are rotated 1½ turns and the torque measured using a torque wrench.

The test is run from the ground surface and is predominantly used for the rapid establishment of the soil profile through a recorded plot of number of blows versus depth. If required, correction factors to the N_{10} values may be applied to allow for energy losses in the hammer.

The N_{10} value can be used to give an estimate of the density index, I_D at that depth for granular soils, depending on the uniformity coefficient, C_u of the soil and the test performed – light (DPL) or heavy (DPH):

$(C_U < 3)$ above GWL:

$I_D = 0.15 + 0.260 \log (N_{10}) (\text{DPL})$
$I_D = 0.10 + 0.435 \log (N_{10}) (\text{DPH})$

Table 7.11   Estimates of ϕ' from I_D and C_u.

Soil type	Grading	Range of I_D		ϕ' (°)
Sands and sandy gravel	Poorly graded ($C_u < 6$)	15–35	Loose	30
		35–65	Medium dense	32.5
		>65	Dense	35
Sands, sandy gravel and gravel	Medium graded ($6 \leq C_u \leq 15$)	15–35	Loose	30
		35–65	Medium dense	34
		>65	Dense	38

Data from EN 1997-2:2007 (BSI, 2007a).

($C_U < 3$) below GWL:

$$I_D = 0.21 + 0.230 \log (N_{10})(DPL)$$
$$I_D = 0.23 + 0.380 \log (N_{10})(DPH)$$

($C_U > 6$) above GWL

$$I_D = -0.14 + 0.550 \log (N_{10})(DPH)$$

These relationships apply for the range $3 \leq N_{10} \leq 50$.

From the density index, values of the effective angle of shearing resistance, ϕ' can be estimated (Table 7.11).

More details on the test procedure and the interpretation of results are given in EN ISO 22476-2:2011 and in EN 1997-2:2007. A variation of the test, used at the base of a borehole and known as *borehole dynamic probing*, is described in EN ISO 22476-14:2020.

Example 7.3: Dynamic probing test

The results of a heavy dynamic probing test through a deposit of gravelly sand are shown in Fig. 7.13. A sample of the sand was taken for particle size distribution and the results are given in Example 1.2. The groundwater level was 3.0 m.

Estimate the angle of shearing resistance, ϕ' of the sand at a depth of 2.0 m.

Solution:

From Example 1.2, $C_u = 8.3$.
From Fig. 7.13, N_{10} @ 2.0 m = 20
$I_D = -0.14 + 0.55 \log(N_{10}) = -0.14 + 0.55 \times \log(20) = 0.57 = 57\%$
From Table 7.11, ϕ' is estimated to be 34°

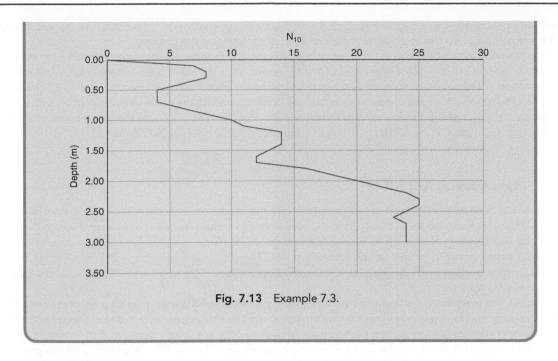

Fig. 7.13   Example 7.3.

7.6.4   Pressuremeter test

This test utilises a cylindrical probe containing an inflatable, flexible membrane operating within a borehole. The probe is lowered into the borehole to the required depth and the membrane is expanded under pressure so that it fills the void between the probe and the borehole sides (Fig. 7.14). Measurements of pressure and

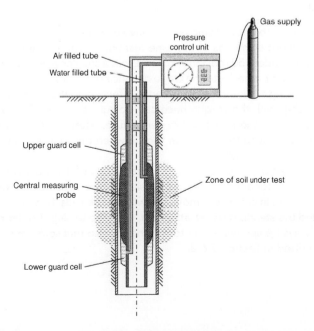

Fig. 7.14   Pressuremeter test.

membrane volumetric expansion are recorded until the maximum expansion is reached. The test results can then be used to derive the strength and deformation parameters of the ground.

The equipment described above is known as the Ménard pressuremeter, named after Louis Ménard who developed the test in the 1950s. Other types of pressuremeter exist, including *self-boring* and *full-displacement* pressuremeters. The various pressuremeter test procedures are described in different parts of EN ISO 22476.

The results from the Ménard pressuremeter test are used to establish the pressuremeter modulus, (E_M), and limit pressure, (p_{LM}) which can be used to determine the bearing resistance and the settlements of spread foundations. Examples of methods for doing this are given in EN 1997-2:2007, Annex E.

7.6.5   Plate loading test (PLT)

Plate loading tests can be carried out at ground level or at the base of an excavation. Their purpose is to retrieve data which can be used to calculate the settlement of spread foundations in a sand and in the estimation of the bearing resistance in foundation design work. They can also be used to determine the undrained shear strength properties of a cohesive soil. The data obtained from a PLT is a record of the load-settlement behaviour of the plate as it is loaded in successive increments and allowed to settle before the next load is applied.

As stated, the loading is applied in increments (usually one-fifth of the proposed bearing pressure) and is increased up to two or three times the proposed loading. Additional increments are only added when there is no continuing detectable settlement of the plate. Measurements are usually taken to 0.01 mm, and where there is no definite failure point, the ultimate bearing capacity (see Chapter 10) is assumed to be the pressure causing a settlement equal to 5–10 or 10 – 20% of the plate width, depending on soil strength and density.

The test procedure is described in BS1377-9 and EN ISO 22476-13, and examples on the determination of undrained shear strength and the settlement of spread foundations in a sand are given in EN 1997-2:2007, Annex K.

7.6.6   Field vane test

In soft sensitive clays it is difficult to obtain samples that have only a slight degree of disturbance, and *in situ* shear tests are usually carried out by means of the vane test (Fig. 7.15). The apparatus consists of a 75 mm diameter vane, with four small blades 150 mm long. For stiff soils a smaller vane, 50 mm diameter and 100 mm high, may be used. The vanes are pushed into the clay a distance of not less than three times the diameter of the borehole ahead of the boring to eliminate disturbance effects, and the undrained strength of the clay is obtained from the relationship with the torque necessary to turn the vane. The rate of turning the rods, throughout the test, is kept within the range 6–12° per minute. After the soil has sheared, its remoulded strength can be determined by noting the minimum torque when the vane is rotated rapidly. The test procedure is described in EN ISO 22476-9:2020.

Figure 7.15 indicates that the torque head is mounted at the top of the rods. This is standard practice for most site investigation work but, for deep bores, it is possible to use apparatus in which the torque motor is mounted down near to the vane, in order to remove the whip in the rods. Because of this development, the vane has largely superseded the standard penetration test, for deep testing. The latter test has the disadvantage that the load must always be applied at the top of the rods so that some of the energy from a blow is dissipated in them as described in Section 7.6.2.

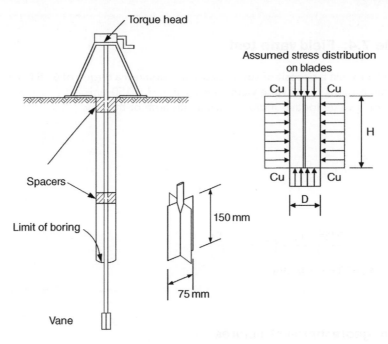

Fig. 7.15 Field vane test.

The actual stress distribution generated by a cylinder of soil being rotated by the blades of a vane which has been either jacked or hammered into it is a matter for conjecture. The most common approach used adopts the simplifying assumption that the resistance to shear is equivalent to a uniform shear stress, equal to the undrained strength of the soil, c_u, acting on both the perimeter and the ends of the cylinder (see Fig. 7.15).

For equilibrium, the applied torque, T is equal to the moment of resistance of the vane blades. The torque due to the ends can be obtained by considering an elemental annulus and integrating over the whole area:

$$\text{End torque} = 2 \times c_u \int_0^r 2\pi r \, dr \times r = 2c_u \left[2\pi \frac{r^3}{3} \right]_0^{D/2} = c_u \frac{\pi D^3}{6}$$

$$\text{Side torque} = c_u \pi DH \times \frac{D}{2} = c_u \frac{\pi D^2 H}{2}$$

$$T = c_u \frac{\pi D^2 H}{2} + c_u \frac{\pi D^3}{6} = c_u \frac{\pi D^2 H}{2} \left(1 + \frac{D}{3H} \right) \tag{7.5}$$

where

D = width of vane
H = height of vane.

Example 7.4: Field vane test

A vane, used to test a deposit of soft alluvial clay, required a torque of 67.5 Nm to fail the soil. The dimensions of the vane were: D = 75 mm; H = 150 mm.
   Determine a value for the undrained shear strength of the clay.

Solution:

$$T = c_u \frac{\pi D^2 H}{2} \left(1 + \frac{D}{3H}\right)$$

i.e.

$$67.5 = c_u \frac{\pi 0.075^2 \times 0.15}{2} \left(1 + \frac{0.075}{0.45}\right)$$

$$\Rightarrow c_u = 43\,653\ \text{Pa} = 44\ \text{kPa}$$

7.6.7   Testing of geotechnical structures

In addition to the tests described before, testing is often performed on geotechnical structures to assess their performance against design requirements. The testing of geotechnical structures, such as reinforced soil, pile foundations and ground anchorage systems, involves sophisticated and specialised processes performed by specialist contractors. Specifications for the testing processes are covered in EN ISO 22477:2018.

7.7   Geotechnical reports

All of the information retrieved during a site investigation must be compiled in written reports and submitted to the client. The reports are the end product of all the investigation work and provide the client and their representative consultants with the information relevant to enable an efficient foundation design and construction plan to be evolved.
   Eurocode 7 stipulates the requirements of two reports:

- *Geotechnical Design Report* (GDR), described in Eurocode 7 Part 1
- *Ground Investigation Report* (GIR), described in Eurocode 7 Part 2

In accordance with the design process set out in the second generation of Eurocode 7, it is apparent that the information retrieved throughout the site investigation develops both the ground model and the geotechnical model. The ground model is the output of the ground investigation report.
   The ground investigation report is the final product of the exploration programme encompassing all of the subject areas described throughout this chapter. It comprises an account of the desk study, the series of laboratory testing reports, the field investigation, sampling and measurement reports, the field testing reports and any other relevant reports, together with an interpretation of the ground conditions across the site. The report will also include any limitations and constraints of the various test results so that the designer can assess the relevance of the test results to the geotechnical design. The report can also include any derived values of geotechnical properties (see Section 6.12.2).

The geotechnical design report contains the GIR along with the results from the calculations performed to verify the safety and serviceability during the geotechnical design (see Chapter 6). Along with the calculations, relevant drawings and foundation design recommendations are included, as is a plan of supervision and monitoring for the site.

The GIR is generally prepared in sections, and typically will include the following:

Preamble

This introductory section consists of a brief summary which gives the location of the site, the date of the investigation and name of the client, the types of bores put down and the equipment used.

Description of site

Here a general description of the site is given: whether it is an open field or a redevelopment of a site where old foundations, cellars and walls, etc., remain. Some mention is made of the general geology of the area, whether there are old mineral workings at depth and, if so, whether the report has considered their possible effects or not. A map, showing the site location and the positions of any investigation points, is usually included in the report.

Description of subsoil conditions encountered

This section should consist of a short, and readable, description of the general subsoil conditions over the site with reference to the borehole logs. The relevance and significance of any *in situ* testing carried out is also included.

Vertical sections (soil profiles) are generally prepared, showing to scale the sequence and thickness of the strata. Design engineers are mainly interested in the materials below the subsoil, and with stratified sedimentary deposits conditions may be more or less homogeneous. Glacial clay deposits can also be homogeneous although unstratified, but they often have an erratic structure in which pockets of different soils are scattered through the main deposit and make it difficult to obtain an average value for the deposit's characteristics. Furthermore, the clay itself may vary considerably, and at certain levels it can even decrease in strength with increasing depth.

Besides the primary structure of stratification, many clays contain a secondary structure of hair cracks, joints and *slickensides*. The cracks (often referred to as macroscopic fissures) and joints generally occurred with shrinkage when at some stage in its development the deposit was exposed to the atmosphere and dried out. Slickensides are smooth, highly polished surfaces probably caused by movement along the joints. If the effect of these fissures is ignored in the testing programme, the strength characteristics obtained may bear little relationship to the properties of the clay mass.

With the application of a foundation load there is little chance of the fissures opening up, but in cuttings (due to the expansion caused by stress relief) some fissures may open and allow the ingress of rainwater which will eventually soften the upper region of the deposit and lead to local slips. Fissures are more prevalent in overconsolidated clays, where stress relief occurs, than in normally consolidated clays, but any evidence of fissuring should be reported in the boring record.

Borehole logs

A borehole log is a list of all the materials encountered during the boring. A log is best shown in sectional form so that the depths at which the various materials were met can be easily seen. A typical borehole log is shown in Fig. 7.16. It should include a note of all the information that was found: groundwater conditions, numbers and types of samples taken, list of *in situ* tests, time taken by boring, etc.

Record of borehole 109

Ref. no.   886
Ground level   3.40 m

Dia. of boring   150 mm to 13.70 m LCP
76 mm to 17.40 m DD

Progress			Sample/test		Strata			
Hole	Casing	Water	Depth	Type	Legend	Depth	Level	Description
28/8/98			0.30	D1		0.30	3.10	Brown sandy topsoil
			1.88–2.13	S(29)				Medium dense to dense red brown silty sand and fine gravel
				D2				
		Met at	0.30–2.18	B1				
		2.00	2.20	W				
			2.50	D3		2.6	0.80	
			3.15–3.45	C(6)				Medium dense brown silty sand with clayey layers, containing occasional gravel
			3.45–3.75	(17)				
				D4				
			2.70–3.90	B2				
			4.10–4.55	U-/120				
4.56	4.56	4.00	4.10	D5		4.56	−1.16	
29/8/98		1.50						Stiff light brown laminated silty clay, with layers of sand
			5.50–5.95	U1/80				
			6.00	D6				
			6.90	D7				
						7.10	−3.70	
			7.45–7.75	S(29)				Medium dense becoming dense brown sand
				D8				
			8.45–8.75	S(22)				
				D9				
			9.45–9.75	S(26)				
				D10				
			10.45–10.75	S(46)				
10.73	10.73	8.00		D11				
30/8/98		1.50	11.20	D12		11.20	−7.80	
			11.30–11.75	U2/120				Compact brown silty sand with layers of silty clay
			11.80	D13		11.80	−8.40	
								Compact brown sand and gravel
13.40	12.90	9.90	12.90–13.35	U3/150		12.80	−9.40	
31/8/98		8.00	13.35	D14				Dense grey-brown clayey sand with occasional gravel
13.70	13.44	13.00	13.70	D15		13.70	−10.30	
4/9/98		1.50	13.70–15.10	1.40•				Hard mottled red-brown, grey and green coarse grained basaltic tuff
		Water	15.10–16.34	1.24				
		Flush						
16.34	13.44	−0.36				16.13	−12.73	
5/9/98		1.04	16.34–17.40	1.06				Soft and medium hard weathered mottled red-brown, grey-green basaltic tuff
						17.19	−13.79	
								Hard mottled grey-green basaltic tuff
17.40	18.41	−0.36				17.40	−14.00	

Remarks: Penetration test continued beyond normal drive from 3.45 m. 40 mm diameter perforated standpipe 18.00 m long inserted, surrounded by gravel filter with bentonite seal and screw cap at surface.

Key:

D	Disturbed sample	S(30)	Standard penetration test	U1/70	Undisturbed sample 100 mm dia
B	Bulk sample	C(27)	Cone penetration test	/70	Blows to drive sample 450 mm
W	Water sample	(27)	No blows for 300 mm pentn.	U-/70	U/d sample - no recovery
•	Core recovery	V	In situ vane shear test	LCP	Light cable percussion
				DD	Rotary diamond drilled

Fig. 7.16   Borehole log.

Laboratory and in situ tests results

This is a list of the tests carried out together with a set of laboratory sheets showing all tests results, e.g. particle size distribution curves, liquid limit plots, Mohr circle plots, etc.

Evaluation of geotechnical information

It is in this section that firm recommendations as to possible foundation types and modes of construction should be given. Unless specified otherwise, it is the responsibility of the architect or consultant to decide on the actual structure and the construction. For this reason, the GIR should endeavour to list possible alternatives: whether strip foundations are possible, if piling is a sensible proposition, etc. In the accompanying GDR, design calculations of each type of foundation are presented. If correlations were used to derive geotechnical parameters or coefficients, the correlations and their applicability are also recorded.

If the investigation has been limited by specification or finance and the ground interpretation has been based on limited information, it is important that this is recorded.

Ground model

The ground model is the culmination of all site investigation information and may be simply the GIR itself, or it may be a sophisticated three-dimensional software model, or anything in between. The geotechnical design model is developed from the ground model as the design evolves through calculation, analysis and further investigation. Thus, the nature of the ground model required by the design team will depend on the needs of the designers and the nature of the project. The ground model may continue to be developed after the site investigation is complete and the GIR published.

Exercises

Exercise 7.1

A thin-walled sample tube is used to recover reasonably undisturbed samples of soft clay. The tube has an internal diameter of 75 mm and the thickness of the sample tube is 1.6 mm.
   Determine the area ratio of the tube.

Answer   8.7%

Exercise 7.2

The internal and external diameter for a standard split-spoon sampler are 35.05 and 50.8 mm respectively while the internal and external diameter for a thin-walled tube sampler are 47.63 and 50.8 mm respectively. Determine the degree of disturbance for both sampling methods.
   Which one will provide a better quality of sample for testing?

Answer   110%; 13.8%
   The sample obtained by the thin-walled sampler will provide a better quality of sample.

Exercise 7.3

A sample of silty sand was tested in the laboratory to determine its maximum, minimum and natural void ratio. The results were:

$$e_{max} = 0.770$$
$$e_{min} = 0.332$$
$$e = 0.551$$

The soil deposit from where the sample was taken was saturated throughout with unit weight = 19.6 kN/m^3. The GWL was encountered at a depth of 2.2 m below the ground surface. The soil was subjected to an SPT test at a depth of 4 m with the N count = 22. The anvil of the equipment was 1 m above ground surface.

If, during calibration of the test equipment, the energy applied to the top of the driving rods was measured as 362 J, determine the $(N_1)_{60}$ value for the soil.

Answer 30

Exercise 7.4

From a particle size distribution test on a sandy gravel which was taken at a depth of 3 m from a site where the ground water table is at the ground surface, the following was obtained:

$$D_{10} = 1.4; \quad D_{60} = 4.1$$

A light dynamic probing test was performed at the site and the N_{10} value at 3 m was measured as 23. Estimate the angle of shearing resistance, ϕ' of the gravel at this same depth.

Answer $\phi' \approx 32.5°$

Exercise 7.5

A soft clay was tested at the base of a borehole using the vane test. Determine the undrained shear strength of the soil where the measured torque was 85 Nm, if the vane diameter was 75 mm and height was 150 mm.

Answer 55 kPa

Part III

Advanced Soil Mechanics and Applications

Chapter 8

Lateral Earth Pressure

Learning objectives:

By the end of this chapter, you will have been introduced to:

- lateral stresses in the ground leading to lateral pressures acting on the back of soil retaining structures;
- earth pressure at rest, and the active and passive states of earth pressure;
- Rankine's theory of earth pressure for granular and cohesive soils;
- Coulomb's wedge method of analysis for granular and cohesive soils;
- the effect of surface surcharges and compaction pressures on retaining walls;
- drainage systems in retaining wall backfill.

8.1 Earth pressure at rest

Consider an element of soil at some depth, z below ground surface. We saw in Chapter 3 that the vertical total stress, σ_1 acting at that point is caused by the total weight acting above. In the case of a homogenous soil with no surface surcharge, σ_1 is due to the weight of the material above ($=\gamma z$) as shown in Fig. 8.1.

A lateral stress, σ_3 also acts at the point and is equal to the vertical stress multiplied by the coefficient of earth pressure, K. In this case, the coefficient is the *coefficient of earth pressure at rest*, denoted by K_0. It was shown experimentally that for granular soils and normally consolidated clays, $K_0 \approx 1 - \sin \phi'$ (Jaky, 1944). Eurocode 7 however, recommends that the effect of any overconsolidation and of a sloping surface be considered:

$$K_0 = (1 - \sin\phi') \times \sqrt{R_o} \times (1 + \sin\beta) \tag{8.1}$$

where R_o = overconsolidation ratio; β = angle of inclination of ground surface to the horizontal.

Smith's Elements of Soil Mechanics, 10th Edition. Ian Smith.
© 2021 John Wiley & Sons Ltd. Published 2021 by John Wiley & Sons Ltd.
Companion website: www.wiley.com/go/smith/soilmechanics10e

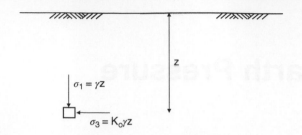

Fig. 8.1 Vertical and lateral stresses.

8.2 Active and passive earth pressure

Let us consider the simple case of a retaining wall with a vertical back (details of wall design and construction are given in Chapter 9) supporting a cohesionless soil with a horizontal surface (Fig. 8.2). Let the angle of shearing resistance of the soil be ϕ' and let its unit weight, γ, be of a constant value. Then the vertical stress, σ_1 acting at a point at depth below the ground surface will be equal to $\gamma \times$ the depth:

Behind the wall, at depth h, the vertical stress, $\sigma_1 = \gamma h$.

In front of the wall, at depth d, the vertical stress, $\sigma_1 = \gamma d$.

If the wall is allowed to yield (i.e. move forward slightly), it is clear that the soil behind the wall will experience a reduction in lateral stress (it will expand slightly), whilst the soil in front will compress slightly and thus experience an increase in lateral stress. This shows that, in addition to the *at rest* state, soil can exist in two states. In expansion, the soil is in an *active* state, and in compression, the soil is in a *passive* state.

We can say therefore (in this example), that the soil behind the wall is in an active state and thus, the pressure that the soil is exerting on the wall is *active pressure*. By contrast, the soil in front of the wall is in a passive state and so the pressure that the soil is exerting on the wall is *passive pressure*.

As with the *at rest* condition described in Section 8.1, the lateral earth pressure acting at some depth is equal to the vertical stress (pressure) multiplied by the appropriate coefficient of earth pressure. We can now introduce the *coefficient of active earth pressure*, K_a and the *coefficient of passive earth pressure*, K_p.

Behind the wall (active), $\sigma_3 = K_a \gamma h$

In front of the wall (passive), $\sigma_3 = K_p \gamma d$

The active earth pressure is the minimum value of lateral pressure, σ_3 the soil can withstand. The passive earth pressure is the maximum value.

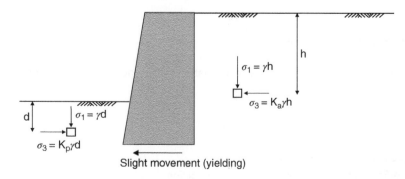

Slight movement (yielding)

Fig. 8.2 Active and passive states.

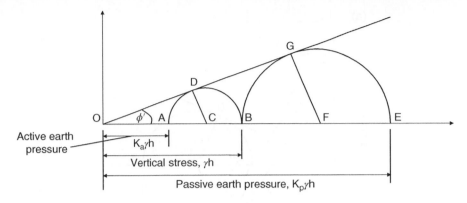

Fig. 8.3   Active and passive earth pressures.

Consider now a single element of soil at depth, h below the ground surface. The vertical stress, σ_1 at this depth is equal to γh. Referring to the previous paragraph, the two extreme values of σ_3 can be obtained from the Mohr circle diagram for the soil, as shown in Fig. 8.3. The vertical stress, σ_1 is represented by point B.

From Fig. 8.3, it is seen that the lateral pressure can reduce to a minimum value at which the stress circle is tangential to the strength envelope of the soil. This minimum value (point A) is the active earth pressure and equals $K_a \gamma h$. The lateral pressure can rise to a maximum value (with the stress circle again tangential to the strength envelope) known as *the passive earth pressure*, which equals $K_p \gamma h$ (point E). From the figure, it is clear that when considering active pressure, the vertical pressure due to the soil weight, γh, is a *major* principal stress and that when considering passive pressure, the vertical pressure due to the soil weight, γh, is a *minor* principal stress.

The two major theories to estimate active and passive pressure values are those by Rankine (1857) and Coulomb (1776). Both theories are very much in use today and both are described below.

8.3   Rankine's theory: granular soils, active earth pressure

8.3.1   Horizontal soil surface

Imagine a smooth (i.e. no friction exists between wall and soil), vertical retaining wall holding back a granular, cohesionless soil with an angle of shearing resistance ϕ'. The top of the soil is horizontal and level with the top of the wall. Consider a point in the soil at a depth, h, below the top of the wall, (Fig. 8.2) and assume that the wall has yielded sufficiently to satisfy active earth pressure conditions.

From the Mohr circle diagram (Fig. 8.3), it is seen that:

$$K_a = \frac{K_a \gamma h}{\gamma h} = \frac{OA}{OB} = \frac{OC - AC}{OC + CB} = \frac{OC - DC}{OC + DC} = \frac{1 - \frac{DC}{OC}}{1 + \frac{DC}{OC}} = \frac{1 - \sin \phi'}{1 + \sin \phi'}$$

It can be shown by trigonometry that:

$$\frac{1 - \sin \phi'}{1 + \sin \phi'} = \tan^2 \left(45° - \frac{\phi'}{2} \right)$$

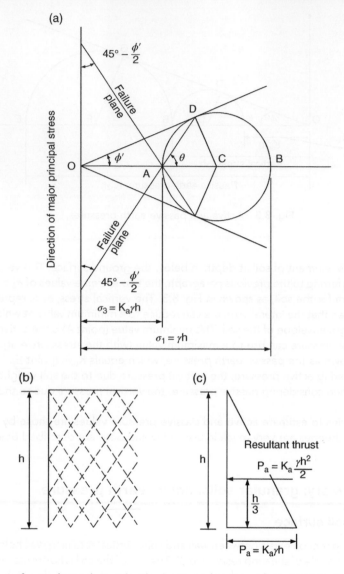

Fig. 8.4   Active pressure for a cohesionless soil with a horizontal upper surface. (a) Mohr diagram. (b) Failure plane network. (c) Pressure distribution on back of wall.

hence

$$K_a = \frac{1 - \sin\phi'}{1 + \sin\phi'} = \tan^2\left(45° - \frac{\phi'}{2}\right)$$

(8.2)

The Mohr circle diagram can be extended to identify the direction of the major principal stress, using the geometry indicated in Fig. 8.4a. The angle that the failure plane makes with the back of the wall is clear from the figure and the network of shear planes formed behind the wall is illustrated in Fig. 8.4b.

The lateral pressure acting on the wall at any depth, σ'_3, is equal to $K_a\sigma'_1$. Since the vertical effective stress $\sigma'_1 = \gamma h$ and K_a is a constant, the lateral pressure increases linearly with depth (Fig. 8.4c). This lateral pressure is of course the active pressure. This pressure is given the symbol p_a, and is defined:

$$p_a = K_a\gamma h$$

The magnitude of the resultant thrust, P_a, acting on the back of the wall is the area of the pressure distribution diagram. This force is a line load that acts through the centroid of the pressure distribution. In the case of a triangular distribution, the thrust acts at a third of the height of the triangle.

8.3.2 Sloping soil surface

The problem of the ground surface behind the wall sloping at an angle, β to the horizontal is illustrated in Fig. 8.5. The evaluation of K_a may be carried out in a similar manner to the previous case, but the vertical pressure will no longer be a principal stress. The pressure on the wall is assumed to act parallel to the surface of the soil, i.e. at angle β to the horizontal.

The active pressure, p_a, is still given by the expression:

$$p_a = K_a\gamma h$$

but in this case, K_a is defined:

$$K_a = \cos\beta\,\frac{\cos\beta - \sqrt{\cos^2\beta - \cos^2\phi'}}{\cos\beta + \sqrt{\cos^2\beta - \cos^2\phi'}} \tag{8.3}$$

For both horizontal and sloping surfaces, the total active thrust, P_a acting on the back of the wall is the area of the pressure distribution diagram. For triangular pressure distributions, as in the example of Fig. 8.5, the active thrust is given by:

$$P_a = \frac{1}{2}K_a\gamma h^2 \tag{8.4}$$

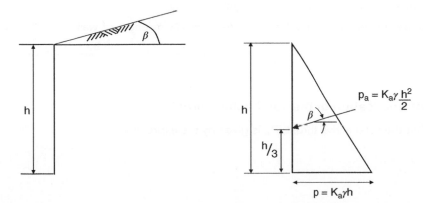

Fig. 8.5 Active pressure for a cohesionless soil with its surface sloping upwards at angle β to the horizontal.

Example 8.1: Rankine active thrust

A vertical retaining wall, 5 m high, retains a soil with a horizontal surface level with the top of the wall with the following properties: $\phi' = 35°$; $\gamma = 19$ kN/m^3.

(a) Determine K_a using Rankine's theory and establish the active thrust, P_a placed on the back of the wall.
(b) Determine the increase in horizontal thrust placed on the back of the wall if the soil surface slopes at an angle of 35° to the horizontal.

Solution:

(a) Soil surface horizontal:

$$K_a = \frac{1 - \sin 35°}{1 + \sin 35°} = 0.271$$

Maximum $p_a = 0.271 \times 19 \times 5 = 25.75$ kPa

Thrust = area of pressure distribution diagram

$$= \frac{25.75 \times 5}{2} = 64 \text{ kN}$$

(b) Soil surface sloping at 35°:
In this case $\beta = \phi'$. When this is the case, the expression for K_a reduces to $K_a = \cos \phi'$. Hence,

$$K_a = \cos 35° = 0.82$$

$$\text{Thrust} = K_a \gamma \frac{h^2}{2} = 0.82 \times 19 \times \frac{5^2}{2} = 195 \text{ kN}$$

This thrust is assumed to be parallel to the slope, i.e. at 35° to the horizontal.

Horizontal component = 195 × cos 35° = 160 kN
Increase in horizontal thrust = (160 − 64) = 96 kN/m length of wall

8.3.3   Point of application of the total active thrust

We have seen that the total active thrust, P_a, is given by the expression:

$$P_a = \frac{1}{2} K_a \gamma h^2$$

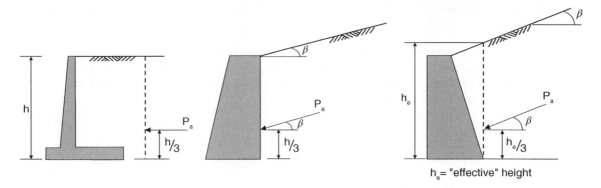

Fig. 8.6 Point of application of total active thrust (Rankine theory).

where K_a is the respective value of the coefficient of active earth pressure, h = height of wall, and γ = unit weight of retained soil.

The position of the centre of pressure on the back of the wall, i.e. the point of application of P_a, is largely indeterminate. Locations suitable for design purposes are shown in Fig. 8.6, and are based on the Rankine theory (with its assumption of a triangular distribution of pressure).

Example 8.2: Rankine active thrust; more than one soil

Details of the soil retained behind a smooth wall are given in Fig. 8.7a. Draw the diagram of the pressure distribution on the back of the wall and determine the total horizontal active thrust acting on the back of the wall using the Rankine theory.

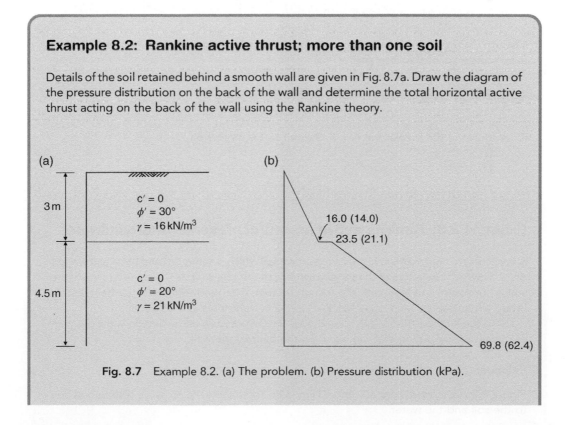

Fig. 8.7 Example 8.2. (a) The problem. (b) Pressure distribution (kPa).

Solution:

For this kind of problem, the values of lateral pressure at the salient depths are determined. The salient depths are those where a change in ground conditions occur. Since K_a and γ are constant for a particular soil, the active pressure increases linearly with depth. Active pressure at the top of the wall ($z = 0$), $P_0 = 0$.
Consider the upper soil layer:

$$K_a = \frac{1 - \sin 30°}{1 + \sin 30°} = 0.33$$

$$P_3 = 0.33 \times 16 \times 3 = 16 \, \text{kPa}$$

Consider the lower soil layer:

$$K_a = \frac{1 - \sin 20°}{1 + \sin 20°} = 0.49$$

$$P_3 = 0.49 \times 16 \times 3 = 23.5 \, \text{kPa}$$

There is a step change in active pressure at a depth of 3 m from 16 to 23.5 kPa.

$$P_{7.5} = 0.49 \times 21 \times 4.5 + 23.5 = 69.8 \, \text{kPa}$$

The active pressure diagram is shown in Fig. 8.7b and the value of the total active thrust is simply the area of this diagram:

$$\left(16 \times \frac{3}{2}\right) + (23.5 \times 4.5) + \left(46.3 \times \frac{4.5}{2}\right) = 234 \, \text{kN}$$

(The figures in the brackets in Fig. 8.7b refer to Example 8.6.)

Example 8.3: Rankine active pressure; presence of groundwater

A vertical retaining wall 6 m high is supporting soil which is saturated and has a unit weight of 22.5 kN/m³. The angle of shearing resistance of the soil, ϕ', is 35°, and the surface of the soil is horizontal and level with the top of the wall. A groundwater level has been established within the soil and occurs at a level of 2 m from the top of the wall.
   Using the Rankine theory, calculate the significant pressure values and draw the diagram of pressure distribution that will occur on the back of the wall.

Solution:

Figure 8.8a illustrates the problem and Fig. 8.8b and c show the pressure distribution due to the soil and the water.

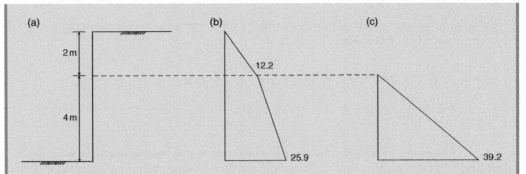

Fig. 8.8 Example 8.3. (a) The problem. (b) Earth pressure (kPa). (c) Water pressure (kPa).

$$K_a = \frac{1 - \sin 35°}{1 + \sin 35°} = 0.27$$

Although there is the same soil throughout, there is a change in unit weight at a depth of 2 m as the unit weight of the soil below the GWL is equal to the effective unit weight. The problem can therefore be regarded as two layers of different soil: the upper having a unit weight of 22 kN/m³ and the lower (22.5–9.81) = 12.7 kN/m³.

Consider the upper soil:

At depth = 2 m : $p_2 = K_a \gamma h = 0.27 \times 22.5 \times 2 = 12.2$ kPa

Consider the lower soil:

At depth = 6 m : $p_6 = 12.2 + (0.27 \times 12.7 \times 4) = 25.9$ kPa

(Note that at the interface of two cohesionless soil layers, the pressure values are the same if the ϕ' values are equal.)

Water pressure

At depth = 2 m, the water pressure = 0
At depth = 6 m, the water pressure = 9.81 × 4 = 39.2 kPa

The two pressure diagrams are shown in Fig. 8.8b and c, and the resultant pressure acting on the back of the wall is the sum of these two drawings:

$$= \left(\frac{1}{2} \times 12.2 \times 2\right) + (12.2 \times 4) + \left(\frac{1}{2} \times 13.7 \times 4\right) + \left(\frac{1}{2} \times 39.2 \times 4\right) = 167 \text{ kPa}$$

Had no groundwater table been present and the soil remained saturated throughout, the active thrust would have been:

$$= \left(\frac{1}{2} \times 0.27 \times 22.5 \times 6^2\right) = 109 \text{ kPa} \quad (\text{i.e. a lower value})$$

Example 8.3 illustrates the significant increase in lateral pressure that the presence of a water table causes on a retaining wall. Except in the case of quay walls, a situation in which there is a water table immediately behind a retaining wall should not be allowed to arise. Where such a possibility is likely an adequate drainage system should be provided (see Section 8.10.1).

For both Examples 8.2 and 8.3, the area of the resulting active pressure diagram will give the magnitude of the total active thrust, P_a. If required, its point of application can be obtained by taking moments of forces about some convenient point on the space diagram. If this approach is not practical, then the assumptions of Fig. 8.6 should generally be sufficiently accurate.

8.4   Rankine's theory: granular soils, passive earth pressure

8.4.1   Horizontal soil surface

In this case, the vertical pressure due to the weight of the soil, γh, is acting as a minor principal stress. Figure 8.9a shows the Mohr circle diagram representing these stress conditions and drawn in the usual position, i.e. with the axis OX (the direction of the major principal plane) horizontal. Figure 8.9b shows the same diagram correctly orientated with the major principal stress, $K_p\gamma h$, horizontal and the major principal plane vertical. The Mohr diagram, it will be seen, must be rotated through 90°.

In the Mohr diagram:

$$\frac{\sigma_1}{\sigma_3} = \frac{OB}{OA} = \frac{OC + DC}{OC - DB} = \frac{1 + \sin\phi'}{1 - \sin\phi'} = \tan^2\left(45° + \frac{\phi'}{2}\right)$$

hence

$$K_p = \frac{1 + \sin\phi'}{1 - \sin\phi'} = \tan^2\left(45° + \frac{\phi'}{2}\right) \tag{8.5}$$

As with active pressure, there is a network of shear planes inclined at $(45° - \phi'/2)$ to the direction of the major principal stress, but this time the soil is being compressed as opposed to expanded.

8.4.2   Sloping soil surface

The directions of the principal stresses are not known, but we assume that the passive pressure acts parallel to the surface of the slope. The analysis gives:

$$K_p = \cos\beta \frac{\cos\beta + \sqrt{\cos^2\beta - \cos^2\phi'}}{\cos\beta - \sqrt{\cos^2\beta - \cos^2\phi'}} \tag{8.6}$$

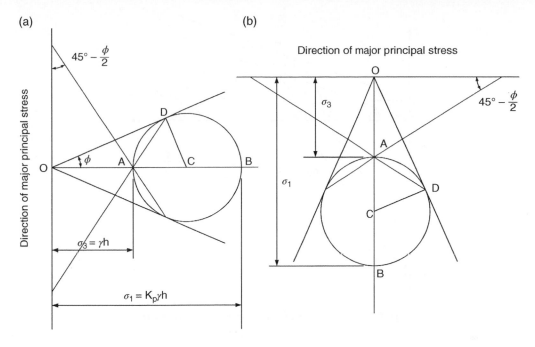

Fig. 8.9 Passive earth pressure for a cohesionless soil with a horizontal upper surface. (a) Mohr diagram drawn in usual position. (b) Diagram correctly orientated.

8.4.3 Rankine's assumption on wall friction

The amount of friction developed between a retaining wall and the soil can be of a high magnitude (particularly in the case of passive pressure). The Rankine theory's assumption of a smooth wall with no frictional effects can therefore lead to a significant underestimation (up to about a half) of the true K_p value. The theory can obviously lead to conservative design which, although safe, might at times be over-safe and lead to an uneconomic structure.

8.5 Rankine's theory: cohesive soils

8.5.1 Effect of cohesion on active pressure

Consider two soils of the same unit weight, one acting as a purely frictional soil with an angle of shearing resistance, ϕ', and the other acting as a cohesive-frictional soil with the same angle of shearing resistance, ϕ', and an effective cohesion, c'. The Mohr circle diagrams for the two soils are shown in Fig. 8.10.

At depth, h, both soils are subjected to the same major principal stress $\sigma'_1 = \gamma h$. The minor principal stress for the cohesionless soil is σ'_3 but for the cohesive soil it is only σ'_{3c}, the difference being due to the cohesive strength, c', that is represented by the lengths AB or EF.

Consider triangle HGF:

$$\frac{HF}{GH} = \frac{HF}{c'} = \frac{\sin(90° - \phi')}{\sin\left(45° + \frac{\phi'}{2}\right)} = 2\frac{\sin\left(45° - \frac{\phi'}{2}\right)\cos\left(45° - \frac{\phi'}{2}\right)}{\cos\left(45° - \frac{\phi'}{2}\right)}$$

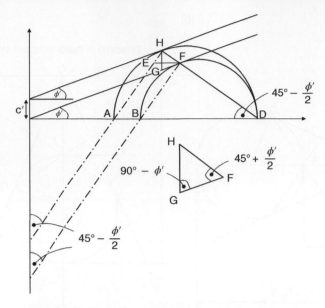

Fig. 8.10   The effect of cohesion on active pressure.

or

$$HF = 2c' \sin\left(45° - \frac{\phi'}{2}\right)$$

The difference between σ'_3 and σ'_{3c}

$$= EF = \frac{HF}{\cos\left(45° - \dfrac{\phi'}{2}\right)}$$

$$= 2c' \frac{\sin\left(45° - \dfrac{\phi'}{2}\right)}{\cos\left(45° - \dfrac{\phi'}{2}\right)} = 2c' \tan\left(45° - \frac{\phi'}{2}\right)$$

Hence, the active pressure, p_a at depth h in a soil exhibiting both frictional and cohesive strength and having a horizontal ground surface, is given by:

$$p_a = K_a\gamma h - 2c' \tan\left(45° - \frac{\phi'}{2}\right)$$

Recall that $K_a = \tan^2\left(45° - \frac{\phi'}{2}\right)$, so

$$p_a = K_a\gamma h - 2c'\sqrt{K_a} \tag{8.7}$$

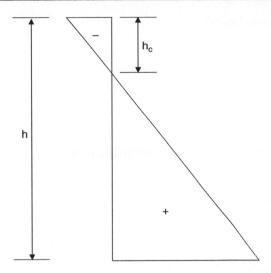

Fig. 8.11 Active pressure diagram for a soil with both cohesive and frictional strength.

This expression was formulated by Bell (1915) and is often referred to as *Bell's solution*. The active pressure diagram for such a soil is shown in Fig. 8.11. The negative values of p_a extending down from the top of the wall to a depth of h_c indicate that this zone of soil is in a state of suction. However, soils cannot really withstand tensile stress, and cracks may form within the soil. It is therefore unwise to assume that any negative active pressures exist within the depth h_c.

Since cohesive soils have low permeability, the tension crack will likely fill with water. For design purposes therefore, it is assumed that a hydrostatic water pressure is experienced in the tension zone.

8.5.2 Depth of the tension zone

In Fig. 8.11, the depth of the tension zone is given the symbol h_c. It is possible for cracks to develop over this depth, and a value for h_c is often required.

If p_a in the expression

$$p_a = K_a\gamma h - 2c' \tan\left(45° - \frac{\phi'}{2}\right)$$

is put equal to zero, we can obtain an expression for h_c:

$$h_c = \frac{2c'}{\gamma K_a} \tan\left(45° - \frac{\phi'}{2}\right)$$

$$h_c = \frac{2c'}{\gamma} \tan\left(45° + \frac{\phi'}{2}\right)$$

(8.8)

h_c may also be expressed:

$$h_c = \frac{2c'}{\gamma\sqrt{K_a}}$$

When $\phi' = 0°$ (i.e. in the undrained state):

$$h_c = \frac{2c_u}{\gamma} \qquad\qquad (8.9)$$

Example 8.4: Lateral pressure distribution

A retaining wall supports a soil as shown in Fig. 8.12. Sketch the lateral pressure distribution acting on the back of the wall.

Solution:

$$K_a = \frac{1 - \sin\phi'}{1 + \sin\phi'} = \frac{1 - \sin 25°}{1 + \sin 25°} = 0.41$$

$$p_a = K_a\gamma h - 2c'\sqrt{K_a}$$

Pressure distribution:

$$P_{0m} = 0 - (2 \times 4) \times \sqrt{0.41} = -5.1\,\text{kPa}$$

$$P_{10m} = (0.41 \times 18 \times 10) - (2 \times 4) \times \sqrt{0.41} = 68.7\,\text{kPa}$$

Tension crack:

$$h_c = \frac{2c'}{\gamma}\tan\left(45° + \frac{\phi'}{2}\right) = 0.70\,\text{m}$$

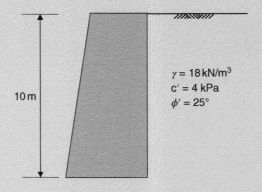

10 m

$\gamma = 18\,\text{kN/m}^3$
$c' = 4\,\text{kPa}$
$\phi' = 25°$

Fig. 8.12 Example 8.4.

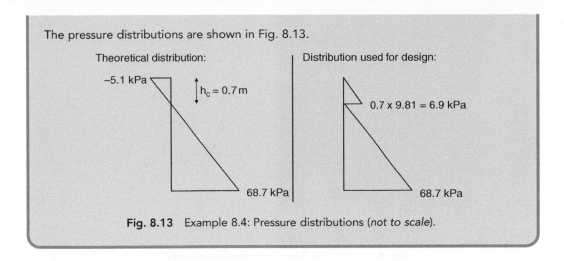

The pressure distributions are shown in Fig. 8.13.

Fig. 8.13 Example 8.4: Pressure distributions (*not to scale*).

8.5.3 The occurrence of tensile cracks

A tension zone, and therefore tensile cracking, can only occur when the soil exhibits cohesive strength. Gravels, sands, and most silts generally operate in a drained state and, having no cohesion, do not experience tensile cracking.

Clays, when undrained, can have substantial values of c_u but, when fully drained, almost invariably have effective cohesive intercepts that are either zero or, have a small enough value to be considered negligible.

It is therefore apparent that tensile cracks can only occur in clays and are only important in undrained conditions. The value of h_c, as determined from the formula derived above, is seen to become smaller as the value of the apparent cohesion reduces. This illustrates that, as a clay wets up and its cohesive intercept reduces from c_u to c', any tensile cracks within it tend to close.

If, there is a uniform surcharge acting on the surface of the retained soil such that its *equivalent height* is h_e (see Section 8.8.1) then the depth of the tension zone becomes equal to z_0 where $z_0 = h_c - h_e$. If, of course, the surcharge value is such that h_e is greater than h_c, then no tension zone will exist (as illustrated in Example 8.9).

8.5.4 Effect of cohesion on passive pressure

Rankine's theory was developed by Bell (1915) for the case of a frictional/cohesive soil. His solution for a soil with a horizontal surface is:

$$p_p = \gamma h \tan^2\left(45° + \frac{\phi'}{2}\right) + 2c' \tan\left(45° + \frac{\phi'}{2}\right)$$
$$= K_p \gamma h + 2c' \sqrt{K_p}$$

(8.10)

8.6 Coulomb's wedge theory: active earth pressure

Instead of considering the equilibrium of an element in a stressed mass, Coulomb's theory considers the soil as a whole.

8.6.1 Granular soils

If a wall supporting a granular, cohesionless soil is suddenly removed, the soil will slump down to its angle of shearing resistance, ϕ', on the plane BC in Fig. 8.14a. It is therefore, reasonable to assume that if the wall only

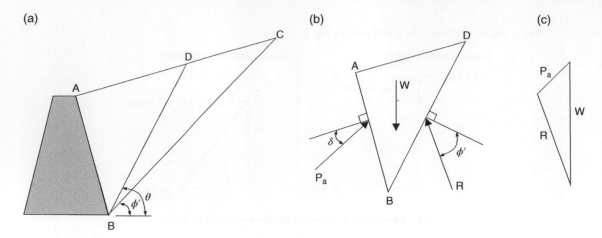

Fig. 8.14   Wedge theory for cohesionless soils.

moved forward slightly, a rupture plane BD would develop somewhere between AB and BC: the wedge of soil ABD would then move down the back of the wall AB, and along the rupture plane BD. The rupture plane is inclined at an angle θ to the horizontal. These wedges do in fact exist and have failure surfaces approximating to planes. Coulomb analysed this problem analytically in 1776 on the assumption that the surface of the retained soil was also a plane.

Consider the *trial* wedge ABD shown in Fig. 8.14b. The wedge is in plastic equilibrium, this being maintained by the three forces acting:

W, the weight of the wedge;
P_a, the reaction from the wall;
and R, the reaction on the plane of failure.

At failure, the reaction on the failure plane will be inclined at maximum obliquity, ϕ', to the normal to the plane. If the angle of wall friction is δ, then the reaction from the wall will be inclined at δ to the normal to the wall (δ cannot be greater than ϕ'). As active pressures are being developed, the wedge is tending to move downwards, and both R and P_a will consequently be on the downward sides of the normal lines (Fig. 8.14b). The weight of the wedge is known (W = area ABD × unit weight) and acts vertically downwards. Since the directions of R and P_a are also known, the triangle of forces can now be completed, and the magnitude of P_a determined (Fig. 8.14c).

The value of P_a found this way is specific to the particular trial wedge considered and is not necessarily the maximum value acting on the back of the wall. To establish the maximum value of P_a, a number of trial wedges are considered, each with its own triangle of forces. The wedge yielding the maximum value of P_a is then the actual failure wedge and thus, the angle of the failure surface θ is determined. Example 8.10 illustrates the procedure.

The maximum value of P_a is given by

$$P_a = \tfrac{1}{2}\,K_a \gamma h^2$$

where h is the height of the wall and K_a is Coulomb's coefficient of active earth pressure:

$$K_a = \left\{ \frac{\operatorname{cosec} \psi \, \sin(\psi - \phi')}{\sqrt{\sin(\psi + \delta)} + \sqrt{\dfrac{\sin(\phi' + \delta)\sin(\phi' - \beta)}{\sin(\psi - \beta)}}} \right\}^2 \tag{8.11}$$

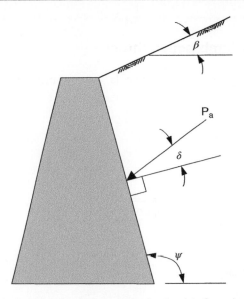

Fig. 8.15 Symbols used in Coulomb's formula.

where ψ = angle of back of wall to the horizontal;
δ = angle of wall friction;
β = angle of inclination of surface of retained soil to the horizontal;
and ϕ' = angle of shearing resistance.

The active thrust, P_a, is assumed to act at angle δ to the normal to the wall. Refer to Fig. 8.15 for explanation of all the symbols used in Coulomb's K_a expression.

It is interesting to note that Coulomb's expression for K_a reduces to the Rankine formula when $\psi = 90^\circ$ and $\delta = \beta$, i.e.

$$K_a = \cos\beta \times \frac{\cos\beta - \sqrt{(\cos^2\beta - \cos^2\phi')}}{\cos\beta + \sqrt{(\cos^2\beta - \cos^2\phi')}}$$

and further reduces to

$$K_a = \frac{1 - \sin\phi'}{1 + \sin\phi'}$$

when $\psi = 90^\circ$ and $\delta = 0$.

The value of the angle of wall friction, δ, can be obtained from large shear box testing (the lower half of the box contains the wall material, the upper half contains the soil), but if test values are not available, design codes provide guidance on the value of δ to use for the particular wall-soil materials used (typically in the region, $\delta = 0.5$ to $1.0\ \phi'$).

Example 8.5: Coulomb active thrust

Solve Example 8.1 Using the Coulomb formula. Assume that $\delta = \frac{1}{2}\,\phi'$.

Solution:

Coulomb's formula for K_a is:

$$K_a = \left\{ \frac{\operatorname{cosec}\,\psi\,\sin\,(\psi - \phi')}{\sqrt{\sin\,(\psi + \delta)} + \sqrt{\dfrac{\sin\,(\phi' + \delta)\,\sin\,(\phi' - \beta)}{\sin\,(\psi - \beta)}}} \right\}^2$$

(a) *Soil surface horizontal*

$\delta = 17.5°; \psi = 90°; \phi' = 35°; \beta = 0.$

$$K_a = \left\{ \frac{\sin 55° / \sin 90°}{\sqrt{\sin 107.5°} + \sqrt{\sin 52.5°\,\sin 35° / \sin 90°}} \right\}^2$$

$$K_a = \left\{ \frac{0.819}{0.976 + 0.675} \right\}^2 = 0.25$$

$P_a = 0.5 K_a \gamma H^2 = 0.5 \times 0.25 \times 19 \times 5^2 = 59.4\ \text{kN}$

This value is inclined at 17.5° to the normal to the back of the wall so that the total horizontal active thrust, according to Coulomb, is $59.4 \times \cos 17.5° = 56.7\ \text{kN}$.

(b) *Soil surface sloping at 35°*
   Substituting $\delta = 17.5°$; $\psi = 90°$; $\phi' = 35°$ and $\beta = 35°$ into the formula gives $K_a = 0.70$. Hence

Total active thrust $= 0.5 \times 0.70 \times 19 \times 5^2 = 166.3\ \text{kN}$

Total horizontal thrust $= 166.3 \times \cos 17.5° = 158.6\ \text{kN}$

Increase in horizontal thrust $= 158.6 - 56.7 = 101.9\ \text{kN}$

Example 8.6: Coulomb active thrust; more than one soil

Determine the total horizontal active thrust acting on the back of the wall of Example 8.2 by the Coulomb theory. Take $\delta = \phi'/2$.

Solution:

Active pressure at the top of the wall, $p_0 = 0$.
  Consider the upper soil layer:

    For $\phi' = 30°, \delta = \phi'/2 = 17.5°, \psi = 90°$ and $\beta = 0$, $K_a = 0.301$

  Hence active pressure at a depth of 3 m $= 0.301 \times 16 \times 3 = 14.5$ kPa.
  But this pressure acts at 15° to the horizontal (as $\delta = 15°$).
  Horizontal pressure at depth $= 3$ m $= p_3 = 14.5 \cos 15° = 14.0$ kPa.
  Consider the lower soil layer:

    For $\phi' = 20°, \delta = \phi'/2 = 10°, \psi = 90°$ and $\beta = 0$, $K_a = 0.447$
    $p_3 = 0.447 \times 16 \times 3 \times \cos 10° = 21.1$ kPa
    $p_{7.5} = [(0.447 \times 21 \times 4.5) + 21.1] \times \cos 10° = 62.4$ kPa

These values are shown in brackets on the pressure diagram in Fig. 8.7b.

8.6.2 The effect of cohesion

In this case, Coulomb's theory is extended by assuming that, at the top of the wall, there is a zone of soil within which there are no friction or cohesive effects along both the back of the wall and on the plane of rupture (Fig. 8.16). Thus, tension cracks are assumed to develop over the extent of the zone, to a depth z_0. Beneath

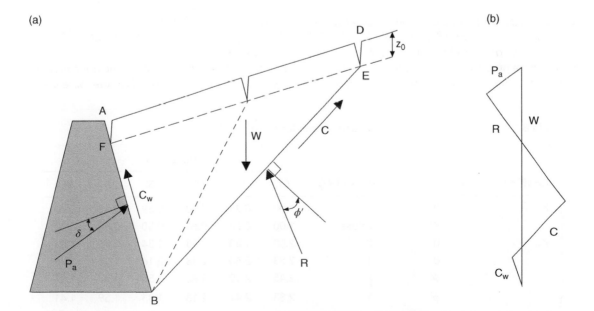

Fig. 8.16 Consideration of the effects of cohesion.

the tension zone, wall adhesion can exist between wall and the soil, which is quantified as a proportion of the soil cohesion, c. This applies for both the undrained state ($c = c_u$) and where $c' > 0$ ($c = c'$) in the drained state.

There are now five forces acting on the wedge, which form the polygon of forces (Fig. 8.16b):

R, the reaction on plane of failure;
W, the weight of whole wedge ABED;
P_a, the resultant thrust on wall;
C_w, the adhesive force along length BF of wall ($C_w = a \times BF$);
a, the wall adhesion; and
C, the cohesive force along rupture plane BE ($C = c \times BE$).

The unit wall adhesion, a, cannot be greater than the apparent cohesion, and Eurocode 7 advises that a and δ should be chosen such that

$$\frac{a}{c} = \frac{\tan \delta}{\tan \phi}$$

The value of W is obtained as before, so there are only two unknown forces: R and P_a which are simply measured from the polygon. As with the cohesionless case, a number of trial wedges can be analysed to find the angle of the failure surface that yields the maximum value of P_a.

Kerisel and Absi (1990) published values of the horizontal components of K_a and K_p for a range of values of ϕ, β, δ, and ψ to ease calculation when using the Coulomb theory. In addition, the spreadsheet *earth pressure coefficients.xls* is available for download, which can be used to determine the horizontal components of K_a and K_p. In this section, we are concerned with the horizontal component of K_a (i.e. $K_a \cos \delta$) only. The horizontal component of K_p is given in Section 8.7.2.

The active pressure acting normally to the wall at a depth, h can be defined by:

$$p_{ah} = K_a \gamma h - c K_{ac}$$

where c = operating value of cohesion and K_{ac} = coefficient of active earth pressure.

Various values of K_a and K_{ac} are given in Table 8.1 for the straightforward case of $\beta = 0$ and $\psi = 90°$. Intermediate values of K_a and K_{ac} can be obtained from the spreadsheet.

It should be noted that the values in Table 8.1 are for pressure components acting in the horizontal direction, not at an angle δ to the horizontal as in the original Coulomb theory. The values in the table are from

Table 8.1 Values of K_a and K_{ac} for a cohesive soil for $\beta = 0$, $\psi = 90°$.

Coefficient	Values of δ	Values of a/c	Values of ϕ'					
			0°	5°	10°	15°	20°	25°
K_a	0	All	1.00	0.85	0.70	0.59	0.48	0.40
	ϕ'	values	1.00	0.78	0.64	0.50	0.40	0.32
K_{ac}	0	0	2.00	1.83	1.68	1.54	1.40	1.29
	0	1	2.83	2.60	2.38	2.16	1.96	1.76
	ϕ'	$\frac{1}{2}$	2.45	2.10	1.82	1.55	1.32	1.15
	ϕ'	1	2.83	2.47	2.13	1.85	1.59	1.41

Kerisel and Absi's original work. The values in the spreadsheet are derived from the following widely recognised formulae, which are sufficiently accurate for most purposes:

$$K_a = \text{Coulomb's value} \times \cos \delta$$

$$K_{ac} = 2\sqrt{K_a\left(1 + \frac{a}{c}\right)}$$

EN 1997-1:2004 (BSI, 2004) recommends K_{ac} be limited to $2.56\sqrt{K_a}$

Example 8.7: Coulomb K_a and K_{ac}

Determine Coulomb's K_a and K_{ac} values for $\phi' = 20°$, $\delta = 10°$, $\beta = 0$, $\psi = 90°$, $c' = 10$ kPa, $c_w = 10$ kPa.

Solution:

$$K_a = \left\{ \frac{\text{cosec } 90° \sin 70°}{\sqrt{\sin 100°} + \sqrt{\dfrac{\sin 30° \sin 20°}{\sin 90°}}} \right\}^2 = 0.447$$

Hence the K_a value for horizontal pressure $= 0.447 \times \cos 10° = 0.44$.

$$K_{ac} = 2\sqrt{0.44\left(1 + \frac{10}{10}\right)} = 1.90$$

8.6.3 Point of application of total active thrust

The magnitude of P_a is obtained directly from the force diagram. Its point of application may be assumed to be where a line drawn through the centroid of the failure wedge, and parallel to the failure plane, intersects the back of the wall. (See Fig. 8.17.)

8.7 Coulomb's wedge theory: passive earth pressure

8.7.1 Granular soils

With the assumption of a plane failure surface leading to a wedge failure, Coulomb's expression for K_p for a granular soil is:

$$K_p = \left\{ \frac{\text{cosec } \psi \sin (\psi - \phi')}{\sqrt{\sin (\psi - \delta)} - \sqrt{\dfrac{\sin (\phi' + \delta) \sin (\phi' + \beta)}{\sin (\psi - \beta)}}} \right\}^2 \tag{8.12}$$

the symbols having the same meanings as previously.

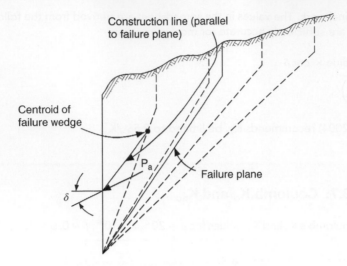

Fig. 8.17   Determination of line of action P_a.

The expression reduces to:

$$K_p = \frac{1 + \sin \phi'}{1 - \sin \phi'}$$

when $\psi = 90°$, $\delta = 0$ and $\beta = 0°$.

With passive pressure, the failure surface only approximates to a plane surface when the angle of wall friction is small. The situation arises because the behaviour of the soil is not only governed by its weight, but also by the compression forces induced by the wall tending to push into the soil. These forces, unlike the active case, do not act on only one plane within the soil, resulting in a non-uniform strain pattern and the development of a curved failure surface (Fig. 8.18).

It is apparent that in most cases, the assumption of a Coulomb wedge for a passive failure can lead to a serious overestimation of the resistance available. Terzaghi (1943) first analysed this problem and concluded that, provided the angle of friction developed between the soil and the wall, δ is not more than $\phi'/3$, where ϕ' is the operative value of the angle of shearing resistance of the soil, the assumption of a plane failure surface generally gives reasonable results. For values of δ greater than $\phi'/3$, the errors involved can be very large.

Adjusted values for K_p, that allow for a curved failure surface, are given in Table 8.2. These values apply to a vertical wall and a horizontal soil surface, and include the multiplier $\cos \delta$ as the values in the table give the components of pressure that will act normally to the wall.

It is seen therefore that for a smooth wall, where $\delta = 0$, the Rankine theory can be used for the evaluation of passive pressure. If wall friction is mobilised, then $\delta \neq 0$, and the coefficients of Table 8.2 should be used (unless $\delta \leq \phi'/3$ in which case the Coulomb equation can be used directly).

8.7.2   The effect of cohesion

As shown in Chapter 4, a clay has a non-linear stress–strain relationship and its shear strength depends upon its previous stress history. Add to this the complications of non-uniform strain patterns within a passive resistance zone, and it is obvious that any design approach must be rigorous in nature: either numerically modelled or an empirical approach based on experimental work.

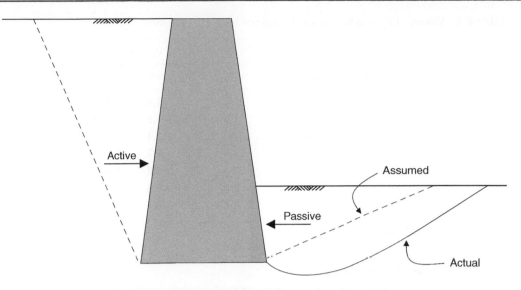

Fig. 8.18   Departure of passive failure surface from a plane.

Table 8.2   Values of K_p for cohesionless soils.

Values of δ	Values of ϕ'			
	25°	30°	35°	40°
	Values of K_p			
0°	2.5	3.0	3.7	4.6
10°	3.1	4.0	4.8	6.5
20°	3.7	4.9	6.0	8.8
30°	–	5.8	7.3	11.4

Based on Kerisel and Absi (1990).

A similar equation to that of Bell can be used for passive pressure values, when the effects of wall friction and adhesion are taken into account.

The passive pressure acting normally to the wall at a depth h can be defined as:

$$P_{ph} = K_p\gamma h - cK_{pc}$$

where c = operating value of cohesion and K_{pc} = coefficient of active earth pressure.

Various values of K_p and K_{pc} are given in Table 8.3 for the straightforward case of $\beta = 0$, $\psi = 90°$. As with the active pressure coefficients given in Table 8.1, they give the value of the pressure acting normally to the wall.

An alternative to using the values set out in Table 8.3 is that of the work of Sokolovski (1960), part of which is presented in Table 8.4. This offers a more realistic set of values than those listed in Tables 8.2 and 8.3.

Table 8.3 Values of K_p and K_{pc} for a cohesive soil for $\beta = 0$; $\psi = 90°$.

Coefficient	Values of δ	Values of a/c	Values of ϕ'					
			0°	5°	10°	15°	20°	25°
K_p	0	All	1.0	1.2	1.4	1.7	2.1	2.5
	ϕ'	values	1.0	1.3	1.6	2.2	2.9	3.9
K_{pc}	0	0	2.0	2.2	2.4	2.6	2.8	3.1
	0	½	2.4	2.6	2.9	3.2	3.5	3.8
	0	1	2.6	2.9	3.2	3.6	4.0	4.4
	ϕ'	½	2.4	2.8	3.3	3.8	4.5	5.5
	ϕ'	1	2.6	2.9	3.4	3.9	4.7	5.7

Based on Sokolovski (1960).

Table 8.4 Values of K_p and K_{pc}.

Values of δ	Values of a/c	Values of ϕ'			
		10°	20°	30°	40°
		Values of K_p			
0	All	1.42	2.04	3.00	4.60
$\phi'/2$	values	1.55	2.51	4.46	9.10
ϕ'		1.63	2.86	5.67	14.10
		Values of K_{pc}			
0	0	2.38	2.86	3.46	4.29
$\phi'/2$	0	2.49	3.17	4.22	6.03
ϕ	0	2.55	3.38	4.76	7.51
0	0.5	2.92	3.50	4.24	5.25
$\phi'/2$	0.5	3.05	3.88	5.17	7.39
ϕ	0.5	3.13	4.14	5.83	9.20
0	1.0	3.37	4.04	4.90	6.07
$\phi'/2$	1.0	3.52	4.48	5.97	8.53
ϕ'	1.0	3.61	4.78	6.73	10.62

Based on Sokolovski (1960).

The K_{pc} values were obtained from the approximate relationship:

$$K_{pc} = 2\sqrt{K_p\left(1 + \frac{a}{c}\right)}$$

EN 1997-1:2004 (BSI, 2004) recommends K_{pc} be limited to $2.56\sqrt{K_p}$

This relationship has been used in the *earth pressure coefficients.xls* spreadsheet which can be used to determine intermediate values of K_p and K_{pc}.

It should be noted that, in the case of passive earth pressure, the amount of wall movement necessary to achieve the ultimate value of ϕ' can be large, particularly in the case of a loose sand where it cannot reasonably be expected that more than one half the value of the ultimate passive pressures will be developed.

As mentioned earlier, δ is a function of ϕ' and typically will fall into the range $0.5\,\phi' < \delta < \phi'$. EN1997-1:2004 (BSI, 2004) quantifies this as

$$\delta = k.\phi_{cv;\,d}$$

where $\phi_{cv;\,d}$ is the design constant volume value of ϕ' and k is a constant depending on the roughness of the soil structure interface, and recommends the following for granular soils:

for concrete cast against soil, k = 1.0, (i.e. $\delta = \phi_{cv;\,d}$)
for precast concrete and steel sheet piling (see Section 9.6), k = $\frac{2}{3}$

EN1997-3:202x develops this and refers to k as k_δ and states that:

- k_δ considers disturbance during construction, and
- for stone-filled gabions and cribs, k_δ may be assumed to equal 1.0.
- k_δ should not exceed $\frac{2}{3}$ for smooth surface.

8.8   Surcharges

The extra loading carried by a retaining wall is known as a *surcharge* and can be a uniform load (roadway, stacked goods, etc.), a line load (trains running parallel to a wall), an isolated load (column footing), or a dynamic load (traffic).

Even when a retaining wall is not intended to support a surcharge over its design life, it should be remembered that it may be subject to surface loadings due to plant movement during construction. It is at this time that the wall will be at its weakest state and, therefore, transient surcharges must be considered in retaining wall design.

8.8.1   Uniform surcharge

Soil surface horizontal

When the surface of the soil behind the wall is horizontal, the pressure acting on the back of the wall due to a wide and long surcharge, q, is uniform with depth and has magnitude equal to K_aq (Fig. 8.19).

Soil surface sloping at angle β to horizontal

When the surface of the soil is not horizontal, the surcharge can be considered as equivalent to an extra height of soil, h_e, placed on top of the soil:

$$h_e = \frac{q\sin\psi}{\gamma\sin(\psi+\beta)} \tag{8.13}$$

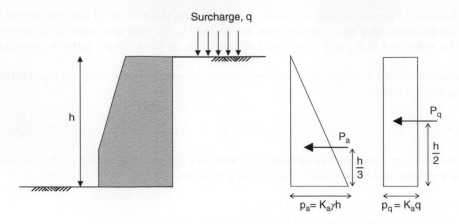

Fig. 8.19   Effect of uniform surcharge on a retaining wall.

where

γ = unit weight of soil (kN/m^3)
q = magnitude of surcharge (kPa)
ψ = angle of back of wall to horizontal
β = angle of inclination of retained soil.

Once again, the pressure acting on the back of the wall due to the surcharge is considered uniform, but this time is of magnitude $K_a\gamma h_e$.

Example 8.8: Uniform surcharge (i)

A smooth-backed vertical wall is 6 m high and retains a soil with a bulk unit weight of 20 kN/m^3 and $\phi' = 20°$. The top of the soil is level with the top of the wall and its surface is horizontal and carries a uniformly distributed surcharge of 50 kPa. Using the Rankine theory, determine the total active thrust per linear metre of wall and its point of application.

Solution:

Figure 8.20a shows the problem and Fig. 8.20b shows the resultant pressure diagram. Using the Rankine theory:

$$K_a = \frac{1 - \sin\phi'}{1 + \sin\phi'} = \frac{1 - \sin 20°}{1 + \sin 20°} = 0.49$$
$$p_a = K_a\gamma h = 0.49 \times 20 \times 6 = 58.8 \text{ kPa}$$

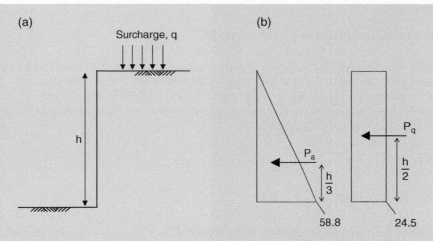

Fig. 8.20 Example 8.8. (a) The problem. (b) Pressure distribution (kPa).

Since soil surface behind wall is horizontal,

$$p_q = K_a q \times 0.49 \times 50 = 24.5 \text{ kPa}$$

The pressure diagram is now plotted (Fig. 8.20b).

Total thrust = Area of pressure diagram
$$= P_a + P_q$$
$$= (\tfrac{1}{2} \times 58.8 \times 6) + (24.5 \times 6)$$
$$= 176.4 + 147 = 323.4 \text{ kN}$$

The point of application of this thrust is obtained by taking moments of forces about the base of the wall, i.e.

$$323.4 \times h = (147 \times 3) + \left(176.4 \times \frac{6}{3}\right)$$
$$\Rightarrow h = \frac{793.8}{323.4} = 2.45 \text{ m}$$

Resultant thrust acts at 2.45 m above base of wall.

Example 8.9: Uniform surcharge (ii)

A vertical retaining wall is 5 m high and supports a soil; the surface of which is horizontal and level with the top of the wall and carrying a uniform surcharge of 75 kPa.

The properties of the soil are $\phi' = 20°$; $c' = 10$ kPa; $\gamma = 20$ kN/m³; $\delta = \phi'/2$.

Determine the value of the maximum horizontal thrust on the back of the wall using the K_a and K_{ac} coefficients of Table 8.1.

Solution:

$$h_c = \frac{2c'}{\gamma} \tan\left(45° + \frac{\phi'}{2}\right) = 1.43\,\text{m}$$

$$h_e = \frac{q}{\gamma} = \frac{75}{20} = 3.75\,\text{m}\ (\text{since }\psi = 90°\text{ and }\beta = 0°)$$

$$\Rightarrow z_o = h_c - h_e = -2.32\,\text{m}, \quad \text{i.e. take } z_o = 0;\ \text{no tension crack will form.}$$

$$\frac{a}{c} = \frac{\tan\delta}{\tan\phi} \Rightarrow a = 10 \times \frac{\tan 10°}{\tan 20°} = 5\,\text{kPa}$$

The coefficients K_a and K_{ac} (Table 8.1) can be obtained by linear interpolation, or using the *coefficients of earth pressure* spreadsheet:

$$K_a = 0.44$$
$$K_{ac} = 1.62$$

Active pressure at top of wall,

$$P_0 = \gamma h_e K_a - c'K_{ac}$$
$$= (20 \times 3.75 \times 0.44) - (10 \times 1.62) = 16.8\,\text{kPa}$$

Active pressure at base of wall,

$$P_5 = \gamma(H + h_e)K_a - c'K_{ac}$$
$$= 20(5 + 3.75)0.44 - 16.2 = 60.8\,\text{kPa}$$

The pressure diagram on the back of the wall is shown in Fig. 8.21. Remembering that these are the values of pressure acting normal to the wall, the maximum horizontal thrust will be the area of the diagram.

$$\text{Maximum horizontal thrust} = \frac{16.8 + 60.8}{2} \times 5 = 194\,\text{kN/m run of wall}$$

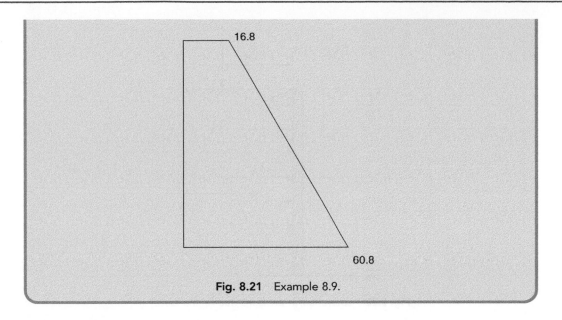

16.8

60.8

Fig. 8.21 Example 8.9.

8.8.2 Line load

The lateral thrust acting on the back of the wall as a result of a line load surcharge is best estimated by plastic analysis.

It is possible to use a Boussinesq analysis (see Chapter 3) to determine the vertical stress increments due to the surface load and then to use these values in the plastic analysis combined with the design value of K_a (see Chapter 9). Figure 8.22 illustrates the theoretical stress distribution placed on a rigid wall caused by a vertical line load running parallel to the wall, first established by Terzaghi (1943).

For values of $m > 0.4$, it can be shown that

$$\Delta\sigma_z = \frac{4Q}{\pi} \frac{x^2 z}{(x^2 + z^2)^2}$$

where x and z are dimensions as illustrated in Fig. 8.22 and Q is the applied surface line load (kN/m).

Setting x and z as proportions of the wall height, H ($x = mH$ and $z = nH$) we have:

$$\Delta\sigma_z = \frac{4Q}{\pi H} \frac{m^2 n}{(m^2 + n^2)^2}$$

The resultant horizontal thrust,

$$P_h = \frac{0.64Q}{m^2 + 1}$$

For cases where $m \leq 0.4$, $P_h = 0.55Q$

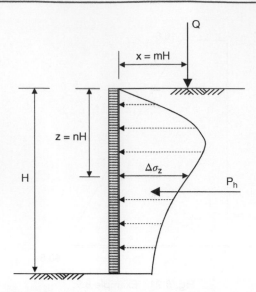

Fig. 8.22   Lateral earth pressure due to line load.

8.8.3   Compaction effects

During the construction of gravity retaining walls, layers of fill are compacted behind the wall, and this compaction process induces lateral stresses within the fill which can act against the back of the wall. If the stresses are high enough, they can lead to movement or deformation of the wall, and so the effect of the compaction is taken into account during the design of the wall.

Typically, the compaction plant used are vibratory rollers (described in Chapter 15) and as the vehicle compacts the soil, the increase in vertical stress induces the increased lateral stress, both of which reduce once the vehicle moves on. However, as some yielding of the wall will have occurred, some of the induced lateral stress remains. The amount of this increased lateral stress decreases with depth, as successive layers are compacted above. Thus, the compaction-induced lateral pressure in a layer remains constant as successive layers are compacted above. However, close to the surface, the increase in horizontal stress does not remain after compaction, and over this extent, the lateral pressure is simply the active pressure with no contribution from compaction.

These findings yield the pressure distribution shown in Fig. 8.23. It is seen that the lateral stress in the shallow soil ($z < z_c$) is equal to $K_p\sigma'_v$, and the lateral stress in the deeper soil ($z > h_c$) is equal to $K_a\sigma'_v$, where σ'_v is the effective vertical stress. The shaded area on the figure indicates the lateral earth pressures due to the compaction.

Based on the work of Ingold (1979), the depths z_c and h_c are defined:

$$z_c = K_a\sqrt{\frac{2p}{\pi\gamma}}$$

$$h_c = \frac{1}{K_a}\sqrt{\frac{2p}{\pi\gamma}}$$

where p is the applied line load per metre width of roller (kN).

Clayton and Symons (1992) provide additional information on determining compaction pressures.

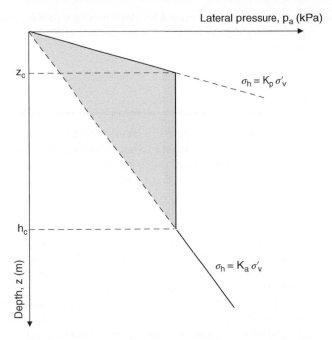

Fig. 8.23   Compaction induced lateral pressure.

8.9   Choice of method for determination of active pressure

The main criticism of the Rankine theory is that it assumes conditions that are unrealistic in soils. There will invariably be friction and/or adhesion developed between the soil and the wall as it will have some degree of roughness and will never be perfectly smooth. Hence, in many cases, the Rankine assumption that no shear forces develop on the back of the wall is simply not true and it may be appropriate to use the Coulomb theory.

As noted earlier, it is not easy to obtain measurements of the value of wall friction, δ, and the value of the wall adhesion, a, so both of these are usually estimated. δ is obviously a function of the angle of shearing resistance, ϕ', of the retained soil immediately adjacent to the wall and can have any value from virtually zero up to some maximum value, which cannot be greater than ϕ'. Similarly, the operative value of a is related to the value of cohesion of the soil immediately adjacent to the wall.

Just what will be the actual operating value of δ depends upon the amount of relative movement between the soil and the wall. A significant downward movement of the soil relative to the wall will result in the development of the maximum δ value.

Cases of significant relative downward movement of the soil are not necessarily all that common. Often there are cases in which there is some accompanying downward movement of the wall resulting in a smaller relative displacement. Examples of such cases can be gravity and sheet pile walls, and a value of δ less than the maximum should therefore be used. (Descriptions of different wall types are given in Chapter 9.)

When the retained soil is supported on a foundation slab, as with a reinforced concrete cantilever or counterfort wall, there will be virtually no movement of the soil relative to the back of the wall. In this case, the adoption of a 'virtual plane' in the design procedure, as illustrated in Example 9.3, justifies the use of the Rankine approach.

The use of the Rankine method affords a quick means for determining a conservative value of active pressure, which can be useful in preliminary design work. Full explanations of the procedures used in retaining wall design are given in Chapter 9.

8.10 Backfill material

The examples used to illustrate lateral earth pressure in this chapter have all been based on gravity walls, i.e. walls which are constructed using a 'bottom-up' process and backfilled with soil after construction, or as construction progresses. The ideal backfill material for such walls is granular, such as suitably graded quarried fill, gravel, or clean sand with a small percentage of fines. Such a soil is free draining and of good durability and strength but can be expensive, even when obtained locally.

Economies can sometimes be achieved using granular material in retaining wall construction in the form of a wedge as shown in Fig. 8.24. The wedge separates the finer material making up the bulk of the backfill from the back of the wall. With such a wedge, lateral pressures exerted on to the back of the wall can be evaluated with the assumption that the backfill is made up entirely of the granular material.

Recycled construction waste, slag, clinker, and other manufactured materials that approximate to a granular soil will generally prove satisfactory as backfill material provided that they do not contain harmful chemicals. Inorganic silts and clays can be used as backfills but require special drainage arrangements and can give rise to swelling and shrinkage problems that are not encountered in granular material. Peat, organic soil, chalk, shale, pulverised fuel ash, and other unsuitable materials should not be used as backfill if at all possible.

8.10.1 Drainage systems

No matter what material is used as backfill, its drainage is of great importance. A retaining wall is designed generally to withstand only lateral pressures exerted by the soil that it is supporting. In any design, the possibility of a groundwater level occurring in the material behind a retaining wall must be examined and an appropriate drainage system decided upon.

For a granular backfill, the only drainage often necessary is the provision of weep holes that go through the wall and are spaced at approximately 3 m centres, both horizontally and vertically. The holes can vary in diameter from 75 to about 150 mm and are protected against clogging by the provision of gravel pockets or geosynthetic filters placed in the backfill immediately behind each weep hole (Fig. 8.25a).

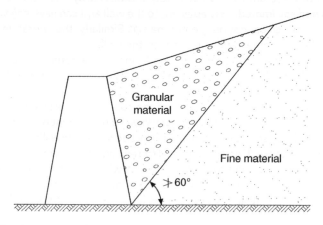

Fig. 8.24 Use of granular material in retaining wall construction.

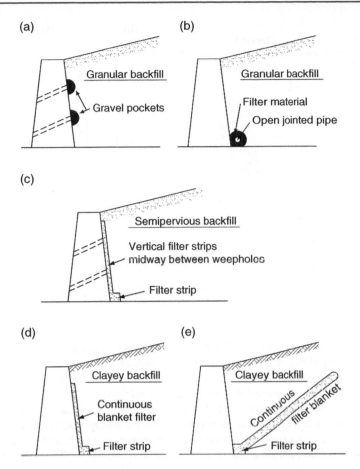

Fig. 8.25   Common drainage systems for retaining walls. (a) Weepholes only. (b) Backdrain. (c) Vertical strips of filter material. (d) Vertical drainage blanket. (e) Inclined drainage blanket.

Generally, weep holes can only be provided in outside walls and an alternative arrangement for granular backfill is illustrated in Fig. 8.25b. It comprises a continuous longitudinal back drain placed at the foot of the wall consisting of open jointed pipes packed around with gravel or some other suitable filter material. The design of filters is discussed in Chapter 2 and geosynthetic filters are mentioned in Chapter 15. Provision for rodding out during service should be provided to ensure that the pipes do not become blocked through a build-up of fines.

If the backfill material is granular, but has more than 5% fine sand, silt, or clay particles mixed within it, then it is only semi-pervious. For such a material, the provision of weep holes on their own will provide inefficient drainage, with the further complication of there being a much greater tendency for clogging to occur. The solution can be to provide additional drainage in the form of vertical strips of geosynthetic or granular filter material placed midway between the weep holes and led down to a continuous longitudinal strip of the same filter material as shown in Fig. 8.25c.

For clayey materials, blanket drains of suitable filter material are necessary. These blankets should be about 0.33 m thick and typical arrangements are shown in Fig. 8.25d and e. Generally, the vertical drainage blanket of Fig. 8.25d will prove satisfactory, especially if the surface of the retained soil can be protected with some form of impervious covering. If this protection cannot be given, then there is the chance of high seepage

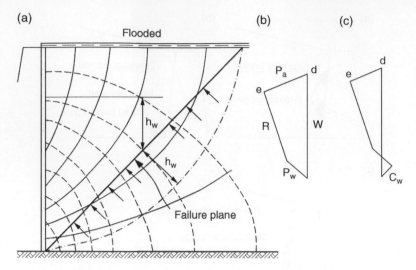

Fig. 8.26 Seepage forces behind a retaining wall with a vertical drain during heavy rain. (a) Flow net. (b, c) Force diagrams.

pressure being created during heavy rain (see Example 8.10). In such a situation, the alternative arrangement of the inclined filter blanket of Fig. 8.25e can substantially reduce such seepage pressures.

The reason for the different effects of the two drainage systems can be seen when we consider the respective seepage flow nets that are generated during flooded conditions.

The flow net for the vertical drain is shown in Fig. 8.26a. It must be appreciated that the drain is neither an equipotential nor a flow line. It is a drained surface and therefore the only head of water that can exist along it is that due to elevation. Hence, if a square flow net has been drawn, the vertical distances between adjacent equipotentials entering the drain will be equal to each other (in a manner similar to the upstream slope of an earth dam).

Owing to the seepage forces, an additional force, P_w, now acts upwards and at right angles to the failure plane. From the flow net, it is possible to determine values of excess hydrostatic pressure, h_w, at selected points along the failure plane (see Fig. 8.26a). If a smooth curve is drawn through these h_w values (when plotted along the failure plane), it becomes possible to evaluate P_w (see Example 8.10).

The resulting force diagram is shown in Fig. 8.26b. In theory, the polygon of forces will be as shown in Fig. 8.26c but, as seepage will only occur once the soil has achieved a drained state, the operative strength parameter is ϕ', with c' generally being assumed to be zero.

The seepage flow net for the inclined drain in Fig. 8.26e is shown in Fig. 8.27. Such a drain induces vertical drainage of the rainwater and it is seen that the portion of the flow net above the drain is absolutely regular and, more importantly, that the equipotentials are horizontal. This latter fact means that, within the soil above the drain, the value of excess hydrostatic head at any point must be zero. The failure plane will not be subjected to the upward force P_w and the pressure exerted on the back of the wall can only be from the saturated soil.

8.10.2 Differential hydrostatic head

When there is a risk of a groundwater level developing behind the wall then the possible increase in lateral pressure due to submergence must be allowed for. This problem will occur in tidal areas, and quay walls must be designed to withstand the most adverse difference created by tidal lag between the water level in front of

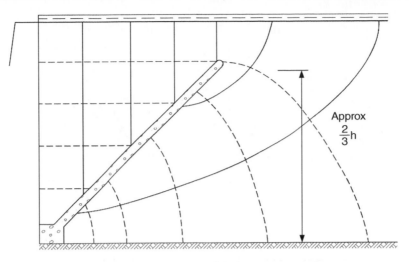

Fig. 8.27 Effect of an inclined drain on seepage forces.

and the groundwater level behind the wall. As there is no real time for steady seepage conditions to develop between the two head levels, the effect of possible seepage forces can safely be ignored.

Example 8.10: Thrust due to saturated soil

A vertical 4 m high wall is founded on a relatively impervious soil and is supporting soil with the properties: $\phi' = 40°$, $c' = 0$, $\delta = 20°$, $\gamma_{sat} = 20$ kN/m³.

The surface of the retained soil is horizontal and is level with the top of the wall. If the wall is subjected to heavy and prolonged rain such that the retained soil becomes saturated and its surface flooded, determine the maximum horizontal thrust that will be exerted on to the wall:

(i)   if there is no drainage system;
(ii) if there is the drainage system of Fig. 8.25d; and
(iii) if there is the drainage system of Fig. 8.25e.

Solution:

(i) No drainage
   The wall is not smooth so we must use the Coulomb wedge analysis method. The total thrust on the back of the wall will be the sum of the active thrust from the submerged soil and the thrust from the water pressure.

   Four trial wedges have been chosen and are shown in Fig. 8.28a, and the corresponding force diagram, drawn to scale, in Fig. 8.28b. Since the diagram is drawn to scale, we can read the magnitude of the maximum value of P_a directly.

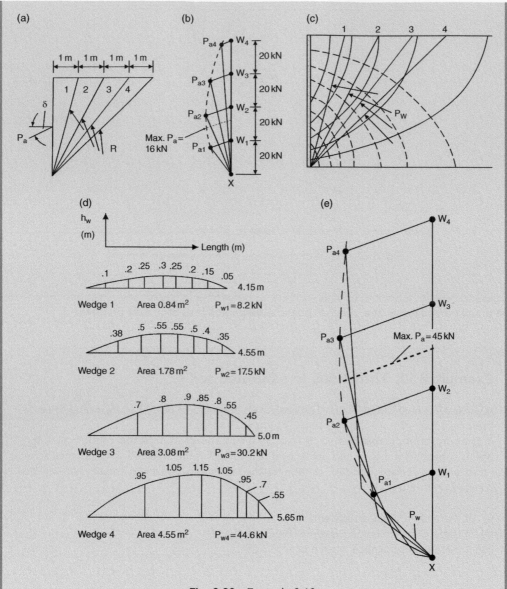

Fig. 8.28 Example 8.10.

Note: It does not matter which dimensions for each wedge are chosen: the final value of maximum P_a will be the same.

e.g. wedge 1:

Weight = ½ × 1 × 4 × (20 − 9.81) = 20 kN

The directions of R and P_a are known, so these are drawn to complete the triangle of forces of the wedge. The weight of wedge 2 (also = 20 kN) is then drawn above W_1 as shown in Fig. 8.28a and its triangle of forces completed. This approach continues for wedges 3 and 4. The curved broken line (referred to as the *Culmann line*) is drawn through the P_a values of each wedge. The maximum P_a value is then easily established and measured from the Culmann line to the vertical weight line.

Maximum P_a due to submerged soil (measured from force diagram) = 16 kN

Horizontal component of P_a = 16.0 × cos 20° = 15 kN

Horizontal thrust from water pressure = $9.81 \times \dfrac{4^2}{2} = 78.5$ kN

Total horizontal thrust = 15 + 78.5 = 93.5 kN/m run of wall

(ii) With vertical drain on back of wall

The flow net for steady seepage from the flooded surface of the soil into the drain is shown in Fig. 8.28c. From this diagram, it is possible to determine the distribution of the excess hydrostatic head, h_w, along the length of the failure surface of each of the four trial wedges. These distributions are shown in Fig. 8.28d and the area of each diagram multiplied by the unit weight of water gives the upward force, P_w, acting at right angles to each failure plane.

The tabulated calculations are:

Wedge	P_w (kN) (=wedge pwp area × γ_w)
1	8.2
2	17.5
3	30.2
4	44.6

The force diagram is shown in Fig. 8.28e.
From the diagram, maximum P_a = 45 kN

⇒ Maximum horizontal thrust on wall = 45 × cos 20° = 42 kN/m

(iii) With inclined drain

As has been shown earlier, for all points in the soil above the drain there can be no excess hydrostatic heads. The force diagram is therefore identical to Fig. 8.28e except that, as P_w is zero for all wedges, it is removed from each polygon of forces. When this is done, it is found that the maximum value of P_a is 30 kN.

⇒ Maximum horizontal thrust on back of wall = 30 × cos 20° = 28 kN/m

This example clearly demonstrates the importance of installing a drainage system behind a retaining wall, as its presence significantly reduces the horizontal thrust being placed on the wall from any anticipated, or unexpected, water pressures.

8.11 Influence of wall yield on design

A wall can yield in one of two ways: either by rotation about its lower edge (Fig. 8.29b) or by sliding forward (Fig. 8.29c). Provided that the wall yields sufficiently, a state of active earth pressure is reached and the thrust on the back of the wall is in both cases about the same (P_a).

The pressure distribution that gives this total thrust value can be very different in each instance. For example, consider a wall that is unable to yield (Fig. 8.29a). The pressure distribution is triangular and is represented by the line AC.

Consider that the wall now yields by rotation about its lower edge until the total thrust = P_a (Fig. 8.29b). This results in conditions that approximate to the Rankine theory and is known as the *totally active case*.

Suppose, however, that the wall yields by sliding forward until active thrust conditions are achieved (Fig. 8.29c). This hardly disturbs the upper layers of soil so that the top of the pressure diagram is similar to the earth pressure at rest diagram. As the total thrust on the wall is the same as in rotational yield, it means that the pressure distribution must be roughly similar to the line AE in Fig. 8.29c.

This type of yield gives conditions that approximate to the wedge theory, with the centre of pressure moving up to between 0.45 and 0.55 h above the wall base and is referred to as the *arching-active case*.

The differences between the various pressure diagrams can be seen in Fig. 8.29d where the three diagrams have been superimposed. It has been found that if the top of a wall moves 0.1% of its height, i.e. a movement of 10 mm in a 10 m high wall, an arching-active case is attained. This applies whether the wall rotates or slides. To achieve the totally active case the top of the wall must move about 0.5%, or 50 mm in a 10 m high wall.

It can therefore be seen that if a retaining wall with a cohesionless backfill is held so rigidly that little yield is possible (e.g. if it is joined to an adjacent structure), it must be designed to withstand earth pressure values much larger than active pressure values.

If such a wall is completely restrained, it must be designed to take earth pressure at rest values, although this condition does not often occur: if a wall is so restrained that only a small amount of yielding can take place, arching-active conditions may be achieved, as in the strutting of trench supports. In this case, the

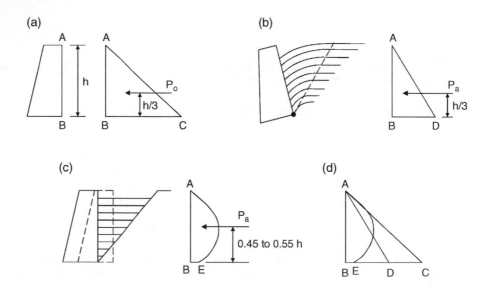

Fig. 8.29 Influence of wall yield on pressure distribution. (a) Wall unable to yield. (b) Wall yield by rotation. (c) Wall yield by sliding forward. (d) Diagrams superimposed.

assumption of a triangular pressure distribution is incorrect, the actual pressure distribution being indeterminate but roughly parabolic.

If the wall yields 0.5% of its height, then the totally active case is attained, and the assumption of a triangular pressure distribution is satisfactory. Almost all retaining walls, unless propped at the top, can yield a considerable amount with no detrimental effects and attain this totally active state.

In the case of a wall with a cohesive backfill, the totally active case is reached as soon as the wall yields, but due to plastic flow within the clay there is a slow build-up of pressure on the back of the wall, which will eventually yield again to re-acquire the totally active pressure conditions. This process is repetitive and, over a number of years, the resulting movement of the wall may be large. For such soils, we can either design for higher pressure or, if the wall is relatively unimportant, design for the totally active case bearing in mind that the useful life of the wall may be short.

8.12   Design parameters for different soil types

8.12.1   Active earth conditions

Owing to various self-compensating factors, the operative values of the strength parameters that determine the value of the active earth pressure are close to the peak values obtained from the triaxial test, even though a retaining wall operates in a state of plane strain. As has been discussed in Chapter 4, the values of these strength parameters vary with both the soil type and the drainage conditions. For earth pressure calculations, attention should be paid to the following.

Sands and gravels

For all stages of construction and for the period after construction the appropriate strength parameter is ϕ'. It is appropriate to take c' as being equal to zero.

Clays

The manner in which a clay soil behaves during its transition from an undrained to a drained state depends upon the previous stress history of the soil and has been described in Chapter 4.

Soft or normally consolidated clay

During and immediately after construction of a wall supporting this type of soil, the vertical effective stress is small, the strength of the soil is at a minimum, and the value of the active earth pressure exerted on to the back of the wall is at a maximum. After construction and after sufficient time has elapsed, the soil will achieve a drained condition. The effective vertical stress will then be equal to the total vertical stress and the soil will have achieved its greatest strength. At this stage therefore, the back of the wall will be subjected to the smallest possible values of active earth pressure (if other factors do not alter).

It is possible to use effective stress analyses to estimate the value of pressure on the back of the wall for any stage of the wall's life. A designer is interested mainly in the maximum pressure values, which occur during and immediately after construction. As it is not easy to predict accurate values of pore water pressures for this stage, an effective stress analysis can be difficult and it is simplest to use the undrained strength parameters in any earth pressure calculations, i.e. ϕ_u (=0) and c_u.

As mentioned in Chapter 4, the sensitivity of a normally consolidated clay can vary from 5 to 10. If it is considered that the soil will be severely disturbed during construction, then the c_u value used in the design calculations should be the undrained strength of the clay remoulded to the same density and at the same water content as the *in situ* values.

If required, the final pressure values on the back of the wall, which apply when the clay is fully drained, can be evaluated in terms of effective stresses using the effective stress parameter ϕ' (N.B. c′ = 0 for a normally consolidated clay). Soft clays usually have to be supported by an embedded wall (see Chapter 9) and water pressures acting on the wall must be considered in the design.

Overconsolidated clay

In the undrained state, negative pore water pressures are generated during shear. This simply means that this type of clay is at its strongest and the pressure on the wall is at its minimum value, during and immediately after construction. The maximum value of active earth pressure will occur when the clay has reached a fully drained condition and the retaining wall should be designed to withstand this value, obtained from the effective stress parameters ϕ' and c′.

With an overconsolidated clay, c′ has a finite value (Fig. 4.39) but, for retaining wall design, this value cannot be regarded as dependable as it could well decrease. It is therefore safest to assume that c′ = 0 and to work with ϕ' only in any earth pressure calculations involving overconsolidated clay. The assumption also helps to allow for any possible increase in lateral pressure due to swelling in an expansive clay as its pore water pressures change from negative (in the undrained state) to zero (when fully drained).

Silts

In many cases, a silt can be assumed to be either purely granular, with the characteristics of a fine sand, or purely cohesive, with the characteristics of a soft clay. When such a classification is not possible then the silt must be regarded as a c − ϕ soil. The undrained strength parameter c_u should be used for the evaluation of active earth pressures which will be applicable to the period of during, and immediately after, construction. The final active earth pressure to which the wall will be subjected can be determined from an effective stress analysis using the parameters ϕ' and c′.

Rainwater in tension cracks

If tension cracks develop within a retained soil and if the surface of the soil is not rendered impervious, then rainwater can penetrate into them. If the cracks become full of water, we can consider that we have a hydro-static, triangular distribution of water pressure acting on the back of the wall over the depth of the cracks, z_0. The value of this pressure will vary from zero at the top of the wall to $9.81 \times z_0$ kPa at the base of the crack. This water pressure should be allowed for in design calculations, see Section 8.5.1 and Example 8.4.

The ingress of water, if prolonged, can lead eventually to softening and swelling of the soil. Swelling could partially close the cracks, but would then cause swelling pressures that could act on the back of the wall. The prediction of values of lateral pressure due to soil swelling is quite difficult.

Shrinkage cracks may also occur and, in the United Kingdom, can extend downwards to depths of about 1.5 m below the surface of the soil. If water can penetrate these shrinkage cracks, then the resulting water pressures should be allowed for as for tension cracks.

8.12.2   Passive earth conditions

Granular soils

It is generally agreed that, for passive pressures in a granular soil, the operative value of ϕ is lower than ϕ_t, the peak triaxial angle obtained from drained tests, particularly for high values of ϕ_t.

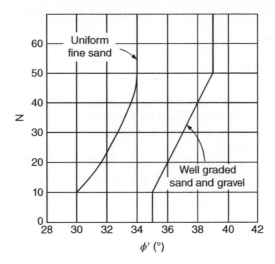

Fig. 8.30 Relationship between N and ϕ'.

With a granular soil, ϕ_t is most often estimated from the results of an *in situ* test such as the standard penetration test (see Chapter 7). It is suggested therefore that values for ϕ', to be used in the determination of passive pressure values, can be obtained from Fig. 8.30 (which is a modified form of Fig. 4.41). The corrected N value can be used in place of the direct blow count N.

Normally consolidated clays

As with the active state, this type of clay is at its weakest when in its undrained state, i.e. during and immediately after construction. For a normally consolidated clay, the operative strength parameter is c_u.

Overconsolidated clays

With this soil, its weakest strength occurs once the soil has reached its drained state. The operative parameters are therefore c' and ϕ', although this is an over-simplification for the case when the level of soil in front of the wall has been reduced by excavation. In this instance, there will be a relief of overburden pressure which could result in softening occurring within the soil. When this happens, some estimation of the strength reduction of the soil must be made, possibly by shear tests on samples of the softened soil.

Silts

As for active pressure, the passive resistance of a silt can be estimated either from the results of *in situ* penetration tests or from a drained triaxial test. For passive pressure, the appropriate strength parameters are c' and ϕ'.

Exercises

Exercise 8.1

A 6 m high retaining wall with a smooth vertical back retains a mass of dry cohesionless soil that has a horizontal surface level with the top of the wall and carries a uniformly distributed surcharge of 10 kPa. The soil weighs 20 kN/m³ and has an angle of shearing resistance of 36°.

Determine the active thrust on the back of the wall per metre length of wall (i) without the uniform surcharge and (ii) with the surcharge.

Answers　(i) 93.6 kN; (ii) 109.2 kN

Exercise 8.2

The back of a 10.7 m high wall slopes away from the soil it retains at an angle of 10° to the vertical. The surface of the soil slopes up from the top of the wall at an angle of 20°. The soil is cohesionless with a weight density of 17.6 kN/m³ and $\phi' = 33°$.

If the angle of wall friction, $\delta = 19°$ determine the maximum thrust on the wall: (a) graphically and (b) analytically using the Coulomb theory.

Answer　(a) and (b) P_a = 476.3 kN

Exercise 8.3

The soil profile acting against the back of a retaining wall is shown in Fig. 8.31. Assuming that Rankine's conditions apply, determine both the theoretical lateral earth pressure distribution and the distribution that would be used in design, and sketch the two pressure distribution diagrams.

From the pressure distribution that would be used in design, determine the magnitude of the total thrust that acts on the wall.

Determine the point of application of the total thrust and express it as the distance from the base of the wall.

Answer　P = 183.6 kN; x = 2.56 m above base.

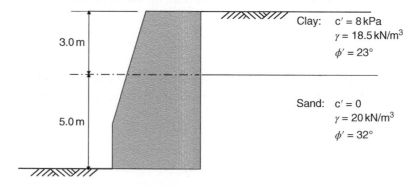

3.0 m

Clay:　$c' = 8\,kPa$
$\gamma = 18.5\,kN/m^3$
$\phi' = 23°$

5.0 m

Sand:　$c' = 0$
$\gamma = 20\,kN/m^3$
$\phi' = 32°$

Fig. 8.31　Exercise 8.3.

Exercise 8.4

A soil has the following properties: $\gamma = 18$ kN/m^3, $\phi' = 30\,°$, $c' = 5$ kPa. The soil is retained behind a 6 m high vertical wall and has a horizontal surface, level with the top of the wall.

If the wall adhesion, a = 5 kPa and $\delta = 15°$ determine the total active horizontal thrust acting on the back of the wall:

(i) with no surcharge acting on the retained soil;
(ii) when the surface of the soil is subjected to a vertical uniformly distributed pressure of 30 kPa.

(Use the values of K_a and K_{ac} from the *earth pressure coefficients* spreadsheet)

Answers   (i) 59.6 kN; (ii) 100.5 kN ($K_a = 0.29$ and $K_{ac} = 1.53$)

Chapter 9

Retaining Structures

9.1 Main types of retaining structures

Various types of retaining structures are used in civil engineering, the main ones being:

- mass construction gravity walls;
- reinforced concrete walls;
- gabion walls;
- crib walls;
- sheet pile walls;
- diaphragm walls;
- reinforced soil walls;
- soil nail walls.

The last two structures are different from the rest in that the soil itself forms part of these structures. Because of this fundamental difference, reinforced soil and soil nail walls are discussed separately at the end of this chapter.

Retaining structures are used to support soils and structures to maintain a difference in elevation of the ground surface and are normally grouped into gravity walls or embedded walls.

Smith's Elements of Soil Mechanics, 10th Edition. Ian Smith.
© 2021 John Wiley & Sons Ltd. Published 2021 by John Wiley & Sons Ltd.
Companion website: www.wiley.com/go/smith/soilmechanics10e

9.2 Gravity walls

9.2.1 Mass construction gravity walls

This type of wall depends upon its weight for its stability and is built of such a thickness that the overturning effect of the lateral earth pressure that it is subjected to does not induce tensile stresses within it.

The walls are built in mass concrete or cemented precast concrete blocks, brick, stone, etc., and are generally used for walls of low height; they become uneconomic for high walls.

The cross-section of the wall is trapezoidal with a base width between 0.3 and 0.5h, where h is the height of the wall. This base width includes any projections of the heel or toe of the wall, which are usually not more than 0.25 m each and are intended to reduce the bearing pressure between the base of the wall and the supporting soil. If the wall is built of poured concrete, then its width at the top should be not less than 0.2 m, and preferably 0.3 m, to allow for the proper placement of the concrete.

9.2.2 Reinforced concrete walls

Cantilever wall

This wall has a vertical, or inclined, stem monolithic with a base slab and is suitable for heights of up to about 7 m. Typical dimensions for the wall are given in Fig. 9.1. Its slenderness is possible as the tensile stresses within its stem and base are resisted by steel reinforcement. If the face of the wall is to be exposed, then general practice is to provide it with a small backward batter of about 1 in 50 in order to compensate for any slight forward tilting of the wall.

Counterfort wall

This wall can be used for heights greater than about 6 m. Its wall stem acts as a slab spanning between the counterfort supports which are usually spaced at about 0.67H but not less than 2.5 m, because of construction considerations. Details of the wall are given in Fig. 9.1b.

A form of the counterfort wall is the buttressed wall, where the counterforts are built on the face of the wall and not within the backfill. There can be occasions when such a wall is useful but, because of the exposed buttresses, it can become unsightly and is not very popular.

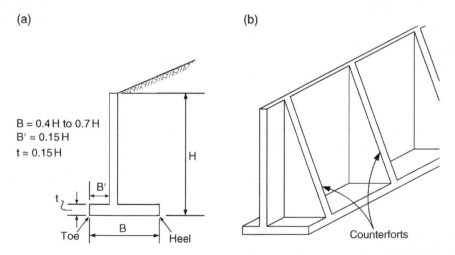

(a) (b)

$B = 0.4H$ to $0.7H$
$B' \approx 0.15H$
$t \approx 0.15H$

Fig. 9.1 Types of reinforced concrete retaining walls. (a) Cantilever wall. (b) Counterfort wall.

Relieving platforms

A retaining wall is subjected to both shear and bending stresses caused by the lateral pressures induced from the soil that it is supporting. A mass construction gravity wall can take such stresses in its stride, but this is not so for the vertical stem of a reinforced concrete retaining wall. If structural failure of the stem is to be avoided, then it must be provided with enough steel reinforcement to resist the bending moment and to have a sufficient thickness to withstand the shear stresses, for all sections throughout its height.

It is this situation that imposes a practical height limitation of about 7 m on the wall stem of a conventional retaining wall. As a wall is increased in dimensions it becomes less flexible and the lateral pressures exerted on it by the soil will tend to be higher than the active values assumed in the design. It is possible therefore to enter a sort of upwards spiral – if a wall is strengthened to withstand increased lateral pressures then its rigidity is increased, and the lateral pressures are increased – and so on.

A way out of the problem is the provision of one or more horizontal concrete slabs, or platforms, placed within the backfill and rigidly connected to the wall stem. A platform carries the weight of the material above it (up as far as the next platform if there are more than one). This vertical force exerts a cantilever moment on to the back of the wall in the opposite direction to the bending moment caused by the lateral soil pressure. The resulting bending moment diagram becomes a series of steps and the wall is subjected to a maximum bending moment value that is considerably less than the value when there are no platforms (Fig. 9.2).

With the reduction of bending moment values to a manageable level, the wall stem can be kept slim enough for the assumption of active pressure values to be realistic, with a consequential more economic construction.

9.2.3 Crib walls

Details of the wall are shown in Fig. 9.3a. It consists of a series of pens made up from prefabricated timber, precast concrete, or steel members which are filled with granular soil. It acts like a mass construction gravity wall with the advantage of quick erection and, due to its flexible nature, the ability to withstand relatively large differential settlements. A crib wall is usually tilted so that its face has a batter of about 1 in 6. The width of the wall can vary from 0.5H to 1.0H and the wall is suitable for heights up to about 6.5 m. It is important to note that, apart from earth fill, a crib wall should not be subjected to surcharge loadings.

9.2.4 Gabion walls

A gabion wall is built of cuboid metal cages or baskets made up from a square grid of galvanised steel fabric, usually 3–5 mm in diameter and spaced 75 mm apart. These baskets are usually 2 m long and 1 m^2 in

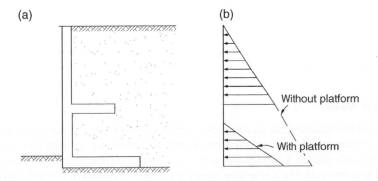

Fig. 9.2 Moment relief platforms. (a) Typical arrangement. (b) Pressure diagram.

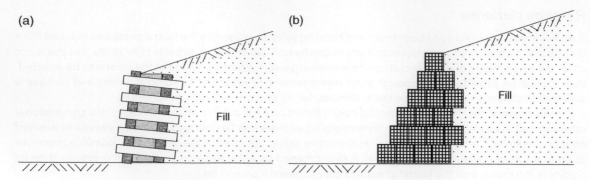

Fig. 9.3   Crib and gabion walls. (a) Crib wall. (b). Gabion wall.

cross-section, filled with stone particles. A central diaphragm fitted in each metal basket divides it into two equal 1 m³ sections, which adds stability. During construction, the stone-filled baskets are secured together with steel wire of 2.5 mm in diameter. The base of a gabion wall is usually about 0.5H, and a typical wall is illustrated in Fig. 9.3b. It is seen that a front face batter can be provided by slightly stepping back each succeeding layer.

9.3   Embedded walls

Embedded walls rely on the passive resistance of the soil in front of the lower part of the wall to provide stability. Anchors or props, where incorporated, provide additional support.

9.3.1   Sheet pile walls

These walls are made up from a series of interlocking piles individually driven into the foundation soil. Most modern sheet pile walls are made of steel, but earlier, walls were also made from timber or precast concrete sections and may still be encountered. There are two main types of sheet pile walls: *cantilever and anchored.*

Cantilever wall

This wall is held in the ground by the active and passive pressures that act on its lower part (Fig. 9.15a).

Anchored wall

This wall is fixed at its base, as is the cantilever wall, but it is also supported by a row, or two rows, of ties or struts placed near its top (Fig. 9.18).

9.3.2   Diaphragm walls

A diaphragm wall could be classed either as a reinforced concrete wall or as a sheet pile wall, but it really merits its own classification. It consists of a vertical reinforced concrete slab fixed in position in the same manner as a sheet pile in that the lower section is held in place by the active and passive soil pressures that act upon it.

A diaphragm wall is constructed by a machine digging a trench in panels of limited length, filled with bentonite slurry as the digging proceeds to the required depth. This slurry has thixotropic properties, i.e. it forms

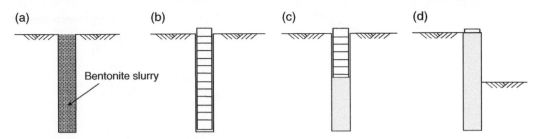

Fig. 9.4 The construction stages of a diaphragm wall. (a) Trench dug. (b) Reinforcement cage inserted. (c) Bentonite displaced by concrete. (d) Soil excavated in front of wall.

into a gel when left undisturbed but becomes a liquid when disturbed. There is no penetration of the slurry into clays, and in sands and silts, water from the bentonite slurry initially penetrates into the soil and creates a virtually impervious skin of bentonite particles, only a few millimetres thick, on the sides of the trench. The reason for the slurry is that it creates lateral pressures which act on the sides of the short trench panel and thus prevents collapse. When excavation is complete, the required steel reinforcement is lowered into position. The trench is then filled with concrete by means of a tremie pipe, the displaced slurry being collected for cleaning and further use.

The wall is constructed in alternating short panel lengths. When the concrete has developed sufficient strength, the remaining intermediate panels are excavated and constructed to complete the wall. The length of each panel is limited to the amount that the soil will arch, in a horizontal direction, to support the ground until the concrete has been placed.

The various construction stages are shown in a simplified form in Fig. 9.4.

9.3.3 Contiguous and secant bored pile walls

Contiguous bored pile walls

This type of wall is constructed from a single or double row of concrete piles placed beside each other. Alternate piles are cast first, and the intermediate piles are then installed. The construction technique allows gaps to be left between piles which can permit an inflow of water in granular conditions. The secant bored pile wall offers a watertight alternative.

Secant bored pile walls

The construction technique is similar to that of the contiguous bored pile wall, except that the alternate piles are drilled at a closer spacing. Then, while the concrete is still green, the intermediate holes are drilled along a slightly offset line so that the holes cut into the first piles. These holes are then concreted to create a watertight continuous wall.

9.4 Design of retaining structures

9.4.1 Failure modes of retaining structures

Retaining structures are designed such that when constructed they will remain stable and support the ground that they are retaining. To enable the design to proceed, an understanding of the potential failure modes of the structure must be known. Common modes of failure, and how they are assessed using Eurocode 7, have

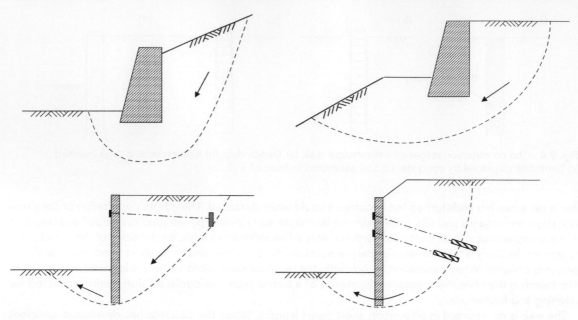

Fig. 9.5 Examples of limit modes for overall stability of retaining structure (based on Figure 9.1, EN 1997-1:2004 (BSI, 2004)).

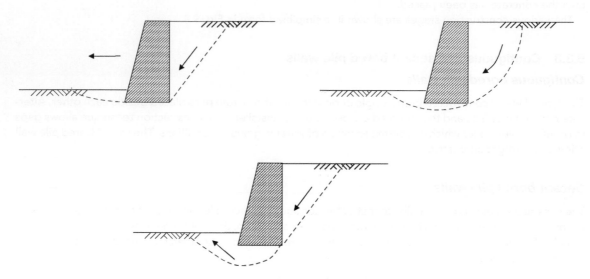

Fig. 9.6 Examples of limit modes for foundation failures of gravity walls (based on Figure 9.2, EN 1997-1:2004 (BSI, 2004)).

been illustrated in Fig. 6.6. Additional examples of how different retaining structures might fail when considering: (i) their overall stability, (ii) failure of their foundation, and (iii) their failure by rotation (embedded walls) are illustrated in Figs. 9.5–9.7.

The traditional approach for the design of retaining structures involved establishing the ratio of the restoring moment (or force) to the disturbing moment (or force) and declaring this ratio as a factor of safety, for any

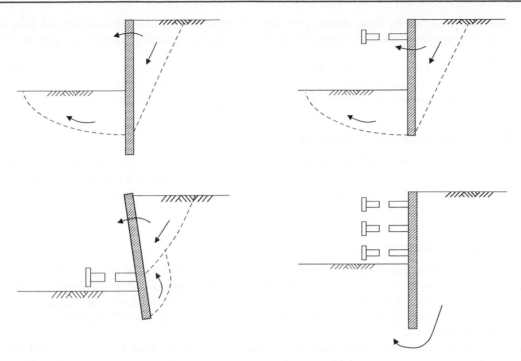

Fig. 9.7 Examples of limit modes for rotational failures of embedded walls (based on Figure 9.3, EN 1997-1:2004 (BSI, 2004)).

of the situations illustrated in Figs. 9.5–9.7. This factor had to be high enough to allow for any uncertainties in the soil parameters used in the analysis, and the approach was generally referred to as the *factor of safety* or *gross pressure* approach. Example 9.2 illustrates the use of this method for a cantilever retaining wall. The limit state design approach set out in Eurocode 7 is of course now used instead.

As explained in Chapter 6, the second generation of Eurocode 7 is under development and is anticipated to be published in the mid-2020s. The differences in the design methods between the two generations of the code are captured in the following sections and the worked examples in this chapter.

9.4.2 Limit states

The relevant limit states for the wall under design must be considered during the design process. These limit states include:

1. Toppling/overturning (Figs. 6.6a and b). Earth pressures acting on the back of the wall induce an overturning moment about the toe of the wall. For a wall to be stable, the resisting moment due to the weight of the wall must be adequate to resist overturning. Further, the resultant thrust must act within the middle third of the base to avoid tilting of the wall.
2. Bearing failure of the soil beneath the structure (Fig. 6.6c). The overturning moment from the lateral thrust causes high bearing pressures at the toe of the wall. These values must be kept within safe limits and less than the design bearing resistance of the soil.
3. Forward sliding (Fig. 6.6d). Caused by insufficient base friction and lack of any passive resistance or supports in front of the wall.
4. Slip of the surrounding soil (Fig. 6.6e). This effect can occur in cohesive soils where the overall mass is at risk of failure. This can be analysed as for a slope stability problem.

5. Structural failure caused by faulty design, poor workmanship, deterioration of materials, etc. (Fig. 6.6f).
6. Excessive deformation of the wall or ground such that, adjacent structures or services reach their ultimate limit state.
7. Rotation of the wall – for embedded walls such as sheet pile walls. Rotation may occur near the toe (cantilevered sheet pile walls) or at the anchor/prop (anchored/propped walls).
8. Unfavourable seepage effects and the adequacy of any drainage system provided.

9.4.3 Design to Eurocode 7 (first generation)

The design of retaining structures is covered in Section 9 of EN 1997-1:2004. The design process adopts the GEO limit state for each check, the only exception being if a gravity wall is founded on rock or a very hard soil where the EQU limit state may be appropriate for the toppling check. The process is illustrated through the worked examples in this chapter.

9.4.4 Design to Eurocode 7 (second generation)

In the second generation, the design of retaining walls is covered in Section 7 of EN1997-3:202x. In accordance with the procedure for establishing the Geotechnical Category (described in Chapter 6), the Geotechnical Complexity Class must be established at an early stage. Most routine retaining wall design work will fall into Geotechnical Complexity Class 2. Higher complexity structures will fall into GCC3, and lower complexity structures will fall into GCC1.

GCC3 retaining structures include situations where: the ground includes weak layers, the ground is experiencing persistent movement or is unstable, the retaining structure is adjacent to existing and sensitive structures, and/or interacts with adjacent structures, or where the retaining structure itself contains complex geometries.

GCC1 retaining structures cover situations where there is negligible risk of the occurrence of an ultimate or serviceability limit state, where the ground conditions are fully known and understood, or where there is negligible risk of ground movement.

As with all geotechnical structures, the limit states against ground failure or rupture are verified using either the material factor approach (MFA) or the resistance factor approach (RFA). The ultimate resistances to sliding and bearing are thus verified using either approach, and the combinations of Design Case and material partial factor set (M1 or M2 – as defined in Table 6.7) are the same as for spread foundations. These are shown in Table 9.1.

Table 9.1 Partial factors for the verification of ground resistance for persistent and transient design situations – gravity retaining structures.

Verification check	Partial factor on	Symbol	Material factor approach*		Resistance factor approach
			(a)	(b)	
Bearing and sliding resistance	Actions and effects of actions	γ_F and γ_E	DC1	DC3	DC4
	Ground properties	γ_M	M1	M2	M1
	Bearing resistance	γ_{Rv}	Not factored		1.4
	Sliding resistance	γ_{Rh}	Not factored		1.1

*Both combinations should be verified:
(a) for structural resistance checks
(b) for geotechnical resistance checks

9.5   Design of gravity retaining walls

The design of gravity retaining walls involves an assessment of the effects of the lateral earth pressures and water pressures acting on the back of the wall. The lateral pressures have a tendency to slide the wall forward or to topple the wall about its toe (Fig. 9.8).

The required dimensions and mass of the wall to resist these effects are determined in the design process. The greater the mass of the wall, the greater the resistance to toppling and sliding. However, the greater the mass, the greater the bearing pressure placed on the soil beneath the wall. Hence, typical geotechnical limit states to be checked for gravity retaining walls are toppling, forward sliding, and bearing failure.

9.5.1   Toppling

The method of analysis for toppling, or overturning, involves a moment equilibrium approach. The wall tends to rotate about its toe. Moments, therefore, are taken about the toe to assess stability: the *destabilising moment* is due to the active and water thrusts acting on the back of the wall (i.e. $M_d = \Sigma (P \times l)$), and the *stabilising, or restoring, moment* is due to the self-weight of the wall ($M_r = W \times l_w$). Lever arms are determined as indicated in Fig. 9.8.

Since the strength of the retained soil affects the destabilising moment, verification against toppling using EN1997: 202x involves using the material factor approach.

9.5.2   Base resistance to sliding

Granular soils and drained clays

The base resistance to sliding is equal to $R_v \tan \delta$, where δ is the angle of friction between the base of the wall and its supporting soil, and R_v is the vertical reaction on the wall base. In limit state design, the sliding limit state will be satisfied if the base resistance to sliding is greater than, or equal to, R_h, the horizontal component of the resultant force acting on the base. In the factor of safety approach, the ratio $(R_v \tan \delta)/R_h$ is determined to establish the factor of safety against sliding. It is common practice to take the passive resistance from any soil in front of a gravity wall as equal to zero, since this soil will be small in depth and in a disturbed state following construction of the wall.

In the case of a drained clay any value of effective cohesion, c' will be so small that it is best ignored.

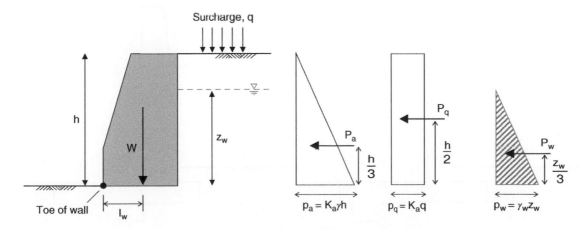

Fig. 9.8 Forces and lever arms: gravity retaining wall.

Undrained clays

The adhesion between the supporting soil and the base of a gravity wall can be taken as equal to the value a used in the determination of the active pressure values (see Section 8.6.2) and based on the value of c_u:

Resistance to sliding = a × Area of base of wall

Shear key

As we have seen, the sliding resistance depends on both the weight of the wall and the coefficient of friction between the base of the wall and the soil. If the resistance to sliding needs to be enhanced to satisfy the limit state requirement, a cost-effective solution is to include a *shear key* in the base of the wall – see Fig. 9.9. This avoids the alternative need to design a larger, and therefore heavier, wall.

The shear key contributes to sliding resistance through two mechanisms: an increase in passive resistance in front of the key, and the relocation of the failure sliding surface from between the soil and wall base, to wholly within the soil. A failure surface wholly within the soil experiences a greater frictional resistance than that experienced along the soil/wall interface.

9.5.3   Bearing pressures on soil

The resultant of the forces due to the pressure of the retained soil and the weight of the wall subject the foundation to both direct and bending effects.

Let R be the resultant force on the foundation, per unit length, and let R_v be its vertical component (Fig. 9.10a).

Considering unit length of wall:

Section modulus of foundation = $\dfrac{B^2}{6}$

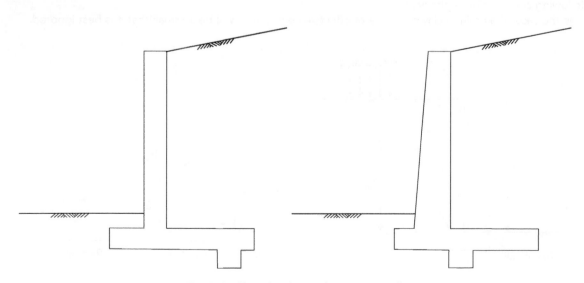

Fig. 9.9   Shear key beneath retaining wall.

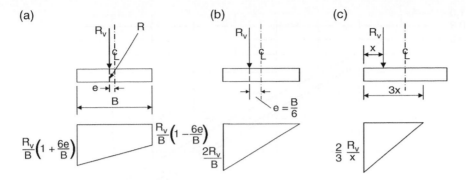

Fig. 9.10 Bearing pressures due to a retaining wall foundation. (a) Within middle third. (b) On middle third. (c) Outside middle third.

Maximum pressure on base = Direct pressure + pressure due to bending

$$= \frac{R_v}{B} + \frac{R6_v e}{B^2} \tag{9.1}$$

$$= \frac{R_v}{B}\left(1 + \frac{6e}{B}\right)$$

$$\text{Minimum pressure on base} = \frac{R_v}{B}\left(1 - \frac{6e}{B}\right) \tag{9.2}$$

The formulae only apply when R_v is within the middle third; when R_v is on the middle third (Fig. 9.10b), then:

$$e = \frac{B}{6}$$

$$\Rightarrow \text{Maximum pressure} = \frac{2R_v}{B}, \quad \text{Minimum pressure} = 0$$

If the resultant R lies outside the middle third (Fig. 9.10c) the formulae become:

$$\text{Maximum pressure} = \frac{2R_v}{3x}, \quad \text{Minimum pressure} = 0$$

9.5.4 Earth pressure coefficients

During the design of retaining walls, it is often appropriate to use Rankine's K_a and K_p, such as in the case of cantilever gravity walls (see Example 9.2). However, when Rankine's conditions do not apply (e.g. where friction exists between wall and soil), Annex C of EN 1997-1:2004 and Annex D of EN 1997-3:202x provide guidance and a set of charts that may be used to determine the horizontal components of K_a and K_p for a given δ/ϕ' ratio (the corrigendum to EN 1997-1:2004, published in 2009, or the UK National Annex to EN 1997-1:2004 (BSI, 2007b) should be used as the original EN 1997-1:2004, contained some published errors). Charts for both horizontal and inclined retained surfaces are given, and the chart to determine the horizontal component of K_a for a horizontal ground surface behind the wall is redrawn in Fig. 9.11. The data on the charts are based on the work of Kerisel and Absi (1990), see Section 8.6.2.

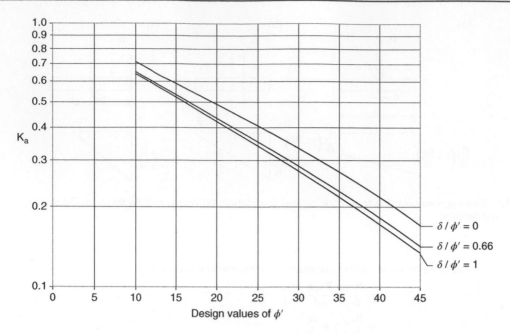

Fig. 9.11 Coefficients K_a (horizontal component) for horizontal retained surface. Based on EN1997-1:2004 (BSI, 2004).

Example 9.1: Mass concrete wall; toppling and sliding by Eurocode 7 (first generation)

Check the proposed design of the mass concrete retaining wall shown in Fig. 9.12. The wall is to be cast into the foundation soil to a depth of 1.0 m and will retain granular fill to a height of 4 m as shown. Take the unit weight of concrete as $\gamma_c = 24\,kN/m^3$ (from EN 1991-1-1:2002) and ignore any passive resistance from the soil in front of the wall. Check the overturning and sliding limit states, using Design Approach 1.

Solution:

(a) Toppling:

Since the wall is founded into soil, the ground will contribute to the stability and therefore toppling is checked using the GEO limit state. For Design Approach 1, we must check both partial factor sets combinations.

1. *Combination 1 (partial factor sets A1 + M1 + R1)*

From Table 6.1: $\gamma_{G;\,unfav} = 1.35$; $\gamma_{G;\,fav} = 1.0$; $\gamma_Q = 1.5$; $\gamma_{\phi}' = 1.0$

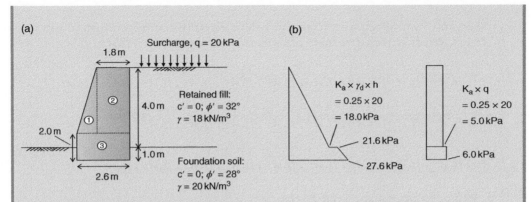

Fig. 9.12 Example 9.1. (a) Retaining wall. (b) Earth pressure diagram (DA1-1).

First, we determine the design material properties and the design actions:

(i) *Design material properties:*
Retained fill:

$$\phi'_d = \tan^{-1}\left(\frac{\tan\phi'}{\gamma_{\phi'}}\right) = \tan^{-1}\left(\frac{\tan 32°}{1.0}\right) = 32°$$

Eurocode 7 states that for concrete walls cast into the soil, δ should be taken as equal to the design value of ϕ', i.e. $\frac{\delta}{\phi'_d} = 1$. From Fig. 9.11, the horizontal component of $K_a = 0.25$.
Foundation soil:

$$\phi'_d = \tan^{-1}\left(\frac{\tan\phi'}{\gamma_{\phi'}}\right) = \tan^{-1}\left(\frac{\tan 28°}{1.0}\right) = 28°$$

From Fig. 9.11, the horizontal component of $K_a = 0.3$.

(ii) *Design actions*
The self-weight of the wall is a permanent, favourable action. Consider the wall as comprising three areas as indicated in Fig. 9.12a. The design weight of each area is determined:

Area 1: $G_{W1;d} = \frac{1}{2} \times 0.8 \times 3 \times \gamma_{concrete} \times \gamma_{G;fav} = 1.2 \times 24 \times 1.0 = 28.8$ kN
Area 2: $G_{W2;d} = 1.8 \times 3 \times \gamma_{concrete} \times \gamma_{G;fav} = 5.4 \times 24 \times 1.0 = 129.6$ kN
Area 3: $G_{W3;d} = 2.6 \times 2 \times \gamma_{concrete} \times \gamma_{G;fav} = 5.2 \times 24 \times 1.0 = 124.8$ kN

The thrust from the active earth pressure behind the wall is a permanent, unfavourable action. Values of the active earth pressure are indicated in Fig. 9.12b.

$$P_{a;d}(fill) = \frac{1}{2} \times 18.0 \times 4 \times \gamma_{G;unfav} = 48.6 \text{ kN}$$

$$P_{a;d}(foundation\ soil) = \frac{1}{2} \times (21.6 + 27.6) \times 1.0 \times \gamma_{G;unfav} = 33.2 \text{ kN}$$

The lateral thrust from the surcharge is a variable, unfavourable action:

$$P_{q;d}(fill) = 5.0 \times 4 \times \gamma_Q = 30.0 \text{ kN}$$

$$P_{q;d}(foundation\ soil) = 6.0 \times 1.0 \times \gamma_Q = 9.0 \text{ kN}$$

(iii) *Design effect of actions and design resistance*

The effect of the actions is to cause the overturning moment about the toe of the wall. This is resisted by the stabilising moment from the self-weight of the wall.

Action	Magnitude of action (kN)	Lever arm (m)	Moment (kNm)
Stabilising:			
Area 1	28.8	$\frac{2}{3} \times 0.8 = 0.53$	15.3
Area 2	129.6	$0.8 + \frac{1.8}{2} = 1.7$	220.3
Area 3	124.8	$\frac{2.6}{2} = 1.3$	162.2
		Total:	397.8
Destabilising:			
P_a (fill)	48.6	$1 + \frac{4}{3} = 2.33$	113.0
P_a (foundation soil)	33.2	$\frac{1(2 \times 21.6 + 27.6)}{3(21.6 + 27.6)} = 0.48$	15.9
P_q (fill)	30.0	$1 + \frac{4}{2} = 3.0$	90.0
P_q (foundation soil)	9.0	$\frac{1.0}{2} = 0.5$	4.5
		Total:	223.4

From the results, it is seen that the limit state is satisfied since the sum of the design destabilising actions and effects (223.4 kNm) is less than the sum of the design stabilising actions and effects (397.8 kNm).

The result may be presented by the *over-design factor*, Γ:

$$\Gamma = \frac{397.8}{223.4} = 1.78$$

2. *Combination 2 (partial factor sets A2 + M2 + R1)*

The partial factors now are: $\gamma_{G;\,fav} = 1.0$; $\gamma_{G;\,unfav} = 1.0$; $\gamma_Q = 1.3$; $\gamma_{\phi}' = 1.25$. The calculations are the same as for Combination 1 except that this time these partial factors are used.

Fill : $\quad \phi_d' = \tan^{-1}\left(\dfrac{\tan\phi'}{\gamma_{\phi'}}\right) = \tan^{-1}\left(\dfrac{\tan 32°}{1.25}\right) = 26.6°; \quad K_a = 0.31$

Foundation soil : $\quad \phi_d' = \tan^{-1}\left(\dfrac{\tan\phi'}{\gamma_{\phi'}}\right) = \tan^{-1}\left(\dfrac{\tan 28°}{1.25}\right) = 23°; \quad K_a = 0.37$

Stabilising moments:

$M_{A1} = 15.3\,\text{kNm}$
$M_{A2} = 220.3\,\text{kNm}$
$M_{A3} = \underline{162.2\,\text{kNm}}$
Total = $397.8\,\text{kNm}$

Destabilising moments:

$M_{Pfill} \quad\quad = 104.8\,\text{kNm}$
$M_{Pfoundation} = 14.6\,\text{kNm}$
$M_{Qfill} \quad\quad = 97.3\,\text{kNm}$
$M_{Qfoundation} = \underline{4.8\,\text{kNm}}$
Total = $221.6\,\text{kNm}$

Thus, the GEO limit state is satisfied and the over-design factor, $\Gamma = \dfrac{397.8}{221.6} = 1.80$

(b) Sliding:

As before, we must check both partial factor sets combinations.

1. *Combination 1 (partial factor sets A1 + M1 + R1)*

From Table 6.1: $\gamma_{G;\,unfav} = 1.35$; $\gamma_{G;\,fav} = 1.0$; $\gamma_Q = 1.5$; $\gamma_{\phi'} = 1.0$

(i) *Design material properties:*
The design values are determined as before:
Retained fill: $\phi'_d = \phi' = 32°$; $K_a = 0.25$
Foundation soil: $\phi'_d = \phi' = 28°$; $K_a = 0.30$

(ii) *Design actions:*
The design total weight of the wall is determined:

$$R_{v;d} = 28.8 + 129.6 + 124.8 = 283.2 \text{ kN}$$

The lateral thrusts are as before.

(iii) *Design effect of actions and design resistance:*
The effect of the actions is to cause forward sliding of the wall. This is resisted by the friction on the underside of the wall.

Total horizontal thrust, $R_{h;d} = 48.6 + 33.2 + 30.0 + 9.0 = 120.8 \text{ kN}$
Design resistance $= R_{v;d} \tan \delta = 283.2 \times \tan 28° = 150.6 \text{ kN}$ (since $\delta = \phi'$)

Thus, the GEO limit state requirement is satisfied and the over-design factor,

$$\Gamma = \frac{150.6}{120.8} = 1.25$$

2. *Combination 2 (partial factor sets A2 + M2 + R1)*
The partial factors are: $\gamma_{G; \text{ fav}} = 1.0$; $\gamma_{G; \text{ unfav}} = 1.0$; $\gamma_Q = 1.3$; $\gamma_{\phi'} = 1.25$. The calculations are the same as for Combination 1 except that this time these partial factors are used.

K_a (fill) = 0.31
K_a (foundation soil) = 0.37
$R_{v;d} = 283.2 \text{ kN}$
$R_{h;d} = 44.9 + 30.4 + 32.5 + 9.7 = 117.5 \text{ kN}$
$R_{v;d} \tan\delta = 283.2 \times \tan 23° = 120.2 \text{ kN}$

Thus, the GEO limit state is satisfied and the over-design factor, $\Gamma = \dfrac{120.2}{117.5} = 1.03$

Overview

The GEO limit state is satisfied for both checks and thus the proposed design of the wall is satisfactory. The lowest value of Γ obtained (in this case 1.03) governs the design.
Annex C of EN 1997-1:2004 (BSI, 2004) also gives formulae which may be used to determine separate active earth pressure coefficients for surcharge loadings (K_{aq}) and for cohesion in the soil (K_{ac}). Example 9.1 has no cohesion ($K_{ac} = 0$) but does have a surcharge. But, as the surface of the soil is horizontal, K_{aq} is equal to K_a.

Example 9.2: Mass concrete wall; toppling and sliding by Eurocode 7 (second generation)

Consider again the retaining wall of Example 9.1. Check the toppling and sliding limit states, if the wall falls into Consequence Class 2.

Solution:

(a) Toppling verification:

Toppling is verified using the MFA. In this approach, we must check both combinations DC1–M1 and DC3–M2.

DC1–M1:

First, we determine the design material properties and the design actions:

(i) *Design material properties:*

The partial factors for materials in set M1 are all equal to 1.0 (see Table 6.7).

Thus, $\gamma_{\tan \phi} = 1.0$.

Retained fill:

$$\phi'_d = \tan^{-1}\left(\frac{\tan \phi'}{\gamma_{\tan \phi}}\right) = \tan^{-1}\left(\frac{\tan 32°}{1.0}\right) = 32°$$

EN 1997-3:202x states that the angle of soil/wall friction, δ should satisfy the inequality, $\tan \delta \leq k_\delta \times \tan \phi$. For concrete walls cast into the soil, k_δ should be taken as equal to 1.0, i.e. $\delta = \phi$.

From Fig. 9.11, the horizontal component of $K_a = 0.25$.

Foundation soil:

$$\phi'_d = \tan^{-1}\left(\frac{\tan \phi'}{\gamma_{\tan \phi}}\right) = \tan^{-1}\left(\frac{\tan 28°}{1.0}\right) = 28°$$

From Fig. 9.11, the horizontal component of $K_a = 0.3$.

(ii) *Design actions*

As wall is in Consequence Class 2, $K_F = 1.0$ (see Table 6.2).

For Design Case 1, the following partial factors apply to actions (see Table 6.6):

Permanent unfavourable, $\gamma_G = 1.35 \, K_F = 1.35$

Permanent favourable, $\gamma_{G, \, fav} = 1.0$

Variable unfavourable, $\gamma_Q = 1.5 \, K_F = 1.5$

The self-weight of the wall is a permanent, favourable action. Consider the wall as comprising three areas as indicated in Fig. 9.12a. The design weight of each area is determined:

Area 1: $G_{W1;d} = \frac{1}{2} \times 0.8 \times 3 \times \gamma_{concrete} \times \gamma_{G;fav} = 1.2 \times 24 \times 1.0 = 28.8\, kN$
Area 2: $G_{W2;d} = 1.8 \times 3 \times \gamma_{concrete} \times \gamma_{G;fav} = 5.4 \times 24 \times 1.0 = 129.6\, kN$
Area 3: $G_{W3;d} = 2.6 \times 2 \times \gamma_{concrete} \times \gamma_{G;fav} = 5.2 \times 24 \times 1.0 = 124.8\, kN$

The thrust from the active earth pressure behind the wall is a permanent, unfavourable action. Values of the active earth pressure are the same as DA1-1 in Example 9.1 as indicated in Fig. 9.12b.

$P_{a;d}$ (fill) $= \frac{1}{2} \times 18.0 \times 4 \times \gamma_G = 48.6\, kN$
$P_{a;d}$ (foundation soil) $= \frac{1}{2} \times (21.6 + 27.6) \times 1.0 \times \gamma_G = 33.2\, kN$

The lateral thrust from the surcharge is a variable, unfavourable action:

$P_{q;d}$ (fill) $= 5.0 \times 4 \times \gamma_Q = 30.0\, kN$
$P_{q;d}$ (foundation soil) $= 6.0 \times 1.0 \times \gamma_Q = 9.0\, kN$

It is noted therefore that the design values of all actions are the same as for DA1-1 in Example 9.1. Therefore, the rest of the verification calculations will be the same too, and thus,

$$\Gamma = \frac{397.8}{223.4} = 1.78$$

The toppling limit state verification is satisfied, since $\Gamma > 1$.
DC3–M2:
For DC3 and M2, the following partial factors apply: $\gamma_G = 1.0$; $\gamma_{G;\, fav} = 1.0$; $\gamma_Q = 1.3$; $\gamma_{\tan\phi} = 1.25\, K_M$.
As wall is in Consequence Class 2, $K_M = 1.0$ (see Table 6.8).
The calculations are the same as for DC1 & M1 above, except that this time these partial factors are used.
It is noted therefore that the design values of all actions and material properties are the same as for DA1-2 in Example 9.1. Therefore, the rest of the verification calculations will be the same too, and thus,

$$\Gamma = \frac{397.8}{221.6} = 1.80$$

Again, the toppling limit state verification is satisfied, since $\Gamma > 1$.
(b) Sliding verification
Sliding is checked by the RFA, using DC4 (see Table 6.6). In this approach, the permanent actions and material properties are not factored, and the safety is applied to the resistance, the variable actions and the effects of the actions.
The partial factor on sliding resistance, $\gamma_{Rh} = 1.1$ (from Table 9.1) and on variable actions, $\gamma_Q = \dfrac{\gamma_{Q,1}}{\gamma_{G,1}}$ (from Table 6.6). Also, the partial factor on the effects of the unfavourable actions, $\gamma_E = 1.35\, K_F$.

$$\gamma_Q = \frac{\gamma_{Q,1}}{\gamma_{G,1}} = \frac{1.5 \times K_F}{1.35 \times K_F} = 1.11$$

Thus,

$$\gamma_{Rh} = 1.1; \quad \gamma_Q = 1.11; \quad \gamma_E = 1.35 \text{ (since } K_F = 1.0); \quad \gamma_{\tan\phi} = 1.0$$

Drawing on the values determined earlier in this example using some of these partial factors, we have:

$$\text{Fill}: \quad \phi'_d = \tan^{-1}\left(\frac{\tan\phi'}{\gamma_{\tan\phi}}\right) = \tan^{-1}\left(\frac{\tan 32°}{1.0}\right) = 32°; \quad K_a = 0.25$$

$$\text{Foundation soil}: \quad \phi'_d = \tan^{-1}\left(\frac{\tan\phi'}{\gamma_{\tan\phi}}\right) = \tan^{-1}\left(\frac{\tan 28°}{1.0}\right) = 28°; \quad K_a = 0.3$$

$P_{a;d}$ (fill) $= \frac{1}{2} \times 18.0 \times 4 = 36.0$ kN
$P_{a;d}$ (foundation soil) $= \frac{1}{2} \times (21.6 + 27.6) \times 1.0 = 24.6$ kN
$P_{q;d}$ (fill) $= 5.0 \times 4 \times \gamma_Q = 22.2$ kN
$P_{q;d}$ (foundation soil) $= 6.0 \times 1.0 \times \gamma_Q = 6.7$ kN
Design total sliding force, $P_d = (36 + 24.6 + 22.2 + 6.7) \times \gamma_E = 120.8$ kN
Also, design total weight of the wall, $R_{v;d} = 28.8 + 129.6 + 124.8 = 283.2$ kN
Now, since $\delta = \phi'$ and $\gamma_{\tan\delta} = 1.0$:

$$\text{Design resistance} = \frac{R_{v;d} \tan\delta}{\gamma_{Rh}} = \frac{283.2 \times \tan 28°}{1.1} = 136.9 \text{ kN}$$

Hence, the over-design factor, $\Gamma = \dfrac{136.9}{120.8} = 1.13$.
The sliding limit state verification is satisfied, since $\Gamma > 1$.

Example 9.3: Strength and stability checks by traditional and Eurocode 7 (first generation) approaches

The proposed design of a cantilever retaining wall is shown in Fig. 9.13. The unit weight of the concrete is 25 kN/m³ (EN 1991-1-1:2002), and the soil has weight density 18 kN/m³. The soil peak strength parameters are $\phi' = 38°$, $c' = 0$ and the ultimate bearing capacity of the soil beneath the wall has been calculated to be 750 kPa.

The retained soil carries a variable surcharge, approximated to a uniform surcharge of intensity 10 kPa. Ignore any passive resistance from the soil in front of the wall.

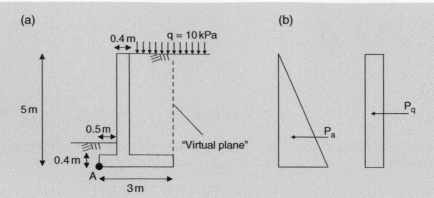

Fig. 9.13   Example 9.3. (a) Wall geometry. (b) Pressure distributions.

Check the safety of the proposed design:

(a) by the traditional (gross pressure) method, assuming that the coefficient of friction between base of wall and soil is equal to tan ϕ';

(b) against the GEO (Design Approach 1) limit state of Eurocode 7.

Solution:

(a) Traditional method

Note: When the retained soil is supported by a heel, the design assumes a virtual plane as shown in Fig. 9.13a provided that the heel width, b satisfies the inequality

$$b \geq h \ \tan\left(45° - \frac{\phi'}{2}\right)$$

If the heel width satisfies the above inequality (it does in this example), Rankine's conditions apply along this face and the earth pressures acting here are established in the design.

Sliding:

Using Rankine's theory (with $\phi' = 38°$), $K_a = 0.238$

Active thrust from soil, $P_a = \frac{1}{2} K_a \gamma h^2 = \frac{1}{2} \times 0.238 \times 18 \times 5^2 = 53.6$ kN

Thrust due to surcharge, $P_q = K_a q h = 0.238 \times 10 \times 5 = 11.9$ kN

$$\sum H = 65.5 \text{ kN}$$

Since the surcharge is variable, it cannot be relied upon to contribute to the resistance to sliding (through the weight of the wall) so is excluded in the determination of R_v.

Vertical reaction, R_v = weight of base + weight of stem + soil on heel (excl.surcharge)

$$= 25(0.4 \times 3.0) + 25(0.4 \times 4.6) + 18(2.1 \times 4.6)$$
$$= 30.0 + 46.0 + 173.9$$
$$= 249.9 \text{ kN}$$

Total force causing sliding, $R_h = 65.5$ kN

Force resisting sliding = $R_v \tan \delta = 249.9 \times \tan 38° = 195.2$ kN

Factor of safety against sliding, $F_s = \dfrac{195.2}{65.5} = 3.0$

Overturning:

Taking moments about point A, the toe of the wall.

Disturbing moment, M_D:

$$M_D = P_a \times \left(\frac{5}{3}\right) + P_q \times \left(\frac{5}{2}\right)$$

$$= 89.3 + 29.8$$

$$= 119.1 \text{ kNm}$$

Resisting moment, M_R

Due to base = $30.0 \times 1.5 = 45.0$ kNm

Due to stem = $46.0 \times 0.7 = 32.2$ kNm

Due to soil on heel = $173.9 \times 1.95 = 339.1$ kNm

$M_R = 45.0 + 32.2 + 339.1 = 416.3$ kNm

Factor of safety against overturning, $F_o = \dfrac{416.3}{119.1} = 3.5$

Bearing:

Refer to Fig. 9.14 and consider both the weight of wall, R_v and effect of the surcharge, q_s over heel.

Take moments about point A.

First, establish lever arm of R_v:

For moment equilibrium, the net moment must be equal to $M_R - M_D$

$$= 416.3 - 119.1 = 297.2 \text{ kNm}$$

$\Rightarrow$ The lever arm, x_1 of R_v = 297.2/249.9 = 1.19 m

Next, determine the lever arm, x_2 of the surcharge force acting over the heel:

$$x_2 = 1.95 \text{ m}$$

Knowing that the total vertical force, $V = R_v + q_s = 249.9 + (10 \times 2.1) = 270.9$ kN

we can find the lever arm of the equivalent resultant vertical force about A:

$$R_v \cdot x_1 + q_s \cdot x_2 = V \cdot x$$

$$x = \frac{(249.9 \times 1.19) + (10 \times 2.1 \times 1.95)}{270.9} = 1.25 \text{ m (within middle } 1/3 \Rightarrow \text{ok)}$$

$$e = B/2 - x = 1.5 - 1.25 = 0.25 \text{ m}$$

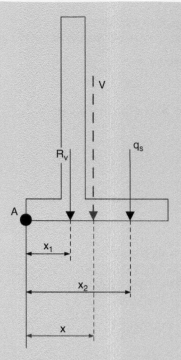

Fig. 9.14 Example 9.3: vertical forces.

Max bearing pressure $= \dfrac{V}{B}\left(1 + \dfrac{6e}{B}\right)$

$= \dfrac{270.9}{3}\left(1 + \dfrac{6 \times 0.25}{3}\right)$

$= 135.6 \text{ kPa}$

Factor of safety against bearing failure, $F_b = \dfrac{750.0}{135.6} = 5.5$

(b) Eurocode 7

For the overall stability of the wall, once again we check safety against the GEO limit state.

1. *Combination 1 (partial factor sets A1 + M1 + R1)*

From Table 6.1: $\gamma_{G;\ unfav} = 1.35$; $\gamma_{G;\ fav} = 1.0$; $\gamma_Q = 1.5$; $\gamma_{\phi'} = 1.0$; $\gamma_{Rv} = 1.0$

(i) Design material properties

$\phi'_d = 38°$: Using Rankine's theory, $K_a = \dfrac{1 - \sin\phi_d}{1 + \sin\phi_d} = 0.238$

(ii) Design actions

In sliding and toppling, the weight of the wall is a permanent, favourable action.

The influence of the surcharge on the soil on the heel is ignored since transient (variable) actions can never be considered favourable.

Stem: $W_d = 0.4 \times 4.6 \times \gamma_{concrete} \times \gamma_{G;\ fav} = 1.84 \times 25 \times 1.0 = 46.0\,kN$

Base: $W_d = 0.4 \times 3.0 \times \gamma_{concrete} \times \gamma_{G;\ fav} = 1.2 \times 25 \times 1.0 = 30.0\,kN$

Soil on heel: $W_d = 21 \times 4.6 \times \gamma \times \gamma_{G;\ fav} = 9.66 \times 18 \times 1.0 = 173.9\,kN$

Total, $R_{v;d}$: 249.9 kN

The thrust from the active earth pressure is a permanent, unfavourable action.

$$P_{a;d} = \frac{1}{2} \times 0.238 \times 18 \times 5^2 \times \gamma_{G;\ unfav} = 72.3\ kN$$

The lateral thrust from the surcharge is a variable, unfavourable action.

$$P_{q;d} = 0.238 \times 10 \times 5 \times \gamma_Q = 17.8\ kN$$

(iii) Design effect of actions and design resistance

Sliding:

Total horizontal thrust, $R_{h;d} = 72.3 + 17.8 = 90.1\,kN$

Design resistance $= R_{v;d} \tan \delta = 249.9 \times \tan 38^\circ = 195.2\,kN$ (since $\delta = \phi'$)

The GEO limit state requirement for sliding is satisfied and the over-design factor,

$$\Gamma = \frac{195.2}{90.1} = 2.17$$

Toppling:

Destabilising moment, $M_{dst;d} = 72.3 \times \left(\frac{5}{3}\right) + 17.8 \times \left(\frac{5}{2}\right) = 165.0\ kNm$

Stabilising moment, $M_{stb;d} = 46.0 \times 0.7 + 30.0 \times 1.5 + 173.9 \times 1.95$
$$= 416.3\ kNm$$

The GEO limit state requirement for toppling is satisfied since $M_{dst;d} < M_{stb;d}$, and the over-design factor,

$$\Gamma = \frac{416.3}{165.0} = 2.52$$

Bearing:

From above, destabilising moment = 165.0 kNm

Stabilising moment:

The weight of the wall, R_v is now considered as an *unfavourable* action and includes the surcharge acting on the soil on the heel:

$$M_R = M_{stem} + M_{base} + M_{heel}$$
$$= (46.0 \times 0.7 + 30.0 \times 1.5 + 173.9 \times 1.95) \times \gamma_{G;\ unfav} + (21.0 \times 1.95) \times \gamma_Q$$
$$= 623.4 \text{ kNm}$$

$$R_{v;d} = (46.0 + 30.0 + 173.9) \times \gamma_{G;\ unfav} + 21.0 \times \gamma_Q = 368.9 \text{ kN}$$

Lever arm of $R_{v;d}$, $x = \dfrac{623.4 - 165.0}{368.9} = 1.24$ m (within middle third of base)

Eccentricity, $e = 1.5 - 1.24 = 0.26$ m

Maximum bearing pressure $= \dfrac{R_{v;d}}{B}\left(1 + \dfrac{6e}{B}\right) = \dfrac{368.9}{3.0}\left(1 + \dfrac{6 \times 0.26}{3.0}\right) = 186.9$ kPa

Design bearing resistance, $R_{v;d} = \dfrac{R_{v;d}}{\gamma_{Rv}} = \dfrac{750}{1.0} = 750$ kPa

The GEO limit state requirement for bearing is satisfied and the over-design factor

$$\Gamma = \frac{750}{186.9} = 4.0$$

2. *Combination 2 (partial factor sets A2 + M2 + R1)*

   The calculations are the same as for Combination 1 except that this time the following partial factors (from Table 6.1) are used:

   $\gamma_{G;\ unfav} = 1.0$; $\gamma_{G;\ fav} = 1.0$; $\gamma_Q = 1.3$; $\gamma_\phi' = 1.25$; $\gamma_{Rv} = 1.0$

   The favourable actions are the same as for Combination 1 since $\gamma_{G;\ fav} = 1.0$.

$$\phi_d' = \tan^{-1}\left(\frac{\tan\phi'}{\gamma_{\phi'}}\right) = \tan^{-1}\left(\frac{\tan 38°}{1.25}\right) = 32°; \quad K_a = 0.307$$

Sliding:

$R_{v;d} = 249.9$ kN

$R_{h;d} = 69.1 + 20.0 = 89.1$ kN

$R_{v;d} \tan\delta = 249.9 \times \tan 32° = 156.2$ kN

The GEO limit state requirement for sliding is satisfied and the over-design factor,

$$\Gamma = \frac{156.2}{89.1} = 1.75.$$

Toppling:

Destabilising moment, $M_{dst;d} = 69.1 \times \left(\dfrac{5}{3}\right) + 20.0 \times \left(\dfrac{5}{2}\right) = 165.2$ kNm

Stabilising moment, $M_{stb;d} = 416.3$ kNm

The GEO limit state requirement for toppling is satisfied and $\Gamma = \dfrac{416.3}{165.2} = 2.52$

Bearing:

$M_{stb} = M_{stem} + M_{base} + M_{heel}$
$$= (46.0 \times 0.7 + 30.0 \times 1.5 + 173.9 \times 1.95) \times \gamma_{G;\ unfav} + (21.0 \times 1.95) \times \gamma_Q$$
$$= 469.5 \text{ kNm}$$

$R_{v;d} = (46.0 + 30.0 + 173.9) \times \gamma_{G;\ unfav} + 21.0 \times \gamma_Q = 277.2 \text{ kN}$

Lever arm of $R_{v;d}$, $x = \dfrac{469.5 - 165.2}{277.2} = 1.1$ m   (within middle third of base)

Eccentricity, $e = 1.5 - 1.1 = 0.4$ m

Maximum bearing pressure $= \dfrac{R_{v;d}}{B} \left(1 + \dfrac{6e}{B}\right) = \dfrac{277.2}{3.0} \left(1 + \dfrac{6 \times 0.4}{3.0}\right) = 166.3$ kPa

Design bearing resistance, $R_{v;d} = \dfrac{R_{v;d}}{\gamma_{Rv}} = \dfrac{750}{1.0} = 750$ kPa

The GEO limit state requirement for bearing is satisfied and the over-design factor

$\Gamma = \dfrac{750}{166.3} = 4.5$

Example 9.4: Strength and stability checks by Eurocode 7 (second generation)

Check the retaining wall of Example 9.3 using the approaches set out in the second generation of Eurocode 7. The wall, as a routine earth retaining structure, falls into Consequence Class 2 (so $K_F = K_M = 1.0$).

Solution:

The calculations are the same as for Example 9.3(b), only this time we adopt the different partial factors combinations as listed for each verification check below.

(i) Toppling – Material Factor Approach
   DC1-M1:
   $\gamma_G = 1.35\ K_F;\ \gamma_{G;\ fav} = 1.0;\ \gamma_Q = 1.5\ K_F;\ \gamma_{\tan\phi} = 1.0$

   $\phi'_d = 38°$: Using Rankine's theory, $K_a = \dfrac{1 - \sin\phi_d}{1 + \sin\phi_d} = 0.238$

   In sliding and toppling, the weight of the wall is a permanent, favourable action. The influence of the surcharge on the soil on the heel is ignored since transient (variable) actions can never be considered favourable.

   The calculations are the same as those of Example 9.3(b), DA1-1 (ii). Hence,

Design weight of wall, $R_{v;d} = 249.9\,kN$
Design active thrust, $P_{a;d} = 72.3\,kN$
Design surcharge thrust, $P_{q;d} = 17.8\,kN$
Destabilising moment, $M_{dst;d} = 165.0\,kNm$
Stabilising moment, $M_{stb;d} = 416.3\,kNm$

The limit state requirement for toppling is satisfied since $M_{dst;d} < M_{stb;d}$, and the over-design factor,

$$\Gamma = \frac{416.3}{165.0} = 2.52$$

DC3-M2:
$\gamma_G = 1.0$; $\gamma_{G;\ fav} = 1.0$; $\gamma_Q = 1.3$; $\gamma_{tan\phi} = 1.25\ K_M$
The calculations are the same as those of Example 9.3(b), DA1-2. Hence,

Destabilising moment, $M_{dst;\,d} = 69.1 \times \left(\dfrac{5}{3}\right) + 20.0 \times \left(\dfrac{5}{2}\right) = 165.2\ kNm$

Stabilising moment, $M_{stb;d} = 416.3\,kNm$

The limit state requirement for toppling is satisfied and $\Gamma = \dfrac{416.3}{165.2} = 2.52$

(ii) Sliding – Resistance Factor Approach
    DC4-M1:
    $\gamma_Q = 1.11$; $\gamma_{tan\phi} = 1.0$; $\gamma_E = 1.35$; $\gamma_{Rh} = 1.1$

    Using Rankine's theory (with $\phi' = 38°$), $K_a = 0.238$
    $P_{a;d} = 53.6\,kN$
    $P_{q;d} = 11.9 \times \gamma_Q = 13.2\,kN$
    $\sum P = 53.6 + 13.2 = 66.8\,kN$
    Design total sliding force, $P_d = 66.8 \times \gamma_E = 90.2\,kN$

As with Example 9.3, since the surcharge is variable, it cannot be relied upon to contribute to the resistance to sliding (through the weight of the wall) so is excluded in the determination of R_v.

    From Example 9.3, design weight of wall, $R_{v;d} = 249.9\,kN$
    Force resisting sliding, $R_h = R_v \tan \delta = 249.9 \times \tan 38° = 195.2\,kN$
    Design resistance, $R_{h;\,d} = \dfrac{195.2}{\gamma_{Rh}} = \dfrac{195.2}{1.1} = 177.5\ kN$

The sliding limit state requirement is satisfied since $R_{h;d} > P_d$ and the over-design factor,

$$\Gamma = \frac{177.5}{90.2} = 1.97$$

(iii) Bearing – Resistance Factor Approach
    DC4-M1:
    $\gamma_Q = 1.11$; $\gamma_{tan\phi} = 1.0$; $\gamma_E = 1.35$; $\gamma_{Rv} = 1.4$
    Refer to Fig. 9.14 and consider both weight of wall, R_v and effect of the surcharge, q_s over heel.

Take moments about point A and establish the lever arm of R_v:

For moment equilibrium the net moment, M must be equal to $M_{stb;d} - M_{dst;d}$

Since G_k is not factored in DC4, effectively $\gamma_{G;\ fav} = 1.0$ (same as MFA).

Therefore, $M_{stb;d} = 416.3$ kNm.

From (ii) $P_{a;d} = 53.6$ kN

$P_{q;d} = 0.238 \times 10 \times 5 \times \gamma_Q = 13.2$ kN

$$M_{dst;d} = 53.6 \times \left(\frac{5}{3}\right) + 13.2 \times \left(\frac{5}{2}\right) = 122.3 \text{ kNm}$$

$$M = 416.3 - 122.3 = 294.0 \text{ kNm}$$

$\Rightarrow$ The lever arm, x_1 of $R_v = 294.0/249.9 = 1.18$ m

Next, determine the lever arm, x_2 of the surcharge force acting over the heel:

$$x_2 = 1.95 \text{m}$$

It is seen that $V = R_v + q_s = 249.9 + (10 \times 2.1) = 270.9$ kN

Now we can find the lever arm of the equivalent resultant vertical force about A:

$$R_v \cdot x_1 + q_s \cdot x_2 = V \cdot x$$

$$x = \frac{(249.9 \times 1.18) + (10 \times 2.1 \times 1.95)}{270.9} = 1.24 \text{ m (within middle } 1/3 \Rightarrow \text{ ok)}$$

$$e = B/2 - x = 1.5 - 1.24 = 0.26 \text{ m}$$

$$\text{Max bearing pressure} = \frac{V}{B}\left(1 + \frac{6e}{B}\right)$$

$$= \frac{270.9}{3}\left(1 + \frac{6 \times 0.26}{3}\right)$$

$$= 137.3 \text{ kPa}$$

Design bearing resistance, $R_{v;d} = \dfrac{R_{v;d}}{\gamma_{Rv}} = \dfrac{750}{1.4} = 535.7$ kPa

The bearing limit state requirement is satisfied since $R_{v;d} >$ the maximum bearing pressure and the over-design factor,

$$\Gamma = \frac{535.7}{137.3} = 3.9$$

9.6 Design of sheet pile walls

A sheet pile wall is a flexible structure which depends for stability upon the passive resistance of the soil in front of and behind the lower part of the wall. Stability also depends on the anchors when incorporated.

Retaining walls of this type differ from other walls in that their weight is negligible compared with the remaining forces involved. Design methods usually neglect the effect of friction between the soil and the wall, but this omission is fairly satisfactory when determining active pressure values. This is because sheet piles are produced under factory conditions that ensures smooth sections roll off the rolling and shaping mills. It should be remembered, however, that the effect of wall friction can almost double the Rankine value of K_p.

9.6.1 Cantilever walls

Sheet pile walls are flexible, and sufficient yield will occur in a cantilever wall to give totally active earth pressure conditions in the retained soil (Fig. 9.15).

Let the height of the wall be h, and suppose that it is required to find the depth of embedment, d, that will make the wall stable. For equilibrium, the active pressure on the back of the wall must be balanced by the passive pressure both in front of and behind the wall. If an arbitrary point C is chosen and it is assumed that the wall will rotate outwards about this point, the theoretical pressure distribution on the wall is as shown in Fig. 9.15c. The toe of the wall (point D, Fig. 9.15a) is deep enough such that the conditions that prevail are known as *fixed earth* conditions.

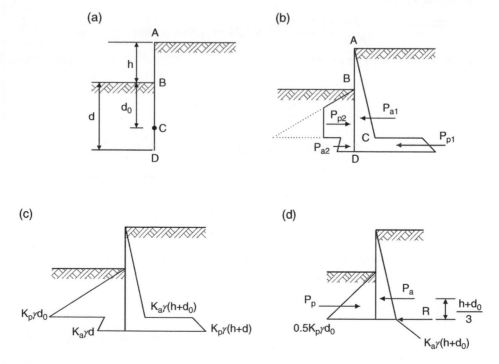

Fig. 9.15 Pressure distribution on sheet pile wall. (a) Wall geometry. (b) Gross pressure: theoretical pressure distribution. (c) Eurocode 7 method. (d) Gross pressure: distribution assumed for design.

Table 9.2 Acceptable F_p values for ranges of ϕ'.

ϕ' (degrees)	F_p
>30	2.0
20–30	1.5–2.0
<20	1.5

Data from Padfield and Mair (1984).

Limit state design method to establish required design depth

The depth, d, is obtained by balancing the disturbing and restoring moments about C, together with the horizontal forces established using the pressure distribution shown in Fig. 9.15c. In Eurocode 7 (first generation), the GEO limit state is applied to assess the rotational stability using this theoretical pressure distribution. The method generates two equations containing the unknowns d and d_0, which are solved by repeated iteration until the correct values are obtained.

Example 9.5 illustrates the design of a cantilever sheet pile wall using the GEO limit state.

Traditional methods

Various traditional methods of design exist, and some are still in use for small projects and temporary works. Each involves the determination of an overall factor of safety, F_p for passive resistance based on different lateral earth pressure distributions. The methods were described in earlier editions of this book: the most common of which is known as the *gross pressure method*, the pressure distribution of which is shown in Fig. 9.15b.

In this method, the total theoretical value of thrust $K_p\gamma d^2/2$ is divided by a factor of safety, traditionally taken as $F_p = 2.0$. The effective passive resistance in front of the wall is therefore assumed to have a magnitude of $K_p\gamma d^2/4$ and is of trapezoidal distribution, the centre of pressure of this trapezium lying between d/2 and d/3 above the base of the pile (for ease of calculation the value is generally taken as d/2). It is common to use lower values of F_p for low values of ϕ'. Padfield and Mair (1984) recommended the values given in Table 9.2.

Calculations are considerably simplified if it is assumed that the passive resistance on the back of the wall, P_{p1}, acts as a concentrated load, R, on the foot of the pile, leading to the pressure distribution shown in Fig. 9.15d, from which d can be obtained by taking moments of thrusts about the base of the pile. The value of d obtained by this method is more nearly the value of d_0 in Fig. 9.15a, the customary practice being to increase the value of d by 20–25% to allow for this effect.

During the design of sheet pile walls, an allowance for a *future unplanned excavation* in front of the wall is made: the retained height is increased by 10%, to a maximum of 0.5 m.

9.6.2 Dealing with passive earth pressure

Eurocode 7 – first generation

There is a question over how passive pressure should be treated when using the Eurocode 7 GEO limit state, as it could be regarded as either a *favourable action* or as a *resistance*.

i.e. to establish the design passive resistance, we have either:

$$P_{p;d} = P_{p;k} \times \gamma_{G;fav}$$

or

$$P_{p;d} = \frac{P_{p;k}}{\gamma_{Re}}$$

The values of both $\gamma_{G; fav}$ and γ_{Re} for each design approach, taken from Table 6.1, are presented in Table 9.3. From the table it is clear that the question is only of concern when using Design Approach 2. For the other design approaches both $\gamma_{G; fav}$ and γ_{Re} are 1.0 and thus $P_{p;d}$ will be equal to $P_{p;k}$.

The prevailing view (Schuppener, 2008) is that the passive thrust for gravity walls should be treated as a resistance, thus even the concern when using Design Approach 2 is addressed. Further, it could be argued that once consideration has been given for an unplanned excavation, the contribution to stability from any passive resistance will be small and could, conservatively, be ignored.

However, the question gains a bit of additional confusion when considering embedded walls. Recalling that the purpose of a partial factor of safety is to allow for the uncertainty in the derived characteristic value, then any design action (e.g. active thrust) determined from a soil property (e.g. angle of shearing resistance, ϕ') will contain the same degree of uncertainty in its value as any other action (e.g. passive thrust) which is determined from the same soil property. In the case of an embedded wall where the same soil exists on both sides of the wall, this means that the uncertainty in the active thrust (taken up by the application of the permanent, unfavourable partial factor, $\gamma_{G,unfav}$) must be the same as that in the passive thrust. We must therefore consider the passive thrust also to be a permanent, unfavourable action. This follows from the *Single Source Principle* in which actions coming from the same source must be combined with a single partial factor of safety. Indeed EN 1997-1:2004 (BSI, 2004) states the following:

NOTE: Unfavourable (or destabilising) and favourable (or stabilising) permanent actions may in some situations be considered as coming from a single source. If they are considered so, a single partial factor may be applied to the sum of these actions or to the sum of their effects.

Table 9.4 updates Table 9.3 with the inclusion of the permanent, unfavourable partial factors, $\gamma_{G; unfav}$.

Table 9.3   Values of $\gamma_{G; fav}$ and γ_{Re} for each design approach.

| | Design approach | | | |
| | 1 | | 2 | 3 |
	Combination 1	Combination 2		
$\gamma_{G; fav}$	1.0	1.0	1.0	1.0
γ_{Re}	1.0	1.0	1.4	1.0

Table 9.4   Values of $\gamma_{G; fav}$, $\gamma_{G; unfav}$, and γ_{Re} for each design approach.

| | Design approach | | | |
| | 1 | | 2 | 3 |
	Combination 1	Combination 2		
$\gamma_{G; fav}$	1.0	1.0	1.0	1.0
$\gamma_{G; unfav}$	1.35	1.0	1.35	1.0
γ_{Re}	1.0	1.0	1.4	1.0

Now, we see that treating the passive thrust as a permanent, unfavourable action affects Design Approach 1 (Combination 1) and Design Approach 2. In the United Kingdom, where Design Approach 1 is adopted, the earth resistance partial factor $\gamma_{Re} = 1.0$ and thus it is appropriate to treat passive resistance as a permanent, unfavourable action since the thrust derives from the same source as the active pressure and the level of uncertainty in its value is the same (i.e. application of single source principle).

Eurocode 7 – second generation

In the second generation of Eurocode 7, the limit state against rotational failure is verified using either the material factor approach (MFA) or the resistance factor approach (RFA). The design case and partial factors combinations are shown in Table 9.5. Since both the MFA and RFA approaches are available, the question of how to deal with passive earth resistance described above is not relevant in the second generation.

Table 9.5 Partial factors for the verification of passive resistance against embedded retaining structures for persistent and transient design situations.

| Verification check | Partial factor on | Symbol | Material factor approach* | | Resistance factor approach |
			(a)	(b)	
Rotational resistance	Actions and effects of actions	γ_F and γ_E	DC4	DC3	DC4
	Ground properties	γ_M	M1	M2	Not factored
	Passive resistance	γ_{Re}	Not factored		1.4

*Both combinations should be verified.

Example 9.5: Cantilever sheet pile wall

Calculate the minimum depth of embedment, d, to provide stability to a cantilever sheet pile wall, retaining an excavated depth of 5 m using:

(a) Eurocode 7 (first generation) – GEO limit state, Design Approach 1;
(b) Gross pressure method;
(c) Eurocode 7 (second generation) – using the RFA (take wall as a Consequence Class 2 structure).

The soil properties are $\phi'_{peak} = 30°$, $c' = 0$, $\gamma = 20 \, kN/m^3$.

Solution:

The problem is illustrated in Fig. 9.16a.

(a) Eurocode 7, GEO Limit State, Design Approach 1
Allowance is made for a future unplanned excavation Δa equal to 10% of the clear height (=0.5 m). The pressure distribution is shown in Fig. 9.16b.

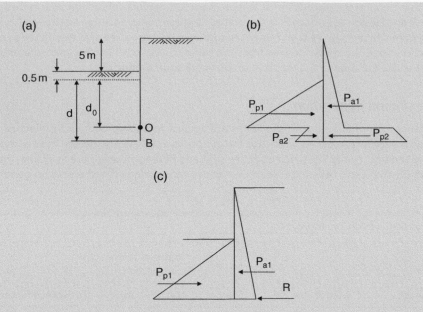

Fig. 9.16 Example 9.5. (a) Wall geometry. (b) Pressure distribution. (c) Simplified pressure distribution.

(i) *Combination 1 (partial factor sets A1 + M1 + R1)*
From Table 6.1: $\gamma_{G;\ unfav} = 1.35$; $\gamma_{\phi'} = 1.0$

$$\phi'_d = \tan^{-1}\left(\frac{\tan\phi'}{\gamma_{\phi'}}\right) = \tan^{-1}\left(\frac{\tan 30°}{1.0}\right) = 30°$$

Using Rankine's theory, $K_a = 0.33$, $K_p = 3.0$
(It is appropriate to use Rankine's theory as the steel sheet piling can be assumed to be smooth and thus no friction exists between wall and soil. Also, the back of the wall is vertical of course.)
The earth pressures acting at the salient points in the distribution are established.
Behind wall:

$$p_{a,O} = 0.33 \times 20 \times (d_0 + 5.5) = 6.67(d_0 + 5.5)\ kPa$$
$$p_{p,O} = 3.0 \times 20 \times (d_0 + 5.5) = 60(d_0 + 5.5)\ kPa$$
$$p_{p,B} = 3.0 \times 20 \times (d + 5.5) = 60(d + 5.5)\ kPa$$

In front of wall:
$$p_{p,O} = 3.0 \times 20 \times d_0 = 60d_0\ kPa$$
$$p_{a,O} = 0.33 \times 20 \times d_0 = 6.67d_0\ kPa$$
$$p_{a,B} = 0.33 \times 20 \times d = 6.67d\ kPa$$

Design actions:
The active thrust due to the earth pressure is a permanent, unfavourable action:

$$P_{a1;d} = \frac{1}{2} \times 6.67(d_0 + 5.5) \times (d_0 + 5.5) \times \gamma_{G;\ unfav} = 3.33(d_0 + 5.5)^2 \times 1.35$$

$$= 4.5(d_0 + 5.5)^2$$

$$P_{a2;d} = \frac{1}{2} \times (6.67d_0 + 6.67d) \times (d - d_0) \times \gamma_{G;\ unfav} = 4.5(d_0 + d) \times (d - d_0)$$

$$= 4.5\left(d^2 - d_0^2\right)$$

The passive reaction is also a permanent, unfavourable action, since the thrust derives from the same source as the active pressure, and the level of uncertainty in its value should be the same. (This follows from the single source principle – see Section 9.6.2.)

$$P_{p1;d} = \frac{1}{2} \times 60d_0 \times d_0 \times \gamma_{G;\ unfav} = 40.5d_0^2$$

$$P_{p2;d} = \frac{1}{2} \times (60(d + 5.5) + 60(d_0 + 5.5)) \times (d - d_0) \times \gamma_{G;\ unfav}$$

$$= 40.5(d - d_0)[(d + 5.5) + (d_0 + 5.5)]$$

$$= 40.5\left(d^2 - d_0^2\right) + 445.5(d - d_0)$$

Effect of actions:
Consider moments about point O.

	Force (kN)	Lever arm (m)	Moment (kNm)
$P_{a1;d}$	$4.5(d_0 + 5.5)^2$	$\dfrac{(d_0 + 5.5)}{3}$	$1.5(d_0 + 5.5)^3$
$P_{a2;d}$	$4.5(d^2 - d_0^2)$	$\dfrac{(d - d_0)(2d + d_0)^*}{3(d + d_0)}$	$\dfrac{1.5 \times \left(d^2 - d_0^2\right)(d - d_0)(2d + d_0)}{(d + d_0)}$
$P_{p1;d}$	$40.5d_0^2$	$\dfrac{d_0}{3}$	$13.5d_0^3$
$P_{p2;d}$	$40.5(d^2 - d_0^2) +$ $445.5(d - d_0)$	$\dfrac{(d - d_0)(2d + d_0 + 16.5)^*}{3(d + d_0 + 11)}$	$\left[40.5\left(d^2 - d_0^2\right) + 445.5(d - d_0)\right]$ $\times \left[\dfrac{(d - d_0)(2d + d_0 + 16.5)}{3(d + d_0 + 11)}\right]$

* These formulae are established from the fact that the thrust acts through the centroid of a trapezoidal shaped part of the pressure distribution. The lever arm is equal to $\dfrac{(d - d_0)\left(2p_{p,B} + p_{p,O}\right)}{3\left(p_{p,O} + p_{p,B}\right)}$. For a simplistic approach, the thrust could be considered as acting at mid-height, i.e. lever arm $\approx \dfrac{(d - d_0)}{2}$.

d and d_0 are obtained by resolving the moment and force equilibrium equations, which necessitates the use of a programmable calculator or spreadsheet such as the spreadsheet *Example 9.5.xls*, available for download:

$$\sum M_O = 0$$

$$M_{Pp1} + M_{Pp2} - M_{Pa1} - M_{Pa2} = 0$$

$$13.5d_0^3 + \left[40.5\left(d^2 - d_0^2\right) + 445.5(d - d_0)\right] \times \left[\frac{(d - d_0)(2d + d_0 + 16.5)}{3(d + d_0 + 11)}\right] \qquad (9.3)$$

$$- 1.5(d_0 + 5.5)^3 - \left[\frac{1.5 \times \left(d^2 - d_0^2\right)(d - d_0)(2d + d_0)}{(d + d_0)}\right] = 0$$

$$\sum H = 0$$

$$P_{p1;d} + P_{a2;d} - P_{a1;d} - P_{p2;d} = 0$$

$$40.5d_0^2 + 4.5\left(d^2 - d_0^2\right) - 4.5(d_0 + 5.5)^2 \qquad (9.4)$$

$$- \left[40.5\left(d^2 - d_0^2\right) + 445.5(d - d_0)\right] = 0$$

Equations (9.3) and (9.4) solve for:

$$d_0 = 4.8 \text{ m}$$

$$d = 5.4 \text{ m}$$

(ii) *Combination 2 (partial factor sets A2 + M2 + R1)*

The calculations are the same as for Combination 1 except that this time the following partial factors are used: $\gamma_{G; \text{ unfav}} = 1.0$; $\gamma_\phi' = 1.25$.

The following expressions are then derived ($K_a = 0.41$; $K_p = 2.44$):

$$\sum M_O = 0$$

$$8.13d_0^3 + 8.13(d - d_0)^2[(d_0 + 5.5) + 2(d + 5.5)] \qquad (9.5)$$

$$- 1.36(d_0 + 5.5)^3 - 0.172(d - d_0)^2(7.9d_0 + 15.8d) = 0$$

$$\sum H = 0$$

$$24.4d_0^2 + 4.09\left(d^2 - d_0^2\right) - 4.09(d_0 + 5.5)^2 \qquad (9.6)$$

$$- \left[24.4\left(d^2 - d_0^2\right) + 268.4(d - d_0)\right] = 0$$

Equations (9.5) and (9.6) solve for:

$$d_0 = 6.4 \text{ m}$$

$$d = 7.2 \text{ m}$$

(b) Gross pressure method

In the gross pressure method, the net passive resistance below the point of rotation is replaced by the horizontal force R, as shown in Fig. 9.17.

Using Rankine's theory (with $\phi' = 30°$) $K_a = \dfrac{1}{3}$; $K_p = 3.0$.

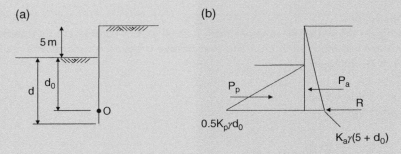

Fig. 9.17 Example 9.5. (a) Wall geometry. (b) Pressure distribution.

	Force (kN)	Lever arm (m)	Moment (kNm)
P_a	$\dfrac{20}{2 \times 3}(5 + d_0)^2$	$\dfrac{(5 + d_0)}{3}$	$\dfrac{10}{9}(5 + d_0)^3$
P_p	$15d_0^2$	$\dfrac{d_0}{3}$	$5d_0^3$

Minimum depth is required, and since $F_p = 2.0$ has already been applied to the pressure distribution (see Section 9.6.1),

$$\frac{5d_0^3}{\frac{10}{9}(5 + d_0)^3} = \frac{9d_0^3}{2(5 + d_0)^3} = 1$$

$$d_0 = 7.7 \text{ m}$$

To obtain the design depth, d, d_0 is increased by an amount equal to the extent required to generate a net passive resistance force below the point of rotation at least as large as R. (R is obtained from simple horizontal force equilibrium). This demands additional calculations, and it is common practice to avoid this by simply increasing d_0 by 20–25% to give d. In this example, we shall take 20%.
i.e. $d = d_0 \times 1.2 = 7.7 \times 1.2 = 9.24$ m

(c) Eurocode 7 (second generation)
Resistance Factor Approach
Allowance is made for a future unplanned excavation Δa equal to 10% of the clear height (=0.5 m). The pressure distribution is shown in Fig. 9.16b.
The RFA uses the partial factor combination of DC4 and $\gamma_{Ro} = 1.4$.
$\gamma_Q = 1.11$; $\gamma_E = 1.35$; $\gamma_{Re} = 1.4$
Since material properties are not factored in the RFA, K_a is the same as Part (a) (i) above, i.e. $K_a = 0.33$. Therefore, the earth pressures are also the same as in Part (a) (i).

Design actions:
Drawing on the earlier calculations from Part (a) (i), but considering that permanent actions are not factored in DC4 and that passive pressure is treated as a resistance in the RFA, we have:

$$P_{a1;d} = 3.33(d_0 + 5.5)^2$$
$$P_{a2;d} = 3.33(d^2 - d_0^2)$$

$$P_{p1;d} = \frac{30\,d_0^2}{\gamma_{Re}} = \frac{30\,d_0^2}{1.4} = 21.4d_0^2$$

$$P_{p2;d} = \frac{30\left(d^2 - d_0^2\right) + 330(d - d_0)}{\gamma_{Re}} = 21.4\left(d^2 - d_0^2\right) + 235.7(d - d_0)$$

Effect of actions:
Consider moments about O. The lever arms have been determined in Part (a) (i). The design effects of the *active actions* are considered by applying the partial factor to their moments:
$$M_{a1;d} = 1.11(d_0 + 5.5)^2 \times \gamma_E = 1.5(d_0 + 5.5)^3$$

$$M_{a2;d} = \frac{1.11 \times \left(d^2 - d_0^2\right)(d - d_0)(2d + d_0) \times \gamma_E}{(d + d_0)} = \frac{1.5 \times \left(d^2 - d_0^2\right)(d - d_0)(2d + d_0)}{(d + d_0)}$$

$$M_{p1;d} = 7.1d_0^3$$

$$M_{p2;d} = \left[21.4\left(d^2 - d_0^2\right) + 235.7(d - d_0)\right]\frac{(d - d_0)}{2}$$
$$= \left[10.7\left(d^2 - d_0^2\right) + 117.9(d - d_0)\right](d - d_0)$$

d and d_0 are obtained by resolving the moment and force equilibrium equations, which necessitates the use of a programmable calculator, symbolic maths solver software or spreadsheet such as the *Example 9.5.xls*, available for download:

$$\sum M_O = 0$$
$$M_{Pp1} + M_{Pp2} - M_{Pa1} - M_{Pa2} = 0$$

$$\sum H = 0$$
$$P_{p1;d} + P_{a2;d} - P_{a1;d} - P_{p2;d} = 0$$

which solve for
$$d_0 = 7.4\,m$$
$$d = 8.6\,m$$

9.6.3   Anchored and propped walls

When the top of a sheet pile wall is anchored, a considerable reduction in the embedment depth can be obtained (Fig. 9.18a). Due to this anchorage, the lateral yield in the upper part of the wall is similar to the yield in a braced excavation (see Section 9.7), whereas in the lower part, the yield is similar to that of a retaining wall yielding by rotation. As a result, the pressure distribution on the back of an anchored sheet pile is a combination of the totally active and the arching-active cases, the probable pressure distribution is indicated in Fig. 9.18b. In practice, the pressure distribution behind the wall is assumed to be totally active.

The anchor or prop force required can be obtained by equating horizontal forces: $T = P_a - P_p$, from which a value is obtained per metre run of wall. The resulting value of T is increased by 25% to allow for flexibility in the piling and arching in the soil. Anchors are usually spaced at 2–3 m intervals and secured to stiffening wales.

Anchorage can be obtained by the use of additional piling or by anchor blocks (large concrete blocks in which the tie is embedded). Any anchorage block must be outside the possible failure plane (Fig. 9.19a), and when space is limited piling becomes necessary (Fig. 9.19b). If bending is to be avoided in the anchorage pile, then a pair of raking piles can be used (Fig. 9.19c).

9.6.4   Depth of embedment for anchored walls

As the depth of embedment is not as great as for the cantilever wall, the toe of the wall is not rigidly fixed into the ground and is free to move slightly. The analysis of this condition is thus referred to as the *free earth support method*. With this method it is assumed that rotation occurs about the anchor point and that sufficient yielding occurs for the development of active and passive pressures. The pressure distribution assumed in design is shown in Fig. 9.18b, and the wall is considered free to move at its base. By taking moments about the anchor at A, an expression for the embedment depth, d, can be obtained. The traditional methods of assessing the ratio of restoring moments to overturning moments described for cantilever walls are also used for anchored walls.

The design of anchored walls to Eurocode 7 (first generation) involves the use of the GEO limit state to assess the rotational stability, as illustrated in Example 9.6. The design to the second generation of Eurocode 7 follows either the MFA or RFA.

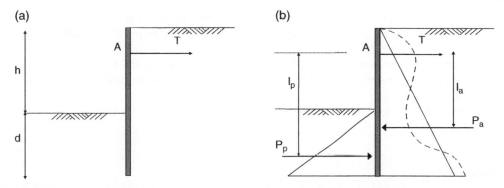

Fig. 9.18   Free earth support method for anchored sheet piled walls. (a) Anchored sheet pile wall. (b) Distribution assumed for design.

(a)

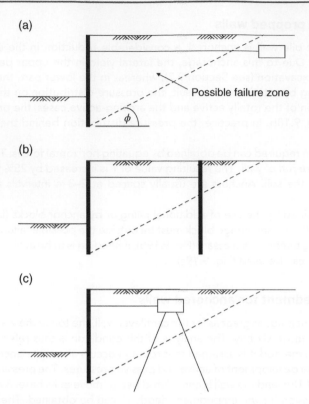

Possible failure zone

ϕ

(b)

(c)

Fig. 9.19 Anchorage systems for sheet pile walls. (a) Anchor block. (b) Parallel row of sheet piles. (c) Raking piles.

Example 9.6: Anchored sheet pile wall

If an anchor is placed 1 m below the ground level behind the sheet pile wall described in Example 9.5, calculate the minimum depth of embedment, d, to provide stability using:

(a) Eurocode 7 (first generation) – GEO limit state, Design Approach 1;
(b) gross pressure method, taking F_p, the factor of safety on passive resistance, as equal to 2.0.
(c) Eurocode 7 (second generation) – using the MFA (take wall as a Consequence Class 2 structure).

Solution:

(a) Eurocode 7, GEO Limit State, Design Approach 1
As before, allowance is made for a future unplanned excavation Δa equal to 10% of the clear height. This is the height between the ground surface in front of the wall and the anchor (=0.4 m). The pressure distribution is shown in Fig. 9.20.

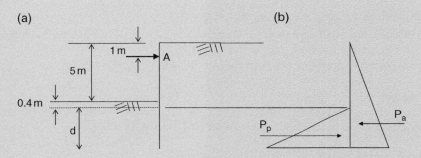

Fig. 9.20 Example 9.6 part (a). (a) Wall geometry. (b) Pressure distribution.

(i) *Combination 1 (partial factor sets A1 + M1 + R1)*
From Table 6.1: $\gamma_{G;\ unfav} = 1.35$; $\gamma_{\phi'} = 1.0$
From before, $K_a = 0.33$, $K_p = 3.0$
Design actions:
The active thrust due to the earth pressure is a permanent, unfavourable action:

$$P_{a;d} = \frac{1}{2} \times K_a \times \gamma \times (d + 5.4)^2 \times \gamma_{G;\ unfav} = 4.5(d + 5.4)^2$$

The passive resistance is also considered a permanent, unfavourable action:

$$P_{p;d} = \frac{1}{2} \times 60 \times d \times d \times \gamma_{G;\ unfav} = 40.5d^2$$

Effect of actions:

d is obtained by resolving the moment equilibrium equation. The lever arms about A are:

$$l_a : \frac{2}{3}(d + 5.4) - 1$$

$$l_p : \frac{2}{3}d + 4.4$$

$$\sum M = 0$$

i.e. $M_{Pa} - M_{Pp} = 0$

$$4.5(d + 5.4)^2 \times \left[\frac{2}{3}(d + 5.4) - 1\right] - 40.5d^2 \times \left(\frac{2}{3}d + 4.4\right) = 0$$

Using Example 9.6.xls, d = 2.1 m
Design depth = 2.1 + 0.4 = 2.5 m

(ii) *Combination 2 (partial factor sets A2 + M2 + R1)*

The calculations are the same as for Combination 1 except that this time the following partial factors are used: $\gamma_{G;\ unfav} = 1.0$; $\gamma_{\phi'} = 1.25$.

The following expressions are then derived ($K_a = 0.41$; $K_p = 2.44$):

$$\sum M = 0$$

$$4.09(d + 5.4)^2 \times \left[\frac{2}{3}(d + 5.4) - 1\right] - 24.4d^2 \times \left(\frac{2}{3}d + 4.4\right) = 0$$

Using Example 9.6.xls, d = 2.9 m
Design depth = 2.9 + 0.4 = 3.3 m

(b) Gross pressure method

The pressure distribution is shown in Fig. 9.21.
Using Rankine's theory (with $\phi' = 30°$) $K_a = 0.33$; $K_p = 3.0$:

	Force (kN)	Lever arm (m)	Moment (kNm)
P_a	$\dfrac{10}{3}(d + 5)^2$	$\dfrac{2}{3}(d + 5) - 1$	$\dfrac{10}{3}(d + 5)^2\left[\dfrac{2}{3}(d + 5) - 1\right]$
P_p	$15d^2$	$\dfrac{2}{3}d + 4$	$15d^2\left[\dfrac{2}{3}d + 4\right]$

Minimum depth is required, and since $F_p = 2.0$ has already been applied to the pressure distribution,

$$\frac{15d^2\left[\dfrac{2}{3}d + 4\right]}{\dfrac{10}{3}(d + 5)^2\left[\dfrac{2}{3}(d + 5) - 1\right]} = 1$$

by trial and error, d = 3.4 m

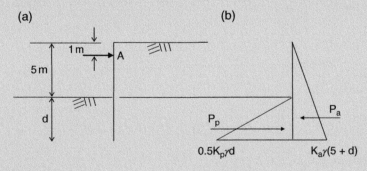

Fig. 9.21   Example 9.6 part (b). (a) Wall geometry. (b) Pressure distribution.

(c) Eurocode 7 (second generation)
Material Factor Approach

(i) DC4 and M1

$\gamma_Q = 1.11$; $\gamma_{tan\phi} = 1.0$; $\gamma_E = 1.35$

Drawing from calculations made in Part (a) (i), $K_a = 0.33$, $K_p = 3.0$.

Design actions (permanent, unfavourabe):

$$P_{a;d} = \frac{1}{2} \times K_a \times \gamma \times (d + 5.4)^2 = 3.33(d + 5.4)^2$$

$$P_{p;d} = \frac{1}{2} \times 60 \times d \times d = 30d^2$$

Effect of actions:

d is obtained by resolving the moment equilibrium equation. The lever arms are:

$$l_a : \frac{2}{3}(d + 5.4) - 1$$

$$l_p : \frac{2}{3}d + 4.4$$

$$\sum M = 0$$

i.e. $(M_{Pa} \times \gamma_E) - (M_{Pp} \times \gamma_E) = 0$

$$\left(3.33(d + 5.4)^2 \times \left[\frac{2}{3}(d + 5.4) - 1\right] \times 1.35\right) - \left(30d^2 \times \left(\frac{2}{3}d + 4.4\right) \times 1.35\right) = 0$$

Using Example 9.6.xls, d = 2.1 m
Design depth = 2.1 + 0.4 = 2.5 m

(ii) DC3 and M2

$\gamma_G = 1.0$; $\gamma_{G; fav} = 1.0$; $\gamma_Q = 1.3$; $\gamma_{tan\phi} = 1.25\, K_M$

Drawing from calculations made in Part (a) (ii), ($K_a = 0.41$; $K_p = 2.44$):

$$\sum M = 0$$

$$4.09(d + 5.4)^2 \times \left[\frac{2}{3}(d + 5.4) - 1\right] - 24.4d^2 \times \left(\frac{2}{3}d + 4.4\right) = 0$$

Using Example 9.6.xls, d = 2.9 m
Design depth = 2.9 + 0.4 = 3.3 m

9.6.5   Reduction of design moments in anchored sheet pile walls

Rowe (1952) conducted a series of model tests in which he showed that the bending moments that actually occur in an anchored sheet pile wall are less than the values computed by the free earth support method. This difference in values is due mainly to arching effects within the soil which create a passive pressure distribution in front of the wall that is considerably different from the theoretical triangular distribution assumed for the analysis. Because of this phenomenon, the point of application of the passive resistive force occurs at a much shallower depth than the generally assumed value of d/3 (where d = depth of penetration of the pile).

Rowe later extended his work to cover clay soils (1957, 1958) and suggested a semi-empirical approach, covering the main soil types, whereby the values computed by the free earth support method for both the moments in the pile and the tension in the tie can be realistically reduced. The method involves the use of two coefficients, r_d and r_t, and worked examples illustrating the use of the method have been prepared by Barden (1974). Numerical studies performed by Potts and Fourie (1984) have confirmed Rowe's findings for normally consolidated clays. However, their studies showed that Rowe's results do not stand for overconsolidated clays, and they have produced separated design charts for this case.

9.6.6   Treatment of groundwater conditions

In order to carry out the stability analysis of a retaining wall involving groundwater, it is necessary to know the values of the water pressures acting on both sides of the wall.

If there is a water level on one side of the wall only, the problem is simple to analyse and was illustrated in Example 8.3. If there are water levels on both sides of the wall but at the same elevation, then the two water pressure diagrams are equal and therefore balance out. Hence, apart from allowing for the fact that the soil below the water table is submerged, no special treatment is necessary.

With different water levels on both sides of the pile, flow can occur:

Clay soils: because of the low permeability of the clay, seepage cannot take place around the toe and hydrostatic water pressures act on each side of the wall.

Granular soils: seepage pressures can be determined using a flow net as indicated in Fig. 9.22a. A more convenient approximation, however, is to assume that the excess head causing the flow is distributed linearly around the length of the pile that is within the water zone, i.e. (2d + h − i − j), as shown in Fig. 9.22b.

It can be shown that the pore water pressure at the toe of the wall,

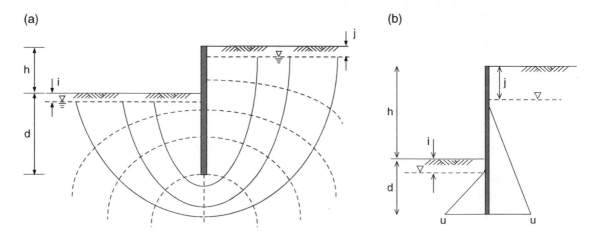

Fig. 9.22   Granular soils: water pressures. (a) Flow net. (b) Simplified distribution.

$$u_t = \frac{2(h + d - j)(d - i)\gamma_w}{(2d + h - i - j)} \tag{9.7}$$

For both hydrostatic and simplified seepage conditions, the *net* water pressure moment (i.e. consideration of water moments on both sides of wall) is determined and used in the moment equilibrium equation.

Example 9.7: Water pressure distribution

Determine an approximation for the water pressure distribution on each side of a sheet pile wall installed in a granular soil if $h = 8\,$m, $d = 6\,$m and $i = j = 0$ (i.e. GWL is at ground surface on both sides of the wall). Take $\gamma_w = 9.81\,$kN/m^3.

Solution:

With the assumption that the excess head is linearly distributed around the length of the pile within the water zone, the water pressure on both sides at the pile toe is

$$u_t = \frac{2(h + d - j)(d - i)\gamma_w}{(2d + h - i - j)} = 82.4 \;\text{kPa}$$

The assumed diagrams for water pressure on each side of the wall are shown in Fig. 9.23a, and the net water pressure diagram is shown in Fig. 9.23b.

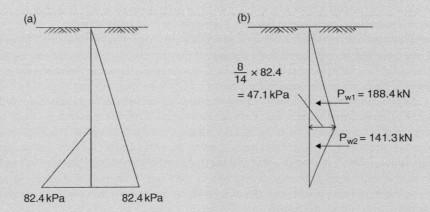

Fig. 9.23   Example 9.7. (a) Simplified water pressure distribution. (b) Net water pressures.

Eurocode 7 – first generation

EN 1997-1:2004 (BSI, 2004) states that in silts and clays, water pressures must be considered to act and, unless a reliable drainage system is installed, the groundwater table should be taken as a coincident with the ground surface of the retained soil. The resulting water pressures are considered as geotechnical actions (permanent, unfavourable), and the appropriate partial factors of safety are selected and applied to the net water pressure acting to yield the design water pressure.

Thenault (2012) looked into the effect that passive pressure has on the design depth of embedment when considered as an unfavourable action, as a favourable action or as a resistance. Her results quantified the significance that the depth of the GWT has on the design depths achieved between the three approaches and showed that careful consideration of how to deal with passive pressure is required when the GWT is taken, as Eurocode 7 states, at the ground surface.

Eurocode 7 – second generation

In the second generation of Eurocode 7, actions resulting from water pressures are directly factored as explained in Section 6.12.5. The partial factors for both permanent and variable water actions for each design case are listed in Table 6.6.

9.7   Braced excavations

When excavating a deep trench, the insertion of *shoring* to support the sides becomes necessary to avoid collapse of the side walls during the trench-based work activities. The shoring forms part of the temporary works and is not a permanent structure.

Modern *shoring systems* and *trench boxes* are employed as a rapid and very effective means of providing the required support. These are reusable systems which, in the main nowadays, use hydraulic rams or adjustable props to provide the required pressure to the internal face of the shoring panels to resist the inwards collapsing pressures of the supported soil, as shown in Fig. 9.24. In all cases, the system is removed at the end of the trench works to be reused further along the trench or on the next project. The shoring system is eased deeper into the soil simultaneously as the base of the trench is excavated deeper.

The vertical shores used with the hydraulic rams or adjustable props may be fabricated from either steel or aluminium and are installed and removed from the top of the trench. The aluminium shores are lightweight and easier to manually handle than steel shores. Once the shores are in place, the props are positioned and adjusted/pressurised to provide the required support.

A trench box is a pre-assembled single unit, lowered into position using lifting chains secured to a crane or excavator boom. For deep trenches (greater than approximately 2 m), the trench boxes can be stacked on top of each other and fixed together to provide side support over the extent of the excavation.

Traditionally, when timbering or sheet piles were used as the side support system, the excavation was carried down first to some point X, and rigidly strutted supports inserted between the levels D to X (Fig. 9.25a). With reference to this figure, as further excavation is carried out, timbering and strutting are inserted in stages, but before the timbering is inserted, the soil yields by an amount that tends to increase with depth (it is relatively small at the top of the trench).

In Fig. 9.25b, the shape A'B'C'D' represents, to an enlarged scale, the original form of the surface that has yielded to the position ABCD of Fig. 9.25a; the resulting pressure on the back of the wall is roughly parabolic and is indicated in Fig. 9.25c.

For design purposes, a trapezoidal distribution of the form developed by Terzaghi and Peck (1967), since revised by Terzaghi *et al.* (1996) is assumed. The design procedure for the struts is semi-empirical. For sands, the pressure distribution is assumed to be uniform over the full depth of the excavation (Fig. 9.25d). For clays, the pressure distribution depends on the stability number, N:

$$N = \frac{\gamma H}{c_u}$$

(a)

(b)

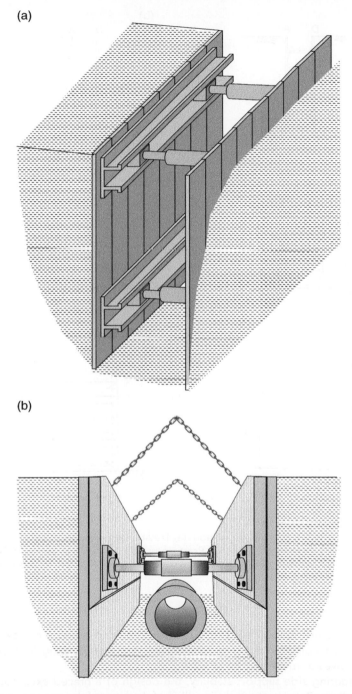

Fig. 9.24 Trench support systems: (a) vertical shores with hydraulic rams, (b) single unit trench box.

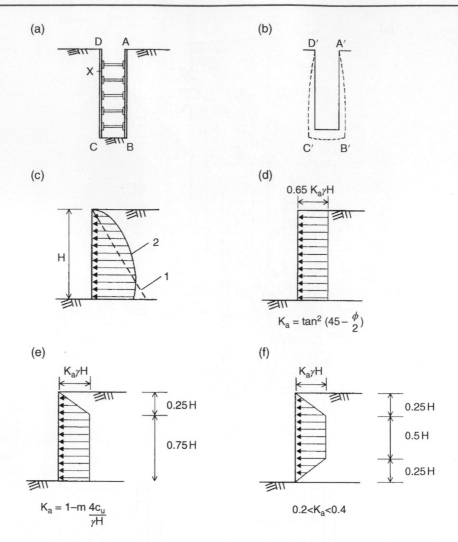

Fig. 9.25 Pressure distribution in strutted excavation. (a) Traditional side support and bracing. (b) Yielding of original excavation surfaces (exaggerated scale). (c) Pressure profile on trench support. (d) Design pressure distribution (sands). (e) Design pressure distribution (clays) when N > 4 and K_a > 0.4. (f) Design pressure distribution (clays) when N < 4 or 0.2 < K_a < 0.4.

If N is greater than 4, the distribution in Fig. 9.25e is used, provided that K_a is greater than 0.4. If N is less than 4, or if 0.2 < K_a < 0.4, the distribution in Fig. 9.25f is used. With respect to Fig. 9.25e, m is generally taken as 1.0. For soft clays, however, m can reduce to ≈ 0.4.

In addition to considering side support stability, the design of a braced excavation should consider failure through *basal heave*. Basal heave is the swelling of the base of the excavation due to the removal of the soil above through excavation. In granular soils, hydraulic gradients can affect the basal stability, and these too should be considered. Guidance on the verification of basal heave is given in EN 1997-3:202x.

9.8 Reinforced soil

The principle of reinforced soil is that a mass of soil can be given tensile strength in a specific direction, if lengths of a material capable of carrying tension are embedded within it in the required direction.

This idea has been known for centuries, and the use of straw to strengthen unburnt clay bricks and fascine mattresses used to strengthen soft soil deposits prior to road construction have been described in ancient texts. The ancient ziggurats, built in Iraq, consisted of dried earth blocks, reinforced across the width of the structure with tarred ropes. However, the full potential of reinforced soil was never realised until Vidal, who coined the term 'reinforced earth', demonstrated its wide potential and produced a rational design approach (Vidal, 1966). There is no doubt that the present-day use of reinforced soil structures stems directly from the pioneering work of Vidal.

Reinforced soil can be used in many geotechnical applications but, in this chapter, we are only concerned with earth retaining structures.

A reinforced soil retaining wall is a gravity structure and a simple form of such a wall is illustrated in Fig. 9.26. Brief descriptions of the components listed in the figure are set out below.

Soil fill

The soil should be granular and free draining with not more than 10% passing the 63 μm sieve.

Reinforcing elements

Metals: Originally, many reinforced soil structures used thin metallic strips usually 50–100 mm wide and 3–5 mm thick. Metals used were aluminium alloy, copper, stainless steel, and galvanised steel, the latter being the most common. The common property of these materials is that they all have high moduli of elasticity so that negligible strains are created within the soil mass.

Geosynthetics: Since the mid-1970s there has been an increasing use of geosynthetics as reinforcement in reinforced soil, either in strip form or in grid form, such as Tensar geogrid. Geosynthetics have the advantage of greater durability than metal in corrosive soil, and their tensile strength can approach that of steel. In grid form, plastic reinforcement can achieve high frictional properties between itself and the surrounding soil. The main disadvantage of plastic reinforcement is that it experiences plastic deformation when subjected to tensile forces, which can lead to relatively large strains within the soil mass.

Another type of polymer reinforcement material is when it is reinforced with glass fibres. Known as *glass fibre-reinforced plastic* (GRP), this material has a tensile strength similar to mild steel with the advantage that it does not experience plastic deformation.

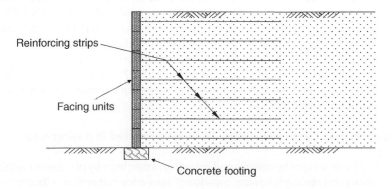

Reinforcing strips

Facing units

Concrete footing

Fig. 9.26 Typical reinforced soil retaining wall.

Facing units

At the free boundary of a reinforced soil structure, it is necessary to provide a barrier in order that the fill is contained. This is provided by a thin weatherproof facing which in no way contributes structural strength to the wall. The facing is usually built up from prefabricated units small and light enough to be manhandled. The units are generally made of precast concrete although steel, aluminium, and plastic units are sometimes encountered. In order to form a platform from which the facing units can be built up, a small mass concrete foundation is required.

Design of reinforced soil retaining structures

In the United Kingdom, the current design standard is BS8006 Part 1: BS 8006-1:2010 (BSI, 2010) *Code of practice for strengthened/ reinforced soils and other fills*. This code adopts a limit state design approach using partial factors; however, it is not fully compatible with Eurocode 7, EN 1997-1:2004 (BSI, 2004). BS 8006-1:2010 describes which design methods are acceptable for reinforced soil slopes and walls, but it does not explain how to actually design these geotechnical structures. However, reinforced soil has been included in the second generation of Eurocode 7, and the design sequence (i.e. consequence class, design case, limit states verifications, etc.) aligns with the design procedures described in Chapter 6. Limit states for both external and internal rupture and stability are verified.

Reinforced soil can provide a method for retaining soil when existing ground conditions do not allow construction by other, more conventional, methods. For example, a compressible soil may be perfectly capable of supporting a reinforced soil retaining structure, whereas it would probably require some form of piled foundation if a more conventional retaining wall were to be constructed. The technique can also be used when there is insufficient land space to construct the sloping side of a conventional earth embankment.

However, reinforced soil should not be thought of as only a form of alternative construction as it is often the first choice of design engineers when considering an earth retaining structure.

9.9 Soil nailing

Soil nailing is an *in situ* reinforcement technique used to stabilise slopes and retain excavations but, in this chapter, we are concerned only with earth retaining structures. The technique uses steel bars fully bonded into the soil mass. The bars are inserted into the soil either by direct driving or by drilling a borehole, inserting the bar and then filling the annulus around the bar with grout. The face of the exposed soil is sprayed with concrete to produce a zone of reinforced soil. The zone then acts as a homogeneous unit supporting the soil behind in a similar manner to a conventional retaining wall. The construction phases of a soil nailed wall are shown in Fig. 9.27, and the specification for soil nailing is given in BS 8006 Part 2: BS 8006-2:2011 *Code of practice for strengthened/reinforced soils, Part 2: soil nail design*. As with reinforced soil, soil nailing has been (briefly) included in the second generation of Eurocode 7 and the design sequence (i.e. consequence class, design case, limit states verifications, etc.) aligns with the design procedures described in Chapter 6.

The construction stages of a soil nailed wall are shown in Fig. 9.27. Although the completed soil structure may be expected to behave similarly to a conventional reinforced soil structure, there are notable differences between the two construction methods:

- natural soil properties may be greatly inferior to those permitted in a reinforced soil structure where selected fill is used;
- soil nails are installed by driving or by drilling and grouting rather than by placement within compacted fill;
- the construction process for nailing follows a 'top-down' sequence rather than a 'bottom-up' sequence for reinforced soils;

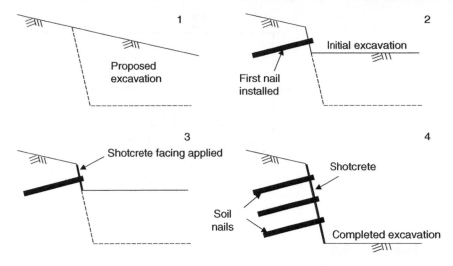

Fig. 9.27 Construction stages of a soil nailed wall.

- the facing to a nailed structure is usually formed from sprayed concrete (shotcrete) or geosynthetics rather than precast units;
- nails are commonly installed at an inclination to the horizontal in contrast to reinforced soil where the reinforcements are placed horizontally.

There are two methods of forming the nail: *drill and gout* and *driving*. With the drill and grout method, steel bars are installed into pre-drilled holes and grout injected around them to bond them fully into the soil mass. This generates a reasonably large contact area between the grout and the soil, thereby providing a high pull-out resistance. With the driving method, nails are either driven into the soil using a hydraulic or pneumatic hammer or fired into the soil from a nail launcher which uses an explosive release of compressed air. This method of installation requires the nails to be relatively robust and to have a reasonably small cross-sectional area.

Exercises

Exercise 9.1

A reinforced concrete cantilever retaining wall, supporting a granular soil, has dimensions shown in Fig. 9.28.

Using a gross factor of safety approach, calculate the factors of safety against sliding and overturning and check the bearing pressure on the soil beneath the wall if the allowable bearing pressure is 300 kPa. Take the unit weight of concrete as 23.5 kN/m^3 and assume that the friction between the base of the wall and the soil is equal to ϕ'. Ignore any passive resistance from the soil in front of the wall.

Answer $F_s = 2.19$; $F_o = 3.63$; $p_{max} = 135.5$ kPa {$<300 \Rightarrow$ OK}

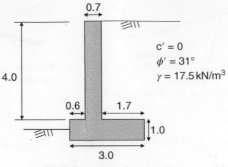

Fig. 9.28   Exercise 9.1.

Exercise 9.2

Check the safety of the mass concrete retaining wall shown in Fig. 9.29 in terms of sliding and overturning using a gross factor of safety approach.

You may assume that Rankine's conditions apply to the soil behind the wall, and that the friction between the base of the wall and the soil, $\delta = \frac{2}{3}\phi'$. Ignore any passive resistance from the soil in front of the wall.

Answer $F_o = 3.53$; $F_s = 1.85$.

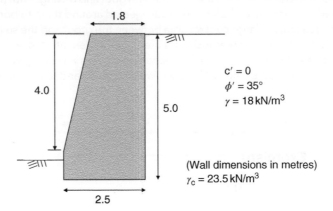

Fig. 9.29   Exercise 9.2.

Exercise 9.3

Check each situation of Exercises 9.1 and 9.2 in accordance with Eurocode 7, first generation, using Design Approach 1 in all cases.

Answer   Exercise 9.1:
   Overturning (GEO), DA1-1, $\Gamma = 2.69$; DA1-2, $\Gamma = 2.94$

Sliding (GEO) DA1-1, $\Gamma = 1.62$; DA1-2, $\Gamma = 1.42$
Exercise 9.2:
Overturning (GEO), DA1-1, $\Gamma = 3.26$; DA1-2, $\Gamma = 3.35$
Sliding (GEO) DA1-1, $\Gamma = 2.76$; DA1-2, $\Gamma = 2.27$

Exercise 9.4

A cantilever sheet pile wall is to be constructed in a granular soil with the following properties:

$\gamma = 19.2 \, kN/m^3$
$\phi' = 29°$
$c' = 0$

The depth of the excavation is to be 4 m. Determine the minimum depth of embedment, d, to provide stability in accordance with EN 1997-1:2004 Design Approach 1.
If the wall is now to be designed to carry a surcharge of 20 kPa, what would be the required depth of embedment?
Note: since a surcharge, by definition, is a variable action, it cannot be relied upon to provide any contribution to stability. Therefore, a surcharge will only contribute to the destabilising moment.

Answer Surcharge = 0: DA1-1, $d_0 = 4.1$ m, d = 4.6 m
  DA1-2, $d_0 = 5.4$ m, d = 6.0 m
  Surcharge = 20 kPa: DA1-1, $d_0 = 5.0$ m, d = 5.7 m
  DA1-2, $d_0 = 6.8$ m, d = 7.8 m

Exercise 9.5

Repeat Exercise 9.4 following the material factor approach of EN 1997-3:202x, taking the wall as a Consequence Class 2 structure.

Answer Surcharge = 0: DC4-M1, $d_0 = 4.1$ m, d = 4.6 m
  DC3-M2, $d_0 = 5.4$ m, d = 6.0 m
  Surcharge = 20 kPa: DC4-M1, $d_0 = 5.0$ m, d = 5.7 m
  DC3-M2, $d_0 = 6.8$ m, d = 7.8 m

Chapter 10

Bearing Capacity and Shallow Foundations Design

Learning objectives:

By the end of this chapter, you will have been introduced to:

- the theory of bearing capacity and methods for the determination of the ultimate and safe bearing capacity;
- the general bearing capacity equation and related factors;
- designing shallow foundations using a factor of safety approach;
- designing shallow foundations using both generations of Eurocode 7;
- the estimation of allowable bearing capacity from field tests.

10.1 Bearing capacity terms

The following terms are used in bearing capacity problems.

Ultimate bearing capacity

The value of the average contact pressure between the foundation and the soil which will produce shear failure in the soil.

Safe bearing capacity

The maximum value of contact pressure to which the soil may be subjected without risk of shear failure. This is based solely on the strength of the soil and is simply the ultimate bearing capacity divided by a suitable factor of safety.

Allowable bearing pressure

The maximum allowable net loading intensity on the soil allowing for both shear and settlement effects.

Smith's Elements of Soil Mechanics, 10th Edition. Ian Smith.
© 2021 John Wiley & Sons Ltd. Published 2021 by John Wiley & Sons Ltd.
Companion website: www.wiley.com/go/smith/soilmechanics10e

10.2   Types of foundation

Strip foundation

Often termed a *continuous footing,* this foundation has a length significantly greater than its width. It is generally used to support a series of columns or a wall.

Pad footing

Generally, an individual foundation designed to carry a single column load although there are occasions when a pad foundation supports two or more columns.

Raft foundation

This is a generic term for all types of foundations that cover large areas. A raft foundation is also called a *mat foundation* and can vary from a fascine mattress supporting a farm road to a large reinforced concrete basement supporting a high-rise block.

Pile foundation

Piles are used to transfer structural loads to either the foundation soil or the bedrock underlying the site. They are usually designed to work in groups, with the column loads they support transferred into them via a capping slab. Pile foundations are covered in Chapter 11.

Pier foundation

This is a large column built up either from the bedrock or from a slab supported by piles. Its purpose is to support a large load, such as that from a bridge. A pier operates in the same manner as a pile, but it is essentially a short squat column whereas a pile is relatively longer and more slender.

Shallow foundation

A foundation whose depth below the surface, z, is equal to or less than its least dimension, B. Most strip and pad footings fall into this category.

Deep foundation

A foundation whose depth below the surface is greater than its least dimension. Piles and piers fall into this category.

10.3   Ultimate bearing capacity of a foundation

The ultimate bearing capacity of a foundation is given the symbol q_u and there are various analytical methods by which it can be evaluated. As will be seen, some of these approaches are not all that suitable but they still form a very useful introduction to the study of the bearing capacity of a foundation.

10.3.1 Earth pressure theory

This is a theoretical model and results in too unrealistic values of q_u to be used in design because the method does not fully replicate the bearing failure mechanism in soils.

Consider an element of soil under a foundation (Fig. 10.1). From the theory covered in Chapter 8, we see that the vertical downward pressure of the footing, q_u, is a major principal stress causing a corresponding Rankine active pressure, p. For particles beyond the edge of the foundation, this lateral stress can be considered as a major principal stress (i.e. passive resistance) with its corresponding vertical minor principal stress γz (the weight of the soil).

Now,

$$p = q_u K_a = q_u \frac{1 - \sin \phi'}{1 + \sin \phi'}$$

Also,

$$p = \gamma z K_p = \gamma z \frac{1 + \sin \phi'}{1 - \sin \phi'}$$

$$\Rightarrow q_u = \gamma z \left[\frac{1 + \sin \phi'}{1 - \sin \phi'} \right]^2$$

This is the formula for the ultimate bearing capacity, q_u. It is seen that it is not satisfactory for shallow footings because when $z = 0$ then, according to the formula, q_u also $= 0$. Bell's development of the Rankine solution for $c - \phi$ soils gives the following equation:

$$q_u = \gamma z \left[\frac{1 + \sin \phi'}{1 - \sin \phi'} \right]^2 + 2c' \sqrt{\left(\frac{1 + \sin \phi'}{1 - \sin \phi'} \right)^3} + 2c' \sqrt{\frac{1 + \sin \phi'}{1 - \sin \phi'}} \tag{10.1}$$

For, the undrained state, $\phi_u = 0°$,

$$q_u = \gamma z + 4c_u$$

or $q_u = 4c_u$ for a surface footing.

10.3.2 Slip circle methods

With slip circle methods, the foundation is assumed to fail by rotation about some slip surface, usually taken as the arc of a circle. Almost all foundation failures exhibit rotational effects, and Fellenius (1927) showed that

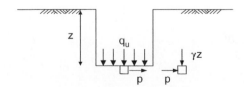

Fig. 10.1 Earth pressure conditions immediately below a foundation.

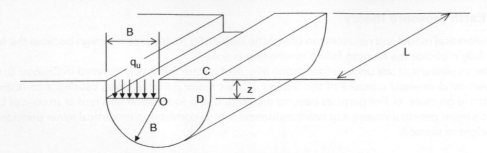

Fig. 10.2   Foundation failure rotation about one edge.

the centre of rotation is slightly above the base of the foundation and to one side of it. He found that in a saturated cohesive soil, the ultimate bearing capacity for a surface footing is,

$$q_u = 5.52\ c_u$$

To illustrate the method, consider a foundation failing by rotation about one edge and founded at a depth z below the surface of a saturated clay of unit weight γ and undrained strength c_u (Fig. 10.2).

Disturbing moment about O:

$$q_u \times LB \times \frac{B}{2} = \frac{q_u LB^2}{2} \tag{10.2}$$

Resisting moments about O:
    Cohesion along cylindrical sliding surface = $c_u \pi LB$

$$\Rightarrow \text{Moment} = c_u \pi LB^2 \tag{10.3}$$

Cohesion along CD = $c_u zL$

$$\Rightarrow \text{Moment} = c_u zLB \tag{10.4}$$

Weight of soil above foundation level = γzLB

$$\Rightarrow \text{Moment} = \frac{\gamma zLB^2}{2} \tag{10.5}$$

For limit equilibrium (10.2) = (10.3) + (10.4) + (10.5)
    i.e.

$$\frac{q_u LB^2}{2} = c_u \pi LB^2 + c_u zLB + \frac{\gamma zLB^2}{2}$$

$$\Rightarrow q_u = 2\pi c_u + \frac{2c_u z}{B} + \gamma z$$

$$= 2\pi c_u \left(1 + \frac{1}{\pi}\frac{z}{B} + \frac{1}{2\pi}\frac{\gamma z}{c_u}\right) \tag{10.6}$$

$$= 6.28 c_u \left(1 + 0.32\frac{z}{B} + 0.16\frac{\gamma z}{c_u}\right)$$

Cohesion of end sectors

The above formula (10.6) only applies to a strip footing, and if the foundation is of finite dimensions then the effect of the ends must be included.

To obtain this, it is assumed that when the cohesion along the perimeter of the sector has reached its maximum value, c_u, the value of cohesion at some point on the sector at distance r from O is $c_r = c_u r/B$, as shown in Fig. 10.3.

Rotational resistance of an elemental ring, dr thick

$$= \frac{c_u r}{B} \times \pi r \; dr$$

$$\text{Moment about O} = \frac{c_u r}{B} \times \pi r \; dr \times r = \pi \frac{c_u}{B} r^3 \; dr$$

$$\text{Total moment of both ends} = 2 \int_0^B \pi \frac{c_u}{B} r^3 \; dr$$

$$= 2\pi \frac{c_u}{B} \times \frac{B^4}{4} = \frac{\pi c_u B^3}{2}$$

$$(10.7)$$

This analysis ignores the cohesion of the soil above the base of the foundation at the two ends, but unless the foundation is very deep this will have little effect on the value of q_u. The term (10.7) should be added into the original equation.

For a surface footing, the formula for q_u is:

$$q_u = 6.28 \; c_u$$

This value is high because the centre of rotation is actually above the base, but in practice a series of rotational centres are chosen, and each circle is analysed (as for a slope stability problem) until the lowest q_u value has been obtained. The method can be extended to allow for frictional effects but is considered most satisfactory when used for cohesive soils; it was extended by Wilson (1941), who prepared a chart (Fig. 10.4) which gives the centre of the most critical circle for cohesive soils (his technique is not applicable to other categories of soil or to surface footings).

The slip circle method is useful when the soil properties beneath the foundation vary, since an approximate position of the critical circle can be obtained from Fig. 10.4 and then other circles near to it can be analysed. When the soil conditions are uniform, Wilson's critical circle gives,

$$q_u = 5.52 \; c_u \text{ for a surface footing.}$$

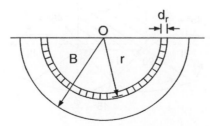

Fig. 10.3 Cohesion of end sectors.

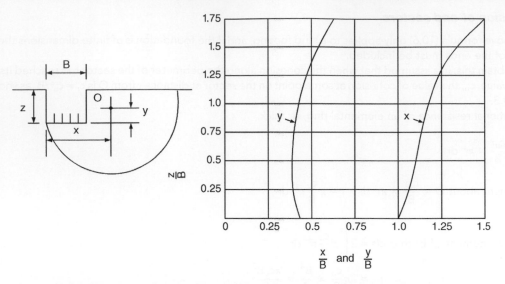

Fig. 10.4 Location of centre of critical circle for use with Fellenius' method.

10.3.3 Plastic failure theory

Forms of bearing capacity failure

Terzaghi (1943) stated that the bearing capacity failure of a foundation is caused by either a general soil shear failure or a local soil shear failure. Vesic (1963) listed punching shear failure as a further form of bearing capacity failure.

1. *General shear failure*
   This is the most commonly used theory in bearing capacity work. The form of this failure is illustrated in Fig. 10.5, which shows a strip footing. The failure pattern is clearly defined, and it can be seen that definite failure surfaces develop within the soil. A wedge of compressed soil (I) goes down with the footing, creating slip surfaces and areas of plastic flow (II). These areas are initially prevented from moving outwards by the passive resistance of the soil wedges (III). Once this passive resistance is overcome, movement takes place and bulging of the soil surface around the foundation occurs. With general shear failure, collapse is sudden and is accompanied by a tilting of the foundation.

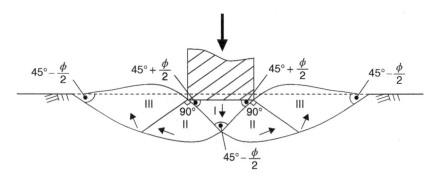

Fig. 10.5 General shear failure.

2. *Local shear failure*

The failure pattern developed is of the same form as for general shear failure but only the slip surfaces immediately below the foundation are well defined. Shear failure is local and does not create the large zones of plastic failure which develop with general shear failure. Some heaving of the soil around the foundation may occur but the actual slip surfaces do not penetrate the surface of the soil and there is no tilting of the foundation.

3. *Punching shear failure*

This is a downward movement of the foundation caused by soil shear failure only occurring along the boundaries of the wedge of soil immediately below the foundation. There is little bulging of the surface of the soil and no slip surfaces can be seen.

For both punching and local shear failure, settlement considerations are invariably more critical than those of bearing capacity so that the evaluation of the ultimate bearing capacity of a foundation is usually obtained from an analysis of general shear failure.

Prandtl's analysis

Prandtl (1921) was interested in the plastic failure of metals and one of his solutions (for the penetration of a punch into metal) can be applied to the case of a foundation penetrating downwards into a soil with no attendant rotation.

The analysis gives solutions for various values of ϕ, and for a surface footing with $\phi = 0$, Prandtl obtained: $q_u = 5.14c_u$.

Terzaghi's analysis

Working on similar lines to Prandtl's analysis, Terzaghi (1943) produced a formula for q_u which allows for the effects of cohesion and friction between the base of the footing and the soil, and is also applicable to shallow ($z/B \leq 1$) and surface foundations. His solution for the ultimate bearing capacity, q_u of a *strip footing* is a three-term expression incorporating the *bearing capacity factors*: N_c, N_q, and N_γ.

$$q_u = cN_c + \gamma zN_q + 0.5\gamma BN_\gamma \tag{10.8}$$

where

c = apparent cohesion;
z = founding depth;
B = width of foundation; and
γ = unit weight of the soil removed.

The coefficients N_c, N_q, and N_γ depend upon the soil's angle of shearing resistance and can be obtained from Fig. 10.6.

When $\phi = 0°$, $N_c = 5.7$, $N_q = 1.0$, and $N_\gamma = 0 \Rightarrow q_u = 5.7c_u + \gamma z$

or $q_u = 5.7c_u$ for a surface footing.

The increase in the value of N_c from 5.14 to 5.7 is due to the fact that Terzaghi allowed for frictional effects between the foundation and its supporting soil.

The coefficient N_q allows for the surcharge effects due to the soil above the foundation level, and N_γ allows for the size of the footing, B. The effect of N_γ is of little consequence with clays, where the angle of shearing resistance is usually assumed to be the undrained value, ϕ_u, (=0°), but it can become significant with wide foundations supported on cohesionless soil.

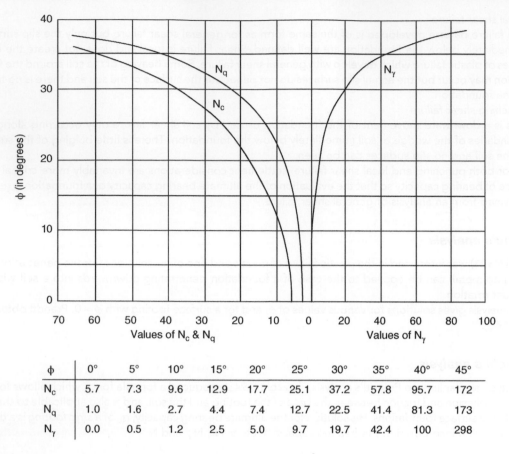

ϕ	0°	5°	10°	15°	20°	25°	30°	35°	40°	45°
N_c	5.7	7.3	9.6	12.9	17.7	25.1	37.2	57.8	95.7	172
N_q	1.0	1.6	2.7	4.4	7.4	12.7	22.5	41.4	81.3	173
N_γ	0.0	0.5	1.2	2.5	5.0	9.7	19.7	42.4	100	298

Fig. 10.6 Terzaghi's bearing capacity coefficients.

Terzaghi's solution for a circular footing is:

$$q_u = 1.3cN_c + \gamma z N_q + 0.3\gamma B N_\gamma \quad \text{(where B = diameter)} \tag{10.9}$$

For a square footing:

$$q_u = 1.3cN_c + \gamma z N_q + 0.4\gamma B N_\gamma \tag{10.10}$$

and for a rectangular footing:

$$q_u = cN_c\left(1 + 0.3\frac{B}{L}\right) + \gamma z N_q + 0.5\gamma B N_\gamma\left(1 - 0.2\frac{B}{L}\right) \tag{10.11}$$

Skempton (1951) showed that for a cohesive soil ($\phi = 0°$) the value of the coefficient N_c increases with the value of the foundation depth, z. His suggested values for N_c, applicable to circular, square, and strip footings, are given in Fig. 10.7. In the case of a rectangular footing on a cohesive soil, a value for N_c can either be estimated from Fig. 10.7 or obtained from the formula:

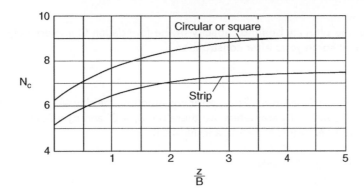

Fig. 10.7 Variation of the coefficient N_c with depth. Modified from Skempton (1951).

$$N_c = 5\left(1 + 0.2\frac{B}{L}\right)\left(1 + 0.2\frac{z}{B}\right)$$

(10.12)

with a limiting value for N_c of $N_c = 7.5(1 + 0.2B/L)$, which corresponds to a z/B ratio greater than 2.5 (Skempton, 1951).

10.3.4 Summary of bearing capacity formula

It can be seen that Rankine's theory does not give satisfactory results and that, for variable subsoil conditions, the slip surface analysis of Fellenius provides the best solution. For normal soil conditions, Equations (10.8)–(10.11) can generally be used and may be applied to foundations at any depth in $c - \phi$ soils and to shallow foundations in cohesive soils. For deep footings in cohesive soil, the values of N_c suggested by Skempton may be used in place of the Terzaghi values.

10.3.5 Choice of soil parameters

As with earth pressure equations, bearing capacity equations can be used with either the undrained or the drained soil parameters. As granular soils operate in the drained state at all stages during and after construction, the relevant soil strength parameter is ϕ'.

Saturated cohesive soils operate in the undrained state during and immediately after construction and the relevant parameter is c_u. If required, the long-term stability can be checked with the assumption that the soil will be drained, and the relevant parameters are c' and ϕ' (with c' generally taken as equal to zero).

Example 10.1: Ultimate bearing capacity (Terzaghi) in short term and long term

A rectangular foundation, 2 m × 4 m, is to be founded at a depth of 1 m below the surface of a deep stratum of soft saturated clay (unit weight = 20 kN/m³).

Quick undrained and consolidated undrained triaxial tests established the following soil parameters: $\phi_u = 0°$, $c_u = 24$ kPa; $\phi' = 25°$, and $c' = 0$.

Determine the ultimate bearing capacity of the foundation (i) immediately after construction and (ii) some years after construction.

Solution:

(i) It may be assumed that immediately after construction, the clay will be in an undrained state. The relevant soil parameters are therefore $\phi_u = 0°$ and $c_u = 24$ kPa.
From Fig. 10.6: $N_c = 5.7$, $N_q = 1.0$, and $N_\gamma = 0.0$.

$$q_u = cN_c(1 + 0.3B/L) + \gamma z N_q$$
$$= 24 \times 5.7(1 + 0.3 \times 2/4) + 20 \times 1 \times 1$$
$$= 177.3 \text{ kPa}$$

(ii) It can be assumed that, after some years, the clay will be fully drained so that the relevant soil parameters are $\phi' = 25°$ and $c' = 0$.
From Fig. 10.6: $N_c = 25.1$, $N_q = 12.7$, and $N_\gamma = 9.7$.

$$q_u = \gamma z N_q + 0.5\gamma B N_\gamma(1 - 0.2B/L)$$
$$= 20 \times 1 \times 12.7 + 0.5 \times 20 \times 2 \times 9.7(1 - 0.2 \times 2/4)$$
$$= 428.6 \text{ kPa}$$

Example 10.2: Ultimate bearing capacity (Terzaghi); effect of ϕ'

A continuous foundation is 1.5 m wide and is founded at a depth of 1.5 m in a deep layer of sand of unit weight 18.5 kN/m^3.

Determine the ultimate bearing capacity of the foundation if the soil strength parameters are $c' = 0$ and $\phi' = $ (i) 35°, (ii) 30°.

Solution:

(i) From Fig. 10.6: for $\phi' = 35°$, $N_c = 58$, $N_q = 41$, $N_\gamma = 42$. For a continuous footing:

$$q_u = c'N_c + \gamma z N_q + 0.5\gamma B N_\gamma$$
$$= 18.5 \times 1.5 \times 41 + 0.5 \times 18.5 \times 1.5 \times 42$$
$$= 1720 \text{ kPa}$$

(ii) From Fig. 10.6: for $\phi' = 30°$, $N_c = 37$, $N_q = 22$, $N_\gamma = 20$

$$q_u = 18.5 \times 1.5 \times 22 + 0.5 \times 18.5 \times 1.5 \times 20$$
$$= 888 \text{ kPa}$$

The ultimate bearing capacity is reduced by about 50% when the value of ϕ' is reduced by about 15%.

10.4   Determination of the safe bearing capacity

The value of the safe bearing capacity is simply the value of the net ultimate bearing capacity divided by a suitable factor of safety, F. The value of F is usually not less than 3.0, except for a relatively unimportant structure, and sometimes can be as much as 5.0. At first glance, these values for F appear high but the necessity for them is illustrated in Example 10.2, which demonstrates the effect on q_u of a small variation in the value of ϕ'.

The net ultimate bearing capacity is the increase in vertical pressure, above that of the original overburden pressure, that the soil can just carry before shear failure occurs. The original overburden pressure is γz and this term should be subtracted from the bearing capacity equations, i.e. for a strip footing:

$$q_{u\ net} = cN_c + \gamma z(N_q - 1) + 0.5\gamma BN_\gamma \tag{10.13}$$

The safe bearing capacity is therefore the above expression divided by F plus the term γz:

$$\text{Safe bearing capacity} = \frac{cN_c + \gamma z(N_q - 1) + 0.5\gamma BN_\gamma}{F} + \gamma z \tag{10.14}$$

In the case of a footing founded in undrained clay, where $\phi_u = 0°$, the net ultimate bearing capacity is $5.7c_u$, since $N_q = 1$ and $N_\gamma = 0$.

The safe bearing capacity notion is not used during design to Eurocode 7 where, as will be demonstrated in Section 10.7, conformity of the bearing resistance limit state is achieved by ensuring that the design effect of the actions does not exceed the design bearing resistance.

10.5   The effect of groundwater on bearing capacity

10.5.1   Water table below the foundation level

If the water table is at a depth of greater than B below the foundation, the expression for net ultimate bearing capacity is Equation (10.13), but when the water table rises to a depth of less than B below the foundation, the expression becomes:

$$q_{u\ net} = cN_c + \gamma z(N_q - 1) + 0.5\gamma' BN_\gamma \tag{10.15a}$$

where

γ = unit weight of soil above groundwater level and
γ' = effective unit weight.

For cohesive soils, ϕ_u is small and the term $0.5\gamma' BN_\gamma$ is of little account, and the value of the bearing capacity is virtually unaffected by groundwater. With sands, however, the term cN_c is zero and the term $0.5\gamma' BN_\gamma$ is about one half of $0.5\gamma BN_\gamma$, so that groundwater has a significant effect.

10.5.2   Water table above the foundation level

For this case, Terzaghi's expressions are best written in the form:

$$q_{u\,net} = cN_c + \sigma'_v(N_q - 1) + 0.5\gamma'BN_\gamma \qquad (10.15b)$$

where σ'_v = effective overburden pressure removed.

From the expression it is seen that, in these circumstances, the bearing capacity of a cohesive soil can be affected by groundwater.

When designing foundations to Eurocode 7, unless an adequate drainage system and maintenance plan are ensured, the ground water table should be taken as the maximum possible level. This could of course be the ground surface.

10.6   Developments in bearing capacity equations

Terzaghi's bearing capacity equations have been successfully used in the design of numerous shallow foundations throughout the world and are still in use. However, they are viewed by many to be conservative as they do not consider factors that affect bearing capacity, such as inclined loading, foundation depth, and the shear resistance of the soil above the foundation. This section describes developments that have been made to the original equations.

10.6.1   General form of the bearing capacity equation

Meyerhof (1963) proposed the following general equation for q_u. Equation (10.16) is known as the *general bearing capacity equation*.

$$q_u = cN_cs_ci_cd_c + \gamma zN_qs_qi_qd_q + 0.5\gamma BN_\gamma s_\gamma i_\gamma d_\gamma \qquad (10.16)$$

where

s_c, s_q, and s_γ are shape factors;
i_c, i_q, and i_γ are inclination factors; and
d_c, d_q, and d_γ are depth factors.

Other factors, G_c, G_q, and G_γ to allow for a sloping ground surface, and B_c, B_q, and B_γ to allow for any inclination of the base, can also be included when required.

It must be noted that the values of N_c, N_q, and N_γ used in the general bearing capacity equation are not the Terzaghi values. The values of N_c and N_q are obtained from Meyerhof's equations (1963), as they are recognised as the most satisfactory:

$$N_c = (N_q - 1)\cot\phi', \quad N_q = \tan^2\left(45° + \frac{\phi}{2}\right)e^{\pi\tan\phi}$$

The undrained N_c value is $\pi + 2 = 5.14$. For this occasion, the symbol N_{cu} is sometimes used in place of N_c to indicate that it is the undrained condition.

Various workers have proposed expressions for N_γ:

$N_\gamma = (N_q - 1) \tan 1.4\,\phi$ Meyerhof (1963);
$N_\gamma = 1.5(N_q - 1) \tan \phi$ Hansen (1970);
$N_\gamma = 2(N_q + 1) \tan \phi$ Vesic (1973); and
$N_\gamma = 2(N_q - 1) \tan \phi$ where friction exists between foundation base and soil, $\delta \geq \phi/2$ (Chen, 1975; EN 1997-1:2004 (BSI, 2004)).

Through its adoption in Eurocode 7, the latter expression has become the most commonly used. Further examples in this chapter will therefore use the following expressions for the bearing capacity coefficients:

$$N_c = (N_q - 1) \cot \phi',$$

$$N_q = \tan^2\left(45° + \frac{\phi}{2}\right) e^{\pi \tan \phi}$$

$$N_\gamma = 2(N_q - 1) \tan \phi$$

Typical values are shown in Table 10.1.

10.6.2 Shape factors

These factors are intended to allow for the effect of the shape of a rectangular foundation of dimensions B × L on its bearing capacity. For circular-shaped foundations, B = L = diameter. The following expressions are widely used and are adopted in Eurocode 7 (both generations):

Table 10.1 Bearing capacity factors in common use.

$\phi(°)$	N_c	N_q	N_γ
0	5.14	1.00	0.00
5	6.49	1.57	0.10
10	8.34	2.47	0.52
15	10.98	3.94	1.58
20	14.83	6.40	3.93
25	20.72	10.66	9.01
30	30.14	18.40	20.09
35	46.12	33.30	45.23
40	75.31	64.20	106.05
45	133.87	134.87	267.75
50	266.88	319.06	758.09

$$s_{cu} = 1 + 0.2(B'/L') \text{ (undrained conditions)}$$

$$s_c = \left(\frac{s_q N_q - 1}{N_q - 1}\right) \text{ (drained conditions)}$$

$$s_q = 1 + (B'/L') \sin \phi'$$

$$s_\gamma = 1 - 0.3 (B'/L')$$

B' and L' are defined in Section 10.6.4.

10.6.3 Depth factors

These factors are intended to allow for the shear strength of the soil above the foundation. Hansen (1970) proposed the following values:

	z/B ≤ 1.0	z/B > 1.0
d_c	$1 + 0.4(z/B)$	$1 + 0.4 \ \arctan(z/B)$
d_q	$1 + 2 \tan \phi (1 - \sin \phi)^2 (z/B)$	$1 + 2 \tan \phi (1 - \sin \phi)^2 \arctan(z/B)$
d_γ	1.0	1.0

Note: The arctan values must be expressed in radians, e.g. if z = 1.5 and B = 1.0 m then arctan(z/B) = arctan(1.5) = 56.3° = 0.983 radians.

In the first generation of Eurocode 7, depth factors were excluded, but they have been introduced in the second generation. The expressions for d_q and d_γ given above apply, together with the following expression for d_{cu} and d_c' (EN1997-3:202x):

$$d_{cu} = 1 + 0.33 \ \tan^{-1}\left(\frac{D}{B}\right)$$

$$d_{c'} = d_q - \left(\frac{1 - N_q}{N_c \tan \phi'}\right)$$

Example 10.3: Ultimate bearing capacity (Meyerhof) in short term and long term

Recalculate Example 10.1 using Meyerhof's general bearing capacity formula.

Solution:

(i) From Table 10.1, for $\phi_u = 0°$, $N_c = 5.14$, $N_q = 1.0$, and $N_\gamma = 0.0$.

Shape factors:

$s_{cu} = 1 + 0.2(2/4) = 1.1$

$s_q = 1 + (2/4) \sin 0° = 1.0$

$s_\gamma = 1 - 0.3(2/4) = 0.85$

Depth factors:
$z/B = 1/2 = 0.5$. Using Hansen's values for $z/B \leq 1.0$:

$d_c = 1 + 0.4(1/2) = 1.2$, $d_q = 1.0$ (as $\phi_u = 0°$), $d_\gamma = 1.0$

$q_u = cN_c s_c d_c + \gamma z N_q s_q d_q$

$\quad = 24 \times 5.14 \times 1.1 \times 1.2 + 20 \times 1.0 \times 1.0 \times 1.0$

$\quad = 182.8$ kPa

(ii) From Table 10.1, for $\phi' = 25°$, $N_q = 10.66$, and $N_\gamma = 9.01$.
The expressions for s_q and d_q involve ϕ. These two factors will therefore have different values from those in case (i):

$s_q = 1 + (2/4) \sin 25° = 1.21$

$d_q = 1 + 2 \tan 25°(1 - \sin 25°)^2 (1/2) = 1.16$

$q_u = \gamma z N_q s_q d_q + 0.5 \gamma B N_\gamma s_\gamma d_\gamma$

$\quad = 20 \times 1 \times 10.66 \times 1.21 \times 1.16 + 0.5 \times 20 \times 2 \times 9.01 \times 0.85 \times 1.0$

$\quad = 452.5$ kPa

Example 10.4: Safe bearing capacity

Using a factor of safety = 3.0, determine the values of safe bearing capacity for cases (i) and (ii) in Example 10.3.

Solution:

Case (i):

$q_{u\,net} = q_u - \gamma z = 162.8$ kPa

Safe bearing capacity $= \dfrac{162.8}{3} + 20 \times 1$

$\quad = 74.3$ kPa

Case (ii):

$$q_{u\,net} = \gamma z(N_q s_q d_q - 1) + 0.5\gamma B N_\gamma s_\gamma d_\gamma$$
$$= 432.5 \text{ kPa}$$

Safe bearing capacity $= (432.5/3) + 20 \times 1$
$$= 164.2 \text{ kPa}$$

10.6.4 Effect of eccentric and inclined loading on foundations

A foundation can be subjected to eccentric loads and/or to inclined loads, which may be either eccentric or concentric.

Eccentric loads

Let us consider first the relatively simple case of a vertical load acting on a rectangular foundation of width B and length L such that the load has eccentricities e_B and e_L (Fig. 10.8). To solve the problem, we must think in terms of the rather artificial concept of effective foundation width and length. That part of the foundation that is symmetrical about the point of application of the load is considered to be useful, or *effective*, and is the area of the rectangle of effective length $L' = L - 2e_L$ and of effective width $B' = B - 2e_B$.

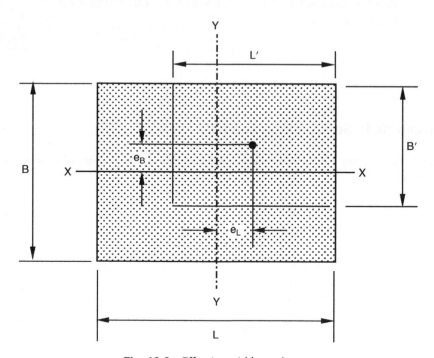

Fig. 10.8 Effective widths and area.

In the case of a strip footing of width B, subjected to a line load with an eccentricity e, then $B' = B - 2e$ and the ultimate bearing capacity of the foundation is found from either Equation (10.8) or, more likely, from the general bearing capacity equation (Equation (10.16)) with the term B replaced by B'.

The overall eccentricity of the bearing pressure, e must consider the self-weight of the foundation and is equal to:

$$e = \frac{P \times e_P}{P + W} \qquad (10.17)$$

where

P = magnitude of the eccentric load;
W = self-weight of the foundation; and
e_p = eccentricity of P.

Inclined loads

The usual method of dealing with an inclined line load, such as P in Fig. 10.9, is to first determine its horizontal and vertical components P_H and P_V and then, by taking moments, determine its eccentricity, e, in order that the effective width of the foundation B' can be determined from the formula $B' = B - 2e$.

The ultimate bearing capacity of the strip foundation (of width B) is then taken to be equal to that of a strip foundation of width B' subjected to a concentric load, P, inclined at an angle α to the vertical.

Various methods of solution have been proposed for this problem, e.g. Janbu (1957), Hansen (1957), but possibly the simplest approach is that proposed by Meyerhof (1953) in which the bearing capacity coefficients N_c, N_q, and N_y are reduced by multiplying them by the factors i_c, i_q, and i_y in his general equation (Equation (10.16)). Meyerhof's expressions for these factors are:

$$i_c = i_q = (1 - \alpha/90°)^2$$
$$i_\gamma = (1 - \alpha/\phi)^2$$

Eurocode 7 offers different expressions for the inclination factors – see the following section.

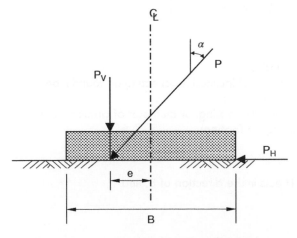

Fig. 10.9 Strip foundation with inclined load.

10.7   Designing spread foundations to Eurocode 7 (first generation)

The design of spread foundations is covered in Section 6 of Eurocode 7, Part 1. The limit states to be checked and the partial factors to be used in the design are the same as we saw when we looked at the bearing and sliding verifications of retaining walls in Section 9.5.

BS8004:2015+A1:2020 *Code of practice for foundations* (BSI, 2020) provides non-contradictory, complementary information which can usefully be used alongside Eurocode 7. Accordingly, reference to BS8004 may identify helpful, complementary information when carrying out a design to Eurocode 7.

10.7.1   Design by calculation

In terms of establishing the bearing resistance, the code states that a commonly recognised method should be used. This would typically be the general bearing capacity equation. Depth factors are excluded in the first generation of Eurocode 7 and for this reason they are excluded too from the solutions to Examples 10.5 and 10.6 in this chapter.

While the design procedure required to satisfy the conditions of Eurocode 7 involves the same methods as we have seen so far in this chapter, there are a few differences listed in Annex D which can be considered for drained conditions. These concern the shape and inclination factors as well as the bearing resistance factor, N_γ, as mentioned earlier. The following expressions are listed in the code:

Drained conditions:

$$N_\gamma = 2(N_q - 1)\tan\phi' \quad \text{(for a rough base, such as a typical foundation);}$$
$$s_q = 1 + (B'/L')\sin\phi' \quad \text{(for a rectangular foundation);}$$
$$s_q = 1 + \sin\phi' \quad \text{(for a square or circular foundation);}$$
$$s_\gamma = 1 - 0.3(B'/L') \quad \text{(for a rectangular foundation);}$$
$$s_\gamma = 0.7 \quad \text{(for a square or circular foundation); and}$$
$$s_c = \frac{s_q N_q - 1}{N_q - 1} \quad \text{(rectangular, square, and circle foundation).}$$

Load inclination factors:

$$i_c = i_q - \left(\frac{1 - i_q}{N_c \tan\phi'}\right); \quad i_q = \left[1 - \frac{H}{V + A'c'\cot\phi'}\right]^m; \quad i_\gamma = i_q^{\left(\frac{m+1}{m}\right)}$$

where

V = vertical load acting on foundation;
H = horizontal load (or component of inclined load) acting on foundation;
A′ = design effective area of foundation;
B = shorter dimension of rectangular footing, or diameter of circular footing;
L = longer dimension of rectangular footing;

$$m = m_B = \frac{2 + \dfrac{B'}{L'}}{1 + \dfrac{B'}{L'}}, \text{ when H acts in the direction of B; and}$$

$$m = m_L = \frac{2 + \dfrac{L'}{B'}}{1 + \dfrac{L'}{B'}}, \text{ when H acts in the direction of L.}$$

Factors for the inclination of the foundation base:

$$b_c = b_q - \left(\frac{1-b_q}{N_c \tan \phi'}\right)$$

$$b_q = b\gamma = (1 - \alpha \tan \phi')^2$$

where α is the inclination of the foundation base to the horizontal.

Undrained conditions:

$$R/_{A'} = (\pi + 2)c_u s_c i_c b_c + q$$

$$s_{cu} = 1 + 0.2(B'/L')$$

$$b_c = 1 - 2\alpha/(\pi + 2)$$

Example 10.5: Traditional, and Eurocode 7 (first generation), approaches (i)

A continuous footing is 1.8 m wide by 0.5 m deep and is founded at a depth of 0.75 m in a clay soil of unit weight 20 kN/m³ and $c_u = 30$ kPa. The foundation is to carry a vertical line load of magnitude 50 kN/m run, which will act at a distance of 0.4 m from the centre line. Take the weight density of concrete as 24 kN/m³.

(i) Determine the safe bearing capacity for the footing, taking F = 3.0.
(ii) Check the Eurocode 7 GEO limit state (Design Approach 1) by establishing the magnitude of the over-design factor.

Solution:

(i) Safe bearing capacity
Self-weight of foundation, $W_f = 0.5 \times 24 \times 1.8 = 21.6$ kN/m run
Weight of soil on top of foundation, $W_s = 0.25 \times 20 \times 1.8 = 9.0$ kN/m run
Total weight of foundation + soil, $W = 21.6 + 9.0 = 30.6$ kN/m run

$$\text{Eccentricity of bearing pressure, } e = \frac{P \times e_P}{P + W} = \frac{50 \times 0.4}{50 + 30.6} = 0.25 \text{ m}$$

Since $e \le \dfrac{B}{6}$, the total force acts within the middle-third of the foundation.
Effective width of footing, $B' = 1.8 - 2 \times 0.25 = 1.3$ m
From Table 10.1, for $\phi_u = 0°$, $N_c = 5.14$, $N_q = 1.0$, and $N_\gamma = 0$.
Footing is continuous (i.e. $L \to \infty$) and conditions are undrained, so $s_c = 1.0$.

$$d_c = 1 + 0.4\left(\frac{0.75}{1.3}\right) = 1.23$$

Safe bearing capacity (per metre run) $= \dfrac{q_{u\,net}}{3} + \gamma z = \dfrac{cN_c s_c d_c}{3} + \gamma z$

$$= \dfrac{30 \times 5.14 \times 1.0 \times 1.23}{3} + 20 \times 0.75$$

$$= 78.2 \text{ kPa}$$

Safe bearing load $= 78.2 \times B' = 101.7 \text{ kN/m run}$

(ii) Eurocode 7 GEO limit state

1. *Combination 1 (partial factor sets A1 + M1 + R1)*
    From Table 6.1: $\gamma_{G;\,unfav} = 1.35$; $\gamma_{cu} = 1.0$; and $\gamma_{Rv} = 1.0$.

    Design material property: $c_{u;d} = \dfrac{c_{u;k}}{\gamma_{cu}} = \dfrac{30}{1} = 30 \text{ kPa}$

    Design actions:

    Weight of foundation, $W_d = W \times \gamma_{G;\,unfav} = 30.6 \times 1.35 = 41.3 \text{ kN/m run}$
    Applied line load, $P_d = P \times \gamma_{G;\,unfav} = 50 \times 1.35 = 67.5 \text{ kN/m run}$

    Effect of design actions:

    Total vertical force, $F_d = 41.3 + 67.5 = 108.8 \text{ kN/m run}$
    Eccentricity, $e = \dfrac{P \times e_P}{P + W} = \dfrac{67.5 \times 0.4}{67.5 + 41.3} = 0.248 \text{ m}$
    Since $e \le \dfrac{B}{6}$, the total force acts within the middle-third of the foundation.
    Effective width of footing, $B' = 1.8 - 2 \times 0.248 = 1.3 \text{ m}$

    Design resistance:

    $s_{cu} = 1.0$ (since $L \to \infty$)
    $R/_{A'} = (\pi + 2)c_u s_c + q = 5.14 \times 30 \times 1.0 \times 1.0 \times 1.0 + (20 \times 0.75)$
    $\qquad = 169.2 \text{ kPa}$

    Characteristic bearing resistance per metre run, $R_k = 169.2 \times 1.3 = 220 \text{ kN/m run}$
    Bearing resistance, $R_d = \dfrac{R_k}{\gamma_{Rv}} = \dfrac{220}{1.0} = 220 \text{ kN/m run}$
    Over-design factor, $\Gamma = \dfrac{R_d}{F_d} = \dfrac{220}{108.8} = 2.03$

    Since $\Gamma > 1$, the GEO limit state requirement is satisfied.

2. *Combination 2 (partial factor sets A2 + M2 + R1)*
    The calculations are the same as for Combination 1 except that this time the following partial factors (from Table 6.1) are used: $\gamma_{G;\,unfav} = 1.0$; $\gamma_{cu} = 1.40$; and $\gamma_{Rv} = 1.0$.

    $c_{u;d} = 21.4 \text{ kPa}$

    $W_d = 30.6 \times \gamma_{G;\,unfav} = 30.6 \text{ kN/m run}$

$P_d = 50.0 \times \gamma_{G; \; unfav} = 50.0 \; kN/m \, run$

$F_d = 30.6 + 50.0 = 80.6 \; kN/m \, run$

$e = 0.248 \; m; \quad B' = 1.3 \; m$

$R_k = ((\pi + 2)c_u s_c + q) \times B' = 125.1 \times 1.3 = 163.1 \; kN/m \, run$

$R_d = \dfrac{R_k}{\gamma_{Rv}} = \dfrac{163.1}{1} = 163.1 \; kN/m \, run$

$\Gamma = \dfrac{R_d}{F_d} = \dfrac{163.1}{80.6} = 2.02$

Since $\Gamma > 1$, the GEO limit state requirement is satisfied.

Example 10.6: Traditional, and Eurocode 7 (first generation), approaches (ii)

A concrete foundation 3 m wide, 9 m long and 0.75 m deep is to be found at a depth of 1.5 m in a deep deposit of dense sand. The angle of shearing resistance of the sand is 35° and its unit weight is 19 kN/m³. The unit weight of concrete is 24 kN/m³.

(a) Using a gross factor of safety approach (take F = 3.0):
   (i) Determine the safe bearing capacity for the foundation.
   (ii) Determine the safe bearing capacity of the foundation if it is subjected to a vertical line load of 220 kN/m at an eccentricity of 0.3 m, together with a horizontal line load of 50 kN/m acting at the base of the foundation as illustrated in Fig. 10.10.
(b) For the situation described in (ii) above, establish the magnitude of the over-design factor for the Eurocode 7 GEO limit state, using Design Approach 1.

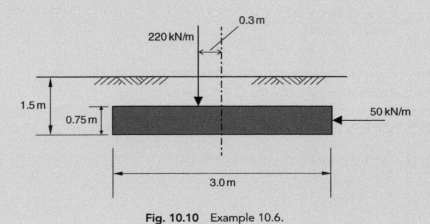

Fig. 10.10   Example 10.6.

Solution:

(a) Gross factor of safety

(i)

$$\text{Safe bearing capacity} = \frac{q_{u\,net}}{3} + \gamma z$$

$$= \frac{\gamma z\left(N_q s_q d_q - 1\right) + 0.5\gamma B N_\gamma s_\gamma d_\gamma}{3} + \gamma z$$

From Table 10.1, for $\phi' = 35°$, $N_q = 33.3$, and $N_\gamma = 45.23$:

$$s_q = 1 + \left(\frac{3}{9}\right)\sin 35° = 1.19, \quad s_\gamma = 1 - 0.3\left(\frac{3}{9}\right) = 0.9$$

$$d_q = 1 + 2\tan 35°(1 - \sin 35°)^2\left(\frac{1.5}{3}\right) = 1.13; \quad d_\gamma = 1$$

Safe bearing capacity

$$= \frac{19 \times 1.5(33.3 \times 1.19 \times 1.13 - 1) + 0.5 \times 19 \times 3 \times 45.23 \times 0.9 \times 1.0}{3} + 19 \times 1.5$$

$$= 831.1 \text{ kPa}$$

(ii) Self-weight of foundation, incl. soil on top, $W = 0.75 \times 9 \times 3 \times (24 + 19) = 871\,\text{kN}$
Total applied vertical load, $P = 220 \times 9 = 1980\,\text{kN}$
Total applied horizontal load, $H = 50 \times 9 = 450\,\text{kN}$
Total vertical load acting on soil, $V = 871 + 1980 = 2851\,\text{kN}$
Eccentricity of bearing pressure

$$e = \frac{P \times e_P}{P + W} = \frac{1980 \times 0.3}{2851} = 0.21 \text{ m}$$

Since $e \leq \dfrac{B}{6}$, the total force acts within the middle-third of the foundation.
Effective width of footing, $B' = 3.0 - 2 \times 0.21 = 2.58\,\text{m}$
The foundation is effectively acted upon by a load of magnitude, F inclined at an angle to the vertical, α:

$$F = \sqrt{V^2 + H^2} = \sqrt{2851^2 + 450^2} = 2886 \text{ kN}$$

$$\alpha = \tan^{-1}\left(\frac{450}{2851}\right) = 9°$$

$$i_q = \left(1 - \frac{9}{90}\right)^2 = 0.81; \quad i_\gamma = \left(1 - \frac{9}{35}\right)^2 = 0.55$$

$$s_q = + \left(\frac{2.58}{9}\right) \sin 35° = 1.2; \quad s_\gamma = 1 - 0.3\left(\frac{2.58}{9}\right) = 0.91$$

$$d_q = 1 + 2\tan 35°(1 - \sin 35°)^2 \left(\frac{1.5}{2.58}\right) = 1.15; \quad d_\gamma = 1$$

Safe bearing capacity

$$= \frac{\gamma z(N_q s_q d_q i_q - 1) + 0.5\ \gamma B' N_\gamma s_\gamma d_\gamma i_\gamma}{3} + \gamma z$$

$$= \frac{19 \times 1.5(33.3 \times 1.2 \times 1.15 \times 0.81 - 1) + 0.5 \times 19 \times 2.58 \times 45.23 \times 0.91 \times 1.0 \times 0.55}{3} + 19 \times 1.5$$

$$= 557.6\ \text{kPa}$$

(b) Eurocode 7

Weight of soil on top of foundation, $W_s = 0.75 \times 9 \times 3 \times 19 = 384.8\ \text{kN}$
Total weight of foundation + soil, $W = 486 + 384.8 = 870.8\ \text{kN}$

1. *Combination 1 (partial factor sets A1 + M1 + R1)*
   From Table 6.1: $\gamma_{G;\ unfav} = 1.35$; $\gamma_Q = 1.5$; $\gamma_{\phi'} = 1.0$; and $\gamma_{Rv} = 1.0$.

$$\text{Design material property: } \phi'_d = \tan^{-1}\left(\frac{\tan\phi'}{\gamma_{\phi'}}\right) = 35°$$

Design actions:
   Weight of foundation, $W_d = W \times \gamma_{G;\ unfav} = 870.8 \times 1.35 = 1175.6\ \text{kN}$
   Applied vertical line load, $P_d = P \times \gamma_{G;\ unfav} = 1980 \times 1.35 = 2673\ \text{kN}$
   Applied horizontal line load, $H_d = H \times \gamma_{G;\ unfav} = 450 \times 1.35 = 607.5\ \text{kN}$

Effect of design actions:

Total vertical force, $F_d = W_d + P_d = 1175.6 + 2673 = 3848.6\ \text{kN}$

Eccentricity, $e = \dfrac{P_d \times e_p}{P_d + W_d} = \dfrac{2673 \times 0.3}{3848.6} = 0.208\ \text{m}$

Since $e \le \dfrac{B}{6}$, the total force acts within the middle-third of the foundation.
Effective width of footing, $B' = 3.0 - 2 \times 0.208 = 2.58\ \text{m}$
Effective area of footing, $A' = 2.58 \times 9 = 23.2\ \text{m}^2$

Design resistance:
From Table 10.1, $N_c = 46.1$, $N_q = 33.3$, and $N_\gamma = 45.23$.
From Eurocode 7, Annex D,

$$s_q = 1 + \left(\frac{B'}{L}\right)\sin\phi' = 1 + \left(\frac{2.58}{9}\right)\sin 35° = 1.16$$

$$s_c = \frac{s_q N_q - 1}{N_q - 1} = 1.17$$

$$s_\gamma = 1 - 0.3\left(\frac{B'}{L}\right) = 0.91$$

$$m = \frac{2 + \dfrac{B'}{L}}{1 + \dfrac{B'}{L}} = 1.78$$

$$i_q = \left[1 - \frac{H}{V + A'c'\cot\phi'}\right]^m = \left[1 - \frac{607.5}{3848.6 + 0}\right]^{1.78} = 0.74 \ (V = F_d)$$

$$i_c = i_q - \left(\frac{1 - i_q}{N_c \tan\phi'}\right) = 0.74 - \left(\frac{1 - 0.74}{46.1 \tan 35°}\right) = 0.72$$

$$i_\gamma = i_q^{\left(\frac{m+1}{m}\right)} = 0.74^{\left(\frac{2.78}{1.78}\right)} = 0.62$$

Characteristic bearing resistance, per m^2,

$$\begin{aligned}
q_u &= c_d' N_c s_c i_c + \gamma_d z N_q s_q i_q + 0.5 B' \gamma_d N_\gamma s_\gamma i_\gamma \\
&= 0 + (19 \times 1.5 \times 33.3 \times 1.16 \times 0.74) \\
&\quad + (0.5 \times 2.58 \times 19 \times 45.2 \times 0.91 \times 0.62) \\
&= 1439 \ \text{kPa}
\end{aligned}$$

Characteristic bearing resistance, $R_k = q_u \times L \times B' = 1439 \times 9 \times 2.58 = 33\,414\,\text{kN}$

Bearing resistance, $R_d = \dfrac{R_k}{\gamma_{Rv}} = \dfrac{33\,414}{1.0} = 33\,414 \ \text{kN}$

Over-design factor, $\Gamma = \dfrac{R_d}{F_d} = \dfrac{33\,414}{3848.6} = 8.68$

Since $\Gamma > 1$, the GEO bearing limit state requirement is satisfied.

2. *Combination 2 (partial factor sets A2 + M2 + R1)*

   The calculations are the same as for Combination 1 except that this time the following partial factors (from Table 6.1) are used: $\gamma_{G;\,unfav} = 1.0$; $\gamma_Q = 1.3$; $\gamma_{\phi'} = 1.25$; and $\gamma_{Rv} = 1.0$.

$$\phi_d' = \tan^{-1}\left(\frac{\tan\phi'}{\gamma_{\phi'}}\right) = \tan^{-1}\left(\frac{\tan 35°}{1.25}\right) = 29.3°$$

$$W_d = 870.8 \times \gamma_G = 870.8 \ \text{kN}$$

$$P_d = 1980 \times \gamma_G = 1980 \ \text{kN}$$

$$F_d = W_d + P_d = 870.8 + 1980 = 2851 \ kN$$

$$H_d = 450 \times \gamma_G = 450 \ kN$$

$$e = \frac{P_d \times e_p}{P_d + W_d} = \frac{1980 \times 0.3}{1980 + 870.8} = 0.208 \ m \ \text{(within the middle-third)}$$

$$B' = 3.0 - 2 \times 0.208 = 2.58 \ m$$

$$N_q = 16.9, \ N_\gamma = 17.8, \ s_q = 1.14, \ s_\gamma = 0.91, \ i_q = 0.74, \ i_\gamma = 0.62.$$

Bearing resistance, per m^2,

$$q_u = c'_d N_c s_c i_c + \gamma_d z N_q s_q i_q + 0.5 B' \gamma_d N_\gamma s_\gamma i_\gamma$$
$$= 653.5 \ kPa$$

Characteristic bearing resistance, $R_k = 653.5 \times L \times B' = 15 \ 174 \ kN$

Bearing resistance, $R_d = \dfrac{R_k}{\gamma_{Rv}} = \dfrac{15 \ 174}{1.0} = 15 \ 174 \ kN$

Over-design factor, $\Gamma = \dfrac{R_d}{F_d} = \dfrac{15 \ 174}{2851} = 5.32$

Since $\Gamma > 1$, the GEO limit state requirement is satisfied.

It should be noted that Γ for both combinations are high, and that the dimensions of the footing could be safely reduced.

Example 10.7: Bearing resistance – undrained and drained (Eurocode 7 first generation)

It is proposed to place a 2 m × 2 m square footing with a horizontal base at a depth of 1.5 m in a glacial clay soil as shown in Fig. 10.11. The 0.5 m thick footing is to support a 0.4 m × 0.4 m centrally located square column which will carry a vertical representative permanent load of 800 kN and a vertical representative transient load of 350 kN. Take the unit weight of reinforced concrete, $\gamma_{concrete}$ as 25 kN/m^3.

The soil has the following properties:

undrained shear strength, $c_u = 200 \ kPa$;
effective cohesion, $c' = 0 \ kPa$;
angle of shearing resistance, $\phi' = 28°$; and
weight density, $\gamma = 20 \ kN/m^3$.

The ground water table is coincident with the base of the foundation.

Check compliance of the bearing resistance limit state using Design Approach 1 for:

(i) the *short-term* state;
(ii) the *long-term* state;

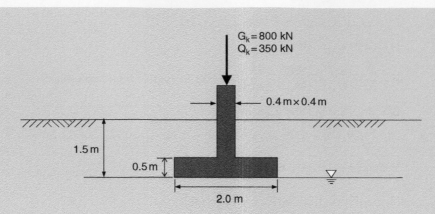

Fig. 10.11 Example 10.7.

Solution:

(i) *Representative actions:*

Self-weight of concrete:
$$W_{concrete} = [(B \times L \times t) + (B_{col} \times L_{col} \times (z-t))] \times \gamma_{concrete}$$
$$= [(2.0 \times 2.0 \times 0.5) + (0.4 \times 0.4 \times 1.0)] \times 25$$
$$= 54.0 \text{ kN}$$

Self-weight of soil:

$$W_{soil} = [(B \times L - B_{col} \times L_{col}) \times (z-t)] \times \gamma_{soil}$$
$$= (2.0 \times 2.0 - 0.4 \times 0.4) \times (1.5 - 0.5) \times 20$$
$$= 76.8 \text{ kN}$$

i/ short-term state

1. *Combination 1 (partial factor sets A1 + M1 + R1)*
   From Table 6.1: $\gamma_{G, unfav} = 1.35$; $\gamma_{G, fav} = 1.0$; $\gamma_Q = 1.5$; $\gamma_{cu} = 1.0$; $\gamma_\gamma = 1.0$; and $\gamma_{Rv} = 1.0$.

   (ii) *Design actions:*
   Design value of self-weight of concrete and soil (unfavourable, permanent action):

   $$W_d = (W_{concrete} + W_{soil}) \times \gamma_{G; \ unfav}$$
   $$= (54.0 + 76.8) \times 1.35 = 176.6 \text{ kN}$$

Design value of the vertical structural (unfavourable) actions:

$$V_d = V_G \times \gamma_{G;\,unfav} + V_Q \times \gamma_Q = (800 \times 1.35) + (350 \times 1.5) = 1605 \text{ kN}$$

Design effect of actions (i.e. sum of vertical forces):

$$F_d = W_d + V_d = 176.6 + 1605 = 1781.6 \text{ kN}$$

Overburden pressure: $q = \gamma_{soil}\, z = 20 \times 1.5 = 30 \text{ kPa}$

(iii) *Design material properties:*

Design cohesion: $c_{u;\,d} = \dfrac{c_{u;k}}{\gamma_{cu}} = \dfrac{200}{1} = 200 \text{ kPa}$

Design weight density of soil: $\gamma_d = \dfrac{\gamma_k}{\gamma_\gamma} = \dfrac{20}{1} = 20 \text{ kN/m}^3$

(iv) *Design geometry:*

No eccentric loading, $\Rightarrow A' = B \times L = 2.0 \times 2.0 = 4.0 \text{ m}^2$
No inclined loading, $\Rightarrow i_c = 1.0$
Foundation base horizontal, $\Rightarrow b_c = 1.0$
$s_c = 1.2$

(v) *Bearing resistance:*

$$R/A' = (\pi + 2)c_u b_c s_c i_c + q$$

$$R_k = 4.0 \times [(5.14 \times 200 \times 1.0 \times 1.2 \times 1.0) + 30] = 5054.4 \text{ kN}$$

$$R_d = \frac{R_k}{\gamma_{Rv}} = \frac{5054.4}{1.0} = 5054.4 \text{ kN}$$

From the results, it is seen that the GEO limit state requirement is satisfied since the design bearing resistance (5054.4 kN) is greater than the design effects of actions (1781.6 kN).

$$\text{Over-design factor, } \Gamma = \frac{R_d}{F_d} = \frac{5054.4}{1781.6} = 2.84$$

2. *Combination 2 (partial factor sets A2 + M2 + R1)*
   From Table 6.1: $\gamma_{G,\,unfav} = 1.0$; $\gamma_{G,\,fav} = 1.0$; $\gamma_Q = 1.3$; $\gamma_{cu} = 1.4$; $\gamma_\gamma = 1.0$; and $\gamma_{Rv} = 1.0$.

   (ii) *Design actions:*
   Design value of self-weight of concrete and soil (unfavourable, permanent action):

$$\begin{aligned} W_d &= (W_{concrete} + W_{soil}) \times \gamma_{G;\,unfav} \\ &= (54.0 + 76.8) \times 1.0 = 130.8 \text{ kN} \end{aligned}$$

Design value of the vertical structural (unfavourable) actions:

$$V_d = V_G \times \gamma_{G;\,unfav} + V_Q \times \gamma_Q = (800 \times 1.0) + (350 \times 1.3) = 1255 \text{ kN}$$

Design effect of actions (i.e. sum of vertical forces):

$$F_d = W_d + V_d = 130.8 + 1255 = 1385.8 \text{ kN}$$

(iii) *Design material properties:*

Design cohesion: $c_{u;d} = \dfrac{c_{u;k}}{\gamma_{cu}} = \dfrac{200}{1.4} = 142.9 \text{ kPa}$

Design weight density of soil: $\gamma_d = \dfrac{\gamma_k}{\gamma_\gamma} = \dfrac{20}{1} = 20 \text{ kN/m}^3$

(iv) *Design geometry:*

No eccentric loading, $\Rightarrow A' = B \times L = 2.0 \times 2.0 = 4.0 \text{ m}^2$
No inclined loading, $\Rightarrow i_c = 1.0$
Foundation base horizontal, $\Rightarrow b_c = 1.0$
$s_c = 1.2$

(v) *Bearing resistance:*

$$R/A' = (\pi + 2)c_u b_c s_c i_c + q$$

$$R_k = 4.0 \times [(5.14 \times 142.9 \times 1.0 \times 1.2 \times 1.0) + 30] = 3645.6 \text{ kN}$$

$$R_d = \frac{R_k}{\gamma_{Rv}} = \frac{3645.6}{1.0} = 3645.6 \text{ kN}$$

GEO limit state requirement satisfied since design bearing resistance (3645.6 kN) > design effects of actions (1385.8 kN).

$$\text{Over-design factor, } \Gamma = \frac{R_d}{F_d} = \frac{3645.6}{1385.8} = 2.63$$

ii/ long-term state

The ground water level is taken as coincident with the ground surface (see Section 10.5.2 and EN 1997-1:2004 §2.4.6.1(11)).

(11) Unless the adequacy of the drainage system can be demonstrated and its maintenance ensured, the design ground-water table should be taken as the maximum possible level, which may be the ground surface.

EN 1997-1:2004 §2.4.6.1(11)

1. **Combination 1** *(partial factor sets A1 + M1 + R1)*
   From Table 6.1: $\gamma_{G, \text{unfav}} = 1.35$; $\gamma_{G, \text{fav}} = 1.0$; $\gamma_Q = 1.5$; $\gamma_{\phi'} = 1.0$; $\gamma_\gamma = 1.0$; and $\gamma_{Rv} = 1.0$.
   (ii) *Design actions:*
       To consider the effects of the buoyant uplift, we can either use the submerged weight of the whole footing or use the total weight and subtract the uplift force due to water pressure under foundation.
       Design value of self-weight of concrete and soil (unfavourable, permanent action):

$$W_d = (W_{\text{concrete}} + W_{\text{soil}}) \times \gamma_{G;\text{ unfav}}$$
$$= (54.0 + 76.8) \times 1.35 = 176.6 \text{ kN}$$

Design value of the vertical structural (unfavourable) actions:

$$V_d = V_G \times \gamma_{G;\ unfav} + V_Q \times \gamma_Q = (800 \times 1.35) + (350 \times 1.5) = 1605\ kN$$

Design value of water pressure under the base (unfavourable (negative) action – from Single Source Principle):

$$U_d = U \times \gamma_{G;\ unfav} = (-1.5 \times 2.0 \times 2.0 \times 9.81) \times 1.35 = -79.5\ kN$$

Design effect of actions (i.e. sum of vertical forces):

$$F_d = W_d + V_d + U_d = 176.6 + 1605 - 79.5 = 1702.1\ kN$$

Overburden pressure: $q = \gamma_{soil}\, z = (20 - 9.81) \times 1.5 = 15.3\ kPa$

(iii) *Design material properties:*

Design angle of shearing resistance: $\phi'_d = \tan^{-1}\left(\dfrac{\tan \phi'}{\gamma_{\phi'}}\right) = \tan^{-1}\left(\dfrac{\tan 28°}{1.0}\right) = 28.0°$

Design weight density of soil: $\gamma_d = \dfrac{\gamma_k}{\gamma_\gamma} = \dfrac{20}{1} = 20\ kN/m^3$

(iv) *Design geometry:*
No eccentric loading, $\Rightarrow A' = B \times L = 2.0 \times 2.0 = 4.0\ m^2$
No inclined loading, $\Rightarrow i_c = i_q = i_\gamma = 1.0$
Foundation base horizontal, $\Rightarrow b_c = b_q = b_\gamma = 1.0$
From EN 1997-1:2004, Annex D:

$$N_q = \tan^2\left(45° + \frac{\phi}{2}\right)e^{\pi \tan \phi} = 14.72$$

$$N_\gamma = 2(N_q - 1)\tan \phi'_d = 14.59$$

$$s_q = 1 + \sin \phi' = 1.47$$

$$s_\gamma = 0.7$$

(v) *Bearing resistance:*

$$R\!/_{A'} = c'N_c b_c s_c i_c + q'N_q b_q s_q i_q + 0.5\gamma B'N_\gamma b_\gamma s_\gamma i_\gamma$$

$$R_k = 4.0 \times [0 + (15.3 \times 14.72 \times 1.0 \times 1.47 \times 1.0) + (0.5 \times (20 - 9.81) \times 2.0 \times 14.59$$
$$\times 1.0 \times 0.7 \times 1.0)]$$

$$= 1738.8\ kN$$

$$R_d = \frac{R_k}{\gamma_{Rv}} = \frac{1738.8}{1.0} = 1738.8\ kN$$

GEO limit state requirement satisfied since design bearing resistance (1738.8 kN) > design effects of actions (1702.1 kN).

Over-design factor, $\Gamma = \dfrac{R_d}{F_d} = \dfrac{1738.8}{1702.1} = 1.02$

2. *Combination 2 (partial factor sets A2 + M2 + R1)*

   From Table 6.1: $\gamma_{G,\,unfav} = 1.0$; $\gamma_{G,\,fav} = 1.0$; $\gamma_Q = 1.3$; $\gamma_{\phi'} = 1.25$; $\gamma_\gamma = 1.0$; and $\gamma_{Rv} = 1.0$.

   (ii) *Design actions:*

   Again, we use the total weight of the foundation and subtract the uplift force due to water pressure under foundation.

   Design value of self-weight of concrete and soil (unfavourable, permanent action):

   $$W_d = (W_{concrete} + W_{soil}) \times \gamma_{G;\,unfav}$$
   $$= (54.0 + 76.8) \times 1.0 = 130.8 \text{ kN}$$

   Design value of the vertical structural (unfavourable) actions:

   $$V_d = V_G \times \gamma_{G;\,unfav} + V_Q \times \gamma_Q = (800 \times 1.0) + (350 \times 1.3) = 1255 \text{ kN}$$

   Design value of water pressure under the base (unfavourable (negative) action – from Single Source Principle):

   $$U_d = U \times \gamma_{G;\,unfav} = (-1.5 \times 2.0 \times 2.0 \times 9.81) \times 1.0 = -58.9 \text{ kN}$$

   Design effect of actions (i.e. sum of vertical forces):

   $$F_d = W_d + V_d + U_d = 130.8 + 1255 - 58.9 = 1326.9 \text{ kN}$$

   Overburden pressure: $q = \gamma_{soil}\, z = (20 - 9.81) \times 1.5 = 15.3 \text{ kPa}$

   (iii) *Design material properties:*

   Design angle of shearing resistance: $\phi'_d = \tan^{-1}\left(\dfrac{\tan 28°}{1.25}\right) = 23.0°$

   Design weight density of soil: $\gamma_d = \dfrac{\gamma_k}{\gamma_\gamma} = \dfrac{20}{1} = 20 \text{ kN/m}^3$

   (iv) *Design geometry:*

   No eccentric loading, $\Rightarrow A' = B \times L = 2.0 \times 2.0 = 4.0 \text{ m}^2$

   No inclined loading, $\Rightarrow i_c = i_q = i_\gamma = 1.0$

   Foundation base horizontal, $\Rightarrow b_c = b_q = b_\gamma = 1.0$

   From EN 1997-1:2004, Annex D:

   $$N_q = \tan^2\left(45° + \frac{\phi}{2}\right)e^{\pi \tan \phi} = 8.7$$

   $$N_\gamma = 2(N_q - 1)\tan \phi'_d = 6.55$$

   $$s_q = 1 + \sin \phi' = 1.39$$

   $$s_\gamma = 0.7$$

   (v) *Bearing resistance:*

   $$R/_{A'} = c'N_c b_c s_c i_c + q'N_q b_q s_q i_q + 0.5\gamma B'N_\gamma b_\gamma s_\gamma i_\gamma$$

$$R_k = 4.0 \times [0 + (15.3 \times 8.7 \times 1.0 \times 1.39 \times 1.0)$$
$$+ (0.5 \times (20 - 9.81) \times 2.0 \times 6.55 \times 1.0 \times 0.7 \times 1.0)]$$
$$= 926.5 \text{ kN}$$

$$R_d = \frac{R_k}{\gamma_{Rv}} = \frac{926.5}{1.0} = 926.5 \text{ kN}$$

GEO limit state requirement NOT satisfied since design bearing resistance (926.5 kN) < design effects of actions (1326.9 kN).

$$\text{Over-design factor, } \Gamma = \frac{R_d}{F_d} = \frac{926.5}{1326.9} = 0.70$$

This example illustrates the need to check the limit state requirements for both combinations and for both the undrained and the drained states. In this case, the footing is inadequately designed, and the dimensions of the footing would have to be increased to ensure the requirements are met in all cases.

Example 10.8: Bearing resistance – vertical and horizontal loading (Eurocode 7 first generation)

In accordance with Eurocode 7, Design Approach 2, check compliance of the bearing resistance limit state under drained conditions of the square pad foundation shown in Fig. 10.12.

The footing is supporting a concentric square column which is carrying both permanent and variable vertical loading, together with a transient horizontal load applied as shown.

The ground water table is coincident with the ground surface, and the following data are provided:

Representative vertical permanent load, V_G = 800 kN;
Representative vertical transient load, V_Q = 400 kN;
Weight density of soil, γ_{soil} = 19.0 kN/m³;
Effective cohesion, c' = 0 kPa;
Angle of shearing resistance, ϕ' = 30°;
Representative horizontal transient load, H_Q = 100 kN; and
Weight density of concrete, $\gamma_{concrete}$ = 24 kN/m³,

Solution:

(i) *Representative actions:*
   Self-weight of concrete:

$$W_{concrete} = (B \times L \times t + B_{col} \times L_{col} \times (z - t)) \times \gamma_{concrete}$$
$$= [(3.0 \times 3.0 \times 0.5) + (0.5 \times 0.5 \times 1.5)] \times 24$$
$$= 117.0 \text{ kN}$$

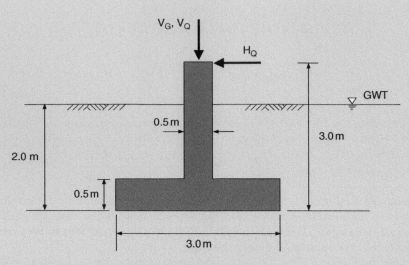

Fig. 10.12 Example 10.8.

Uplift thrust under the base of foundation:

$$U = -d_w \times B \times L \times \gamma_w = -2.0 \times 3.0 \times 3.0 \times 9.81 = -176.6 \ kN$$

Self-weight of soil:

$$
\begin{aligned}
W_{soil} &= (B \times L - B_{col} \times L_{col}) \times (z - t)) \times \gamma_{soil} \\
&= (3.0 \times 3.0 - 0.5 \times 0.5) \times (2.0 - 0.5) \times 19 \\
&= 249.4 \ kN
\end{aligned}
$$

Overburden pressure: $q' = \gamma z = (19 - 9.81) \times 2.0 = 18.4 \ kPa$

Design Approach 2 (partial factor sets A1 + M1 + R2)
From Table 6.1: $\gamma_{G, \ unfav} = 1.35$; $\gamma_{G, \ fav} = 1.0$; $\gamma_Q = 1.5$; $\gamma_\phi = 1.0$; $\gamma_\gamma = 1.0$; and $\gamma_{Rv} = 1.4$.

(ii) *Design actions:*
Design value of self-weight of concrete and soil (unfavourable, permanent action):

$$
\begin{aligned}
W_d &= (W_{concrete} + W_{soil}) \times \gamma_{G; \ unfav} \\
&= (117.0 + 249.4) \times 1.35 = 494.6 \ kN
\end{aligned}
$$

Design value of the vertical structural actions:

$$
\begin{aligned}
V_d &= V_G \times \gamma_{G; \ unfav} + V_Q \times \gamma_Q \\
&= (800 \times 1.35) + (400 \times 1.5) = 1680 \ kN
\end{aligned}
$$

Design value of water pressure under the base (unfavourable (negative) action): (from Single Source Principle)

$$
\begin{aligned}
U_d &= U \times \gamma_{G; \ unfav} \\
&= -176.6 \times 1.35 = -238.4 \ kN
\end{aligned}
$$

Design effect of actions (i.e. sum of vertical forces):

$$F_d = W_d + V_d + U_d = 494.6 + 1680 - 238.4 = 1936.2 \text{ kN}$$

Design value of the horizontal structural transient action:

$$H_d = H_Q \times \gamma_Q$$
$$= 100 \times 1.5 = 150 \text{ kN}$$

(iii) *Material properties:*

Design cohesion: $c'_d = \left(\dfrac{c'_k}{\gamma_{c'}}\right) = 0 \text{ kPa}$

Design angle of shearing resistance: $\phi'_d = \tan^{-1}\left(\dfrac{\tan\phi'}{\gamma_{\phi'}}\right) = \tan^{-1}\left(\dfrac{\tan 30°}{1.0}\right) = 30°$

Design weight density of soil: $\gamma_d = \left(\dfrac{\gamma_k}{\gamma_\gamma}\right) = \dfrac{19}{1} = 19 \text{ kN/m}^3$

(iv) *Design geometry:*

The horizontal force will induce an eccentric loading. The magnitude of e can be determined by considering moments about mid-point of base of footing.

$$V \times e = H_d \times x_H \quad (V \equiv F_d)$$

$$e = \frac{H_d \times x_H}{V} = \frac{150 \times 3.0}{1936.2} = 0.232 \text{ m} \quad \text{(acts within middle-third)}$$

$$B' = B - 2 \times e = 3.0 - 2 \times 0.232 = 2.535 \text{ m}$$
$$\Rightarrow A' = B' \times L = 2.535 \times 3.0 = 7.606 \text{ m}^2$$

From EN 1994-1:2004, Annex D:

$$N_q = \tan^2\left(45° + \frac{\phi}{2}\right)e^{\pi \tan\phi} = 18.4$$

$$N_\gamma = 2(N_q - 1)\tan\phi'_d = 20.1$$

$$m = \frac{2 + \dfrac{B'}{L'}}{1 + \dfrac{B'}{L'}} = 1.542$$

$$i_q = \left[1 - \frac{H}{V + A'c'\cot\phi'}\right]^m = \left[1 - \frac{150.0}{1936.2 + 0}\right]^{1.542} = 0.88 \quad (V = F_d)$$

$$i_\gamma = i_q^{\left(\frac{m+1}{m}\right)} = 0.88^{2.452/1.542} = 0.81$$

$$s_q = 1 + \frac{B'}{L}\sin\phi' = 1.42$$

$$s_\gamma = 1 - 0.3\left(\frac{B'}{L}\right) = 0.75$$

Foundation base horizontal, $\Rightarrow b_q = b_\gamma = 1.0$

(v) *Bearing resistance:*

$$R_k/A' = (c_d' \times N_c \times b_c \times s_c \times i_c + q' \times N_q \times b_q \times s_q \times i_q + 0.5 \times \gamma_d \times B' \times N_\gamma \times b_\gamma \times s_\gamma \times i_\gamma)$$

$$R_k = 7.606 \times [0 + (18.4 \times 18.4 \times 1.0 \times 1.42 \times 0.88) + (0.5 \times (19 - 9.81) \times 2.535 \times 20.1$$
$$\times\ 1.0 \times 0.75 \times 0.81)]$$
$$= 4313 \ \text{kN}$$

$$R_d = \left(\frac{R_k}{\gamma_{Rv}}\right) = \frac{4313}{1.4} = 3081 \ \text{kN}$$

From the results, it is seen that the GEO limit state is satisfied since the design bearing resistance (3081 kN) is greater than the design effects of actions (1936 kN).

Over-design factor, $\Gamma = \dfrac{R_d}{F_d} = \dfrac{3081}{1936} = 1.59$

10.7.2   Design by prescriptive method

As mentioned in Chapter 6, Eurocode 7 Part 1 states that a prescriptive method may be used to check a limit state on occasions where calculation of the soil properties is not possible or necessary, provided that generally conservative rules of design are used. Earlier versions of BS 8004 provided a list of safe bearing capacity values (reproduced in Table 10.2) and this list forms the basis of such a prescriptive approach for the case of the checking of the bearing resistance limit state. The values are based on the following assumptions:

(i) the site and adjoining sites are reasonably level;
(ii) the ground strata are reasonably level;
(iii) there is no softer layer below the foundation stratum; and
(iv) the site is protected from deterioration.

Foundations designed to these values will normally have adequate protection to satisfy the requirements of the bearing resistance limit state, provided that they are not subjected to inclined loading. It should be remembered however that settlement effects would also have to be considered, and thus a serviceability limit state check (see Example 12.8) should be undertaken.

Table 10.2   Presumed safe bearing capacity, q_s, values.

	q_s (kPa)
Rocks	
(Values based on assumption that foundation is carried down to unweathered rock)	
Hard igneous and gneissic	10 000
Hard sandstones and limestones	4000
Schists and slates	3000
Hard shale and mudstones, soft sandstone	2000
Soft shales and mudstones	1000–600
Hard chalk, soft limestone	600
Cohesionless soils	
(Values to be halved if soil submerged)	
Compact gravel, sand and gravel	>600
Medium dense gravel or sand and gravel	600–200
Loose gravel or sand and gravel	<200
Compact sand	>300
Medium dense sand	300–100
Loose sand	<100
Cohesive soils	
(Susceptible to long-term consolidation settlement)	
Very stiff boulder clays and hard clays	600–300
Stiff clays	300–150
Firm clays	150–75
Soft clays and silts	<75
Very soft clays and silts	Not applicable

Based on British Standards Institution (2020).

10.8   Designing spread foundations to Eurocode 7 (second generation)

In the second generation, the design of shallow foundations is covered in Section 5 of EN1997-3:202x. In accordance with the procedure for establishing the Geotechnical Category (described in Chapter 6), the Geotechnical Complexity Class must be established at an early stage. Most routine foundation design work will fall into Geotechnical Complexity Class 2. Higher complexity structures will fall into GCC3 and lower complexity structures will fall into GCC1.

GCC3 shallow foundations include situations where: the ground includes weak layers, the ground is experiencing persistent movement or is unstable, the foundation carries high concentrated loads or situations where the foundation is subject to cyclic, dynamic or seismic loading.

GCC1 foundations cover situations where there is negligible risk of the occurrence of an ultimate or serviceability limit state, where the ground conditions are fully known and understood, or where there is negligible risk of ground movement.

As with all geotechnical structures, the limit states against ground failure or rupture are verified using either the material factor approach (MFA) or the resistance factor approach (RFA). The ultimate resistance to bearing is thus verified using either approach, and the combinations of Design Case and material partial factor set (M1 or M2 – as defined in Table 6.7) are the same as for gravity retaining walls. These were listed in Table 9.1.

Examples 10.9 and 10.10 illustrate the design processes for both the MFA and the RFA.

Depth factors are introduced in the second generation. Also, all the formulae listed in Section 10.7.1 apply in bearing resistance verification.

Example 10.9: Eurocode 7 (second generation) (i)

Consider again Example 10.5. The structure is in Consequence Class 2. Using the resistance factor approach, verify that the bearing limit state will not be exceeded.

Solution:

Resistance Factor Approach: DC4 & M1 & γ_{Rv}
From Table 6.7: $\gamma_{cu} = 1.0$
From Table 6.9: $\gamma_{Rv} = 1.4$; $\gamma_E = 1.35\,K_F$ ($K_F = 1.0$, since CC2)

$c_{u;d} = 30$ kPa;

$W_d = 30.6$ kN/m run; and

$P_d = 50.0$ kN/m run.

$F_d = (30.6 + 50.0) \times \gamma_E = 108.8$ kN/m run

$e = \dfrac{P \times e_P}{P + W} = \dfrac{50.0 \times 0.4}{108.8} = 0.184$ m

Since $e \leq \dfrac{B}{6}$, the total force acts within the middle-third of the foundation.
Effective width of footing, $B' = 1.8 - 2 \times 0.184 = 1.43\,\text{m}$

$s_c = 1.0$

$d_{cu} = 1 + 0.33 \tan^{-1}\left(\dfrac{D}{B}\right) = 1.13$

$R_k = ((\pi + 2)c_u s_c d_{cu} + q) \times B' = 189.3 \times 1.43 = 270.6$ kN/m run

$R_d = \dfrac{R_k}{\gamma_{Rv}} = \dfrac{270.6}{1.4} = 193.3$ kN/m run

$\Gamma = \dfrac{R_d}{F_d} = \dfrac{193.3}{108.8} = 1.8$

Since $\Gamma > 1$, the GEO limit state requirement is satisfied.

Example 10.10: Eurocode 7 (second generation) (ii)

Consider again Example 10.7. The structure is in Consequence Class 2. Using the resistance factor approach, verify that the drained (long-term) bearing limit state will not be exceeded.

Solution:

Resistance Factor Approach: DC4 & M1 & γ_{Rv}
From Table 6.7: $\gamma_{\tan\phi} = 1.0$.
From Table 6.9: $\gamma_{Rv} = 1.4$; $\gamma_E = 1.35\,K_F$ ($K_F = 1.0$, since CC2)
    As with the procedure set out in the first generation of Eurocode 7, the ground water level is taken as coincident with the ground surface.
    Many of the following values have been derived earlier (in Example 10.7).

(i) *Design effect of actions* (i.e. sum of vertical forces):

$$F_d = (W_d + V_d + U_d) \times \gamma_E = (130.8 + 1255 - 58.9) \times 1.35 = 1791 \text{ kN}$$
$$q = 15.3 \text{ kPa}$$

(ii) *Design material properties:*

Design angle of shearing resistance: $\phi'_d = \tan^{-1}\left(\dfrac{\tan 28°}{1.0}\right) = 28.0°$

Design weight density of soil: $\gamma_d = \dfrac{\gamma_k}{\gamma_\gamma} = \dfrac{20}{1} = 20 \text{ kN/m}^3$

(iii) *Design geometry:*
    No eccentric loading, $\Rightarrow A' = B \times L = 2.0 \times 2.0 = 4.0 \text{ m}^2$
    No inclined loading, $\Rightarrow i_c = i_q = i_\gamma = 1.0$
    Foundation base horizontal, $\Rightarrow b_c = b_q = b_\gamma = 1.0$

(iv) *Bearing resistance:*

$$N_q = \tan^2\left(45° + \frac{\phi}{2}\right)e^{\pi \tan\phi} = 14.72$$

$$N_\gamma = 2(N_q - 1)\tan\phi'_d = 14.59$$
$$s_q = 1 + \sin\phi' = 1.47$$
$$s_\gamma = 0.7$$
$$d_q = 1 + 2\tan\phi'(1 - \sin\phi')^2(1.5/2) = 1.22$$
$$d_\gamma = 1.0$$
$$R/A' = c'N_c d_c s_c + q'N_q d_q s_q + 0.5\gamma'B'N_\gamma d_\gamma s_\gamma$$

$$R_k = 4.0 \times [0 + (15.3 \times 14.72 \times 1.22 \times 1.47) + (0.5 \times (20 - 9.81) \times 2.0 \times 14.59 \times 1.0 \times 0.7)]$$
$$= 2032 \text{ kN}$$

$$R_d = \frac{R_k}{\gamma_{Rv}} = \frac{2032}{1.4} = 1451 \text{ kN}$$

Over-design factor, $\Gamma = \dfrac{R_d}{F_d} = \dfrac{1451}{1791} = 0.81$

As with Example 10.7, part (ii), it is seen that the limit state requirement is not satisfied. The dimensions of the footing need to be enlarged to ensure a safe design.

10.9 Non-homogeneous soil conditions

The bearing capacity Equations (10.8)–(10.16) are based on the assumption that the foundation soil is homogeneous and isotropic.

In the case of variable soil conditions, the analysis of bearing capacity is best carried out using a finite element analysis, although it can be carried out using some form of slip circle method, as described earlier in this chapter. This approach can take time and designs based on one of the bearing capacity formulae are consequently quite often used.

For the case of a foundation resting on thin layers of soil, of thicknesses H_1, H_2, H_3, ... H_n and of total depth H, Bowles (1996) suggests that these layers can be treated as one layer with an average c value c_{av} and an average ϕ value ϕ_{av}, where

$$c_{av} = \frac{c_1 H_1 + c_2 H_2 + c_3 H_3 + \cdots + c_n H_n}{H}$$

$$\phi_{av} = \tan^{-1}\left(\frac{H_1 \tan\phi_1 + H_2 \tan\phi_2 + H_3 \tan\phi_3 + \cdots + H_n \tan\phi_n}{H}\right)$$

Vesic (1975) suggested that, for the case of a foundation founded in a layer of soft clay which overlies a stiff clay, the ultimate bearing capacity of the foundation can be expressed as:

$$q_u = c_u N_{cm} + \gamma z$$

where c_u = the undrained strength of the soft clay and N_{cm} = a modified form of N_c, the value of which depends upon the ratio of the c_u values of both clays, the thickness of the upper layer, the foundation depth, and the shape and width of the foundation. Values of N_{cm} are quoted in Vesic's paper.

The converse situation, i.e. that of a foundation founded in a layer of stiff clay which overlies a soft clay, was studied by Brown and Meyerhof (1969), who quoted a formula for N_{cm} based on a punching shear failure analysis.

At first glance a safe way of determining the bearing capacity of a foundation might be to base it on the shear strength of the weakest soil below it, but such a procedure can be uneconomical, particularly if the weak soil is overlain by much stronger soil. A more suitable method is to calculate the safe bearing capacity using the shear strength of the stronger material and then to check the amount of overstressing that this will cause in the weaker layers.

10.10 Estimates of bearing capacity from *in situ* testing

10.10.1 The plate loading test

The test procedure was introduced in Section 7.6.5. In the test, an excavation is made to the expected foundation level of the proposed structure and a steel plate, usually from 300 to 750 mm diameter, or square, is placed in position and loaded by means of a hydraulic jack placed beneath a truck or other heavy site vehicle to provide the jacking resistance.

The test measures the settlement of the plate for various increments of applied pressure. The results of settlement against pressure are plotted, as illustrated by the examples in Fig. 10.13.

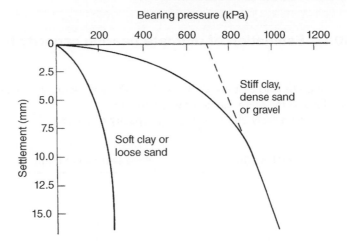

Fig. 10.13   Typical plate loading test results.

The ultimate bearing capacity of the soil can be determined by extrapolating the results.

If the diameter (circular plate) or side length (square plate) of the plate is D, the ultimate bearing capacity, q_u is taken to be equal to:

- *for soft clays and loose sands:* the pressure causing settlement equal to somewhere in the range 10% D (minimum likely q_u) to 20% D (maximum likely q_u).
- *for stiff clays, dense sands, or gravels:* the pressure causing settlement equal to somewhere in the range 5% D (minimum likely q_u) to 10% D (maximum likely q_u).

The extrapolation process has to be performed carefully as the relationship between applied pressure and plate settlement in the test follows a different path for different types of soil, as shown in Fig. 10.13. It can be seen that the extrapolation is not always linear.

On dense sands and gravels and stiff clays after an initial linear response, the relationship between bearing pressure and plate settlement is as shown in Fig. 10.13. The q_u value is then determined by extrapolating backwards (as shown in the figure). For testing carried out within the linear range, simple linear extrapolation can be appropriate to establish q_u using the relationship mentioned above.

With a soft clay or a loose sand, the plate experiences a more or less constant rate of settlement under load and no definite failure point can be established.

In spite of the fact that a plate loading test can only assess a metre or two of the soil layers below the test level, the method can be extremely helpful in gravelly soils where undisturbed sampling is not possible provided that it is accompanied by a borehole investigation programme, to prove that the soil does not exhibit significant variations.

However, the test can give erratic results in sands and soft clays when there is a variation in density over the site, and several tests should be carried out to determine a sensible average. This procedure is costly, particularly if the groundwater level is near the foundation level and groundwater lowering techniques consequently become necessary.

Example 10.11: Estimation of ultimate bearing capacity from plate load test

The results from a plate load test are shown below. The soil tested was a densely compacted granular fill and the plate diameter was 610 mm.

Estimate the minimum likely ultimate bearing capacity of the soil.

Applied load (kN)	Plate settlement (mm)
0	0
5	1.20
10	2.64
15	4.11
25	6.87
35	10.6

Solution:

The first step is to determine the pressure settlement plot:

Area of plate = $0.29\,m^2$

Applied pressure (kPa)	Plate settlement (mm)
0	0
=5/0.29 = 17	1.20
34	2.64
51	4.11
86	6.87
120	10.6

It is seen that the plot is linear (typical for a well-compacted gravelly material), so the equation of the straight line can be used to estimate q_u (Fig. 10.14).

The absolute value of the slope of the line is determined, m = 0.08.
Estimated minimum q_u (at 5% plate diameter) = 30.5/0.08 = 381 kPa
Estimated maximum q_u (at 10% plate diameter) = 61/0.08 = 762 kPa
(Normally it would be the minimum value that the designer would use.)

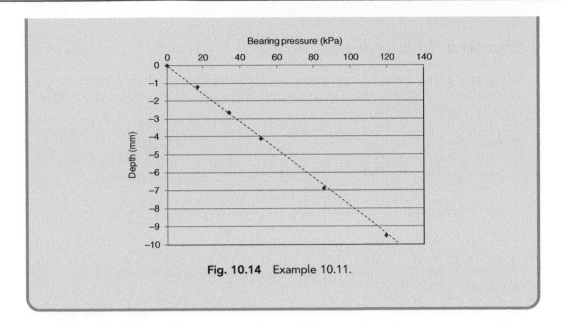

Fig. 10.14   Example 10.11.

As would be expected, the settlement of a square footing applying a constant pressure to the soil increases as the size of the footing increases.

Terzaghi and Peck (1948) investigated this effect and produced the relationship:

$$S = S_1 \left(\frac{2B}{B + 0.3} \right)^2$$

(10.18)

where

S_1 = settlement (mm) of a loaded area 0.305 m square under a given loading intensity p; and
S = settlement (mm) of a square or rectangular footing of width B (in metres) under the same pressure p.

To use plate loading test results, the designer must first decide upon an acceptable value for the maximum allowable settlement. Unless there are other conditions to be taken into account, it is generally accepted that maximum allowable settlement is 25 mm.

The method for determining the allowable bearing pressure for a foundation of width B m is apparent from Equation (10.18). If S is put equal to 25 mm and the numerical value of B is inserted in the formula, S_1 will be obtained. From the plate loading test results, we have the relationship between S_1 and p (Fig. 10.13), so the value of p corresponding to the calculated value of S_1 is the allowable bearing pressure of the foundation, subject to any adjustment that may be necessary for certain groundwater conditions. The adjustment procedure for the presence of groundwater, if the soil is submerged, is the same as that employed to obtain the allowable bearing pressure from the standard penetration test – see following section.

Example 10.11: (continued)

If the proposed foundation is to be a square pad of side length 2 m, with a maximum allowable settlement of 25 mm, establish an estimate of the allowable bearing pressure that could be applied to the foundation.

Solution:

$$S = S_1 \left(\frac{2B}{B + 0.3} \right)^2$$

$$25 = S_1 \left(\frac{2 \times 2}{2 + 0.3} \right)^2$$

$$\Rightarrow S_1 = 8.3 \text{ mm}$$

Referring to Fig. 10.14, anticipated bearing pressure causing a settlement of 8.3 mm is approximately 100 kPa.

Note: As mentioned earlier in this chapter, typically a factor of safety of at least 3.0 is applied to the ultimate bearing capacity to give an indication of the allowable bearing pressure. In this example, the ultimate value was estimated to be 380 kPa, so an allowable value of 100 kPa is well within approximation limits, demonstrating that this approximation method provides acceptable values of allowable bearing pressure.

10.10.2　Standard penetration test

This test was described in Section 7.6.2 and the results from it can be used to establish an approximate allowable bearing pressure, inferred from acceptable settlement values.

Having determined the $(N_1)_{60}$ value, the determination of the allowable bearing pressure is generally based upon an empirical relationship evolved by Terzaghi and Peck (1948) that is based on the measured settlements of various foundations on sand (Fig. 10.15). The allowable bearing pressure for these curves (which are applicable to both square and rectangular foundations) was defined by Terzaghi and Peck as the pressure that will not cause a settlement greater than 25 mm.

When several foundations are involved, the normal design procedure is to determine an average value for $(N_1)_{60}$ from all the boreholes. The allowable bearing pressure for the widest foundation is then obtained with this figure and this bearing pressure is used for the design of all the foundations. The procedure generally leads to only small differential settlements, but even in extreme cases, the differential settlement between any two foundations will not exceed 20 mm.

The curves of Fig. 10.15 apply to unsaturated soils, i.e. when the water table is at a depth of at least 1.0B below the foundation. When the soil is submerged, the value of allowable bearing pressure, q_a obtained from the curves should be reduced according to the expression:

$$q_a = q \times 0.5 \times \left(1 + \frac{D_w}{D + B} \right) \tag{10.19}$$

where

D_w = depth of water table;
D = depth of foundation level; and
B = width of foundation.

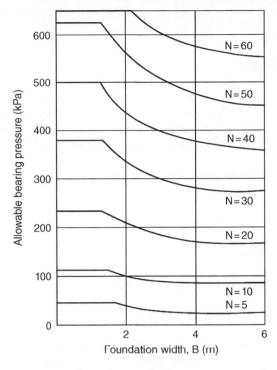

Fig. 10.15 Allowable bearing pressure from the standard penetration test. Based on Terzaghi and Peck (1948).

Example 10.12: Estimate of acceptable bearing capacity from SPT

Refer to the SPT result from Example 7.2. A strip footing, 3 m wide, is to be found at a depth of 3 m in the silty sand. Assuming that the strength characteristics of the sand are constant with depth, estimate the allowable bearing capacity.

Solution:

From Example 7.2, $(N_1)_{60} = 19$
From Fig. 10.15, for a corrected $N = 19$ and $B = 3$ m:
Allowable bearing pressure = 180 kPa

But this value is for *dry* soil and the sand below the foundation is also below groundwater level and is therefore submerged. Therefore,

$$q_a = q \times 0.5 \times \left(1 + \frac{D_w}{D + B}\right) = 180 \times 0.5 \times \left(1 + \frac{1.5}{3 + 3}\right) = 112.5 \text{ kPa}$$

Exercises

Exercise 10.1

A strip footing 3 m wide is to be found at a depth of 2 m in a saturated soil of unit weight 19 kN/m^3. The soil has an angle of shearing resistance, ϕ' of 28° and an effective cohesion, c' of 5 kPa. Groundwater level is at a depth of 4 m. By adopting a factor of safety of 3, determine a value for the safe bearing capacity of the foundation.

If the groundwater level was to rise to the ground surface, determine the new value of safe bearing capacity.

Answer   442 kPa; 242 kPa

Exercise 10.2

A soil layer has the following properties:

$c' = 7$ kPa; $\phi' = 20°$; $\gamma = 20$ kN/m^3.

A footing of dimensions 7.5 m × 2.5 m is to be founded at a depth of 1.5 m into this layer. Determine the ultimate bearing capacity of the soil.

Answer   488 kPa

Exercise 10.3

A continuous concrete footing ($\gamma_c = 24$ kN/m^3) of breadth 2.0 m and thickness 0.5 m is to be founded in a clay soil ($c_u = 22$ kPa; $\gamma = 19$ kN/m^3) at a depth of 1.0 m. The footing will carry an applied vertical load of magnitude 85 kN/m run. The load will act on the centre line of the footing.

Using Eurocode 7 Design Approach 1, determine the magnitude of the over-design factor for the bearing resistance limit state under undrained conditions.

Answer   1.53 (DA1-1); 1.56 (DA1-2)

If you were to include depth factors in the design procedure, what would be the revised value of the over-design factor for each combination?

Answer   1.79 (DA1-1); 1.81 (DA1-2)

Note: Adopting depth factors in the design will invariably lead to higher values of over-design factor.

Exercise 10.4

A rectangular foundation (2.5 m × 6 m × 0.8 m deep) is to be founded at a depth of 1.2 m in a dense sand ($c' = 0$; $\phi' = 32°$; $\gamma = 19.4$ kN/m³). The unit weight of concrete = 24 kN/m³. The foundation will carry a vertical line load of 250 kN/m at an eccentricity of 0.4 m.

By following Eurocode 7, Design Approach 1, establish the proportion of the available resistance that will be used.

Answer   21% (DA1-1); 32% (DA1-2)

Note: The proportion of available resistance that will be used is determined by taking the reciprocal of the over-design factor.

Chapter 11
Pile Foundations

Learning objectives:

By the end of this chapter, you will have been introduced to:
- the different types of pile foundations that exist and their methods of installation;
- methods for the determination of the bearing resistance of piles in different soil types;
- negative skin friction;
- the design of pile foundations using both generations of Eurocode 7;
- piles in tension and experiencing transverse loading.

11.1   Introduction

Pile foundations are used to transfer the load of the structure to the bearing soil or rock located at a significant depth below ground surface. They are long and slender structural elements that transfer the load to soil/rock of high bearing capacity that lie beneath shallower soils of lower bearing capacity. Piles may be made from concrete, steel or timber, or from some composite of these materials and are installed either by driving, drilling, or jacking. A pile cap is fixed to the top of the pile and it is this cap onto which the structural loads are transmitted.

In addition to piles being used to transmit the foundation load to a solid stratum (by end bearing) or throughout a deep mass of soil (through soil-pile friction), pile foundations are also used to resist horizontal or uplift loads where such forces may act.

There are several types of pile and these are described in the coming sections.

11.2   Classification of piles

Piles can be classified by different criteria, such as their material (e.g. concrete, steel, timber), their method of installation (e.g. driven or bored), the degree of soil displacement during installation, or their size (e.g. large diameter, small diameter). However, in terms of pile design, the most appropriate classification criterion is the behaviour of the pile once installed (e.g. end-bearing pile, friction pile, combination pile).

Smith's Elements of Soil Mechanics, 10th Edition. Ian Smith.
© 2021 John Wiley & Sons Ltd. Published 2021 by John Wiley & Sons Ltd.
Companion website: www.wiley.com/go/smith/soilmechanics10e

11.2.1   End bearing

These piles transfer their load to a firm stratum located at a considerable depth below the base of the structure and they derive most of their carrying capacity from the penetration resistance of the soil at the toe of the pile (Fig. 11.1a). The pile behaves as an ordinary column and should be designed as such. Even in weak soil, a pile will not fail by buckling and this effect only needs to be considered if part of the pile is unsupported, i.e. if it is in either air or water.

A further variation of the end-bearing pile is piles with enlarged bearing areas. This is achieved by forcing a bulb of concrete into the soft stratum immediately above the firm layer to give an enlarged base. A similar effect is produced with bored piles by forming a large cone or bell at the bottom with a special reaming tool.

11.2.2   Friction

Here, the carrying capacity is derived mainly from the adhesion or friction of the soil in contact with the shaft of the pile (Fig. 11.1b).

11.2.3   Combination

This is an extension of the end-bearing pile when the bearing stratum is not hard, such as a firm clay. The pile is driven far enough into the lower material to develop adequate frictional resistance (Fig. 11.1c).

11.3   Method of installation

The installation process is every bit as important as the design process for pile foundations. There are two main methods of pile installation: pile driving, and boring by rotating auger. Before pile design can commence the type of pile and its method of installation should be known. In order to avoid damage to the pile during installation, the method of installation should be considered during the site monitoring aspects of the design process.

11.3.1   Driven piles

These are prefabricated piles that are installed into the ground through the use of a pile driver as illustrated in Fig. 11.2. The pile is hoisted into position on the pile driver and aligned against the runners so that the pile is driven into the ground at exactly the required angle, to exactly the required depth. Most commonly the pile is

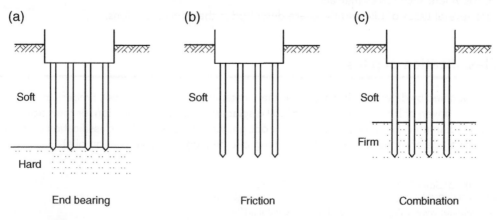

Fig. 11.1   Classification of piles. (a) End bearing. (b) Friction. (c) Combination.

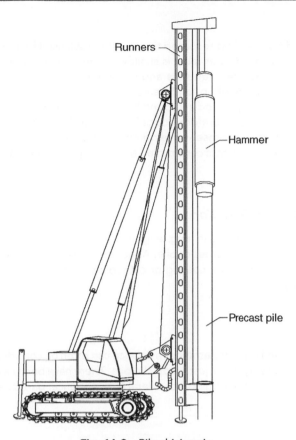

Fig. 11.2 Pile driving rig.

driven into the soil vertically and by striking the top of the pile repeatedly with a pneumatic or percussive hammer. However, this is a noisy method of installation and induces a significant amount of vibration in the ground, so usage is limited to situations where noise and vibration limits permit.

At sites where noise is a concern, vibratory hammers can be used in place of percussive hammers. These hammers are fixed to the top of the pile and employ a vibrating unit to induce vibrations down through the pile to ease its installation into the ground. They are suitable for use in granular soils and generate much less noise than percussive hammers. Water jetting can be used to aid the penetration of piles into dense sands or sandy gravel.

In situations where hammers are not possible or acceptable, piles can be installed by jacking (see *Jacked pile* below). This method uses hydraulic rams to drive the pile into the ground, utilising either adjacent piles or structures to provide the jacking reaction force. The method tends to be used most in micro-piling, where the reaction loads can be provided by the structure being underpinned, or for driving steel sheet piling into the ground.

Most commonly driven piles are made from precast concrete, though steel and timber piles are also available.

Precast concrete

These are usually of square or octagonal section. Reinforcement is necessary within the pile to help withstand both handling and driving stresses. Prestressed concrete piles can also be used on occasions. These piles require less reinforcement than ordinary precast concrete piles.

Timber

Timber piles have been used from earliest recorded times and are still used for permanent works where timber is plentiful and where the groundwater table is shallow. In the United Kingdom, timber piles are used mainly in temporary works, due to their lightness and shock resistance, but they are also used for piers and fenders and can have a design life of up to 25 years or more if kept completely below the water table. However, they can deteriorate rapidly if used in ground in which the water level varies and allows the upper part to come above the water surface. Therefore, special treatments including creosoting are applied to the timber before use. The impregnation of preservative treatment in a pressure chamber is an effective method of protection. Much of central Amsterdam is constructed on timber piles where the groundwater table is very close to the ground surface.

Steel piles: Tubular, box, or H-section

These are suitable for handling and driving in long lengths. They have a relatively small cross-sectional area and penetration is easier than with other types of pile. Although made of steel, the risk of strength reduction due to corrosion is not great, although tar coating or cathodic protection can be employed in permanent work if considered necessary.

Jetted pile

When driving piles in granular soils, the penetration resistance can often be considerably reduced by jetting a stream of high-pressured water into the soil just below the pile. There have been cases where piles have been installed by jetting alone. The method requires considerable experience, particularly when near to existing foundations.

Jacked pile

Generally built up with a series of short sections of precast concrete, this pile is jacked into the ground and progressively increased in length by the addition of a pile section whenever space becomes available. The jacking force is easily measured and the load to pile penetration relationship can be obtained as jacking proceeds. Jacked piles are often used to underpin existing structures where lack of space excludes the use of large pile installation plant and where minimal ground disturbance is essential.

Screw pile

A screw pile consists of a steel, or concrete, cylinder with helical blades attached to its lower end. The pile is made to screw down into the soil by rotating the cylinder with a capstan at the top of the pile. A screw pile, due to the large size of its screw blades, can offer large uplift resistance.

11.3.2   Bored and cast-in-place piles

These piles are formed within a drilled borehole. During the drilling process, the sides of the borehole are supported to prevent the soil from collapsing inwards and temporary sections of steel cylindrical casing are advanced along with the drilling process to provide this required support. As the drilling progresses, the soil is removed from within the casing and brought to the surface. Once the full depth of the borehole has been reached, the casing is gradually withdrawn, the reinforcement cage is placed, and the concrete which forms the pile is pumped into the borehole. For very deep boreholes, the installation of many sections of temporary

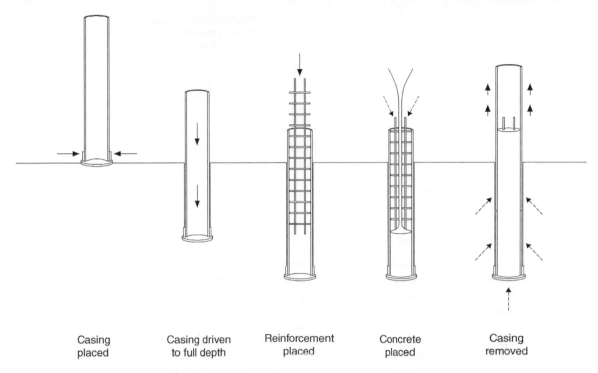

Casing	Casing driven	Reinforcement	Concrete	Casing
placed	to full depth	placed	placed	removed

Fig. 11.3   Driven and cast-in-place pile installation.

casing can be an expensive and slow process, and an alternative means of supporting the sides is through the use of a bentonite slurry in the same manner as for a diaphragm wall (see Section 9.3.2).

An alternative technique which does not use borehole side support is the *continuous flight auger* (CFA) pile. With this technique, a continuous flight auger with a hollow stem is used to create the borehole. The sides of the borehole are supported by the soil on the flights of the auger and so no casing is required. Once the required depth has been reached, the concrete is pumped down the hollow stem and the auger is steadily withdrawn. The steel reinforcement is placed once the auger is clear of the borehole.

11.3.3   Driven and cast-in-place piles

These piles are formed in a relatively similar manner to the bored piles, except that the casing is driven to full depth by a pile driver, rather than advanced in short sections as the hole is formed. The casing is closed at the bottom end by a detachable and sacrificial threaded driving shoe. Once the casing is driven to depth, the reinforcement cage is placed, and the concrete poured to form the pile. The casing is then unthreaded from the driving shoe and brought to the surface as indicated in Fig. 11.3.

A variant on this approach is the *Franki pile*. A steel tube is erected vertically over the place where the pile is to be driven, and about a metre depth of gravel is placed at the end of the tube. A drop hammer, 1500–4000 kg mass, compacts the aggregate into a solid plug which then penetrates the soil and takes the steel tube down with it. When the required set has been achieved, the tube is raised slightly, and the aggregate is broken out. Dry concrete is now added and hammered until a bulb is formed. Reinforcement is placed in position and more concrete is placed and vibrated until the pile top comes up to ground level. The sequence of operations is illustrated in Fig. 11.4.

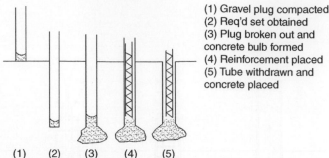

(1) Gravel plug compacted
(2) Req'd set obtained
(3) Plug broken out and concrete bulb formed
(4) Reinforcement placed
(5) Tube withdrawn and concrete placed

(1)   (2)   (3)   (4)   (5)

Fig. 11.4   Installation of a Franki pile.

11.3.4   Large diameter bored piles

The driven or bored and cast-in-place piles discussed above generally have maximum diameters in the order of 0.6 m and are capable of carrying loads of up to about 2 MN. In modern structures, column loads in the order of 20 MN are not uncommon. A column carrying such a load would need about 10 conventional piles, placed in a group and capped by a concrete slab, probably about 25 m² in area.

A consequence of this challenge has been the increasing use of the large diameter bored pile. This pile has a minimum shaft diameter of 0.75 m and may be *underreamed* to give a larger bearing area if necessary. Such a pile is capable of carrying loads in the order of 25 MN and, if taken down through the soft to the hard material, will minimise settlement problems, so that only one such pile is required to support each column of the building. Large diameter bored piles have been installed in depths down to 60 m.

11.4   Pile load testing

The only really reliable means of determining a pile's load capacity is through a pile load test. These tests are expensive, particularly if the ground is variable and a large number of piles must therefore be tested, but they do provide reliable data by which the design of further piles can be based. In the tests, full-scale piles are used, and these are installed in the same manner as those placed for the permanent work.

During pile testing, a load is applied to the top of the pile and the settlement of the pile is recorded against force or time, depending on the test. Tests can be categorised as either *static load* or *dynamic load* tests. In addition, soil test results can be used to aid the determination of the pile load capacity.

11.4.1   Static load tests

Maintained load test (MLT)

In this test, a load is applied to the pile in discrete increments, usually equal to 25% of the designated load that the pile will carry in service, and the resulting pile settlement is monitored. Subsequent load increments are only applied when the rate of induced settlement drops below a specified criterion.

The load is generally applied through a hydraulic jack which uses a beam assembly affixed to adjacent piles to provide the reaction force, see Fig. 11.5c. Alternatively, static weights can be applied to provide the load (Fig. 11.5a) or to provide the jack reaction force (Fig. 11.5b). The test normally lasts between 24 and 48 hours and is the most suitable in determining the load/settlement performance of a pile under serviceability conditions.

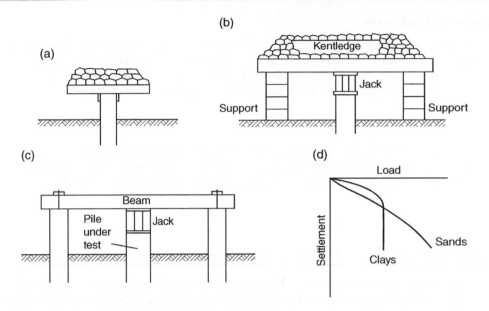

Fig. 11.5 Maintained load test. (a) Static weights. (b) Jack reaction force provision through kentledge. (c) Jack reaction force provision through adjacent piles and beam. (d) Typical load – settlement plots.

The form of load to settlement relationship obtained from a loading test is shown in Fig. 11.5d. Loading is continued until failure occurs, except for large diameter bored piles which, having a serviceability load of about 25 MN, would require massive applied loadings if failure loads were to be achieved. General practice has become to test load these piles to the load they will carry in service, plus 50%.

Guidance on the test procedure is given in the European standard for *Geotechnical investigation and testing: Static compression load testing* (BS EN ISO 22477-1:2018 (BSI, 2018)).

Constant rate of penetration test (CRP)

An alternative to the maintained load test is the constant rate of penetration test. Here, rather than applying a single load and waiting for the settlement to more or less stop, the load is steadily increased so as to induce a continuous rate of settlement of the pile. Thus, the pile is jacked downwards at a constant rate of penetration. The ultimate pile load is considered to be the load at which either a shear failure takes place within the soil or the settlement of the pile equals 10% of its diameter.

Design failure load

The design failure load for a pile is usually taken as the load that causes a settlement equal to 10% of the pile diameter (e.g. for a 250 mm diameter test pile, the design failure load would be the load that causes a settlement of 25 mm). Both the MLT and the CRP tests provide plots of settlement against load, and thus each test can be used to determine the design failure load.

11.4.2   Dynamic load tests

These tests are less popular than static load tests and monitor the response of a pile subjected to hammer blows applied at the pile head. The measured response parameters are analysed to give predictions of the soil resistance that would be mobilised by the pile under static load conditions, based on stress wave theory. The analysis can also provide prediction of the load/settlement performance of the pile.

11.4.3   Soil tests

A ground investigation will invariably be carried out for any foundation project. Results from tests on soil samples can be used to give an indication of the shear strength parameters of the soils acting along the shaft of the pile and at the pile base. These parameters can be used in the design process as will be seen later in this chapter.

11.5   Determination of the bearing resistance of a pile

A pile is supported in the soil by the resistance of the ground beneath the base of the pile plus the frictional or adhesive forces along its embedded length. The ultimate bearing resistance, Q_u is the sum of the base resistance, Q_b and the shaft resistance, Q_s:

$$Q_u = Q_b + Q_s \tag{11.1}$$

11.5.1   Cohesive soils

For piles installed in (non-highly overconsolidated) clays, the critical condition is the undrained state. At the base of the pile, the cohesive component of the general bearing capacity equation (Equation 10.16) applies

$$Q_b = c_b \times N_c \times A_b \tag{11.2}$$

where

c_b is the undisturbed, undrained shear strength at the base of the pile;
A_b is the cross-sectional area of the base of the pile;
N_c is usually taken as equal to 9.0.

The resistance along the shaft of the pile depends on the circumferential area and the amount of cohesion acting between pile and soil:

$$Q_s = c_a \times A_s \tag{11.3}$$

where

c_a is the undrained shear strength along the pile;
A_s is the surface area of the embedded length of the pile.

Ideally, c_a should be determined from a pile load test, but since this is not always possible, c_a is often correlated with the undrained cohesion c_u by an empirical adhesion factor α, thus:

$$Q_s = \alpha \times \overline{c_u} \times A_s \tag{11.4}$$

where

α is the adhesion factor, ranging between 0.3 and 1.0;
$\overline{c_u}$ is the average undisturbed undrained shear strength along the pile.

The ultimate bearing resistance for a pile in cohesive soil is thus:

$$Q_u = c_b N_c A_b + \alpha \overline{c_u} A_s \qquad (11.5)$$

The adhesion factor α

Most of the bearing resistance of a pile in cohesive soil is derived from its shaft resistance, and the problem of determining the ultimate capacity resolves into determining a value for α. The value of α not only depends on the type of clay, but also on the type of pile installation.

For driven piles in soft clays, α can be equal to or greater than 1.0 as, after driving, soft clays tend to increase in strength. However, the value of α tends to reduce rapidly with increasing soil strength and in overconsolidated clays α has been found to vary from 0.6 to 0.3. The usual value assumed for London clay is 0.6.

For bored piles, α may be taken as 0.45.

11.5.2   Granular soils

The ultimate load resistance of a pile installed in granular soil is estimated using only the value of the drained parameter, ϕ', and assuming that there is no contribution due to cohesion. At the base of the pile, the overburden component of the general bearing capacity equation (Equation 10.16) applies

$$Q_b = \gamma z \times N_q \times A_b \qquad (11.6)$$

where

γz is the effective overburden pressure at the base of the pile;
A_b is the cross-sectional area of the base of the pile;
N_q is the bearing capacity coefficient.

The selection of a suitable value for N_q is obviously a crucial part of the design of the pile. The values suggested by Berezantzev *et al.* (1961) are often used and are reproduced in Fig. 11.6. Note that the full value of N_q is used as it is assumed that the weight of soil removed or displaced is equal to the weight of the pile that replaced it.

The resistance along the shaft of the pile depends on the circumferential area and the amount of friction acting between the pile and the soil. This friction is often referred to as *skin friction*:

$$Q_s = f_s \times A_s \qquad (11.7)$$

where

f_s is the average skin friction along the pile;
A_s is the surface area of the embedded length of the pile.

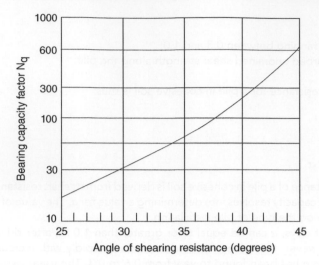

Fig. 11.6 Variation of bearing capacity factor N_q with angle of shearing resistance. Based on Berezantzev et al. (1961).

Clearly, establishing the magnitude of f_s is a critical aspect of the analysis and ground tests results can be used to determine the value. Empirical methods also exist, and Meyerhof (1959) proposed that f_s can be established:

$$f_s = K_s \overline{\gamma z} \tan \delta \tag{11.8}$$

where

K_s is the coefficient of earth pressure;
$\overline{\gamma z}$ is the average effective overburden pressure acting along the embedded length of the pile shaft;
δ is the angle of friction between the pile and soil.

Hence,

$$Q_s = A_s K_s \overline{\gamma z} \tan \delta \tag{11.9}$$

And

$$Q_u = \gamma z N_q A_b + A_s K_s \overline{\gamma z} \tan \delta \tag{11.10}$$

Typical values for δ and K_s were derived by Broms (1966) and are listed in Table 11.1.

Vesic (1973) pointed out that the value of q_b, i.e. $\gamma z N_q$ does not increase indefinitely but has a limiting value at a depth of around 20 times the pile diameter. There is therefore a maximum value of $\gamma z N_q$ that can be used in the calculations for Q_b.

In a similar manner, there is a limiting value that can be used for the average ultimate skin friction, f_s. This maximum value of f_s occurs when the pile has an embedded length between 10 and 20 pile diameters. Vesic (1970) suggested that the maximum value of the average ultimate skin resistance should be obtained from the formula:

$$f_s = 0.08(10)^{1.5(I_D)^4} \tag{11.11}$$

where I_D = the density index of the granular soil.

Table 11.1   Typical values for δ and K_s.

Pile material	δ	K_s Density index of soil Loose	Dense
Steel	20°	0.5	1.0
Concrete	0.75ϕ'	1.0	2.0
Timber	0.67ϕ'	1.5	4.0

Data from Broms (1966).

In practice, f_s is often taken as 100 kPa if the formula gives a greater value. Unlike piles embedded in cohesive soils, the end resistances of piles in granular soils are of considerable significance and short piles are therefore more efficient in granular soils than in cohesive soils.

Example 11.1: Undrained analysis

It is proposed to install a pile of diameter 400 mm and length 6 m into a deep deposit of normally consolidated clay. Laboratory results from a ground investigation have shown the clay to have an average undrained shear strength of 100 kPa over the depth 0–6 m and an undrained shear strength of 160 kPa at a depth of 6 m.

Assuming $N_c = 9.0$ and $\alpha = 0.6$, determine the ultimate bearing resistance of the pile.

Solution:

$$Q_b = c_b \times N_c \times A_b$$

$$= 160 \times 9.0 \times \left(\pi \times \frac{0.4^2}{4} \right)$$

$$= 181.0 \text{ kN}$$

$$Q_s = \alpha \times \overline{c_u} \times A_s$$

$$= 0.6 \times 100 \times \pi \times 0.4 \times 6$$

$$= 452.4 \text{ kN}$$

$$Q_u = Q_b + Q_s$$

$$= 181.0 + 452.4$$

$$= 633.4 \text{ kN}$$

Example 11.2: Drained analysis

A 500 mm diameter bored concrete pile is to be formed in the soil profile shown in Fig. 11.7.
   The ground conditions are as follows:

Granular fill: $\gamma = 20\,kN/m^3$
                $\phi' = 30°$
                $K_s = 1.0$
Dense gravel: $\gamma = 21\,kN/m^3$
                $\phi' = 35°$
                $K_s = 2.0$
Glacial clay: $\gamma = 20\,kN/m^3$
                c_u at 7.0 m = 120 kPa
                c_u at 8.0 m = 145 kPa
                c_u at 11.0 m = 220 kPa
                adhesion factor, $\alpha = 0.6$
                $N_c = 9.0$

Determine the ultimate bearing resistance of the pile for:

(a) embedded length = 8 m
(b) embedded length = 11 m

Solution:

Fill:

$$Q_s = A_s K_s \overline{\gamma z} \tan\delta = (\pi \times 0.5 \times 3) \times 1.0 \times \left(\frac{20 \times 3.0}{2}\right) \times \tan(0.75 \times 30°) = 58.6 \ kN$$

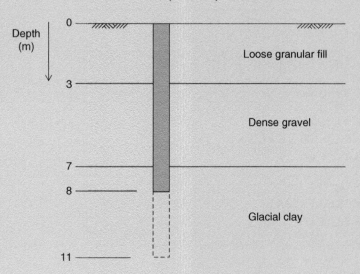

Fig. 11.7   Example 11.2.

Gravel:

$$Q_s = A_s K_s \overline{\gamma z} \tan \delta = (\pi \times 0.5 \times 4) \times 2.0 \times \left(60 + \frac{21 \times 4.0}{2}\right) \times \tan(0.75 \times 35°) = 632.1 \text{ kN}$$

Clay (length = 8 m):

$$Q_s = \alpha \times \overline{c_u} \times A_s = 0.6 \times \left(\frac{120 + 145}{2}\right) \times (\pi \times 0.5 \times 1.0) = 124.9 \text{ kN}$$

$$Q_b = c_b \times N_c \times A_b = 145 \times 9.0 \times \left(\pi \times \frac{0.5^2}{4}\right) = 256.2 \text{ kN}$$

Clay (length = 11 m):

$$Q_s = \alpha \times \overline{c_u} \times A_s = 0.6 \times \left(\frac{120 + 220}{2}\right) \times (\pi \times 0.5 \times 4.0) = 640.9 \text{ kN}$$

$$Q_b = c_b \times N_c \times A_b = 220 \times 9.0 \times \left(\pi \times \frac{0.5^2}{4}\right) = 388.8 \text{ kN}$$

Ultimate bearing resistance:

$$Q_u - Q_b + Q_s$$

(a) 8 m: $Q_u = 256.2 + (58.6 + 632.1 + 124.9) = 1071.8 \text{ kN}$
(b) 11 m: $Q_u = 388.8 + (58.6 + 632.1 + 640.9) = 1720.4 \text{ kN}$

11.5.3   Determination of soil piling parameters from *in situ* tests

With cohesionless soils, it is possible to make reasonable estimates of the values of q_b and f_s from *in situ* penetration tests.

(i) *Standard penetration test results*

Meyerhof (1976) suggests the following formulae to be used in conjunction with the standard penetration test:

Driven piles:

Sands and gravel $q_b \approx \dfrac{40ND}{B} \leq 400 \text{ N kPa}$

Non-plastic silts $q_b \approx \dfrac{40ND}{B} \leq 300 \text{ N kPa}$

Large diameter driven piles $f_s \approx 2\overline{N} \text{ kPa}$

Average diameter driven piles $f_s \approx \overline{N} \text{ kPa}$

Bored piles:

Any type of granular soil $q_b \approx \dfrac{14ND}{B}$ kPa

$f_s \approx 0.67\overline{N}$ kPa

where

N = the *uncorrected* blow count at the pile base;
$\overline{N}$ = the average *uncorrected* N value over the embedded length of the pile;
D = embedded length of the pile in the end-bearing stratum;
B = width, or diameter, of pile.

(ii) *Cone penetration test results*

An alternative method is to use the results of the cone penetration test. Typical results from such a test are shown in Fig. 11.8 and are given in the form of a plot showing the variation of the cone penetrations resistance with depth.

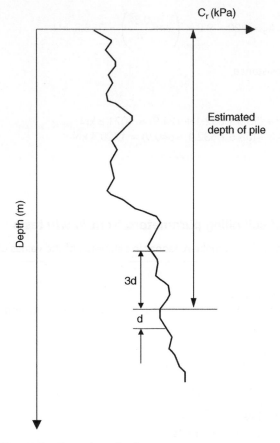

Fig. 11.8   Typical results from a cone penetration test.

For the ultimate base resistance, the cone resistance C_r is taken as the average value of C_r over the depth 4d as shown, where d = diameter of shaft. Then:

$$Q_b = C_r A_b$$

The ultimate skin friction, f_s, can be obtained from one of the following:

$$f_s = \frac{\overline{C_r}}{200} \text{ kPa for driven piles in dense sand}$$

$$f_s = \frac{\overline{C_r}}{400} \text{ kPa for driven piles in loose sand}$$

$$f_s = \frac{\overline{C_r}}{150} \text{ kPa for driven piles in non-plastic silts}$$

where $\overline{C_r}$ = average cone resistance along the embedded length of the pile (De Beer, 1963). Then $Q_s = f_s A_s$ and, as before, $Q_u = Q_b + Q_s$.

Example 11.3: Allowable load from *in situ* testing results

A 5 m thick layer of medium sand overlies a deep deposit of dense gravel. A series of standard penetration tests carried out through the depth of the sand has established that the average blow count, $\overline{N}$, is 22. Further tests show that the gravel has a standard penetration value of N = 40 in the region of the interface with the sand. A precast pile of square section 0.25×0.25 m^2 is to be driven down through the sand and to penetrate sufficiently into the gravel to give good end bearing.

Adopting a safety factor of 3.0 determine the allowable load that the pile will be able to carry.

Solution:

Ultimate bearing resistance of the pile, $Q_u = Q_b + Q_s$

Q_b: All end-bearing effects will occur in the gravel. Now

$$q_b \approx \frac{40ND}{B} \text{ or } 400 \times N \text{ (whichever is the lesser)}$$

i.e.

$$q_b \approx 40 \times 40 \times \frac{D}{0.25} = 400 \times 40 = 16\,000 \text{ kPa}$$

$$\text{Penetration into gravel, D} = \frac{16\,000 \times 0.25}{40 \times 40} = 2.5 \text{ mm}$$

and

$$Q_b = 16\,000 \times 0.25^2 = 1000 \ kN$$

Q_s in sand: $Q_s = f_s A_s = 22 \times 5 \times 0.25 \times 4 = 110 \ kN$

Q_s in gravel: $Q_s = f_s A_s = 40 \times 2.5 \times 0.25 \times 4 = 100 \ kN$

i.e.

$$Q_u = 210 + 1000 = 1210 \ kN$$

$$\text{Allowable load} = \frac{1210}{3} = 400 \ kN$$

Example 11.3 illustrates that, as discussed earlier, the end-bearing effects are much greater than those due to side friction. It can be argued that, in order to develop side friction (shaft resistance) fully, a significant downward movement of the pile is required which cannot occur in this example because of the end resistance of the gravel. As a result of this phenomenon, it is common practice to apply a different factor of safety to the shaft resistance than that applied to the end-bearing resistance. Historically, a factor of safety, F_s of around 1.5 was applied to shaft resistance, and a factor of safety, F_b between 2.5 and 3.0 applied to the end-bearing resistance. These days however, we design pile foundations in accordance with Eurocode 7, which do not use the factor of safety notion – see Sections 11.7 and 11.8.

Returning to Example 11.3 and adopting $F_b = 3$, $F_s = 1.5$, the allowable load now becomes

$$= \frac{1000}{3} + \frac{210}{1.5} = 473 \ kN$$

11.5.4 Negative skin friction, or downdrag

If a soil settles or consolidates around a pile, then the pile will tend to support the soil and there can be a considerable increase in the load on the pile. This effect is known as downdrag and is quantified as the additional shear stress applied to the surface of a pile by the soil as it settles.

The main causes for downdrag are

  (i) bearing piles driven into recently placed fill, which then begins to settle;
 (ii) fill placed around the piles after driving which causes consolidation in the soil below;
(iii) consolidation due to a reduction in the pore water pressure in the soil;
(iv) consolidation due to pile installation (of particular concern in sensitive and normally consolidated clays).

If negative friction effects are likely to occur, then the piles must be designed to carry the additional load. In extreme cases, the value of negative skin friction can equal the positive skin friction. However, the maximum value of negative skin friction cannot act over the entire embedded length of the pile, and it is found to be virtually zero at the top of the pile and reaches the maximum value at its base.

Various methods for determining the magnitude of the negative skin friction exist, but the maximum unit value can be determined in the same way as for the maximum shaft adhesion or friction as described by Equations (11.4) and (11.9). This maximum value should be considered as the *drag force* in pile design.

11.6 Pile groups

11.6.1 Action of pile groups

Piles are usually driven in groups and connected at the top to a *pile cap* onto which the structural load is placed (Fig. 11.9c). The zone of soil or rock which is stressed by the entire group extends to a much greater width and depth than the zone beneath a single pile.

When a pile is installed immediately adjacent to another, the respective bulbs of vertical pressure can overlap (Fig. 11.9b). In the case of a pile group (Fig. 11.9c), not only is the underlying soil stressed to a considerably greater depth than with a single pile, but soil outside the perimeter of the group is also stressed. Because of this superposition of vertical stresses at points within the soil, it is to be expected that the bearing resistance of a pile group can be less than the sum of the individual pile capacities.

Hence, the ultimate load-carrying capacity of a group of piles cannot always be considered as the sum of the individual carrying capacities of each pile. This is due to a phenomenon known as *group action*. Group action effects are more important with friction piles than with end-bearing piles.

Failure of the group may occur either by failure of an individual pile or by failure of the overall mass of soil supporting the group.

Efficiency of a pile group

The ratio of the overall bearing resistance of a pile group to the sum of the individual bearing resistances of all piles in the group is termed the pile group efficiency, E:

$$E = \frac{\text{ultimate bearing resistance of the group}}{n \times \text{ultimate bearing resistance of individual pile}}$$

where n is the number of piles in the group.

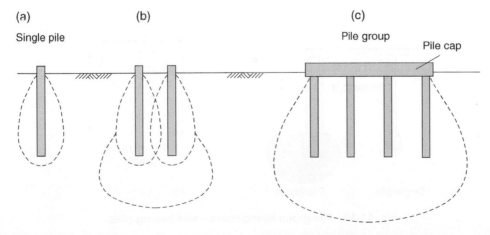

Fig. 11.9 Group action of piles. (a) Single pile bulb of pressure. (b) Adjacent piles overlap of pressure bulbs. (c) Pile group bulb of pressure.

The efficiency of the group becomes significant when establishing the ultimate bearing resistance of the group from a knowledge (measured or calculated) of the bearing resistance of a single pile, as illustrated in Example 11.8.

11.6.2   End-bearing piles

In the case of end-bearing piles, the pressure bulbs of the individual piles will overlap if the spacing between the piles is less than five times the diameter of a single pile. This is the usual condition. Provided that the bearing strata are firm throughout the affected depth of this combined bulb, then the bearing resistance of the group will be equal to the summation of the individual strengths of the piles as shown in Fig. 11.10. However, where a compressible layer exists within or immediately below the shaded zone, checks must be made to ensure that the layer will not be overstressed.

11.6.3   Friction/combination piles

Pile groups in granular soils

Pile installation in sands and gravels causes compaction of the soil between the piles. This densification of the soil leads to an increase in the strength of the soil such that the bearing resistance for the group exceeds the sum of the bearing resistances of the individual piles that comprise the group. However, as a conservative approach in design, it is usual to take the *group* bearing resistance to be equal to the sum of the *individual* bearing resistances. The spacing of the piles is usually around two to three times the diameter, or breadth, of the individual piles.

Pile groups in cohesive soils

By contrast, in clays, the load-carrying capacity of a group of vertically loaded piles is considerably less than the sum of the capacities of individual piles comprising the group, and this phenomenon must be considered in the design else excessive settlement might occur.

An important characteristic of pile groups in cohesive soils is the phenomenon of *block failure*. This is where the entire block of soil containing the piles fails along the perimeter of the group.

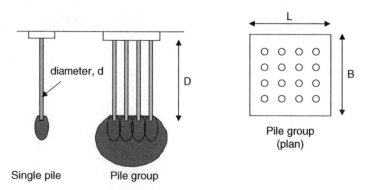

Fig. 11.10   Pile group arrangement – end-bearing piles.

For block failure

$$Q_u = 2D(B + L)\overline{c_u} + 1.3c_bN_cBL \tag{11.12}$$

where

D, B, and L are the dimensions indicated in Fig. 11.10;
$\overline{c_u}$ is the average undrained strength along the sides of the piles;
c_b is the undrained strength at the base of the piles;
N_c is the bearing capacity coefficient (usually taken as 9.0).

In general, block failure can be avoided if the piles are spaced at centres not less than the perimeter of a pile.

As mentioned, in clays, the capacity of an individual pile within a closely spaced group is lower than that for an equivalent 'isolated' pile. This effect is pretty insignificant and so may be ignored in design. Of more concern, however, is the fact that the block capacity of the group is less than the sum of the individual pile capacities. The spacing of the piles is thus influential. If the piles are placed close together (i.e. less than a distance of approximately 1.5d apart), the strength of the group may be governed by the resistance against block failure and thus block failure becomes a likely failure mode. To prevent block failure, the piles should be spaced at least 3d apart.

The base of the pile cap bears onto the soil and will also contribute to the overall bearing resistance of the foundation. The contribution to bearing resistance from the pile cap is simply found using Equations (11.2) or (11.6) and considering the area, A_b as the gross area of the pile cap, less the sum of the cross-sectional areas of all piles.

In such cases:

$$Q_u = E\,n\,Q_{up} + Q_{u,cap} \tag{11.13}$$

where

E = efficiency of pile group (approximately 0.7 for spacing 2d–3d; 1.0 for spacing $\geq$ 5d)
n = number of piles in group.
Q_{up} = ultimate bearing resistance of single pile
$Q_{u,cap}$ = portion of bearing resistance of pile cap

In design situations, both Equations (11.12) and (11.13) are determined and the minimum value taken as the ultimate bearing resistance of the group.

11.6.4   Settlement effects in pile groups

Quite often it is the allowable settlement, rather than the bearing resistance, that decides the design load that a pile group may carry.

For bearing piles, the total foundation load is assumed to act at the base of the piles on a foundation of the same size as the plan of the pile group. With this assumption, it becomes a simple matter to examine settlement effects.

With friction piles, it is virtually impossible to determine the level at which the foundation load is effectively transferred to the soil. An approximate method, often used in design, is to assume that the effective transfer level is at a depth of 2D/3 below the top of the piles. It is also assumed that there is a spread of the total load, one horizontal to four vertical. The settlement of this equivalent foundation (Fig. 11.11) can then be determined by the normal methods.

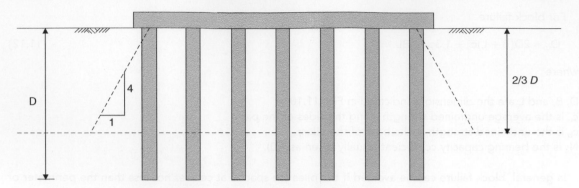

Fig. 11.11 Transfer of load in friction piles.

11.7 Designing pile foundations to Eurocode 7 (first generation)

The principles of Eurocode 7, as described in Chapter 6, apply to the design of pile foundations and the design of pile foundations is covered in Section 7 of Eurocode 7 Part 1, EN 1997-1:2004 (BSI, 2004).

There are several limit states listed that should be considered, though only those limit states most relevant to the particular situation would normally be considered in the design. These include the loss of overall stability, bearing resistance failure of the pile, uplift of the pile, and structural failure of the pile. In this section, we will look only at checking against ground resistance failure through the compressive loading of the pile.

Pile design methods acceptable to Eurocode 7 are in the main based on the results of static pile load tests, and the design calculations should be validated against the test results. When considering the compressive ground resistance limit state, the task is to demonstrate that the design axial compression load on a pile or pile group, $F_{c;d}$, is less than or equal to the design compressive ground resistance, $R_{c;d}$, against the pile or pile group. In the case of pile groups, $R_{c;d}$ is taken as the lesser value of the design ground resistance of an individual pile and that of the whole group.

In keeping with the rules of Eurocode 7, the design value of the compressive resistance of the ground is obtained by dividing the characteristic value by a partial factor of safety. The characteristic value is obtained by one of three approaches: from static load tests, from ground tests results, or from dynamic tests results.

Considering Design Approach 1, the following partial factor sets (see Section 6.7.1) are used for the design of axially loaded piles:

Combination 1: A1 + M1 + R1
Combination 2: A2 + (M1 or M2)* + R4

*M1 is used for calculating pile resistance; M2 is used for calculating unfavourable actions on piles.

11.7.1 Note on the UK National Annex, NA EN 1997-1:2004

As mentioned in Chapter 6, the National Annex allows each nation to provide Nationally Determined Parameters (NDPs) that should be used in place of the values published in the main Eurocode document. For Eurocode 7 Part 1, NDPs include partial factors and rules for use where national choice is permitted in the main Eurocode. The partial factors for the GEO limit state in the UK National Annex are the same as those listed in Eurocode 7 Part 1, with the exception of the partial resistance factors (γ_R) and the correlation factors (ξ) for pile foundations. It is important to note therefore that during pile foundation design in the United Kingdom, the values provided in the National Annex must be used, rather than the values published in the main Eurocode document. In the following sections, both sets of values are provided.

11.7.2 Ultimate compressive resistance from static load tests

The characteristic value of the compressive ground resistance, $R_{c;k}$, is obtained by combining the measured value from the pile load tests with a correlation factor, ξ (related to the number of piles tested). More explicitly, $R_{c;k}$ is taken as the lesser value of:

$$R_{c;k} = \frac{(R_{c;m})_{mean}}{\xi_1} \quad \text{and} \quad R_{c;k} = \frac{(R_{c;m})_{min}}{\xi_2} \qquad (11.14)$$

where

$(R_{c;m})_{mean}$ is the mean measured resistance;
$(R_{c;m})_{min}$ is the minimum measured resistance;
ξ_1, ξ_2 are correlation factors obtained from Table 11.2A (Eurocode 7, Part 1) or Table 11.2B (UK National Annex).

It may be that the characteristic compressive resistance of the ground is more appropriately determined from the characteristic values of the base resistance, $R_{b;k}$ and the shaft resistance, $R_{s;k}$:

$$R_{c;k} = R_{b;k} + R_{s;k} \qquad (11.15)$$

The design compressive resistance of the ground may be derived by either:

$$R_{c;d} = \frac{R_{c;k}}{\gamma_t} \qquad (11.16)$$

Table 11.2A   Correlation factors – static load tests results (from EN 1997-1:2004 Table A9).

	Number of piles tested				
	1	2	3	4	≥5
ξ_1	1.4	1.3	1.2	1.1	1.0
ξ_2	1.4	1.2	1.05	1.0	1.0

Table 11.2B   Correlation factors – static load tests results (from NA to BS EN 1997-1:2004 Table A.NA.9).

	Number of piles tested				
	1	2	3	4	≥5
ξ_1	1.55	1.47	1.42	1.38	1.35
ξ_2	1.55	1.35	1.23	1.15	1.08

Table 11.3A Piles in compression: partial factor sets R1, R2, R3, and R4 (from EN 1997-1:2004 Tables A6–A8).

Partial factor set	R1			R2	R3	R4		
	Driven	Bored	CFA	All	All	Driven	Bored	CFA
Base, γ_b	1.0	1.25	1.1	1.1	1.0	1.3	1.6	1.45
Shaft, γ_s	1.0	1.0	1.0	1.1	1.0	1.3	1.3	1.3
Total, γ_t	1.0	1.15	1.1	1.1	1.0	1.3	1.5	1.4

Table 11.3B Piles in compression: partial factor sets R1, R2, R3, and R4 (from NA to BS EN 1997-1:2004 Tables A. NA.6–A.NA.8).

Partial factor set	R1			R2	R3	R4*		
	Driven	Bored	CFA	All	All	Driven	Bored	CFA
Base, γ_b	1.0	1.0	1.0	1.1	1.0	1.7/1.5	2.0/1.7	2.0/1.7
Shaft, γ_s	1.0	1.0	1.0	1.1	1.0	1.5/1.3	1.6/1.4	1.6/1.4
Total, γ_t	1.0	1.0	1.0	1.1	1.0	1.7/1.5	2.0/1.7	2.0/1.7

*R4: The pairs of values listed in Table 11.3B indicate that there is a choice of value to be used depending on whether or not explicit verification of the serviceability limit state (i.e. settlement) has been carried out. The higher value is to be used when no explicit verification is undertaken, and the lower value is used when there is explicit verification. Explicit verification includes situations where serviceability is verified by load tests on more than 1% of the constructed piles or situations where settlement is of no concern.

or

$$R_{c;d} = \frac{R_{b;k}}{\gamma_b} + \frac{R_{s;k}}{\gamma_s} \tag{11.17}$$

where γ_b, γ_s, and γ_t are partial factors on base resistance, shaft resistance, and the total resistance, respectively. The partial factors for piles in compression recommended in Eurocode 7 are given in Table 11.3A (Eurocode 7, Part 1) or Table 11.3B (UK National Annex).

Example 11.4: Static load tests (to UK National Annex)

A series of static load tests on a set of four bored piles gave the following results:

	Test number			
	1	2	3	4
Measured load (kN)	382	425	365	412

From an understanding of the ground conditions, it is assumed that the ratio of base resistance to shaft resistance is 3 : 1. Determine the design compressive resistance of the ground in accordance with Eurocode 7, Design Approach 1, following the UK National Annex and assume explicit SLS verification has taken place.

Solution:

$$(R_{c;m})_{mean} = \frac{382 + 425 + 365 + 412}{4} = 396 \text{ kN}$$

$$(R_{c;m})_{min} = 365 \text{ kN}$$

From Table 11.2B, $\xi_1 = 1.38$; $\xi_2 = 1.15$

$$R_{c;k} = \frac{(R_{c;m})_{mean}}{\xi_1} = \frac{396}{1.38} = 287 \text{ kN}$$

$$R_{c;k} = \frac{(R_{c;m})_{min}}{\xi_2} = \frac{365}{1.15} = 317 \text{ kN}$$

That is

$$R_{c;k} = 287 \text{ kN (i.e. the minimum value)}$$

Since ratio of base resistance to shaft resistance is 3:1, we have

Characteristic base resistance, $R_{b;k} = 287 \times 0.75 = 215 \text{ kN}$
Characteristic shaft resistance, $R_{s;k} = 287 \times 0.25 = 72 \text{ kN}$

Design Approach 1, Combination 1:
   Partial factor set R1 is used:

$$R_{c;d} = \frac{R_{c;k}}{\gamma_t} = \frac{287}{1.0} = 287 \text{ kN}$$

or

$$R_{c;d} = \frac{R_{b;k}}{\gamma_b} + \frac{R_{s;k}}{\gamma_s} = \frac{215}{1.0} + \frac{72}{1.0} = 287 \text{ kN}$$

Design Approach 1, Combination 2:
   Partial factor set R4 is used:

$$R_{c;d} = \frac{R_{c;k}}{\gamma_t} = \frac{287}{1.7} = 169 \text{ kN}$$

or

$$R_{c;d} = \frac{R_{b;k}}{\gamma_b} + \frac{R_{s;k}}{\gamma_s} = \frac{215}{1.7} + \frac{72}{1.4} = 178 \text{ kN}$$

The design compressive resistance of the ground is thus determined:

$$R_{c;d} = \min \{287, \ 287, \ 169, \ 178\} = 169 \text{ kN}$$

Example 11.5: Maintained load tests (to EN 1997-1:2004)

A structure is to be supported by a series of nine bored piles, arranged symmetrically beneath a pile cap. Each pile has diameter of 600 mm. The pile cap will carry a vertical representative permanent load, P_G of 8 MN (includes self-weight) and a vertical representative transient load, P_Q of 2800 kN applied with no eccentricity.

Results from two maintained load tests carried out on test piles are shown in Fig. 11.12.

Check compliance of the bearing resistance limit state of a single pile by determining the over-design factor for:

(a) Design Approach 1
(b) Design Approach 2

Use the partial and correlation factors from EN 1997-1:2004.

Solution:

From Table 11.2A:

$$\xi_1 = 1.30; \; \xi_2 = 1.20$$

The measured resistance $R_{c,m}$ is taken as the pile load recorded at a settlement equal to 10% of the pile diameter (=60 mm).

$R_{c,m} = 2750$ kN (Test 1), $R_{c;m} = 2900$ kN (Test 2)
$R_{min} = 2750$ kN
$R_{mean} = (2750 + 2900)/2 = 2825$ kN

$$R_{c;k} = \min\left\{\frac{(R_{c;m})_{mean}}{\xi_1}; \frac{(R_{c;m})_{min}}{\xi_2}\right\} = \min\left\{\frac{2825}{1.3}; \frac{2750}{1.2}\right\} = 2173 \text{ kN}$$

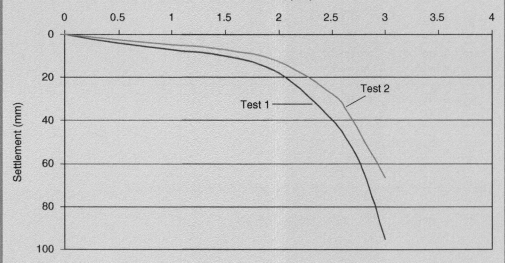

Fig. 11.12 Example 11.5: maintained load test results.

From Table 6.1 and Table 11.3A:

DA1-1: $\gamma_{G,unfav} = 1.35$; $\gamma_Q = 1.50$; $\gamma_t = 1.15$
DA1-2: $\gamma_{G,unfav} = 1.00$; $\gamma_Q = 1.30$; $\gamma_t = 1.50$
DA2: $\gamma_{G,unfav} = 1.35$; $\gamma_Q = 1.50$; $\gamma_t = 1.10$

Consider a single pile:

$$P_G = 8000/9 = 889 \text{ kN}$$

$$P_Q = 2800/9 = 311 \text{ kN}$$

(a) *Design Approach 1:*
Combination 1:

$$V_{G,d} = 889 \times 1.35 = 1200 \text{ kN}$$

$$V_{Q,d} = 311 \times 1.5 = 467 \text{ kN}$$

$$F_d = 1200 + 467 = 1667 \text{ kN}$$

$$R_{c;d} = \frac{R_{c;k}}{\gamma_t} = \frac{2173}{1.15} = 1890 \text{ kN}$$

$$\Gamma = \frac{R_{c;d}}{F_d} = \frac{1890}{1667} = 1.13$$

Combination 2:

$$V_{G,d} = 889 \times 1.0 = 889 \text{ kN}$$

$$V_{Q,d} = 311 \times 1.3 = 404 \text{ kN}$$

$$F_d = 889 + 404 = 1293 \text{ kN}$$

$$R_{c;d} = \frac{R_{c;k}}{\gamma_t} = \frac{2173}{1.5} = 1449 \text{ kN}$$

$$\Gamma = \frac{R_{c;d}}{F_d} = \frac{1449}{1293} = 1.12$$

(b) *Design Approach 2:*

$$F_d = 1667 \text{ kN}$$

$$R_{c;d} = \frac{R_{c;k}}{\gamma_t} = \frac{2173}{1.1} = 1975 \text{ kN}$$

$$\Gamma = \frac{R_{c;d}}{F_d} = \frac{1975}{1667} = 1.18$$

11.7.3   Ultimate compressive resistance from ground tests results

The design compressive resistance can be determined from ground tests results. Here the characteristic compressive resistance, $R_{c;k}$, is taken as the lesser value of:

$$R_{c;k} = \frac{(R_{b;cal} + R_{s;cal})_{mean}}{\xi_3} \quad \text{and} \quad R_{c;k} = \frac{(R_{b;cal} + R_{s;cal})_{min}}{\xi_4} \tag{11.18}$$

where

$(R_{b;cal})_{mean}$ = the mean calculated base resistance
$(R_{s;cal})_{mean}$ = the mean calculated shaft resistance
$(R_{b;cal})_{min}$ = the minimum calculated base resistance
$(R_{s;cal})_{min}$ = the minimum calculated shaft resistance
ξ_3, ξ_4 = correlation factors obtained from Table 11.4A (Eurocode 7, Part 1) or Table 11.4B (UK National Annex).

The calculated base and shaft resistances are determined using the equations set out in Section 11.5.

Table 11.4A   Correlation factors – ground tests results (from EN 1997-1:2004 Table A10).

	Number of test profiles						
	1	**2**	**3**	**4**	**5**	**7**	**10**
ξ_3	1.4	1.35	1.33	1.31	1.29	1.27	1.25
ξ_4	1.4	1.27	1.23	1.20	1.15	1.12	1.08

Table 11.4B   Correlation factors – ground tests results (from NA to BS EN 1997-1:2004 Table A.NA.10).

	Number of test profiles						
	1	**2**	**3**	**4**	**5**	**7**	**10**
ξ_3	1.55	1.47	1.42	1.38	1.36	1.33	1.30
ξ_4	1.55	1.39	1.33	1.29	1.26	1.20	1.15

Example 11.6: Design from ground tests results

A 10 m long × 0.7 m diameter CFA pile is to be founded in a uniform soft clay. The following test results were established in a geotechnical laboratory as part of a site investigation:

Borehole no.	1	2	3	4
Mean undrained strength along shaft, $c_{u;shaft}$ (kPa)	65	62	70	73
Mean undrained strength at base, $c_{u;base}$ (kPa)	90	79	96	100

The pile will carry a permanent axial load of 350 kN (includes the self-weight of the pile) and an applied transient (variable) axial load of 120 kN.

Check the bearing resistance (GEO) limit state in accordance with Eurocode 7, Design Approach 1 (UK) by establishing the magnitude of the over-design factor. Assume $N_c = 9$ and $\alpha = 0.7$ and assume that explicit SLS verification has been undertaken.

Solution:

Area of base of pile, $A_b = \dfrac{\pi D^2}{4} = \dfrac{\pi \times 0.7^2}{4} = 0.385$ m^2

The total resistance is determined from the results from each borehole:

$(R_{c;cal})_1 = (N_c \times c_u \times A_b) + (\pi \times D \times L \times \alpha \times c_u)$
$= (9 \times 90 \times 0.385) + (\pi \times 0.7 \times 10 \times 0.7 \times 65) = 1312$ kN

$(R_{c;cal})_2 = (9 \times 79 \times 0.385) + (\pi \times 0.7 \times 10 \times 0.7 \times 62) = 1228$ kN

$(R_{c;cal})_3 = (9 \times 96 \times 0.385) + (\pi \times 0.7 \times 10 \times 0.7 \times 70) = 1410$ kN

$(R_{c;cal})_4 = (9 \times 100 \times 0.385) + (\pi \times 0.7 \times 10 \times 0.7 \times 73) = 1470$ kN

$(R_{c;cal})_{mean} = \dfrac{1312 + 1228 + 1410 + 1470}{4} = 1355$ kN

$(R_{c;cal})_{min} = 1228$ kN (i.e. Borehole 2)

From Table 11.4B, $\xi_3 = 1.38$; $\xi_4 = 1.29$

$R_{c;k} = \dfrac{(R_{c;cal})_{mean}}{\xi_3} = \dfrac{1355}{1.38} = 982$ kN

$R_{c;k} = \dfrac{(R_{c;cal})_{min}}{\xi_4} = \dfrac{1228}{1.29} = 952$ kN that is, $(R_{c;cal})_{min}$ governs and this lower value of $R_{c;k}$ is taken as the characteristic compressive resistance.

Therefore, using ξ_4;

$$\text{Characteristic base resistance, } R_{b;k} = \frac{9 \times 79 \times 0.385}{1.29} = 212 \text{ kN}$$

$$\text{Characteristic shaft resistance, } R_{s;k} = \frac{\pi \times 0.7 \times 10 \times 0.7 \times 62}{1.29} = 740 \text{ kN}$$

1. *Design Approach 1, Combination 1:*
   Design resistance: partial factor set R1 is used (Table 11.3B):

$$R_{c;d} = \frac{R_{b;k}}{\gamma_b} + \frac{R_{s;k}}{\gamma_s} = \frac{212}{1.0} + \frac{740}{1.0} = 952 \text{ kN}$$

or

$$R_{c;d} = \frac{R_{c;k}}{\gamma_t} = \frac{952}{1.0} = 952 \text{ kN}$$

Design actions: partial factor set A1 is used (Table 6.1):

$$F_{c;d} = 350 \times 1.35 + 120 \times 1.5 = 652.5 \text{ kN}$$

$$\text{Over-design factor, } \Gamma = \frac{952}{652.5} = 1.46$$

2. *Design Approach 1, Combination 2:*
   Design resistance: partial factor set R4 is used (Table 11.3B):

$$R_{c;d} = \frac{R_{b;k}}{\gamma_b} + \frac{R_{s;k}}{\gamma_s} = \frac{212}{1.7} + \frac{740}{1.4} = 653 \text{ kN}$$

or

$$R_{c;d} = \frac{R_{c;k}}{\gamma_t} = \frac{952}{1.7} = 560 \text{ kN}$$

Design actions: partial factor set A2 is used (Table 6.1):

$$F_{c;d} = 350 \times 1.0 + 120 \times 1.3 = 506 \text{ kN}$$

$$\text{Over-design factor, } \Gamma = \frac{560}{506} = 1.11$$

Since $\Gamma \geq 1$, the design of the pile satisfies the GEO limit state requirement.

11.7.4 Ultimate compressive resistance from dynamic tests results

Although the use of static load tests and ground tests are the most common methods of determining the compressive resistance of the pile, the resistance can also be estimated from dynamic tests provided that the test procedure has been calibrated against static load tests.

11.8   Designing pile foundations to Eurocode 7 (second generation)

11.8.1   Geotechnical reliability

The design of piled foundations is covered in Section 6 of Eurocode 7 Part 3, EN 1997-3:202x and involves verification of the same limit states as listed in the first generation of the code. Ultimate limit states to be checked include bearing resistance failure of the ground surrounding the pile, failure of the ground between piles, uplift of the pile, and structural failure of the pile. Serviceability limit states include excessive pile settlement, excessive differential settlement, excessive downdrag, and excessive heave.

Pile foundations fall into *Geotechnical Categories 2* and *3* (refer to Table 6.5) and a pile foundation design will normally fall into *Geotechnical Complexity Class* 2 or 3. Only if there is negligible uncertainty in the ground conditions and if no ground movements are possible, would the design be considered GCC 1. Pile foundations in GCC 3 include situations where difficult ground conditions exist, where friction piles are proposed in low strength ground, where there is ground movement or site instability, or situations where the foundation will be subject to cyclic, dynamic, or repeated loading. Pile foundations, by default, will be in GCC 2 if the design is not in either GCC 1 or GCC 3.

In EN 1997-3:202x, piles are classified as either *displacement piles* (of high displacement or of low displacement) or as *replacement piles* (bored piles or CFA piles). These types of piles have been described in Section 11.3. The partial factors applied in establishing the design pile resistance, depend on this classification (see Section 11.8.2 and Table 11.8).

The ground model, developed from site investigation data including field and laboratory tests results, must contain adequate information on the ground properties to enable the verification checks to be made. The site investigation should therefore include field testing to either enable direct correlation with the pile shaft and base resistance to be made, or to enable the determination of the strength and stiffness properties of the different soils in the profile. Sampling from boreholes should also be carried out to allow laboratory strength and stiffness testing to be performed.

The ground investigation must determine ground conditions over the anticipated full depth of the piles and may also involve pile installation trials and/or installation of test piles for load testing.

During the design, groundwater conditions are considered as those that will be prevalent during the service of the pile, and not necessarily those encountered during the site investigation. Similarly, ground properties used in the design must represent those prevalent after pile installation and should consider any reduction in strength as a result of the installation process.

11.8.2   Design, by calculation, of single axially loaded piles

Note: The values of the various factors listed in this section are from the draft version of EN 1997-3:202x. The National Annex, once published, may instruct different values to be used. Further, the values in the final published version of EN 1997-3 may also differ from those used in this section.

The axial resistance of a single pile can be determined using the results of field and laboratory tests, or from pile load tests results. Ground properties are derived from the field and laboratory tests results, together with additional relevant information and experience. Using these derived ground properties, the design process follows one of two *calculation methods*:

Method A, the Ground Model Method
Method B, the Model Pile Method

In *Calculation Method A*, all the results of the field and laboratory tests are used to derive the ground properties which are then used in the design;

In *Calculation Method B*, the design uses either data obtained by direct correlations with sets of specific field or laboratory ground tests results, or the results of pile load testing.

In the verification of the compressive ground resistance limit state, the task is to demonstrate that the design axial compression load on a pile or pile group, including any downdrag, F_{cd}, is less than or equal to the design axial compressive resistance, R_{cd}.

In the case of pile groups, R_{cd} is taken as the lesser value of the design ground resistance of an individual pile and that of the whole group. Failure of the whole group is referred to as *block failure* – see Section 11.6.

The axial compressive resistance of a single pile, R_c follows from Equation (11.1):

$$R_c = R_b + R_s \tag{11.19}$$

where

R_b = base resistance;
R_s = shaft resistance.

The pile base resistance is obtained from either Equation (11.2) or (11.6).
The pile shaft resistance is obtained from either Equation (11.4) or (11.9).

Calculation Method A

The *representative* axial compressive resistance of the pile, $R_{c,rep}$ is equal to R_c in Equation (11.19). The *design* axial compressive resistance, R_{cd} is then determined:

$$R_{cd} = \frac{R_{c,rep}}{\gamma_{Rc} \times \gamma_{Rd}} \tag{11.20}$$

or

$$R_{cd} = \frac{R_{b,rep}}{\gamma_{Rb} \times \gamma_{Rd}} + \frac{R_{s,rep}}{\gamma_{Rs} \times \gamma_{Rd}} \tag{11.21}$$

where

$R_{c,rep}$ is the representative total resistance of the pile in axial compression;
$R_{b,rep}$ is the representative base resistance;
$R_{s,rep}$ is the representative shaft resistance;
γ_{Rd} is a model factor – see Tables 11.5A and 11.5B;
and γ_{Rc}, γ_{Rb}, and γ_{Rs} are resistance factors – see Table 11.8.

Table 11.5A   Model factor values for the verification of R_{cd} by calculation (Calculation Method A).

Additional information	Model factor, γ_{Rd}
Comprehensive	1.3
Limited	1.55
Minimal	1.8

Table 11.5B   Model factor values for the verification of R_{cd} by calculation (Calculation Method B).

Verification by	Model factor, γ_{Rd}
Pressuremeter test	1.15
CPT	1.2
CPT with comprehensive comparable experience	1.0

Pile design by calculation should be validated using pile load testing results. These results are used to verify that the calculated compressive resistance, R_c, agrees with the measured value in the field. Within this validation process, classification is made of the extent (*comprehensive, limited,* or *minimal*) of additional information gleaned from the pile testing which, in turn, influences the magnitude of the model factor, γ_{Rd}.

The magnitude of γ_{Rd} also depends on the calculation method used, and values are listed in Tables 11.5A (Method A) and 11.5B (Method B).

Calculation Method B

With this method, direct correlation of a profile of field test results is used to establish the ground properties and the model factor, γ_{Rd}, to be used in the verification. A profile is a single data set capturing the ground conditions, from the ground surface to a depth deeper than the anticipated length of pile, at a single location on the site. Values of γ_{Rd} for this calculation method are shown in Table 11.5B.

If appropriate, a number of ground profiles may be used to establish a single data set, from which values of the mean and standard deviation of the calculated bearing resistances for all profiles may be established. Such an approach is only possible if the individual profiles are from the same general area of the site and have the same ground conditions. The *coefficient of variation, CoV,* of this data set is the value of the *standard deviation, s,* divided by the *mean value, R_{av}*:

$$CoV = \frac{s}{R_{av}} \tag{11.22}$$

The representative resistance, R_{rep}, is taken as the minimum of the two values in Equation (11.23):

$$R_{rep} = min\left\{\frac{(R_m)_{mean}}{\xi_{m,mean}} ; \frac{(R_m)_{min}}{\xi_{m, min}}\right\} \tag{11.23}$$

where:

$(R_m)_{mean}$ is the mean calculated or measured pile resistance for a set of profiles;
$(R_m)_{min}$ is the minimum calculated or measured pile resistance for a set of profiles;
$\xi_{m,mean}$ is a correlation factor for the mean of the measured values;
$\xi_{m,min}$ is a correlation factor for the minimum of the measured values.
The values of $\xi_{m,mean}$ and $\xi_{m,min}$ for various numbers of profiles and ranges of coefficient of variation are shown in Tables 11.6 and 11.7.

When following Calculation Method B, an alternative approach to determining R_c by calculation is to use the load testing results directly and then validate the design through calculation. When R_c is determined from static load testing, the correlation factors listed in Table 11.7 are used to determine R_{rep} and the magnitude of the model factor γ_{Rd} is taken as 1.0.

The resistance partial factors and the design cases to be used for both Method A and Method B are listed in Table 11.8.

Table 11.6 Correlation factors for pile design by calculation (Method B).

Correlation factor	Coefficient of variation	Number of profiles						
		1	2	3	4	5	7	10
$\xi_{m,mean}$	≤12%	Use $\xi_{m,min}$ alone		1.30	1.28	1.28	1.27	1.26
	15%			1.40	1.39	1.38	1.37	1.36
	20%			1.67	1.64	1.63	1.61	1.60
	≥25%			1.98	1.95	1.93	1.90	1.89
$\xi_{m,min}$	All	1.40	1.27	1.23	1.20	1.15	1.12	1.08

Table 11.7 Correlation factors for pile design based on results of static load tests.

Correlation factor	Number of static load tests, per 2500 m² area				
	1	2	3	4	5
$\xi_{m,mean}$	1.40	1.30	1.20	1.10	1.00
$\xi_{m,min}$	1.40	1.20	1.05	1.00	1.00

Table 11.8 Partial factors for the verification of ultimate compressive axial resistance of single piles.

Partial factor	Symbol	Pile class	Method A		Method B	
Actions and effects of actions	γ_F and γ_E	All	DC3		DC4	
Downdrag	$\gamma_{F,drag}$		1.00		1.15	
Total resistance	γ_{Rc}	High displacement	1.30		1.10	
		Low displacement	1.35			
		CFA	1.40			
		Bored	1.50			
		Unclassified	1.75		1.30	
			γ_{Rb}	γ_{Rs}	γ_{Rb}	γ_{Rs}
Base and shaft resistance	γ_{Rb} and γ_{Rs}	High displacement	1.30	1.30	1.20	1.00
		Low displacement	1.35	1.30	1.20	1.00
		CFA	1.45	1.30	1.10	1.10
		Bored	1.60	1.30	1.10	1.10
		Unclassified	1.90	1.50	1.35	1.25

11.8.3 Verification of ultimate axial resistance of pile groups

The verification of the ultimate axial resistance of a pile group may be carried out using either the material factor approach or the resistance factor approach. The partial factor combinations are listed in Table 11.9. When using the MFA, both combinations (a) and (b) should be checked.

Table 11.9 Partial factors for the verification of ultimate compressive axial resistance of pile groups.

Partial factor	Symbol	Resistance factor approach (RFA)	Material factor approach (MFA)	
			(a)	(b)
Actions and effects of actions	γ_F and γ_E	DC4	DC4	DC3
Ground properties	γ_M	Not factored	M1	M2
Resistance	γ_R	1.4	Not factored	

Example 11.7: Static load tests (to Eurocode 7 second generation)

A series of static load tests on a set of four large displacement piles on a site gave the following results:

	Test number			
	1	2	3	4
Measured load (kN)	688	657	765	742

From an understanding of the ground conditions, it is approximated that the base resistance is equal to three times the shaft resistance. Determine the single pile design compressive resistance of the ground.

Solution:

$$(R_m)_{mean} = \frac{688 + 657 + 765 + 742}{4} = 713 \text{ kN}$$

$$(R_m)_{min} = 657 \text{ kN}$$

From Table 11.7, $\xi_{m,mean} = 1.10$; $\xi_{m,min} = 1.00$

$$R_{rep} = \frac{(R_m)_{mean}}{\xi_{m,mean}} = \frac{713}{1.10} = 648 \text{ kN}$$

$$R_{rep} = \frac{(R_m)_{min}}{\xi_{m,min}} = \frac{657}{1.00} = 657 \text{ kN}$$

Therefore, $R_{rep} = 648$ kN (i.e. the minimum value)
Since ratio of base resistance to shaft resistance is 3:1, we have

Representative base resistance, $R_{b,rep} = 648 \times 0.75 = 486$ kN
Representative shaft resistance, $R_{s,rep} = 648 \times 0.25 = 162$ kN

Since the ground information is obtained from pile testing, Calculation Method B is used, and partial factors are read from Table 11.8. Also, since static load testing is used, $\gamma_{Rd} = 1.0$.

$$R_{cd} = \frac{R_{c,rep}}{\gamma_{Rc} \times \gamma_{Rd}} = \frac{648}{1.1 \times 1.0} = 589 \text{ kN}$$

or

$$R_{cd} = \frac{R_{b,rep}}{\gamma_{Rb} \times \gamma_{Rd}} + \frac{R_{s,rep}}{\gamma_{Rs} \times \gamma_{Rd}} = \frac{486}{1.2 \times 1.0} + \frac{162}{1.0 \times 1.0} = 567 \text{ kN}$$

The design compressive resistance of the ground is thus determined:

$$R_{cd} = \min\{589, \ 567\} = 567 \text{ kN}$$

Example 11.8: Single, and group, pile verification (to Eurocode 7 second generation)

A structure is to be supported by a series of nine bored piles, arranged symmetrically beneath a square pile cap with side length 5.0 m. Each pile has diameter of 500 mm. The pile cap will carry a vertical representative permanent load, P_G of 1200 kN (includes self-weight), and a vertical representative transient load, P_Q of 300 kN, applied with no eccentricity.

Results from two maintained load tests carried out on test piles are shown in Fig. 11.13.

The proposed foundation arrangement is shown in Fig. 11.14 and is to be formed in a deep deposit of saturated clay, overlain by a thin layer of fill. The base of the pile cap is to

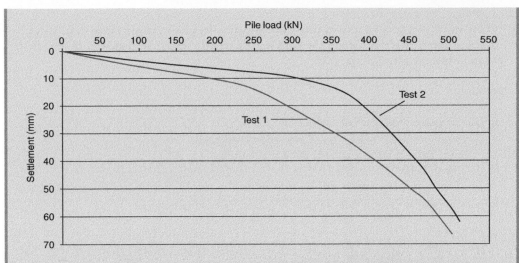

Fig. 11.13 Example 11.8: maintained load tests results.

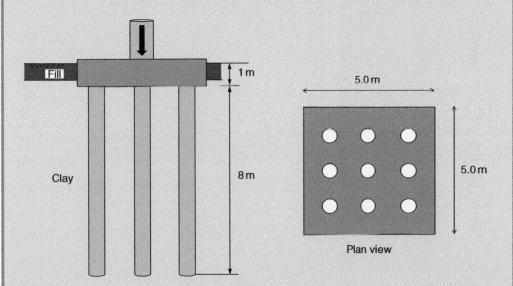

Fig. 11.14 Example 11.8.

be formed at a depth of 1 m below the ground surface and the 8 m long piles will be fully cast in the clay. A single borehole record, together with a limited amount of additional information, has enabled the average undrained shear strength of the soil over the length of the piles to be determined as $\overline{c_u}$ = 75 kPa and the undrained shear strength at the base of the piles was measured as $c_{u,b}$ = 130 kPa. The GWT was encountered at a significant depth below the proposed end of the piles.

If the undrained shear strength of the clay at the formation level of the pile cap is 20 kPa, check compliance of the undrained compressive resistance limit state (a) of a single pile and (b) of the group, using the resistance factor approach.

Assume $N_c = 9.0$, $\alpha = 0.45$ and that the foundation falls into Consequence Class 2. No consolidation of the clay around the piles is anticipated and the efficiency, E, of the pile group based on pile spacing is 0.8.

Solution:

(a) *Compressive resistance of single pile*

(i) Static load tests results: use Calculation Method B.

From Table 11.7: $\xi_{m,mean} = 1.30$; $\xi_{m,min} = 1.20$

From Table 11.8: $\gamma_{Rc} = 1.1$

The measured resistance, R_m is taken as the pile load recorded at a settlement equal to 10% of the pile diameter (= 50 mm).

$R_m = 450\,kN$ (Test 1), $R_m = 480\,kN$ (Test 2)

$(R_m)_{min} = 450$ kN

$(R_m)_{mean} = (450 + 480)/2 = 465$ kN

$$R_{rep} = \min\left\{\frac{(R_m)_{mean}}{\xi_{m,mean}}; \frac{(R_m)_{min}}{\xi_{m,min}}\right\} = \min\left\{\frac{465}{1.3}; \frac{450}{1.2}\right\} = 357.7 \text{ kN}$$

Static load testing used $\Rightarrow \gamma_{Rd} = 1.0$

$$R_{cd} = \frac{R_{c,rep}}{\gamma_{Rc} \times \gamma_{Rd}} = \frac{357.7}{1.1 \times 1.0} = 325.2 \text{ kN}$$

Design Case 4 applies (Table 11.8)

From Table 6.6: $\gamma_Q = 1.5/1.35 = 1.11$; $\gamma_E = 1.35\,K_F$ ($K_F = 1.0$, since CC2)

For a single pile,

$P_{G,rep} = 1200/9 = 133.3$ kN

$P_{Q,rep} = 300/9 = 33.3$ kN

$V_{G,d} = 133.3$ kN

$V_{Q,d} = 33.3 \times 1.11 = 37.0$ kN

As no consolidation occurs around piles, drag force = 0.

$F_d = (133.3 + 37.0) \times \gamma_E = 170.3 \times 1.35 = 229.9$ kN

Since $R_{cd} > F_d$ the limit state verification is satisfied.

(ii) Check by calculation: use Calculation Method A.

From Equation (11.5),

$$R_{c,calc} = c_{u,b}N_cA_b + \alpha\overline{c_u}A_s$$

$$= 130 \times 9 \times \frac{\pi 0.5^2}{4} + 0.45 \times 75 \times \pi \times 0.5 \times 8.0$$

$$= 229.7 + 424.1$$

$$= 653.8 \text{ kN}$$

One profile (borehole) used to determine ground properties $\Rightarrow \xi_{m,min} = 1.40$ (from Table 11.6)

$$R_{c,rep} = \frac{R_{c,calc}}{\xi_{m,min}} = \frac{653.8}{1.40} = 467.0 \text{ kN}$$

Also, $R_{b,rep} = \frac{R_{b,calc}}{\xi_{m,min}} = \frac{229.7}{1.40} = 164.1 \text{ kN}$

$$R_{s,rep} = \frac{R_{s,calc}}{\xi_{m,min}} = \frac{424.1}{1.40} = 302.9 \text{ kN}$$

From Table 11.8: $\gamma_{Rc} = 1.5$; $\gamma_{Rb} = 1.6$; $\gamma_{Rs} = 1.3$
From Table 11.5A: *Limited* additional information $\Rightarrow \gamma_{Rd} = 1.55$

$$R_{cd} = \frac{R_{c,rep}}{\gamma_{Rc} \times \gamma_{Rd}} = \frac{467.0}{1.5 \times 1.55} = 200.9 \text{ kN}$$

or

$$R_{cd} = \frac{R_{b,rep}}{\gamma_{Rb} \times \gamma_{Rd}} + \frac{R_{s,rep}}{\gamma_{Rs} \times \gamma_{Rd}} = \frac{164.1}{1.6 \times 1.55} + \frac{302.9}{1.3 \times 1.55} = 216.5 \text{ kN}$$

$$R_{cd} = \min \{200.9, \ 216.5\} = 200.9 \text{ kN}$$

Design Case 3 applies (Table 11.8)
From Table 6.6: $\gamma_G = 1.0$; $\gamma_Q = 1.3$
For a single pile,

$$V_{G,d} = 133.3 \times 1.0 = 133.3 \text{ kN}$$
$$V_{Q,d} = 33.3 \times 1.3 = 43.3 \text{ kN}$$

$$F_d = 133.3 + 43.3 = 176.6 \text{ kN}$$

Since $R_{cd} > F_d$ the limit state verification is satisfied.

(b) *Compressive resistance of pile group*
Resistance factor approach, Design Case 4 applies (Table 11.9)
From Table 6.6: $\gamma_Q = 1.5/1.35 = 1.11$; $\gamma_E = 1.35 K_F$ ($K_F = 1.0$, since CC2)
From Table 11.9: $\gamma_R = 1.4$
From Table 11.6, $\xi_{m,min} = 1.4$ (Since data based on one profile.)
Check (i) – Determine Q_u using Equation (11.12),

$$Q_u = 2D(B + L)\overline{c_u} + 1.3c_bN_cBL$$
$$= 2 \times 8 \times (5.0 + 5.0) \times 75 + 1.3 \times 130 \times 9 \times 5.0 \times 5.0$$
$$= 50\,025 \text{ kN}$$

$$R_{c,rep} = \frac{(R_m)_{min}}{\xi_{m,min}} = \frac{50\,025}{1.4} = 35\,732 \text{ kN}$$

$$R_{cd} = \frac{R_{c,rep}}{\gamma_R} = \frac{35\,732}{1.4} = 25\,523 \text{ kN}$$

Check (ii) – Determine Q_u using Equation (11.13),
Contribution of pile cap on bearing resistance,

$$Q_{u,cap} = c_u \times N_c \times [(B \times L) - (9 \times \pi \times d)]$$
$$= 20 \times 9 \times [(5 \times 5) - (9 \times \pi \times 0.5)]$$
$$= 1955.3 \text{ kN}$$

$$Q_u = E\,n\,Q_{up} + Q_{u,cap} = 0.8 \times 9 \times 653.8 + 1955.3 = 6662.7 \text{ kN}$$

$$R_{c,rep} = \frac{(R_m)_{min}}{\xi_{m,min}} = \frac{6662.7}{1.4} = 4759 \text{ kN}$$

$$R_{cd} = \frac{R_{c,rep}}{\gamma_R} = \frac{4759}{1.4} = 3399 \text{ kN}$$

$$R_{cd} = \min\{25\,523, 3399\} = 3399 \text{ kN}$$

$$V_{G,d} = 1200 \text{ kN}$$

$$V_{Q,d} = 300 \times 1.11 = 333 \text{ kN}$$

$$F_d = (1200 + 333) \times \gamma_E = 1533 \times 1.35 = 2070 \text{ kN}$$

Since $R_{cd} > F_d$ the block failure limit state verification is satisfied.

Example 11.9: Design from ground tests results (to Eurocode 7 second generation)

Consider again Example 11.6. Check compliance of the compressive limit state using the ground tests results, given that the additional information available to assist the design is considered as *comprehensive*.

Solution:

The design is by calculation from ground properties, so Calculation Method A is used.

From Example 11.6, $(R_{c;cal})_{mean} = 1355$ kN; $(R_{c;cal})_{min} = 1228$ kN

$(R_{b;cal})_1 = (N_c \times c_u \times A_b) = (9 \times 90 \times 0.385) = 312$ kN

$(R_{b;cal})_2 = (9 \times 79 \times 0.385) = 274$ kN

$(R_{b;cal})_3 = (9 \times 96 \times 0.385) = 333$ kN

$(R_{b;cal})_4 = (9 \times 100 \times 0.385) = 347$ kN

$(R_b)_{mean} = 317$ kN

$(R_b)_{min} = 274$ kN

$(R_{s;cal})_1 = (\pi \times D \times L \times \alpha \times c_u) = (\pi \times 0.7 \times 10 \times 0.7 \times 65) = 1001$ kN

$(R_{s;cal})_2 = (\pi \times 0.7 \times 10 \times 0.7 \times 62) = 954$ kN

$(R_{s;cal})_3 - (\pi \times 0.7 \times 10 \times 0.7 \times 70) = 1078$ kN

$(R_{s;cal})_4 = (\pi \times 0.7 \times 10 \times 0.7 \times 73) = 1124$ kN

$(R_s)_{mean} = 1039$ kN

$(R_s)_{min} = 954$ kN

From Table 11.6: $\xi_{m,mean} = 1.28$; $\xi_{m,min} = 1.20$

$R_{rep} = \min\left\{\dfrac{(R_m)_{mean}}{\xi_{m,mean}}; \dfrac{(R_m)_{min}}{\xi_{m,min}}\right\}$

$R_{c,rep} = \dfrac{(R_{c,cal})_{mean}}{\xi_{m,mean}} = \dfrac{1355}{1.28} = 1059$ kN

$R_{c,rep} = \dfrac{(R_{c,cal})_{min}}{\xi_{m,min}} = \dfrac{1228}{1.20} = \underline{1023 \text{ kN}}$

$R_{b,rep} = \dfrac{(R_{b;cal})_{mean}}{\xi_{m,mean}} = \dfrac{317}{1.28} = 248$ kN

$R_{b,rep} = \dfrac{(R_{b;cal})_{min}}{\xi_{m,min}} = \dfrac{274}{1.20} = \underline{228 \text{ kN}}$

$R_{s,rep} = \dfrac{(R_{s;cal})_{mean}}{\xi_{m,mean}} = \dfrac{1039}{1.28} = 812$ kN

$$R_{s,rep} = \frac{(R_{s;cal})_{min}}{\xi_{m,min}} = \frac{954}{1.20} = \underline{795\ kN}$$

The minimum values are underlined.
The additional information available is *comprehensive* $\Rightarrow \gamma_{Rd} = 1.3$
From Table 11.8, $\gamma_{Rc} = 1.4$; $\gamma_{Rb} = 1.45$; $\gamma_{Rs} = 1.3$

$$R_{cd} = \frac{R_{c,rep}}{\gamma_{Rc} \times \gamma_{Rd}} = \frac{1023}{1.4 \times 1.3} = 562\ kN$$

or

$$R_{cd} = \frac{R_{b,rep}}{\gamma_{Rb} \times \gamma_{Rd}} + \frac{R_{s,rep}}{\gamma_{Rs} \times \gamma_{Rd}} = \frac{228}{1.45 \times 1.3} + \frac{795}{1.3 \times 1.3} = 591\ kN$$

$$R_{cd} = min\{562, \ 591\} = 562\ kN$$

Design Case 3 applies (Table 11.8)
From Table 6.6: $\gamma_G = 1.0$; $\gamma_Q = 1.3$

$$V_{G,d} = 350 \times 1.0 = 350\ kN$$
$$V_{Q,d} = 120 \times 1.3 = 156\ kN$$

$$F_d = 350 + 156 = 506\ kN$$

Since $R_{cd} > F_d$ the limit state verification is satisfied.

11.9   Piles subjected to additional, non-compressive loadings

Although piles in most foundation applications act in compression, it is possible that the piles may experience tensile or transverse loading, as illustrated in Fig. 11.15.

11.9.1   Piles in tension

Piles supporting structures such as water towers, electrical transmission pylons, and other tall structures are often required to resist tensile, or pull-out, forces. In vertical piles, the uplift forces can be assumed to be resisted by shaft friction effects only, as indicated in Fig. 11.15a. The uplift resistance of a pile can be calculated in the same manner as for a pile under compression except that reduced values of the unit skin friction or adhesion should be used. Unless data have been obtained from pull-out tests, unit skin friction and adhesion values should be reduced by at least 50% if the piles are short and are liable to significant loading.

A method for maximising the pull-out resistance of a pile is to enlarge its base, through a process known as *underreaming* (Fig. 11.15b). This has the effect of mobilising a torus of soil above the enlarged base, which provides increased resistance against uplift through the tensile force.

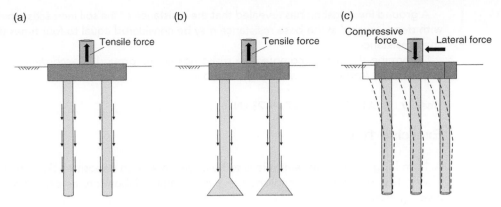

Fig. 11.15   Tensile and transverse loadings in piles. (a) Resistance through shaft friction only. (b) Resistance through underreaming and shaft friction. (c) Transverse loading deflection profile.

11.9.2   Transversely loaded piles

If a pile is subjected to a horizontal force, it is necessary to check that the moments and deflections that will be induced within the pile are acceptable. Various solutions have been proposed for the determination of deflections, moments, and shear force values along the length of a pile. It has been found that the manner in which a pile behaves, under the action of a horizontal load, is largely governed by the length of the pile. With a short pile, failure will be caused by rupture of the surrounding soil, the pile itself remaining undamaged. With a long pile, however, failure will be caused by structural damage to the pile which will occur if the value of its yield moment is less than the applied moment. The deflection profile of a long pile is shown (exaggerated) in Fig. 11.15c.

Both generations of Eurocode 7 provide design guidance for piles in situations where tensile and lateral loading states exist.

Exercises

Exercise 11.1

A single test pile, 300 mm diameter, is driven through a depth of 8 m of clay which has an undrained cohesive strength varying from 10 kPa at its surface to 50 kPa at a depth of 8 m. Estimate the safe load capacity of the pile, for an overall factor of safety = 3.0. Assume $\alpha = 0.65$ and $N_c = 9.0$.

Answer   60 kN

Exercise 11.2

Three static load tests were carried out on CFA piles and gave the following results:

Test no.	1	2	3
Measured load (kN)	1210	1350	1490

A ground investigation has revealed that the resistance of the soil increases rapidly with depth, such that the base resistance may be considered equal to four times the shaft resistance.

Determine the design compressive resistance of the ground in accordance with both Design Approach 1 and Design Approach 2.

Answer   (DA1: 794 kN; DA2: 1023 kN)

Exercise 11.3

A 11 m long × 0.5 m diameter pile is to be driven in a deep deposit of clay. The following test results were established in a geotechnical laboratory as part of a site investigation:

Borehole no.	1	2	3	4
Mean undrained strength along shaft, $c_{u;shaft}$ (kPa)	120	150	200	135

Further tests revealed that the mean undrained strength at the base of the pile can be approximated to 1.5 times the shaft mean strength for each borehole.

The pile will carry a permanent axial load of 600 kN (includes the self-weight of the pile) and an applied transient (variable) axial load of 250 kN.

Check the bearing resistance (GEO) limit state in accordance with Eurocode 7, Design Approach 1 by establishing the magnitude of the over-design factor. Assume $N_c = 9$ and $\alpha = 0.65$.

Answer   (DA1-1: $\Gamma = 1.17$; DA1-2: $\Gamma = 1.15$)

Exercise 11.4

A structure is to be supported by a series of 12 driven piles, arranged symmetrically beneath a pile cap. Pile geometry and loading data are listed below:

$P_G=$	600 kN	$\xi_3=$	1.33
$P_Q=$	300 kN	$\xi_4=$	1.23
$D=$	0.8 m	$L=$	18.5 m

Ground tests results from three boreholes have given the following values of undrained shear strength:

	BH 1	BH 2	BH 3
$c_{u,shaft}$ (kPa)	52	46	51
$c_{u,base}$ (kPa)	33	30	42

Assume $N_c = 9$ and $\alpha = 0.75$.

Check compliance of the bearing resistance limit state by determining the over-design factor for:

(i) Design Approach 1
(ii) Design Approach 2.

Answer   (DA1-1: $\Gamma = 1.12$; DA1-2: $\Gamma = 1.10$; DA2: $\Gamma = 1.02$)

Exercise 11.5

Consider again Exercise 11.2. Determine the design compressive resistance of the ground in accordance with Eurocode 7, second generation.

Answer $R_{cd} = 1023$ kN

Chapter 12

Foundation Settlement and Soil Compression

Learning objectives:

By the end of this chapter, you will have been introduced to:
- methods of analysing immediate settlement, for both cohesive and granular soils;
- the concept of one-dimensional consolidation of clay soils;
- the consolidation test and the oedometer;
- the theory of general consolidation and settlement analysis;
- Eurocode 7 serviceability limit state verification for foundation settlement;
- stress paths in the oedometer, and for general consolidation.

12.1 Settlement of a foundation

Probably the most difficult of the problems that a geotechnical engineer is asked to solve is the accurate prediction of the settlement of a loaded foundation.

The problem is in two distinct parts: (i) the value of the total settlement that will occur and (ii) the rate at which this value will be achieved.

When a soil is subjected to an increase in compressive stress due to a foundation load, the resulting soil compression consists of elastic compression, primary compression, and secondary compression.

Elastic compression

This compression is usually taken as occurring immediately after the application of the foundation load. Its vertical component causes a vertical movement of the foundation, referred to as *immediate settlement*, that in the case of a partially saturated soil is mainly due to the expulsion of air and to the elastic bending reorientation of the soil particles. With saturated soils, immediate settlement effects are assumed to be the result of vertical soil compression before there is any change in volume.

Smith's Elements of Soil Mechanics, 10th Edition. Ian Smith.
© 2021 John Wiley & Sons Ltd. Published 2021 by John Wiley & Sons Ltd.
Companion website: www.wiley.com/go/smith/soilmechanics10e

Primary compression

The sudden application of a foundation load, besides causing elastic compression, creates a state of excess pore water pressure in saturated soil. These excess pore water pressure values can only be dissipated by the gradual expulsion of water through the voids of the soil, which results in a volume change that is time-dependent. A soil experiencing such a volume change is said to be consolidating, and the vertical component of the change is called the *consolidation settlement*.

Secondary compression

Volume changes that are more or less independent of the excess pore water pressure values cause secondary compression. The nature of these changes is not fully understood but they are apparently due to a form of plastic flow resulting in a displacement of the soil particles. Secondary compression effects can continue over long periods of time and, in the consolidation test (see Section 12.3.2), become apparent towards the end of the primary compression stage: due to the thinness of the sample, the excess pore water pressures are soon dissipated and it may appear that the main part of secondary compression occurs after primary compression is completed. This effect is absent in the case of an *in situ* clay layer because the large dimensions involved mean that a considerable time is required before the excess pore pressures drain away. During this time, the effects of secondary compression are also taking place so that, when primary compression is complete, little, if any, secondary effect is noticeable. The terms 'primary' and 'secondary' are therefore seen to be rather arbitrary divisions of the single, continuous consolidation process. The time relationships of these two factors will be entirely different if they are obtained from two test samples of different thicknesses.

12.2   Immediate settlement

12.2.1   Cohesive soils

If a saturated clay is loaded rapidly, the soil will be deformed during the load application and excess pore pressures are set up. This deformation occurs with virtually no volume change and, due to the low permeability of the clay, little water is squeezed out of the voids. Vertical deformation due to the change in shape is the immediate settlement.

This change in shape is illustrated in Fig. 12.1a, where an element of soil is subjected to a vertical major principal stress increase $\Delta\sigma_1$, which induces an excess pore water pressure, Δu. The lateral expansion causes an increase in the minor principal stress, $\Delta\sigma_3$.

The formula for immediate settlement of a flexible foundation was provided by Terzaghi (1943) and is

$$\rho_i = \frac{pB(1-\nu^2)N_p}{E} \tag{12.1}$$

where

p = uniform net contact pressure;
B = width of foundation;
E = Young's modulus of elasticity for the soil (Pa);
ν = Poisson's ratio for the soil (= 0.5 in saturated soil); and
N_p = an influence factor depending upon the dimensions of the flexible foundation.

(a)

Excess pore
pressure = Δu

$\Delta\sigma_1$

$\Delta\sigma_3$

(b)

$\Delta\sigma'_1$

Excess pore
pressure = 0

$\Delta\sigma'_3$

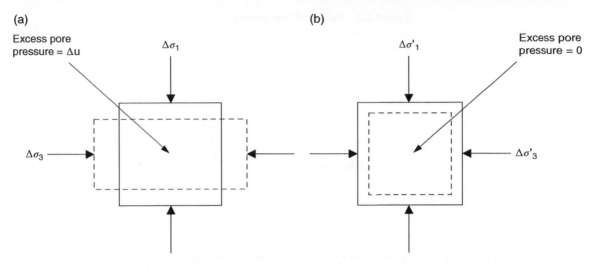

Fig. 12.1   Compressive deformation. (a) Immediate settlement. (b) Consolidation settlement.

This relationship gives the immediate settlement *at the corners* of a rectangular footing, length L and width B. In the case of a uniformly loaded, perfectly flexible square footing, the immediate settlement under its centre is twice that at its corners.

Various values for N_p are given in Table 12.1.

By the principle of superposition, it is possible to determine the immediate settlement under any point of the base of a foundation (for example, see Example 12.2). An earth embankment or spoil heap can be taken as flexible; to determine the immediate settlement of deposits below such a construction, the coefficients of Table 12.1 should be used.

Foundations are generally more rigid than flexible and tend to impose a uniform settlement which is roughly the same value as the mean value of settlement under a flexible foundation. The *mean value of settlement* for a rectangular rigid foundation on the surface of a semi-elastic medium is given by the expression:

$$\rho_i = \frac{pB(1-v^2)I_p}{E} \tag{12.2}$$

Table 12.1   Values of Terzaghi's N_p.

L/B	N_p
1.0	0.56
2.0	0.76
3.0	0.88
4.0	0.96
5.0	1.00

Table 12.2   Values of Skempton's I_p.

L/B	I_p
Circle	0.73
1	0.82
2	1.00
5	1.22
10	1.26

Data from Skempton (1951).

where

I_p = an influence factor depending upon the dimensions of the foundation.

Skempton (1951) suggested the values for I_p given in Table 12.2.

Immediate settlement of a thin clay layer

The coefficients of Tables 12.1 and 12.2 only apply to foundations on deep soil layers. Vertical stresses extend to about four times the breadth (4.0B) below a strip footing and the formulae, strictly speaking, are not applicable to layers thinner than this, although little error is incurred if the coefficients are used for layers of thicknesses greater than 2.0B. A drawback of the method is that it can only be applied to a layer immediately below the foundation.

For cases when the thickness of the layer is less than 4.0B, a solution is possible with the use of coefficients prepared by Steinbrenner (1934), whose procedure was to determine the immediate settlement at the top of the layer (assuming infinite depth) and to calculate the settlement at the bottom of the layer (again assuming infinite depth) below it. The difference between the two values is the actual settlement of the layer.

The total immediate settlement at the corner of a rectangular foundation on an infinite layer is:

$$\rho_i = \frac{pB(1 - v^2)I_p}{E} \tag{12.3}$$

The values of the coefficient I_p (when $v = 0.5$) are given in Fig. 12.2c. To determine the settlement of a point beneath the foundation, the area is divided into rectangles that meet over the point (the same procedure used when determining vertical stress increments by Steinbrenner's method – see Section 3.6.3). The summation of the settlements of the corners of the rectangles gives the total settlement of the point considered.

This method can be extended to determine the immediate settlement of a clay layer which is at some depth below the foundation. In Fig. 12.2b, the settlement of the lower layer (of thickness $H_2 - H_1$) is obtained by first determining the settlement of a layer extending from below the foundation that is of thickness H_2 (using E_2); from this value is subtracted the imaginary settlement of the layer H_1 (again using E_2).

It should be noted that the settlement values obtained by this method are for a perfectly flexible foundation. Usually, the value of settlement at the centre of the foundation is evaluated and reduced by a rigidity factor (generally taken as 0.8) to give a mean value of settlement that applies over the whole foundation.

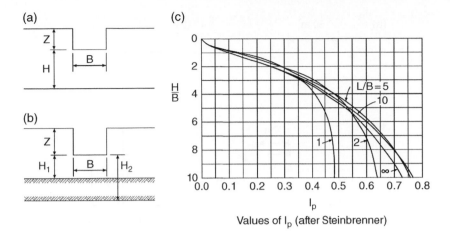

Fig. 12.2 Immediate settlement of thin clay layer. (a) Thin clay layer (H < 4B). (b) Clay layer deep below foundation. (c) Values of I_p (after Steinbrenner).

Example 12.1: Immediate settlement of a rigid foundation

A reinforced concrete foundation, of dimensions 20 m × 40 m exerts a uniform pressure of 200 kPa on a semi-infinite saturated soil layer (E = 50 MPa).

Determine the value of immediate settlement under the foundation using Table 12.2.

Solution:

$$\frac{L}{B} = \frac{40}{20} = 2.0$$

From Table 12.2, I_p = 1.0.

$$\rho_i = \frac{pB(1 - v^2)I_p}{E} = \frac{200 \times 20 \times 0.75 \times 1.0}{50\ 000} = 0.06 \text{ m} = 60 \text{ mm}$$

Example 12.2: Immediate settlement of a flexible foundation

The plan of a proposed spoil heap is shown in Fig. 12.3a. The tip will be about 17 m high and will sit on a thick, soft alluvial deposit (E = 25 MPa). It is estimated that the eventual uniform bearing pressure on the soil will be about 300 kPa. Estimate the immediate settlement under the point A at the surface of the soil.

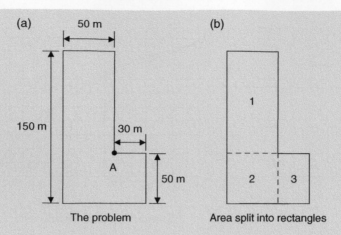

Fig. 12.3 Example 12.2. (a) The problem. (b) Area split into rectangles.

Solution:

The procedure is to divide the plan area into a number of rectangles, the corners of which must meet at the point A. In Fig. 12.3b, it is seen that three rectangles are required. As the structure is flexible and the soil deposit is thick, the coefficients of Table 12.1 should be used:

Rectangle (1): $\quad 100\,\text{m} \times 50\,\text{m}\quad \dfrac{L}{B} = 2.0;\quad N_p = 0.76$

Rectangle (2): $\quad 50\,\text{m} \times 50\,\text{m}\quad \dfrac{L}{B} = 1.0;\quad N_p = 0.56$

Rectangle (3): $\quad 50\,\text{m} \times 30\,\text{m}\quad \dfrac{L}{B} = 1.67;\quad N_p = 0.69$

$$\rho_i = \frac{p(1 - \nu^2)(N_{p1}B_1 + N_{p2}B_2 + N_{p3}B_3)}{E}$$

$$= \frac{300 \times 0.75}{25\,000}(0.76 \times 50 + 0.56 \times 50 + 0.69 \times 30)$$

$$= 0.78\ \text{m} = 78\ \text{cm}$$

The effect of depth

It is found that for deep foundations (z > B), the calculated immediate settlements are more than the actual ones, and a reduction may be applied to the calculated values. If z = B, the reduction is approximately 25%, increasing to about 50% for infinitely deep foundations.

Most foundations are shallow, however, and although this reduction can be allowed for when a layer of soil is some depth below a foundation, the settlement effects in this case are small so it is not customary practice to reduce them further.

Determination of modulus of elasticity

The modulus of elasticity, E, is usually obtained from the results of a consolidated undrained triaxial test carried out on a representative sample of the soil that is consolidated under a cell pressure approximating to the effective overburden pressure at the level from which the sample was taken. The soil is then sheared undrained to obtain the plot of total deviator stress against strain. This is never a straight line and to determine E, a line must be drawn from the origin up to the value of deviator stress that will be experienced in the field when the foundation load is applied. In deep layers, there is the problem of assessing the depth at which the average value of E is encountered. In these situations, the deep layer should be split into thinner layers with a value of E determined for each.

A certain amount of analysis work is necessary to carry out the above procedure. The increments of principal stress $\Delta\sigma_1$ and $\Delta\sigma_3$ must be obtained so that the value of $\Delta\sigma_1 - \Delta\sigma_3$ is known, and a value of safe bearing capacity can be determined for the situation. Skempton (1951) analysed this issue and observed that when a factor of safety of 3.0 is considered, the maximum shear stress induced in the soil is not greater than 65% of the ultimate shear strength, so that a value of E can be obtained directly from the triaxial test results by determining the strain corresponding to 65% of the maximum deviator stress and dividing this value into its corresponding stress. The method produces results that are well within the range of accuracy possible with other techniques.

12.2.2 Granular soils

Owing to the high permeabilities of granular soils, both the elastic and the primary effects occur more or less together. The resulting settlement from these factors is termed the immediate settlement.

The chance of bearing capacity failure in a foundation supported on a granular soil is remote, and for these soils, it is important that settlement is considered in the foundation design, such as through the serviceability limit state verification in a Eurocode 7 design. The allowable bearing pressure, p was historically often taken as the pressure that would cause an average settlement of 25 mm in the foundation. However, the Eurocode 7 foundation design procedures, as described in Chapter 10, perform a more rigorous analysis than simply establishing an allowable bearing pressure.

If the actual bearing pressure is not equal to the value of p then the value of settlement is not known and, since it is difficult to obtain this value from laboratory tests, we can use some specific *in situ* tests to enable us to estimate the likely settlement. The three main tests which we can use are: (i) the cone penetration test (CPT), (ii) the standard penetration test (SPT), and (iii) the plate loading test (PLT). Most methods used require the value of C_r, the penetration resistance of the cone penetration test, which is usually expressed in MPa or kPa.

Meyerhof's method

A quick estimate of the settlement, ρ of a footing on sand was proposed by Meyerhof (1974):

$$\rho = \frac{\Delta p B}{2\overline{C_r}} \tag{12.4}$$

where

B = the least dimension of the footing;
$\overline{C_r}$ = the average value of C_r over a depth below the footing equal to B; and
Δp = the net foundation pressure increase, which is simply the foundation loading less the value of vertical effective stress at foundation level, σ'_{v0}.

The two other methods commonly in use were proposed by De Beer and Martens (1957) and by Schmertmann (1970). Both methods require a value for C_r and, if either is to be used with standard penetration test results, it is necessary to have the correlation between C_r and N.

Perhaps obviously, the value of C_r obtained from the cone penetration test must be related to the number of recorded blows, N obtained from the standard penetration test. Various workers have attempted to find this relationship with moderate success. Meigh and Nixon (1961) showed that, over a number of sites, C_r varied from $(430 \times N)$ to $(1930 \times N)$ kPa.

The relationship most commonly used at the present time is that proposed by Meyerhof (1956):

$$C_r = 400 \times N \text{ kPa} \tag{12.5}$$

where N = actual number of blows recorded in the standard penetration test.

It goes without saying that, whenever possible, C_r values obtained from actual cone tests should be used in preference to values estimated from N values.

De Beer and Martens' method

From the results of the *in situ* tests carried out, a plot of C_r (or N) values against depth is prepared. With the aid of this plot, the profile of the compressible soil beneath the proposed foundation can be divided into a suitable number of layers, preferably of the same thickness, although this is not essential.

In the case of a deep soil deposit, the depth of soil considered as affected by the foundation should not be less than 2.0B, ideally 4.0B, where B = foundation width.

The method proposes the use of a constant of compressibility, C_S, where:

$$C_s = 1.5 \frac{C_r}{p_{o1}} \tag{12.6}$$

where

C_r = static cone resistance (kPa); and
p_{o1} = effective overburden pressure at the point tested.

Total immediate settlement is:

$$\rho_i = \frac{H}{C_s} \ln \left(\frac{p_{o1} + \Delta \sigma_z}{p_{o2}} \right) \tag{12.7}$$

where

$\Delta \sigma_z$ = vertical stress increase at the centre of the settling layer of thickness H; and
p_{o2} = effective overburden pressure at the centre of the layer before any excavation or load application.

Note: Meyerhof (1956) suggested that a more realistic value for C_S is

$$C_s = 1.9 \frac{C_r}{p_{o1}}$$

Such a refinement may be an advantage if the calculations use C_r values which have been determined from cone penetration tests, but if the C_r values used have been obtained from the relationship $C_r = 400$ kPa, such a refinement seems naive.

Schmertmann's method

Originally proposed by Schmertmann in 1970 and modified by Schmertmann *et al.* (1978), the method is generally preferred to De Beer and Martens' approach.

The method is based on two main assumptions:

(i) the greatest vertical strain in the soil beneath the centre of a loaded foundation of width B occurs at depth of B/2 below a square foundation, and at depth of B below a strip foundation;

(ii) stresses caused by the foundation loading can be regarded as insignificant at depths greater than z = 2.0B for a square foundation, and z = 4.0B for a strip foundation.

The method involves the use of a vertical strain influence factor, I_z, whose value varies with depth. Values of I_z, for a net foundation pressure increase, Δp equal to the effective overburden pressure at depth B/2, are shown in Fig. 12.4.

The procedure consists of dividing the sand below the footing into n layers, of thicknesses Δ_{z1}, Δ_{z2}, Δ_{z3}, ... Δ_{zn}. If soil conditions permit, it is simpler if the layers can be made of equal thickness, Δz. The vertical strain of a layer is taken as equal to the increase in vertical stress at the centre of the layer, i.e. Δp, multiplied by I_z, which is then divided by the product of C_r and a factor x. Hence:

$$\rho = C_1 C_2 \Delta p \sum_{1}^{n} \frac{I_z}{x C_r} \Delta z_i \tag{12.8}$$

where

x = 2.5 for a square footing and 3.5 for a long footing;

I_z = the strain influence factor, valued for each layer at its centre, and obtained from a diagram similar to Fig. 12.4 but redrawn to correspond to the foundation loading;

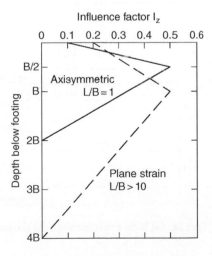

Fig. 12.4 Variation of I_z with depth.

C_1 = a correction factor for the depth of the foundation

$$= 1.0 - 0.5\frac{\sigma'_v}{\Delta p}\ (= 1.0 \ \text{for a surface footing})$$

C_2 = a correction factor for creep

$$= 1 + 0.2\log_{10}10t\ (t = \text{time in years after the application of foundation loading for which the settlement value is required}).$$

As mentioned above, Fig. 12.4 must be redrawn. This is achieved by obtaining a new peak value for I_z from the expression:

$$I_z = 0.5 + 0.1\left(\frac{\Delta p}{\sigma'_{vp}}\right)^{0.5} \tag{12.9}$$

where σ'_{vp} = the effective vertical overburden pressure at a depth of 0.5B for a square foundation and at a depth of 1.0B for a long foundation.

Example 12.3: Settlement on a cohesionless soil

A foundation, 1.5 m square and founded at a depth of 0.75 m in a deep deposit of sand, will apply a pressure of 300 kPa to the soil. The soil may be regarded as saturated throughout with a unit weight of 20 kN/m³, and the approximate N to z relationship is shown in Fig. 12.5a.

If the groundwater level occurs at a depth of 1.5 m below the surface of the soil, determine a value for the settlement at the centre of the foundation, (a) by De Beer and Martens' method and (b) by Schmertmann's method (assume creep occurs within five weeks).

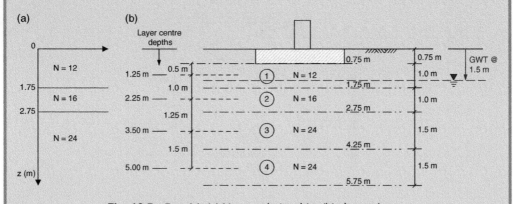

Fig. 12.5   Part (a): (a) N to z relationship; (b) chosen layers.

Solution:

(a) *De Beer and Martens' method*

The soil deposit is deep; therefore, we should investigate to a depth of about 3B–4B below the foundation. In conjunction with Fig. 12.5a it is seen that a depth of 5 m below the foundation can be conveniently divided into four layers of soil, two of 1 m and two of 1.5 m thickness, as shown in Fig. 12.5b and the tabulated workings below.

Net pressure, $p = 300 - (0.75 \times 20) = 285$ kPa

Layer	Thickness, H (m)	C_r (kPa)	p_{o1} at layer centre (kPa)	$C_S = \dfrac{1.5 C_r}{P_{o1}}$
1	1.0	$400 \times 12 = 4800$	$20 \times 1.25 = 25$	288
2	1.0	$400 \times 16 = 6400$	$(20 \times 2.25) - (9.81 \times 0.75) = 37.6$	255
3	1.5	$400 \times 24 = 9600$	$(20 \times 3.5) - (9.81 \times 2.0) = 50.4$	286
4	1.5	9600	$(20 \times 5.0) - (9.81 \times 3.5) = 65.7$	219

Use Fadum's chart (Section 3.6.3) to determine vertical pressure increments. Then $B = L = 1.5/2 = 0.75$ m, and $p_{o1} = p_{o2}$ for each layer.

Layer	B/z = L/z	I_σ	$4 I_\sigma$	$\Delta\sigma_z = 4 p I_\sigma$	$\ln\left(\dfrac{p_0 + \Delta\sigma_z}{p_0}\right)$ (A)	$\dfrac{H}{C_s} \times$ (A) (m)
1	$0.75/0.5 = 1.5$	0.213	0.852	243	2.3721	0.00824
2	$0.75/1.5 = 0.5$	0.088	0.352	100	1.2973	0.00509
3	$0.75/2.75 = 0.27$	0.030	0.120	34	0.5156	0.00270
4	$0.75/4.25 = 0.18$	0.015	0.060	17	0.2301	0.00158
						$\Sigma 0.01761$

Total settlement $= 0.01761 \times 1000 = 17.6$ mm, say 18 mm

(b) *Schmertmann's method*

For a square footing, significant depths extend to 2.0B and σ'_{vp} is taken as the effective vertical overburden pressure at a depth of 0.5B below the foundation, i.e. in this example, 0.75 m, so that $\sigma'_{vp} = 20(0.75 + 0.75) = 30$ kPa.

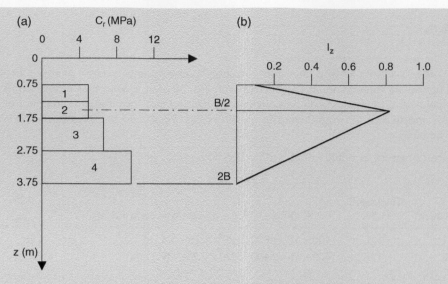

Fig. 12.6 Part (b): (a) C_r to z relationship; (b) variation of I_z.

Net foundation pressure increase $\Delta p = 300 - 20 \times 0.75 = 285$ kPa.
Hence:

$$I_z = 0.5 + 0.1 \left(\frac{\Delta p}{\sigma'_{vp}} \right)^{0.5}$$

$$= 0.5 + 0.1(285/30)^{0.5} = 0.81$$

The variation of I_z for depths up to 2B below the foundation is shown in Fig. 12.6b. The C_r values shown in Fig. 12.6a are obtained from the N values using the relationship $C_r = 0.4N$ MPa. With these C_r values, it is possible to decide upon the number and thicknesses of the layers that the soil can be divided into. For this example, for the purpose of illustration, only four layers have been chosen and these are shown in Fig. 12.6a. (For greater accuracy, the number of layers should be about 8 for a square footing and up to 16 for a long footing.) For a square foundation, Schmertmann recommends that the value of the factor x = 2.5. The calculations are set out below.

Layer	Δz_i (m)	Depth below foundation to centre of layer (m)	C_r (MPa)	I_z	$\frac{I_z \Delta z_i}{x C_r}$
1	0.5	0.25	4.8	0.34	0.014
2	0.5	0.75	4.8	0.81	0.034
3	1.0	1.5	6.4	0.54	0.034
4	1.0	2.5	9.6	0.18	0.008
					$\Sigma 0.09$

$$C_1 = 1.0 - 0.5 \frac{\sigma'_v}{\Delta p} = 1 - \left(0.5 \times \frac{15}{285} \right) = 0.97$$

T = 5 weeks (0.1 year) → $C_2 = 1.0$, so,

$$\rho = 0.97 \times 285 \times 0.09 = 24.9 \text{ mm}$$

Total settlement of centre of foundation = 25 mm

The plate loading test

The results from a plate loading test (see Section 7.6.5) can be used to estimate the average settlement of a proposed foundation on granular soil. The test should be carried out at the proposed foundation level and the soil tested must be relatively homogeneous for some depth.

If ρ_1 is the settlement of the test plate under a certain value of bearing pressure, then the average settlement of the foundation, ρ under the same value of bearing pressure can be obtained by the empirical relationship proposed by Terzaghi and Peck (1948):

$$\rho = \rho_1 \left(\frac{2B}{B + B_1} \right)^2 \tag{12.10}$$

where

B_1 = width or diameter of test plate; and
B = width or diameter of proposed foundation.

One aspect of using the results from a plate loading test for settlement predictions is that it is important to know the position of the groundwater level. It may be that the bulb of pressure from the plate load test is partly or completely above the groundwater level whereas, when the foundation is constructed, the groundwater level will be significantly within the bulb of pressure. Such a situation could lead to actual settlement values as much as twice the values predicted by Equation (12.10).

12.3 Consolidation settlement

This effect occurs in clays where the value of permeability prevents the initial excess pore water pressures from draining away immediately. The design loading used to calculate consolidation settlement must be consistent with this effect.

A large wheel load rolling along a roadway resting on a clay will cause an immediate settlement that is in theory completely recoverable once the wheel has passed, but if the same load is applied permanently there will, in addition, be consolidation. Judgement is necessary in deciding what portion of the superimposed loading carried by a structure will be sustained long enough to cause consolidation, and this involves a quite different procedure from that used in a bearing capacity analysis, which must allow for total permanent and variable loadings.

12.3.1 One-dimensional consolidation

The pore water in a saturated clay will commence to drain away soon after immediate settlement has taken place. The removal of this water leading to the volume change is known as *consolidation* (Fig. 12.1b). The element contracts both horizontally and vertically under the actions of $\Delta\sigma'_3$ and $\Delta\sigma_1$, which gradually increase in magnitude as the excess pore water pressure, Δu, decreases. Eventually, when $\Delta u = 0$, then $\Delta\sigma'_3 = \Delta\sigma_3$ and $\Delta\sigma'_1 = \Delta\sigma_1$, and at this stage consolidation ceases, although secondary consolidation may still be apparent.

If it can be arranged for the lateral expansion due to the change in shape to equal the lateral compression consequent upon the change in volume, and for these changes to occur together, then there will be no immediate settlement and the resulting compression will be one-dimensional with all the strain occurring in the vertical direction. Settlement by one-dimensional strain is by no means uncommon in practice, and most natural soil deposits have experienced one-dimensional settlement during the process of deposition and consolidation.

The consolidation of a clay layer supporting a foundation whose dimensions are much greater than the layer's thickness is essentially one-dimensional as lateral strain effects are negligible save at the edges.

12.3.2 The consolidation test

The apparatus generally used in the laboratory to determine the primary compression characteristics of a soil is known as the consolidation test apparatus (or oedometer) and is illustrated in Fig. 12.7a.

The soil sample (generally 75 mm diameter and 20 mm thick) is encased in a steel cutting ring. Porous discs, saturated with air-free water, are placed on top of and below the sample, which is then inserted in the oedometer.

A vertical load is then applied, and the resulting compression is measured by means of a transducer at intervals of time. Readings are recorded until the sample has achieved full consolidation (usually for a period of 24 hours). Further load increments are then applied, and the procedure repeated until the full stress range expected *in situ* has been covered by the test (Fig. 12.7b).

The test sample is generally flooded with water soon after the application of the first load increment to prevent pore suction.

After the sample has consolidated under its final load increment, the pressure is released in stages at 24-hour intervals and the sample allowed to expand. In this way, an expansion to time curve can also be obtained.

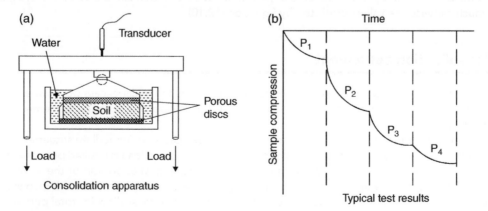

Fig. 12.7　The consolidation test. (a) Consolidation apparatus. (b) Typical test results.

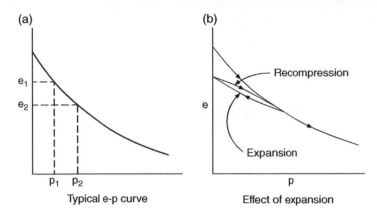

Fig. 12.8   Void ratio to effective pressure curves. (a) Typical e–p curve. (b) Effect of expansion.

After the loading has been completely removed, the final thickness of the sample can be obtained, from which it is possible to calculate the void ratio of the soil for each stage of consolidation under the load increments. The graph of void ratio to consolidation pressure can then be drawn, such a curve generally being referred to as an *e–p curve* (Fig. 12.8a).

It should be noted that the values of p refer to effective stress; for after consolidation, the excess pore pressures become zero and the applied stress increment is equal to the effective stress increment.

If the sample is recompressed after the initial cycle of compression and expansion, the e–p curve for the whole operation is similar to the curves shown in Fig. 12.8b. The recompression curve is flatter than the original compression curve, primary compression being made up of a reversible part and an irreversible part. Once the consolidation pressure is extended beyond the original consolidation pressure value (the preconsolidation pressure), the e–p curve follows the trend of the original compression curve.

All types of soil, whether sand, silt or clay, have the form of compression curves illustrated in Fig. 12.8. The curves shown can be produced quite quickly in the laboratory for teaching purposes, using a dry sand sample, but consolidation problems are mainly concerned with clays and the oedometer is therefore only used to test these types of soil.

12.3.3   Volumetric change

The volume change per unit of original volume constitutes the volumetric change. If a mass of soil of volume V_1 is compressed to a volume V_2, the assumption is made that the change in volume has been caused by a reduction in the volume of the voids. The specific volume, v is the total volume of soil that contains a unit volume of solids and was defined in Equation (5.5) as v = 1 + e. We can use this relationship to quantify the volumetric change.

$$
\begin{aligned}
\text{Volumetric change} &= \frac{V_1 - V_2}{V_1} \\
&= \frac{(1 + e_1) - (1 + e_2)}{1 + e_1} \\
&= \frac{e_1 - e_2}{1 + e_1}
\end{aligned}
\qquad (12.11)
$$

where

e_1 = void ratio at p_1

e_2 = void ratio at p_2.

The slope of the e–p curve is given the symbol a, then:

$$a = \frac{e_1 - e_2}{p_1 - p_2} \ m^2/kN \tag{12.12}$$

i.e.

$$a = \frac{de}{dp}$$

The slope of the e–p curve is seen to **decrease** with increase in pressure; in other words, a is not a constant but will vary depending upon the pressure. Settlement problems are usually only concerned with a range of pressure (that between the initial pressure and the final pressure), and over this range, a is taken as constant by assuming that the e–p curve between these two pressure values is a straight line.

12.3.4 The Rowe cell

An alternative cell to the consolidation cell shown in Fig. 12.7 is the hydraulic cell known as a Rowe cell described by Rowe and Barden (1966) and was previously listed in the former specification for soil testing, BS 1377 (BSI, 1990). The equipment is rarely used in commercial materials testing laboratories and is mainly only found in research laboratories.

The oedometer is hydraulically operated and a various range of cell sizes are available so that test specimens as large as 500 mm diameter and 250 mm thick can be tested. The cell is particularly applicable to testing samples from clay deposits where macrofabric effects are significant.

A constant pressure system applies a hydraulic pressure, via a rubber jack, on to the top of the test specimen. Vertical settlement is measured at the centre of the sample by means of a hollow brass spindle, 10 mm diameter, attached to the jack and passing out through the centre of the top plate to a suitable dial gauge or transducer.

Drainage of the sample can be made to vary according to the nature of the test and can be either vertical or radial, the latter being arranged to be either inwards or outwards. The expelled water exits via the spindle and it is possible to measure pore water pressures during the test, as well as applying a back pressure to the specimen if required. The apparatus can also be used for permeability tests.

12.3.5 Coefficient of volume compressibility, m_v

This value, which is sometimes called the coefficient of volume decrease, represents the compression of a soil, per unit of original thickness, due to a unit increase in pressure, i.e.

m_v = Volumetric change/Unit of pressure increase

Table 12.3 m_v ranges for different soil types.

Soil	m_v (m²/MN)
Peat	10.0–2.0
Plastic clay (normally consolidated alluvial clays)	2.0–0.25
Stiff clay	0.25–0.125
Hard clay (boulder clays)	0.125–0.0625

If H_1 = original thickness and H_2 = final thickness:

$$\text{Volumetric change} = \frac{V_1 - V_2}{V_1} = \frac{H_1 - H_2}{H_1} \quad \text{(as area is constant)}$$

$$= \frac{e_1 - e_2}{1 + e_1}$$

Now,

$$a = \frac{e_1 - e_2}{dp}$$

$$\Rightarrow \text{Volumetric change} = \frac{a \, dp}{1 + e_1}$$

$$\Rightarrow m_v = \frac{a \, dp}{1 + e_1} \frac{1}{dp} = \frac{a}{1 + e_1} \quad \text{m}^2/\text{MN} \tag{12.13}$$

For most practical engineering problems, m_v values can be calculated for a pressure increment of 100 kPa in excess of the present effective overburden pressure at the sample depth.

Once the coefficient of volume decrease has been obtained, we know the compression/unit thickness/unit pressure increase. It is therefore an easy matter to predict the total consolidation settlement of a clay layer of thickness H:

$$\text{Total consolidation settlement,} \quad \rho_c = m_v \, dp \, H \tag{12.14}$$

Typical values of m_v are given in Table 12.3.

In the laboratory consolidation test, the compression of the sample is one-dimensional as there is lateral confinement. The initial excess pore water pressure induced in a saturated clay on loading is equal to the magnitude of the applied major principal stress (due to the fact that there is no lateral yield). This applies no matter what type of soil is tested, provided it is saturated.

One-dimensional consolidation can be produced in a triaxial test specimen by means of a special procedure known as the K_0 test (see Lade, 2016).

Example 12.4: Consolidation test

The following results were obtained from a consolidation test on a sample of saturated clay, each pressure increment having been maintained for 24 hours.

Pressure (kPa)	Thickness of sample after consolidation (mm)
0	20.00
50	19.65
100	19.52
200	19.35
400	19.15
800	18.95
0	19.25

After it had expanded for 24 hours the sample was removed from the apparatus and found to have a water content of 25%. The particle specific gravity of the soil was 2.65.

Plot the void ratio to effective pressure curve and determine the value of the coefficient of volume change for a pressure range of 250–350 kPa.

Solution:

$$w = 0.25; \quad G_s = 2.65$$

Now $e = wG_s$ (as soil is saturated) $= 0.25 \times 2.65 = 0.662$. This is the void ratio corresponding to a sample thickness of 19.25 mm.

$$\frac{dH}{H_1} = \frac{de}{1 + e_1} \Rightarrow de = \frac{1 + e_1}{H_1} dH = \frac{1.662}{19.25} dH = 0.0865 \ dH$$

The values of e at the end of each consolidation can be calculated from this expression.

Pressure (kPa)	H (mm)	dH (mm)	de	e
0	20.0	+0.75	+0.065	0.727
50	19.65	+0.40	+0.035	0.697
100	19.52	+0.27	+0.023	0.685
200	19.35	+0.10	+0.009	0.671
400	19.15	−0.10	−0.009	0.653
800	18.95	−0.30	−0.026	0.636
0	19.25	0	0	0.662

The e–p and H–p plots can now be made and are shown in Fig. 12.9.

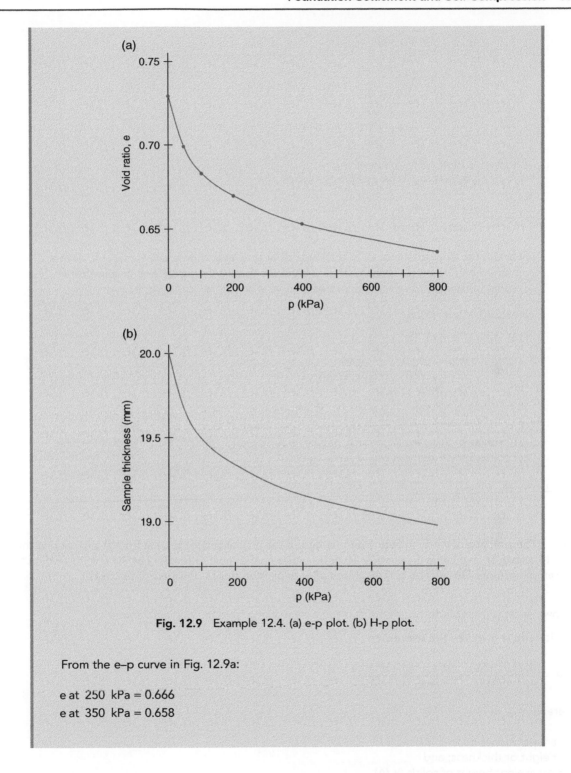

Fig. 12.9   Example 12.4. (a) e-p plot. (b) H-p plot.

From the e–p curve in Fig. 12.9a:

e at 250 kPa = 0.666
e at 350 kPa = 0.658

$$a = \frac{de}{dp} = \frac{0.666 - 0.658}{100}$$

$$= 0.00008 \ m^2/kN$$

$$\Rightarrow m_v = \frac{a}{1 + e_1} = \frac{0.00008}{1.666}$$

$$= 4.8 \times 10^{-5} \ m^2/kN$$

Alternative method for determining m_v
m_v can be expressed in terms of thicknesses:

$$m_v = \frac{dH}{H_1} \frac{1}{dp} = \frac{1}{H_1} \frac{dH}{dp}$$

dH/dp is the slope of the curve of thickness of sample against pressure. Hence, m_v can be obtained by finding the slope of the curve at the required pressure and dividing by the original thickness. The thickness/pressure curve is shown in Fig. 12.9b; from it:

H at 250 kPa = 19.28

H at 350 kPa = 19.19

$$m_v = \frac{19.28 - 19.19}{19.28 \times 100} = \frac{0.09}{19.28 \times 100}$$

$$= 4.7 \times 10^{-5} \ m^2/kN$$

If a layer of this clay, 20 m thick, had been subjected to this pressure increase then the consolidation settlement would have been:

$$\rho_c = m_v H \ dp = 0.000047 \times 20 \times 100 \times 1000 = 96 \ mm$$

Note: The practice of working back from the end of the consolidation test, i.e. from the expanded thickness, to obtain an e–p curve is generally accepted as being the most satisfactory as there is little doubt that the sample is more likely to be fully saturated after expansion than at the start of the test.

However, it is possible to obtain the e–p curve by working from the original thickness.

Void ratio is given by the expression

$$e = \frac{V_v}{V_s} = \frac{V - V_s}{V_s} = \frac{A(H - H_s)}{AH_s} = \frac{H - H_s}{H_s}$$

where:

A = area of sample;
H = height or thickness; and
H_s = equivalent height of solids (V_s/A).

Now, from Equation (1.10),

$$V_s = \frac{M_s}{G_s \rho_w}$$

$$\Rightarrow H_s = \frac{M_s}{G_s \rho_w A}$$

By way of illustration, let us use the test results of Example 12.4 together with the following information:

Original dimensions of test sample: 75 mm diameter, 20 mm thickness
Mass of sample after removing from consolidation apparatus at end of test and drying in oven = 135.6 g.

$$M_s = 135.6 \text{ g}; \quad A = \frac{\pi \times 75^2}{4} = 4418 \text{ mm}^2$$

$$\Rightarrow \quad H_s = \frac{135.6 \times 1000}{2.65 \times 1 \times 4418} = 11.58 \text{ mm}$$

Now, as shown above:

$$e = \frac{H - H_s}{H_s}$$

Hence, the void ratio to pressure relationship can be found.

Pressure (kPa)	Thickness, H (mm)	$e = \dfrac{H - H_s}{H_s}$
0	20.00	$\dfrac{20.00 - 11.58}{11.58} = 0.727$
50	19.65	0.697
100	19.52	0.685
200	19.35	0.671
400	19.15	0.653
800	18.95	0.636

Note: Such close agreement between the two methods for determining the e–p relationship could only happen in a theoretical example. In practice, discrepancies between the two methods would be found.

12.3.6 The virgin consolidation curve

Clay is generally formed by the process of sedimentation from a liquid in which the soil particles were gradually deposited and compressed as more material was placed above them. The e–p curve corresponding to this natural process of consolidation is known as the virgin consolidation curve (Fig. 12.10a).

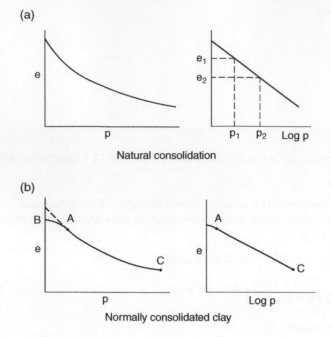

Fig. 12.10   e–p and e–log p curves for natural consolidation and for a normally consolidated clay. (a) Natural consolidation. (b) Normally consolidated clay.

This curve is approximately logarithmic. If the values are plotted to a semi-log scale (e to a natural scale, p to a logarithmic scale), the result is a straight line of equation:

$$e = e_0 - C_c \log_{10} \frac{p_0 + dp}{p_0}$$

Hence, in Fig. 12.10a, e_2 can be expressed in terms of e_1:

$$e_2 = e_1 - C_c \log_{10} \frac{p_2}{p_1} \qquad (12.15)$$

C_C is known as the *compression index* of the clay.

Compression curve for a normally consolidated clay

A normally consolidated clay is one that has never experienced a consolidation pressure greater than that corresponding to its present overburden. The compression curve of such a soil is shown in Fig. 12.10b.

The clay was originally compressed, by the weight of material above, along the virgin consolidation curve to some point A. Owing to the removal of pressure during sampling the soil has expanded to point B. Hence, from B to A, the soil is being recompressed whereas from A to C the virgin consolidation curve is followed.

The semi-log plot is shown in Fig. 12.10b. As before, on the straight-line part:

$$e_2 = e_1 - C_c \log_{10} \frac{p_2}{p_1}$$

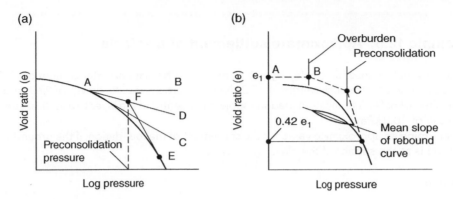

Fig. 12.11 Compression curves for an overconsolidated clay. (a) Graphical determination of preconsolidation pressure (Casagrande). (b) Determination of corrected compression curve (Schmertmann).

Compression curve for an overconsolidated clay

An overconsolidated clay is one which has been subjected to a preconsolidation pressure in excess of its existing overburden (Fig. 12.11a), the resulting compression being much less than for a normally consolidated clay. The semi-log plot is no longer a straight line and a compression index value for an overconsolidated clay is no longer a constant.

From the e–p curve, it is possible to determine an approximate value for the preconsolidation pressure with the use of a graphical method. First, estimate the point of greatest curvature, A, then draw a horizontal line through A (AB) and the tangent to the curve at A (AC). Bisect the angle BAC to give the line AD and locate the straight part of the compression curve (in Fig. 12.11a, the straight part commences at point E). Finally, project the straight part of the curve upwards to cut AD at F. The point F then gives the value of the preconsolidation pressure.

Evaluation of consolidation settlement from the compression index

$$\frac{dH}{H_1} = \frac{e_1 - e_2}{1 + e_1}$$

$$\Rightarrow dH = \frac{e_1 - e_2}{1 + e_1} H_1$$

$$e_1 - e_2 = C_c \log_{10} \frac{p_2}{p_1}$$

$$\Rightarrow \rho_c = dH = \frac{C_C}{1 + e_1} \log_{10} \frac{p_2}{p_1} H_1 \qquad (12.16)$$

Equation (12.16) is only relevant when a clay is being compressed for the first time and therefore cannot be used for an overconsolidated clay.

Determination of compression index C_C

Terzaghi and Peck (1948) showed that there is an approximate relationship between the liquid limit of a normally consolidated clay and its compression index. This relationship has been established experimentally as:

$$C_C \approx 0.009(w_L - 10\%)$$

Example 12.5: Approximate settlement of a soft clay

A soft, normally consolidated clay layer is 15 m thick with a natural water content of 45%. The clay has a saturated unit weight of 17.2 kN/m^3, a particle-specific gravity of 2.68, and a liquid limit of 65%. A foundation load will subject the centre of the layer to a vertical stress increase of 10.0 kPa.

Determine an approximate value for the settlement of the foundation, if the groundwater level is at the surface of the clay.

Solution:

Initial vertical effective stress at centre of layer

$$= (17.2 - 9.81)\frac{15}{2}$$

$$= 55.4 \text{ kPa}$$

Final effective vertical stress $= 55.4 + 10 = 65.4$ kPa

Initial void ratio, $e_1 = wG_s = 0.45 \times 2.68 = 1.21$

$$C_C = 0.009(65 - 10) = 0.009 \times 55 = 0.495$$

$$\rho_c = \frac{0.495}{2.21} \log_{10}\frac{65.4}{55.4} \times 15 = 0.24 \text{ m} = 240 \text{ mm}$$

This method can be used for a rough settlement analysis of a relatively unimportant small structure on a soft clay layer. For large structures, consolidation tests would be carried out.

12.4 Application of consolidation test results

The range of pressure generally considered in a settlement analysis is the increase from p_1 (the existing vertical effective overburden pressure) to p_2 (the vertical effective pressure that will operate once the foundation load has been applied and consolidation has taken place). Hence, in the previous discussion, e_1 represents the void ratio corresponding to the effective overburden pressure and e_2 represents the final void ratio after consolidation. In some textbooks and papers, the initial void ratio, e_1, is given the symbol e_0.

Obtaining a test sample entails removing all of the stresses which are applied to it. This reduction in effective stress causes the sample to either swell or develop negative pore water pressures within it. Owing to the restraining effect of the sampling tube, most soil samples tend to have a negative pore pressure.

In the consolidation test, the sample is submerged in water to prevent evaporation losses, with the result that the negative pore pressures will tend to draw in water and the sample consequently swells. To obviate this effect, the normal procedure is to start the test by applying the first load increment and then to add the water, but if the sample still tends to swell an increased load increment must be added and the test readings started again. The point e_1 is taken to be the position on the test e–p curve that corresponds to the effective overburden pressure at the depth from which the sample was taken. In the case of a uniform deposit, various values of e_1 can be obtained for selected points throughout the layer by reading the test values of void ratio corresponding to the relevant effective overburden pressures. Generally, the test e–p curve lies a little below the actual *in situ* e–p curve, the amount of departure depending upon the degree of disturbance in the test sample. Bearing in mind the inaccuracies involved in any analysis, this departure from the consolidation curve will generally be of small significance (unless the sample is severely disturbed), and most settlement analyses are based on the actual test results.

An alternative method, mainly applicable to overconsolidated clays, was proposed by Schmertmann (1953). He observed that e_1 must be equal to wGs, where w is the *in situ* water content, and that in a consolidation test on an ideal soil with no disturbance, the void ratio of the sample should remain constant at e_1 throughout the pressure range from zero to the effective overburden pressure value. Schmertmann found that the test e–p curve tends to cut the *in situ* virgin consolidation curve at a void ratio value somewhere between 37 and 42% of e_1 and concluded that a reasonable figure for this intersection is $e = 0.42e_1$.

To obtain the corrected curve, with disturbance effects removed, the test sample is either loaded through a pressure range that eventually reduces the void ratio of the sample to $0.42e_1$ or else the test is extended far enough for extrapolated values to be obtained, at least one cycle of expansion and recompression being carried out during the test. The approximate value of the preconsolidation pressure is obtained and the test results are put in the form of a semi-log plot of void ratio to log p (Fig. 12.11b). The value of e_1 is obtained from wGs. The water content, w is found from a separate test sample – usually cuttings obtained during the preparation of the consolidation test sample.

It is now possible to plot on the test curve (point A) and a horizontal line (AB) is drawn to cut the ordinate of the existing overburden pressure at point B. The line BC is next drawn parallel to the mean slope of the laboratory rebound curve to cut the preconsolidation pressure ordinate at point C, and the value of void ratio equal to $0.42e_1$ is obtained and established on the test curve (point D). Finally, points C and D are joined. The corrected curve therefore consists of the three straight lines: AB (parallel to the pressure axis with a constant void ratio value e_1), BC (representing the recompression of the soil up to the preconsolidation pressure), and CD (representing initial compression along the virgin consolidation line).

Apart from the elimination of disturbance effects, the method is useful because it permits the use of a formula similar to the compression index of a normally consolidated clay:

$$\rho_c = \frac{C}{1 + e_1} \log_{10} \frac{p_2}{p_1} H \tag{12.17}$$

where C is the slope of the corrected curve (generally recompression). If the pressure range extends into initial compression, the calculation must be carried out in two parts using the two different C values.

12.5 General consolidation

In the case of a foundation of finite dimensions, such as a pad foundation sitting on a deep deposit of clay, lateral strains will occur, and the consolidation is no longer one-dimensional. If two saturated clays of equal compressibility and thickness are subjected to the same size of foundation and loading, the resulting settlements may be quite different even though the consolidation tests on the clays would give identical results. This is because lateral strain effects in the field may induce unequal pore pressures, whereas in the consolidation test the induced pore pressure is always equal to the increment of applied stress.

From Equation (4.15), we know that for a saturated soil:

$$\Delta u = \Delta\sigma_3 + A(\Delta\sigma_1 - \Delta\sigma_3)$$

Let

p'_1 = initial effective major principal stress;
$\Delta\sigma_1$ = increment of total major principal stress due to the foundation loading; and
Δu = excess pore water pressure induced by the load.

The effective major principal stress on load application will be:

$$p'_1 + \Delta\sigma_1 - \Delta u$$

The effective major principal stress after consolidation will be:

$$p' + \Delta\sigma_1$$

Let

p' = initial effective minor principal stress
$\Delta\sigma_3$ = increment of total minor principal stress due to the foundation loading.

The horizontal effective stress on load application will be:

$$p'_3 + \Delta\sigma_3 - \Delta u$$

If the expression for Δu is examined, it will be seen that Δu is greater than $\Delta\sigma_3$. The horizontal effective stress therefore reduces when the load is applied and there will be a lateral expansion of the soil. Hence, in the early stages of consolidation, the clay will undergo a recompression in the horizontal direction for an effective stress increase of $\Delta u - \Delta\sigma_3$. The strain from this recompression will be small but as consolidation continues the effective stress increases beyond the original value of p'_3 and the strain effects will become larger until consolidation ceases.

12.6   Settlement analysis

The method of settlement analysis most commonly in use is that originally proposed by Skempton and Bjerrum (1957). In this procedure, the lateral expansion and compression effects are assumed to have no significant effect and are therefore ignored. This assumption has been proved satisfactory through trials of measured and predicted settlements.

Ignoring secondary consolidation, the total settlement of a foundation is given by the expression:

$$\rho = \rho_i + \rho_c \tag{12.18}$$

where

ρ_i = immediate settlement
ρ_c = consolidation settlement

In the consolidation test:

$$\rho_{oed} = m_v \Delta\sigma_1 h$$

where h = sample thickness.

Since there is no lateral strain in the consolidation, $\Delta\sigma_1 = \Delta u$. Hence:

$$\rho_{oed} = m_v \Delta u h \tag{12.19}$$

or

$$\rho_{oed} = \int_0^H m_v \Delta u \; dH \tag{12.20}$$

where H = thickness of consolidating layer.

In a saturated soil, $\Delta u = \Delta\sigma_3 + \Lambda(\Delta\sigma_1 - \Delta\sigma_3)$. This may be expressed as:

$$\Delta u = \Delta\sigma_1 \left[A + \frac{\Delta\sigma_3}{\Delta\sigma_1}(1-A) \right]$$

and substituting for Δu in Equation (12.19), we obtain a truer estimation of the consolidation settlement, ρ_c:

$$\rho_C = \int_0^H m_v \; \Delta\sigma_1 \left[A + \frac{\Delta\sigma_3}{\Delta\sigma_1}(1-A) \right] dH \tag{12.21}$$

Equation (12.20) can be expressed in terms of Equation (12.19) by introducing the correction factor μ:

$$\rho_c = \mu\rho = \rho_{oed}$$
$$= \mu \int_0^H m_v \; \Delta\sigma_1 \; dH$$

where

$$\mu = \frac{\int_0^H m_v \; \Delta\sigma_1 \left[A + \frac{\Delta\sigma_3}{\Delta\sigma_1}(1-A) \right] dH}{\int_0^H m_v \Delta\sigma_1 \; dH}$$

If m_v and A are assumed constant with depth, the equation for μ reduces to:

$$\mu = A + (1-A)\alpha \tag{12.22}$$

where

$$\alpha = \frac{\int_0^H \Delta\sigma_3 \; dH}{\int_0^H \Delta\sigma_1 \; dH}$$

Table 12.4 Values of α.

H/B	Circular footing	Strip footing
		α
0	1.00	1.00
0.25	0.67	0.74
0.50	0.50	0.53
1.0	0.38	0.37
2.0	0.30	0.26
4.0	0.28	0.20
10.0	0.26	0.14
∞	0.25	0

Poisson's ratio for a saturated soil is generally taken as 0.5 at the stage when the load is applied, so α is a geometrical parameter which can be determined. Various values for α that were obtained by Skempton and Bjerrum are given in Table 12.4.

The value of the pore pressure coefficient A can now be substituted in Equation (12.22) and a value for μ obtained, typical results being:

Soft sensitive clays > 1.0
Normally consolidated clays < 1.0
Average overconsolidated clays ≈ 0.5
Heavily overconsolidated clays ≈ 0.25

Example 12.6: Total settlement

A sample of the clay of Example 12.4 was subjected to a consolidated undrained triaxial test with the results shown in Fig. 12.12b. The sample was taken from a layer 20 m thick and has a saturated unit weight of 18.5 kN/m³.

It is proposed to construct a reinforced concrete foundation, length 30 m and width 10 m, on the top of the layer. The uniform bearing pressure will be 200 kPa. Determine the total settlement of the foundation under its centre, if the groundwater level occurs at a depth of 5 m below the top of the layer.

Solution:

The vertical pressure increment at the centre of the layer can be obtained by splitting the plan area into four rectangles (Fig. 12.12a) and using Fig. 3.14:

$$\Delta\sigma_1 = 108 \text{ kPa} \quad (I_o = 0.135)$$

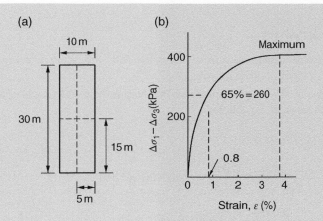

Fig. 12.12   Example 12.6. (a) Foundation plan. (b) Triaxial results.

To obtain the E value for the soil, $\Delta\sigma_3$ should now be evaluated so that the deviator stress $(\Delta\sigma_1 - \Delta\sigma_3)$ can be obtained.

Alternatively the approximate method can be used:

65% of maximum deviator stress = $0.65 \times 400 = 260$ kPa

Strain at this value = 0.8% (from Fig. 12.12b)

Hence:

$$E = \frac{260 \times 100}{0.8} = 32\ 500\ kPa = 32.5\ MPa$$

Immediate settlement

Using the rectangles of Fig. 12.12a and Fig. 12.2:

$$\frac{L}{B} = \frac{15}{5} = 3.0 \qquad \frac{H}{B} = \frac{20}{5} = 4.0$$

From Fig. 12.2, $I_p = 0.48$
But we are analysing four rectangles, hence:

$$I_p = 0.48 \times 4 = 1.92$$

$$\begin{aligned}
\rho_i &= \frac{pB(1 - v^2)I_p}{E} \\
&= \frac{200 \times 5 \times 0.75 \times 1.92 \times 0.8}{32\ 500} \quad (0.8 = \text{rigidity factor}) \\
&= 0.036\ m = 36\ mm
\end{aligned}$$

Consolidation settlement

Initial effective overburden pressure = $18.5 \times 10 - 9.81 \times 5 = 136$ kPa.

Since $\Delta\sigma_1 = 108$ kPa, the range of pressure involved is from 136 to 244 kPa.
Using the e–p curve of Fig. 12.9a:

$e_1 = 0.680;$ $e_2 = 0.666$

$$a = \frac{de}{dp} = \frac{0.680 - 0.666}{108} = \frac{0.014}{108} = 0.000130 \ m^2/kN$$

$$m_v = \frac{a}{1 + e_1} = \frac{0.000130}{1.680} = 7.7 \times 10^{-5} \ m^2/kN$$

$$\rho_c = m_v \ dp \ H = 7.7 \times 108 \times 20 \times 10^{-5} = 0.167 \ m = 167 \ mm$$

Total settlement = 36 + 167 = 203 mm

Some reduction could possibly be applied to the value of ρ_c if the value of μ was known.

Alternative method for determining ρ_c

In one-dimensional consolidation, the volumetric strain must be equal to the axial strain, i.e.

$$\frac{dH}{H} = \frac{\rho_c}{H} = \frac{de}{1 + e_1}$$

hence:

$$\rho_c = \frac{de}{1 + e_1} H$$

In the example:

$$\rho_c = \frac{0.680 - 0.666}{1.680} \times 20$$

$$= 0.008834 \times 20 = 0.167 \ m = 167 \ mm$$

Example 12.7: Total settlement using SPT results

The plan of a proposed raft foundation is shown in Fig. 12.13a. The uniform bearing pressure from the foundation will be 350 kPa and a site investigation has shown that the upper 7.62 m of the subsoil is a saturated coarse sand of unit weight 19.2 kN/m³. Groundwater level occurs at a depth of 3.05 m below the top of the sand.

(a)

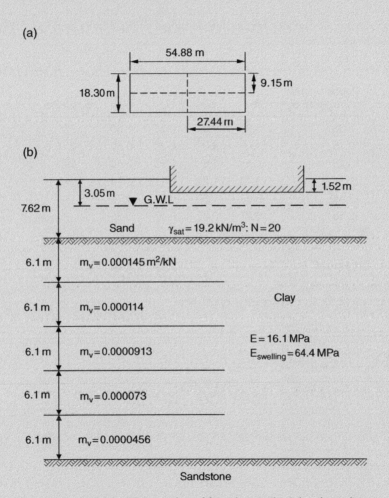

(b)

Fig. 12.13   Example 12.7. (a) Plan of foundation. (b) Subsoil conditions.

The result from a standard penetration test taken at a depth of 4.57 m below the top of the sand gave N = 20. Below the sand, there is a 30.5 m thick layer of clay (A = 0.75, E = 16.1 MPa, $E_{swelling}$ = 64.4 MPa). The clay rests on hard sandstone (Fig. 12.13b).
Determine the total settlement under the centre of the foundation.

Solution:

Using the De Beer and Martens' approach:
  Vertical pressure increments

Gross foundation pressure = 350 kPa
Relief due to excavation of sand = 1.52 × 19.2 = 29 kPa
Net foundation pressure increase, Δp = 350 − 29 = 321 kPa

The foundation is split into four rectangles, as shown in Fig. 12.13a, and Fig. 3.14 is then used to determine values for I_σ.

Depth below foundation (m)	B/z	L/z	I_σ	$4I_\sigma$	$\Delta\sigma_z$ (kPa)
3.05	3.0	9.0	0.25	1.00	321
9.15	1.0	3.0	0.20	0.80	257
15.25	0.6	1.8	0.15	0.60	193
21.35	0.43	1.29	0.11	0.44	141
27.45	0.33	1.00	0.09	0.36	116
33.55	0.27	0.82	0.07	0.28	90

Immediate settlement

Sand: SPT result, N = 20:
From Equation (12.5): C_r = 400 × 20 = 8000 kPa

$$p_0' = 4.57 \times 19.2 - 1.52 \times 9.81 = 73 \text{ kPa}$$

$$C_s = \frac{1.5 \times 8000}{73} = 165$$

From Equation (12.7): $\rho_i = \frac{6.1}{165} \ln\left(\frac{73 + 321}{73}\right) = 0.062 = 62$ mm

As the SPT was carried out on submerged soil, there is no need to increase this value to allow for groundwater effects.
Hence ρ_i in the sand = 62 mm
Clay: In Fig. 12.2, H_1 = 6.1 m and H_2 = 36.6 m

For H_2: $\frac{L}{B} = \frac{27.44}{9.15} = 3.0$; $\frac{H}{B} = \frac{36.6}{9.15} = 4.0$

Hence $I_p = 0.475$

For H_1: $\dfrac{L}{B} = 3.0$; $\quad \dfrac{H}{B} = \dfrac{6.1}{9.15} = 0.67$

Hence $I_p = 0.18$

Settlement under centre of foundation:

(*Note* as heave effects will be allowed for, use the gross contact pressure. If heave is not allowed for, net foundation pressure should be used).

$$\begin{aligned}
\rho_i &= \frac{pB(1 - v^2)4I_p}{E} \times \text{rigidity factor} \\
&= \frac{350 \times 9.15 \times 0.75 \times 4(0.475 - 0.18) \times 0.8}{16\ 100} \\
&= 0.141\ \text{m} = 141\ \text{mm}
\end{aligned}$$

Heave effects: relief of pressure due to sand excavation $= 1.52 \times 19.2 = 29\ \text{kPa}$

$$\Rightarrow \text{Heave} = \frac{29 \times 9.15 \times 0.75 \times 4(0.475 - 0.18) \times 0.8}{64\ 400} = 0.0029\ \text{m} = 3\ \text{mm}$$

Hence, ρ_i in the clay $= 141 - 3 = 138\ \text{mm}$

As can be seen from this example, the effects of heave are usually only significant when a great depth of material is excavated.

Consolidation settlement

The clay layer has been divided into five layers of thickness, H equal to 6.1 m.

m_v (m²/kN)	$\Delta\sigma_z$ (kPa)	$m_v\Delta\sigma_z H$ (m)
0.000145	257	0.227
0.000114	193	0.134
0.0000913	141	0.079
0.000073	116	0.052
0.0000456	90	0.025
		0.517 m = 517 mm

This value of settlement can be reduced by the factor

$$\mu = A + (1 - A)\alpha$$

An approximate value for α can be obtained from Table 12.4, with $\dfrac{H}{B} = \dfrac{36.6}{18.3} = 2.0$

Hence:

$\alpha = 0.26$

$\mu = 0.75 + 0.25 \times 0.26 = 0.82$

$\rho_c = 517 \times 0.82 = 424$ mm

Total settlement = 62 + 138 + 424 = 624 mm

12.7 Eurocode 7 serviceability limit state

12.7.1 First generation: EN 1997-1:2004

As mentioned in Chapter 6, the serviceability limit state (SLS) should be checked in addition to the ultimate limit state (ULS) during a geotechnical design. This is particularly the case where the SLS may be more likely to be exceeded than the ULS. This can be the case with the settlement of shallow foundations and indeed EN 1197-1:2004 (BSI, 2004) states that if the ratio of the undrained bearing capacity to the applied serviceability loading is less than 3 (i.e. the *undrained overdesign factor < 3*), calculations of settlement should be undertaken. If the undrained overdesign factor is less than 2, the settlement calculations should take account of non-linear stiffness effects in the ground.

The provisions for serviceability limit state design with respect to shallow foundations are given in EN 1997-1:2004 (BSI, 2004) Section 6 and these guide the designer to consider foundation displacement (i.e. settlement or heave) and rotation caused by the applied actions. Annex F proposes the use of the adjusted elasticity method (as described earlier in this chapter; Section 12.2.1) to determine the settlement, *s* for a foundation of width *b* resting on a homogeneous soil:

$$s = p \times b \times f/E_m \qquad (12.23)$$

where:

E_m is the design value of the modulus of elasticity;
f is a settlement coefficient; and
p is the bearing pressure, linearly distributed on the base of the foundation.

The settlement coefficient *f* is a function of the size and shape of the foundation, the variation of stiffness with depth, the thickness of the compressible formation, Poisson's ratio, the distribution of the bearing pressure and the point for which the settlement is calculated, and can be derived using any appropriate method such as Skempton's I_p values (Table 12.2). In performing settlement calculations, all partial factors on actions and material properties have value of unity (i.e. $\gamma_F = \gamma_M = 1.0$) as demonstrated (by their exclusion) in Example 12.8.

12.7.2 Second generation: EN 1997-1:202x and EN 1997-3:202x

In both EN 1997-1:202x and EN 1997-3:202x guidance is provided for checking the serviceability limit state on foundation settlement. This guidance is in line with that given in the first generation of the code and guides the designer to check for ground movement through settlement (uniform and differential), heave, tilting, and vibration. As with EN 1997-1:2004 (BSI, 2004), partial factors are not applied to material properties or actions in the verification of serviceability limit states. Instead, the calculated ground movement is compared with a design criterion specific to the structure, as illustrated in Example 12.8.

Example 12.8: Serviceability limit state

If the soil beneath the footing of Example 10.8 is a deep deposit of homogenous clay, check the total settlement against the allowable settlement of 25 mm.

Undisturbed samples of soil taken through the profile were tested in the laboratory and the following results were established from consolidation tests, under applied pressure ranges related to the anticipated stress increase in the profile due to the foundation:

Depth below ground surface (m)	m_v (m²/kN)
0.75	0.000043
2.25	0.000025
3.75	0.000012
5.25	0.000008

It was also determined in the laboratory that the clay deposit has a design modulus of elasticity, $E_m = 36$ MPa.

Solution:

$$\text{Net bearing pressure, } p = (G'_{foundation} + G_k + Q_k)/A$$
$$G'_{foundation} = \text{weight of footing} - \text{weight of overburden removed}$$
$$= (117 + 249.4 - 176.6) - 2 \times (19 - 9.81)$$
$$= 189.8 - 18.4$$
$$= 171.4 \text{ kN}$$
$$\Rightarrow p = (171.4 + 800 + 400)/3^2 = 152.4 \text{ kPa}$$

Immediate settlement:

$$s_0 = \frac{p(1-v^2)Bf}{E_m}$$

$$s_0 = \frac{152.4 \times (1-0.5^2) \times 3.0 \times 0.82}{36\ 000} \quad \text{(using Table 12.2)}$$

$$= 7.8 \text{ mm}$$

Consolidation settlement:

Consider the base plan of the footing as four rectangles (1.5 m × 1.5 m) and use Fig. 3.14 to establish increases in vertical stress at the centres of predefined layers beneath the footing.

Stress increments to depth of 2.0B (=6.0 m) should be established. This can be done by considering the profile as comprising four layers, each 1.5 m thick:

I_σ is determined from Fadum's chart (Fig. 3.14).

Layer	Depth to centre of layer, z (m)	m = n = b/z	I_σ	4 × I_σ	Stress increment at centre of layer, $\Delta\sigma_z$ (kPa)
1	0.75	2.0	0.24	0.96	= 0.96 × 152.4 = 146.3
2	2.25	0.67	0.12	0.48	73.2
3	3.75	0.4	0.06	0.24	36.6
4	5.25	0.29	0.04	0.16	24.4

Consolidation settlement, $s_1 = \Sigma m_v\, h\Delta\sigma_z$

$$= 1.5(146.3 \times 0.000043) + 1.5(73.2 \times 0.000025)$$
$$+ 1.5(36.6 \times 0.000012) + 1.5(24.4 \times 0.000008)$$
$$= 13.1 \text{ mm}$$

$\Rightarrow$ Total settlement = 7.8 + 13.1 = 20.9 mm

The anticipated total settlement is therefore less than the permitted (25 mm) and thus the serviceability limit state requirement is satisfied.

12.8   Stress paths in the oedometer

Fig. 12.14 shows the stress conditions that arise during and after the application of a pressure increment in the consolidation test. Initially, the sample has been consolidated under a previous load and the pore pressure is zero, and the Mohr circle is represented by (p, q) the point X, circle I. As soon as the vertical pressure increase, $\Delta\sigma_1$, is applied, the total stresses move from X to Y (circle 1). As the soil is saturated, $\Delta u = \Delta\sigma_1$ and therefore the effective stress circle is still represented by point X. As consolidation commences, the pore water pressure, Δu, begins to decrease and $\Delta\sigma'_1$ begins to increase. The consolidation is one-dimensional and therefore an increase in the major principal effective stress, $\Delta\sigma'_1$, will induce an increase in the minor principal effective stress $\Delta\sigma'_3 = K_0\Delta\sigma'_1$. Hence, the effective stress circles move steadily towards point Z (circles II, III, and IV), where Z represents full consolidation.

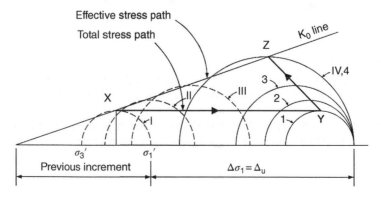

Fig. 12.14   Stress paths in the consolidation test.

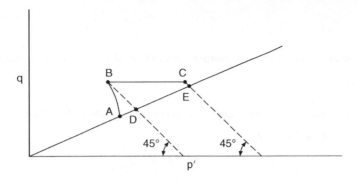

Fig. 12.15   Effective stress path for the general consolidation of a normally consolidated clay.

The total stress circles can be determined from a study of the effective stress circles. For example, the difference between $\Delta\sigma_1$ and $\Delta\sigma_1'$ for circle III represents the pore water pressure within the sample at that time: hence, $\Delta\sigma_3$ at this stage in the consolidation is $\Delta\sigma_3'$ for circle III plus the value of the pore water pressure. It can be seen therefore that Δu decreases with consolidation and the size of the Mohr circle for total stress increases until the point Z is reached (circles 2, 3, and 4). It is clear that circles 4 and IV are coincident.

12.9   Stress path for general consolidation

The effective stress plot of Fig. 12.15 represents a typical case of general consolidation. The soil is normally consolidated and point A represents the initial K_0 consolidation. AB is the effective stress path on the application of the foundation load and BC is the effective stress path during consolidation.

Skempton and Bjerrum's assumption that lateral strain effects during consolidation can be ignored presupposes that the strain due to the stress path BC is the same as that produced by the stress path DE. The fact that the method proposed by Skempton and Bjerrum gives reasonable results indicates that the effective stress path during the consolidation of soil in a typical foundation problem is indeed fairly close to the effective stress path DE of Fig. 12.15. There are occasions when this will not be so, however, and the stress path method of analysis can give a more reasonable prediction of settlement values. The calculation of settlement in a soft soil layer under an embankment by this procedure was discussed by Smith (1968a), and the method is also applicable to spoil heaps.

Example 12.9: Effective stress paths

The sample of clay tested in Example 5.2 was sampled from a layer of soft, normally consolidated clay 9.25 m thick and with an existing effective overburden pressure at its centre of 85 kPa.

It is proposed to construct a flexible foundation on the surface of the clay, and the increases in stresses at the centre of the clay, beneath the centre of the foundation, are estimated to be $\Delta\sigma_1 = 28.8$ kPa and $\Delta\sigma_3 = 19.2$ kPa.

By considering a point at the centre of the clay and below the centre of the foundation, project the effective stress paths for the immediate and consolidation settlements that the foundation will experience onto Fig. 5.4.

Assume that $K_0 = 1 - \sin\phi'$ and determine an approximate value for the immediate settlement of the foundation.

Solution:

Effective stresses at centre of layer before application of foundation load (initial):

$$\sigma'_{1i} = 85 \text{ kPa}$$

From Example 5.2, $K_0 = 0.446$. Clay is normally consolidated, therefore:

$$\sigma'_{3i} = 0.446 \times 85 = 37.9 \text{ kPa}$$

$$\Rightarrow p' = \frac{85 + 37.9}{2} = 61.5 \text{ kPa}; \quad q = \frac{85 - 37.9}{2} = 23.6 \text{ kPa}$$

The coordinates p' and q are superimposed onto Fig. 5.4 to give the point A, the initial state of stress in the soil. This is shown in Fig. 12.16.

Effective stress at centre of clay after application and consolidation of foundation load (final):

$$\sigma'_{1f} = \sigma'_{1i} + \Delta\sigma_1 = 85 + 28.8 = 113.8 \text{ kPa}$$

$$\sigma'_{3f} = \sigma'_{3i} + \Delta\sigma_3 = 37.9 + 19.2 = 57.1 \text{ kPa}$$

$$\Rightarrow p' = \frac{113.8 + 57.1}{2} = 85.5 \text{ kPa}; \quad q = \frac{113.8 - 57.1}{2} = 28.4 \text{ kPa}$$

The coordinates p' and q are plotted in Fig. 12.16 to give the point C, the state of the effective stresses in the soil after consolidation.

As illustrated in Fig. 12.15, the stress path from A to B represents the effect of the immediate settlement, whereas the stress path from B to C represents the effects of the consolidation settlement. The problem is to establish the point B, the point that represents the effective stress state in the soil immediately after the application of the foundation load.

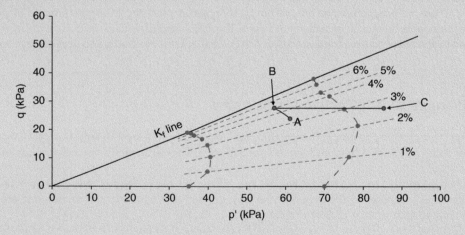

Fig. 12.16 Example 12.9.

During consolidation, at all times,

$$q = \frac{1}{2}(\sigma_1 - \sigma_3) = \frac{1}{2}(\sigma_1' - \sigma_3')$$

Hence, no matter how the individual values of effective stress vary during consolidation, the value of q remains constant. The line BC must be parallel to the horizontal axis. Hence, the point B must lie somewhere along the horizontal line through C.

From A to B, the effective undrained stress path is unknown but it is possible to sketch in an approximate, but sufficiently accurate, path by comparing the two test stress paths on either side of it. This has been done in Fig. 12.16. The immediate settlement can now be found.

On the diagram, the strain contours (lines joining equal strain values on the two test paths) are drawn. It is seen that the point A lies a little above the 3% strain contour (3.2%). Point B lies on the 5% strain contour. Hence, the strain suffered with immediate settlement = 5 − 3.2 = 1.8%.

$$\Rightarrow \rho_i = \frac{1.8}{100} \times 9.25 = 0.167 \text{ m}$$

Exercises

Exercise 12.1

Using the test results from Example 4.10, determine an approximate value for E of the soil and calculate the average settlement of a foundation, 5 m × 1 m, founded on a thick layer of the same soil with a uniform pressure of 600 kPa.

Answer 66 mm

Exercise 12.2

A rectangular, flexible foundation has dimensions L = 4 m and B = 2 m, and is loaded with a uniform pressure of 400 kPa. The foundation sits on a layer of deep saturated clay, E = 10 MPa. Determine the immediate settlement values at its centre and at the central points of its edges.

Answer At centre = 91 mm
At centre of long edge = 67 mm
At centre of short edge = 58 mm

Exercise 12.3

A rectangular foundation, 10 × 2 m, is to carry a total uniform pressure of 400 kPa and is to be founded at a depth of 1 m below the surface of a saturated sand of considerable thickness. The bulk unit weight of the sand is 18 kN/m^3 and standard penetration tests carried out below the water table indicate that the deposit has an average N value of 15.

If the water table occurs at the proposed foundation depth, determine a value for the settlement of the centre of the foundation. (Use De Beer and Martens' method.)

Answer   50 mm

Exercise 12.4

The following data exist for a proposed square footing, which is to be cast into a deep deposit of silty sand.

Width, B = 2.0 m
Founding depth, d = 1.2 m
Depth to ground water table, d_w = 1.5 m
Unit weight of sand, γ = 20 kN/m^3
Applied (gross) pressure = 250 kPa

SPT data for the 6 m of soil, below the founding depth (1.2 m), have been averaged over four layers thus:

Layer	Layer thickness (m)	Average SPT N-value
1	1.0	10
2	1.0	15
3	2.0	20
4	2.0	25

Using de Beer and Marten's method, establish the likely immediate settlement of the footing.

Answer   23.9 mm

Exercise 12.5

Elsewhere on the site as Exercise 12.4, the CPT test was used rather than the SPT. If the same foundation and soil properties apply, determine the likely settlement if the CPT data for the 6 m of soil below the bottom of the foundation were:

Layer	Layer thickness (m)	Average cone resistance, Cr (kPa)
1	1.0	3000
2	1.0	3500
3	2.0	9000
4	2.0	15 000

Answer   24.1 mm

Exercise 12.6

Recalculate Exercises 12.4 and 12.5 using Schmertmann's method, if the time assumed for creep is one year.

Hint: Schmertmann's method only investigates as deep as 2B so consider Layers 1, 2, and 3 only.

Answer   28; 39 mm

Exercise 12.7

A long strip footing (25 m × 2 m) is to cast at a depth of 1.0 m into a loose gravelly sand. The sand has saturated unit weight 19 kN/m^3 and the ground water table is coincident with the ground surface. The gross pressure applied through the foundation will be 120 kPa.

CPT test data for the profile, extending from the base of the foundation is:

Layer	Layer thickness (m)	Average cone resistance, Cr (MPa)
1	1.0	2.5
2	1.0	3.6
3	2.0	4.2
4	2.0	4.6
5	2.0	5.4

Use Schmertmann's method to determine the likely immediate settlement of the foundation.

Hint: Since the foundation is long (L/B > 10), plane strain conditions apply.

Answer   23 mm

Exercise 12.8

A saturated sample of a normally consolidated clay gave the following results when tested in a consolidation apparatus (each loading increment was applied for 24 hours).

Consolidation pressure (kPa)	Thickness of sample (mm)
0	17.32
53.65	16.84
107.3	16.48
214.6	16.18
429.2	15.85
0	16.51

After the sample had been allowed to expand for 24 hours, it was found to have a water content of 30.2%. The particle specific gravity of the soil was 2.65.

(i) Plot the void ratio to effective pressure.
(ii) Plot the void ratio to log effective pressure and hence determine a value for the compression index of the soil.
(iii) A 6.1 m layer of the soil is subjected to an existing effective over burden pressure at its centre of 107.3 kPa, and a foundation load will increase the pressure at the centre of the layer by 80.5 kPa.

Determine the probable total consolidation settlement of the layer (a) by the coefficient of volume compressibility and (b) by the compression index. Explain why the two methods give slightly different answers.

Answer   (a) Settlement by coefficient of volume compressibility = 88 mm
(b) Settlement by compression index = 98 mm

The compression index method is not as accurate as it represents the average of conditions throughout the entire pressure range, whereas the coefficient of volume compressibility applies to the actual pressure range considered.

Chapter 13

Rate of Foundation Settlement

Learning objectives:

By the end of this chapter, you will have been introduced to:
- Terzaghi's theory of consolidation;
- the degree of consolidation and drainage path length;
- the determination of c_v from the consolidation test using the square root of time fitting method;
- consolidation during construction and by drainage in two and three dimensions;
- numerical solutions of consolidation rates;
- using prefabricated vertical drains to accelerate consolidation.

The settlement of a foundation in cohesionless soil and the elastic settlement of a foundation in clay can be assumed to occur as soon as the load is applied. The consolidation settlement of a foundation on clay will only take place as water seeps from the soil at a rate depending upon the permeability of the clay.

13.1 Analogy of consolidation settlement

The model shown in Fig. 13.1 helps to give an understanding of the consolidation process. When load is applied to the piston, it will be carried initially by the water pressure created, but due to the weep hole there will be a slow bleeding of water from the cylinder accompanied by a progressive settlement of the piston until the spring is compressed to its corresponding load. In the analogy, the spring represents the compressible soil skeleton, and the water represents the water in the voids of the soil. The size of the weep hole is analogous to the permeability of the soil.

As consolidation progresses with time as water drains from the soil, we can establish the degree of consolidation that has occurred at any point in time:

$$\text{The degree of consolidation, U,} = \frac{\text{Consolidation attained at time, t}}{\text{Total consolidation}}$$

Smith's Elements of Soil Mechanics, 10th Edition. Ian Smith.
© 2021 John Wiley & Sons Ltd. Published 2021 by John Wiley & Sons Ltd.
Companion website: www.wiley.com/go/smith/soilmechanics10e

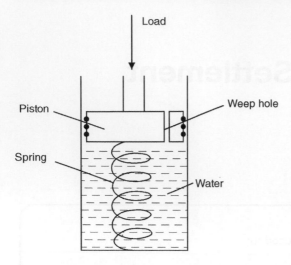

Fig. 13.1 Analogy of consolidation settlement.

13.2 Distribution of the initial excess pore pressure, u_i

If we consider points below the centre of a foundation, it is seen that there are three main forms of possible initial excess pore water pressure, u_i distribution:

- Uniform distribution can occur in thin layers (Fig. 13.2a) so that for all practical purposes, u_i is constant and equals $\Delta\sigma_1$ at the centre of the layer.
- A triangular distribution is found in a deep layer under a foundation, where u_i varies from a maximum value at the top to a negligible value (taken as zero) at some depth below the foundation (Fig. 13.2b(i)). The depth of this variation depends on the dimensions of the footing. Figure 13.2b(ii) shows how a triangular distribution may vary from $u_i = 0$ at the top of a layer to $u_i =$ a maximum value at the bottom. This condition can arise with a newly placed layer of soil, the applied pressure being the soil's weight.
- Trapezoidal distribution results from the quite common situation of a clay layer located at some depth below the foundation (Fig. 13.2c(i)). In the case of a new embankment carrying a superimposed load, a reversed form of trapezoidal distribution is possible (Fig. 13.2c(ii)).

13.3 Terzaghi's theory of consolidation

Terzaghi first presented this theory in 1925 and most practical work on the prediction of settlement rates is based on the differential equation he evolved. The main assumptions in the theory are as follows:

(i) Soil is saturated and homogeneous.
(ii) The coefficient of permeability is constant.
(iii) Darcy's law of saturated flow applies.
(iv) The resulting compression is one-dimensional.
(v) Water flows in one direction.
(vi) Volume changes are due solely to changes in void ratio, which are caused by corresponding changes in effective stress.

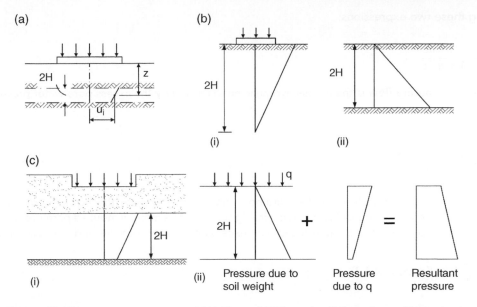

Fig. 13.2 Forms of initial excess pore pressure. (a) Uniform. (b) Triangular. (i) Deep layer. (ii) Newly placed soil. (c) Trapezoidal. (i) Clay layer at depth. (ii) New embankment with superimposed load.

The expression for flow in a saturated soil was established in Chapter 2. The rate of volume change in a cube of volume dx.dy.dz is

$$\left(k_x\frac{\partial^2 h}{\partial x^2} + k_y\frac{\partial^2 h}{\partial y^2} + k_z\frac{\partial^2 h}{\partial z^2}\right)dx.dy.dz \tag{13.1}$$

For one-dimensional flow (assumption (v)), there is no component of hydraulic gradient in the x and y directions, and putting $k_z = k$ the expression becomes

$$\text{Rate of change of volume} = k_z\frac{\partial^2 h}{\partial z^2}dx.dy.dz \tag{13.2}$$

The volume changes during consolidation are assumed to be caused by changes in void ratio.
    We saw earlier (Equation 1.8) that the porosity n is

$$n = \frac{V_v}{V} = \frac{e}{1+e}$$

hence,

$$V_v = dx.dy.dz\frac{e}{1+e}$$

Another expression for the rate of change of volume is therefore:

$$\frac{\partial}{\partial t}\left(dx.dy.dz\frac{e}{1+e}\right)$$

Equating these two expressions:

$$k\frac{\partial^2 h}{\partial z^2} = \frac{1}{1+e}\frac{\partial e}{\partial t} \tag{13.3}$$

The head, h_w, causing flow is the excess hydraulic head caused by the excess pore water pressure, u:

$$h_w = \frac{u}{\gamma_w}$$

$$\Rightarrow \frac{k}{\gamma_w}\frac{\partial^2 u}{\partial z^2} = \frac{1}{1+e}\frac{\partial e}{\partial t} \tag{13.4}$$

With one-dimensional consolidation, there are no lateral strain effects and the increment of applied pressure is therefore numerically equal (but of opposite sign) to the increment of induced pore pressure. Hence, an increment of applied pressure, dp, will cause an excess pore water pressure of du (= –dp). Now:

$$a = -\frac{de}{dp}\text{(see Section 12.3.3)}$$

hence,

$$a = \frac{de}{du}$$

or

$$de = a\,du$$

Substituting for de:

$$\frac{k}{\gamma_w}(1+e)\frac{\partial^2 u}{\partial z^2} = a\frac{\partial u}{\partial t}$$

$$\Rightarrow c_v\frac{\partial^2 u}{\partial z^2} = \frac{\partial u}{\partial t} \tag{13.5}$$

where c_v = the coefficient of consolidation and equals

$$\frac{k}{\gamma_w a}(1+e) = \frac{k}{\gamma_w m_v}$$

In the foregoing theory, z is measured from the top of the clay and complete drainage is assumed at both the upper and lower surfaces, the thickness of the layer being taken as 2H. The initial excess pore pressure, $u_i = -dp$.

The boundary conditions can be expressed mathematically:

when z = 0,   u = 0
when z = 2H,   u = 0
when t = 0,   u = u_i

A solution for

$$c_v \frac{\partial^2 u}{\partial z^2} = \frac{\partial u}{\partial t}$$

that satisfies these conditions can be obtained and gives the value of the excess pore pressure at depth z at time t, u_z:

$$u_z = \sum_{m=0}^{m=\infty} \frac{2u_i}{M}\left(\sin\frac{Mz}{H}\right)e^{-M^2 T} \tag{13.6}$$

where

u_i = the initial excess pore pressure, uniform over the whole depth

$M = \frac{1}{2}\pi(2m+1)$ where m is a positive integer varying from 0 to ∞

$T = \frac{c_v t}{H^2}$ is known as the time factor.

Owing to the drainage at the top and bottom of the layer, the value of u_i will immediately fall to zero at these points. With the mathematical solution, it is possible to determine u at time t for any point within the layer. If these values of pore pressures are plotted, a curve (known as an isochrone) can be drawn through the points (Fig. 13.3b). The maximum excess pore pressure is seen to be at the centre of the layer and, for any point, the applied pressure increment, $\Delta\sigma_1 = u + \Delta\sigma'_1$. After a considerable time, u will become equal to zero and $\Delta\sigma_1$ will equal $\Delta\sigma'_1$.

The plot of isochrones for different time intervals is shown in Fig. 13.3c. For a particular point the degree of consolidation, U_z, will be equal to:

$$\frac{u_i - u_z}{u_i}$$

The mathematical expression for U_z is

$$U_z = 1 - \sum_{m=0}^{m=\infty} \frac{2}{M}\left(\frac{\sin Mz}{H}\right)e^{-M^2 T} \tag{13.7}$$

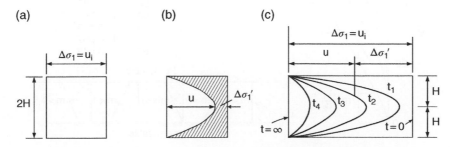

Fig. 13.3 Variation of excess pore pressure with depth and time.

13.4 Average degree of consolidation

Instead of thinking in terms of U_z, the degree of consolidation of a particular point at depth z, we think in terms of U, the average state of consolidation throughout the whole layer. The amount of consolidation still to be undergone at a certain time is represented by the area enclosed under the particular isochrone, and the total consolidation is represented by the area of the initial excess pore pressure distribution diagram (Fig. 13.3a). The consolidation achieved at this isochrone is, therefore, the total consolidation less the area under the curve (shown hatched in Fig. 13.3b):

Average degree of consolidation,

$$U = \frac{2Hu_i - \text{Area under isochrone}}{2Hu_i}$$

The mathematical expression for U is

$$U = 1 - \sum_{m=0}^{m=\infty} \frac{2}{M^2} e^{-M^2 T} \tag{13.8}$$

A theoretical relationship between U and T can therefore be established and is shown in Fig. 13.4, which also gives the relationship for u_i distributions that are not uniform, $m = u_1/u_2$.

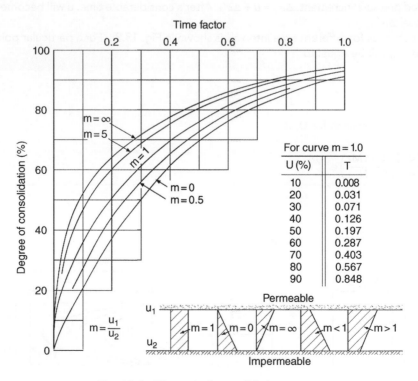

Fig. 13.4 Theoretical consolidation curves.

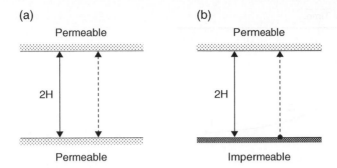

Fig. 13.5 Drainage path length. (a) Two-way drainage. (b) One-way drainage.

13.5   Drainage path length

A consolidating soil layer is usually enclosed, having at its top either the foundation or another layer of soil and beneath it either another soil layer or rock. If the materials above and below the layer are pervious, the water under pressure in the layer will travel either upwards or downwards (a concrete foundation is taken as being pervious compared with a clay layer). This case is known as *two-way drainage* and the drainage path length, i.e. the maximum length that a water particle can travel (Fig. 13.5a) is equal to:

$$\frac{\text{Thickness of layer}}{2} = H$$

If one of the materials is impermeable, water will only travel in one direction. This is known as the *one-way drainage* case, and the length of the drainage path = thickness of layer = 2H (Fig. 13.5b).

The curves of Fig. 13.4 refer to cases of one-way drainage (drainage path length = 2H). Owing to the approximations involved, the curve for m = 1 is often taken for the other cases with the assumption that u_i is the initial excess pore pressure at the centre of the layer. For cases of two-way drainage, the curve for m = 1 should be used and the drainage path length, for the determination of T, is taken as H.

13.6   Determination of the coefficient of consolidation, c_v, from the consolidation test

If, for a particular pressure increment applied during a consolidation test, the compression of the test sample is plotted against the square root of time, the result shown in Fig. 13.6 will be obtained.

The curve is seen to consist of three distinct parts: AB, BC, and CD.

- *AB (initial compression or frictional lag)*
  A small, but rapid, compression sometimes occurs at the commencement of the increment and is probably due to the compression of any air present or to the reorientation of some of the larger particles in the sample. In the majority of tests this effect is absent, and points A and B are coincident. Initial compression is not considered to be due to any loss of water from the soil and should be treated as a zero error for which a correction is made.
- *BC (primary compression)*
  All the compression in this part of the curve is taken as being due to the expulsion of water from the sample, although some secondary compression will also occur. When the pore pressure has been reduced to a negligible amount it is assumed that 100% consolidation has been attained.

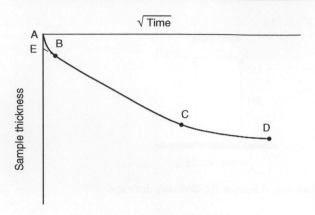

Fig. 13.6   Typical consolidation test results.

- CD (secondary compression)
  The amount by which this effect is evident is a function of the test conditions and therefore cannot be used to determine the *in situ* value.

The square root of time fitting method

It will be appreciated that the curve described above is an actual consolidation curve and would not be obtainable from one of the theoretical curves of Fig. 13.4, which can only be used to plot the primary compression range. To evaluate the coefficient of consolidation, it is necessary to establish the point C, representing 100% primary consolidation, but it is difficult from a study of the test curve to fix C with accuracy and a procedure in which the test curve is 'fitted' to the theoretical curve becomes necessary.

A method was described by Taylor (1948). If the theoretical curve U against $\sqrt{T}$ is plotted for the case of a uniform initial excess pore pressure distribution, the curve will be like that shown in Fig. 13.7a. Up to values of U equal to about 60%, the curve is a straight line of equation $U = 1.13\sqrt{T}$, but if this straight line is extended to cut the ordinate U = 90%, the abscissa of the curve is seen to be 1.15 times the abscissa of the straight line. This fact is used to fit the test and theoretical curves.

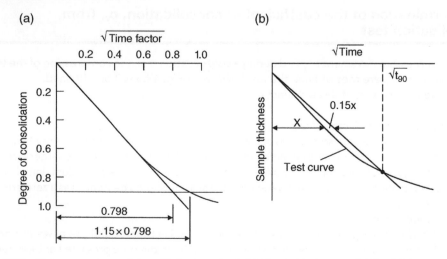

Fig. 13.7   The square root of time 'fitting' method. (a) Theoretical curve. (b) Establishment of t_{90}.

With the test curve, a corrected zero must first be established by projecting the straight line part of the primary compression back to cut the vertical axis at E (Fig. 13.6). A second line, starting through E, is now drawn such that all abscissas on it are 1.15 times the corresponding values on the laboratory curve, and the point at which this second line cuts the laboratory curve is taken to be the point representing 90% primary consolidation (Fig. 13.7b).

To establish c_v, T_{90} is first found from the theoretical curve that fits the drainage conditions (the curve $m = 1$). t_{90} is determined from the test curve:

$$T_{90} = \frac{c_v t_{90}}{H^2} \tag{13.9}$$

i.e.

$$c_v = \frac{T_{90} H^2}{t_{90}} \tag{13.10}$$

It is seen that the point of 90% consolidation, rather than the point for 100% consolidation, is used to establish c_v. This is simply a matter of suitability. A consolidation test sample is always drained on both surfaces and in the formula, H is taken as half the mean thickness of the sample for the pressure range considered. At first glance, it would seem that c_v could not possibly be constant, even for a fairly small pressure range, because as the effective stress is increased the void ratio decreases and both k and m_v decrease rapidly. However, the ratio of k/m_v remains sensibly constant over a large range of pressures, so it is justifiable to assume that c_v is in fact constant.

One drawback of the consolidation theory is the assumption that both Poisson's ratio and the elastic modulus of the soil remain constant, whereas in reality they both vary as consolidation proceeds. Owing to this continuous variation, there is a continuous change in the stress distribution within the soil which, in turn, causes a continuous change in the values of excess pore water pressures. Attempts to prepare theories which allow for this effect of change in applied stress with time have been made. However, because of the necessary approximations made, these theories are not in common use.

13.7   Determination of the permeability coefficient from the consolidation test

Having established c_v, k can be obtained from the formula $k = c_v m_v \gamma_w$. It should be noted that since the mean thickness of the sample is used to determine c_v, m_v should be taken as $a/(1 + \bar{e})$ where $\bar{e}$ is the mean void ratio over the appropriate pressure range.

13.8   Determination of the consolidation coefficient from the triaxial test

It is possible to determine the c_v value of a soil from the consolidation stage of the consolidated undrained triaxial test. In this case, the consolidation is three-dimensional and the value of c_v obtained is greater than would be the case if the soil was tested in the oedometer. Filter paper drains are usually placed around the sample in the triaxial to create radial drainage so that the time for consolidation is reduced. The effect of three-dimensional drainage is allowed for in the calculation of c_v, but the value obtained is not usually dependable as it is related to the relative permeabilities of the soil and the filter paper.

The time taken for consolidation to occur in the triaxial test generally gives a good indication of the necessary rate of strain for the undrained shear part of the test, but it is not advisable to use this time to determine c_v unless there are no filter drains.

The consolidation characteristics of a partially saturated soil are best obtained from the triaxial test, which can give the initial pore water pressures and the volume change under undrained conditions. Having applied the cell pressure and noted these readings, the pore pressures within the sample are allowed to dissipate while further pore pressure measurements are taken. The accuracy of the results obtained is much greater than with the consolidation test as the difficulty of fitting the theoretical and test curves when air is present is largely removed.

Example 13.1: Consolidation test

Results obtained from a consolidation test on a clay sample for a pressure increment of 100–200 kPa were

Thickness of sample (mm)	Time (min)
12.200	0
12.141	0.25
12.108	1
12.075	2.25
12.046	4
11.985	9
11.922	16
11.865	25
11.827	36
11.809	49
11.800	64

(i) Determine the coefficient of consolidation of the soil.
(ii) How long would a layer of this clay, 10 m thick and drained on its top surface only, take to reach 75% primary consolidation?
(iii) If the void ratios at the beginning and end of the increment were 0.94 and 0.82, respectively, determine the value of the coefficient of permeability.

Solution:

(i) The first step is to determine t_{90}. The thickness of the sample is plotted against the square root of time (Fig. 13.8) and, if necessary, the curve is corrected for zero error to establish the point E. The 1.15 line is next drawn from E and where it cuts the test curve (point F) gives $\sqrt{t_{90}} = 6.54$. Hence, $t_{90} = 42.7$ minutes.

To establish c_v, we use the relationship, $T = \dfrac{c_v t}{H^2}$.

From the curve for m = 1 (Fig. 13.4), $T_{90} = 0.85$.

Mean thickness of sample during increment (for corrected initial thickness = 12.168)

$$= \frac{12.168 + 11.800}{2} = 11.984 \, \text{mm}$$

$$\Rightarrow H = \frac{11.984}{2} = 5.992 \, \text{mm}$$

$$c_v = \frac{0.85 \times 5.992^2}{42.7} = 0.715 \, \text{mm}^2/\text{min}$$

(ii) For U = 75 %, T = 0.48 (from Fig. 13.4).
Drainage path length of layer = 10 m = 10 000 mm

$$\text{Time to reach 75\% consolidation} = \frac{0.48 \times 10\ 000^2}{0.715} \times \frac{1}{60} \times \frac{1}{24} \times \frac{1}{365}$$
$$= 128 \, \text{years}$$

(iii) $a = \dfrac{de}{dp} = \dfrac{0.94 - 0.82}{100} = 0.0012$

$$\bar{e} = \frac{0.84 + 0.92}{2} = 0.88$$

$$\text{Average } m_v = \frac{a}{1 + \bar{e}} = \frac{0.0012}{1.88} = 0.000638 \, \text{m}^2/\text{kN}$$

$$k - c_v \gamma_w m_v = \frac{0.715 \times 9.81 \times 0.000638}{1000} = 4.48 \times 10^{-6} \, \text{mm/min}$$

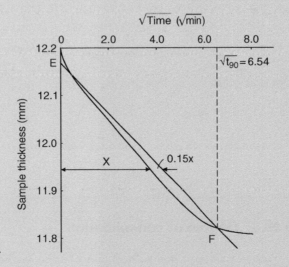

Fig. 13.8 Example 13.1.

13.9 The model law of consolidation

If two layers of the same clay with different drainage path lengths H_1 and H_2 are acted upon by the same pressure increase and reach the same degree of consolidation in times t_1 and t_2, respectively, then theoretically their coefficients of consolidation must be equal as must be their time factors, T_1 and T_2:

$$T_1 = \frac{c_{v1}t_1}{H_1^2}; \quad T_2 = \frac{c_{v2}t_2}{H_2^2} \tag{13.11}$$

Equating:

$$\frac{t_1}{H_1^2} = \frac{t_2}{H_2^2} \tag{13.12}$$

This gives a simple method for determining the degree of consolidation in a layer if the simplifying assumption is made that the compression recorded in the consolidation test is solely due to primary compression.

Example 13.2: Consolidation in the field

During a pressure increment, a consolidation test sample attained 25% primary consolidation in five minutes with a mean thickness of 18 mm. How long would it take a 20 m thick layer of the same soil to reach the same degree of consolidation if (i) the layer was drained on both surfaces and (ii) it was drained on the top surface only?

Solution:

In the consolidation test the sample is drained top and bottom

$$\Rightarrow H_1 = \frac{18}{2} = 9.0 \, \text{mm}$$

(i) With layer drained on both surfaces $H_2 = 10 \, \text{m} = 10\ 000 \, \text{mm}$.

$$t_2 = \frac{t_1}{H_1^2} H_2^2 = \frac{5 \times 10\ 000^2}{9^2} \times \frac{1}{60} \times \frac{1}{24} \times \frac{1}{365} = 11.7 \, \text{years}$$

(ii) With layer drained on top surface only $H_2 = 20 \, \text{m}$.

$$\Rightarrow t_2 = 4 \times 11.7 = 47 \, \text{years}$$

Example 13.3: Degree of consolidation

A 19.1 mm thick clay sample, drained top and bottom, reached 30% consolidation in 10 minutes. How long would it take the same sample to reach 50% consolidation?

Solution:

As U is known (30%) we can obtain T, either from Fig. 13.4 or by using the relationship that $U = 1.13\sqrt{T}$ (up to U = 60%).

$$T_{30} = \left(\frac{0.3}{1.13}\right)^2 = 0.07$$

$$T = \frac{c_v t}{H^2}, \text{ so } c_v = \frac{0.07 \times 9.55^2}{10} = 0.6384 \text{ mm}^2/\text{min}$$

$$T_{50} = \left(\frac{0.5}{1.13}\right)^2 = 0.197 \quad \text{(or obtain from Fig.13.4)}$$

$$t_{50} = \frac{T_{50}H^2}{c_v} = \frac{0.197 \times 9.55^2}{0.6384} = 28.1 \text{ minutes}$$

13.10 Consolidation during construction

A sufficiently accurate solution to the question of how much consolidation will occur during construction is generally achieved by assuming that the entire foundation load is applied halfway through the construction period. For large constructions, spread over some years, it is sometimes useful to know the amount of consolidation that will have taken place by the end of construction, the problem being that whilst consolidating, the clay is subjected to an increasing load.

Figure 13.9 illustrates the loading diagram during and after construction. While excavation is proceeding, swelling may occur (see Example 12.7). If the coefficient of swelling, c_{vs}, is known, it would be fairly straightforward to obtain a solution, first as the pore pressures increase (swelling) and then as they decrease (consolidation). But the assumption is usually made that once the construction weight equals the weight of soil excavated (time t_1 in Fig. 13.9) heave is eliminated and consolidation commences.

By plotting the load–time relationship (with the time in which the net foundation load is applied as t_2) the time t_1 can be found (Fig. 13.9). The settlement curve, assuming instantaneous application of the load at time t_1, is now plotted and a correction is made to the curve by assuming that the actual consolidation settlement at the end of time t_2 has the same value as the settlement on the instantaneous curve at time $t_2/2$. Point A, corresponding to $t_2/2$, is obtained on the instantaneous curve, and point B is established on the corrected curve by drawing a horizontal from A to meet the ordinate of time t_2 at point B. To establish other points on the corrected curve the procedure is to:

(i) select a time, t;
(ii) determine the settlement on the instantaneous curve for t/2 (point C);
(iii) draw a horizontal from C to meet the ordinate for t_2 at D, and
(iv) join OD.

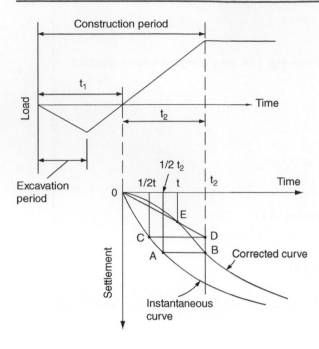

Fig. 13.9 Consolidation during construction.

The point where OD cuts the t ordinate, gives the point E on the corrected curve. This procedure is repeated with different values of t until sufficient points are established for the curve to be drawn. Points beyond B on the corrected curve are displaced horizontally by the distance AB from the corresponding points on the instantaneous curve.

Example 13.4: Settlement versus time relationship

If, in Example 12.7, the excavation is anticipated to take 6 months and the structure is anticipated to be completed after a further 18 months, determine the settlement to time relationship for the central point of the raft during the first 5 years. The clay has a c_v value of 1.86 m²/year and the sandstone may be considered permeable.

Solution:

The initial excess pore water pressure distribution will be roughly trapezoidal. The first step is to determine the values of excess pore pressures at the top and bottom of the clay layer (use Fig. 3.14).

	Depth below foundation (m)	$\frac{B}{z}$	$\frac{L}{z}$	I_σ	$4I_\sigma$	$\Delta\sigma_1$ (kPa)
Top of clay	6.1	1.5	4.5	0.23	0.92	295
Bottom of clay	36.6	0.25	0.75	0.06	0.24	77

Drainage path length, $H = \dfrac{36.6 - 6.1}{2} = 15.25$ m

$m = \dfrac{295}{77} = 3.83$; values of U are obtained from Fig. 13.4.

From Example 12.7, we have $\rho_c = 424$ mm. Thus, we can determine ρ_c for each year interval.

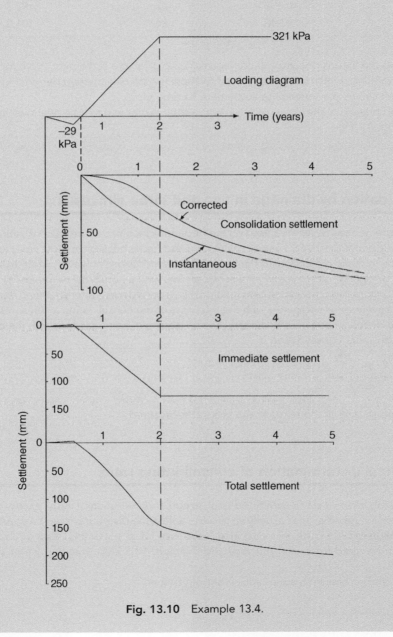

Fig. 13.10 Example 13.4.

t (years)	$T = \dfrac{c_v t}{H^2}$	U (%)	ρ_c (mm)
1	0.008	10	=0.1 × 424 = 42.4
2	0.016	15	63.6
3	0.024	18	76.3
4	0.032	22	93.3
5	0.040	24	101.8

Plotting the values of consolidation against time gives the settlement curve for instantaneous loading, which can be corrected to allow for the construction period (Fig. 13.10, which also shows the immediate settlement to time plot).

The summation of these two plots gives the total settlement to time relationship (lower part of Fig. 13.10).

13.11   Consolidation by drainage in two and three dimensions

The majority of settlement analyses are based on the frequently incorrect assumption that the flow of water in the soil is one-dimensional, partly for ease of calculation and partly because in most cases knowledge of soil compression values in three dimensions is limited. There are occasions however, when this assumption can lead to significant errors, as in the case of an anisotropic soil with a horizontal permeability so much greater than its vertical value that the time–settlement relationship is considerably altered. When dealing with a foundation which is relatively small compared with the thickness of the consolidating layer, some form of analysis allowing for lateral drainage becomes necessary. For an isotropic, homogeneous soil, the differential equation for three-dimensional consolidation is

$$c_v \left(\frac{\partial^2 u}{\partial x^2} + \frac{\partial^2 u}{\partial y^2} + \frac{\partial^2 u}{\partial z^2} \right) = \frac{\partial u}{\partial t} \tag{13.13}$$

For two dimensions, one of the terms in the bracket is dropped.

13.12   Numerical determination of consolidation rates

When a consolidating layer of clay is subjected to an irregular distribution of initial excess pore water pressure, the theoretical solutions are not usually applicable unless the distribution can be approximated to one of the cases considered. In such circumstances, the use of a numerical method is fairly common. A spreadsheet can be used for such a purpose and Example 13.5 illustrates the use of a spreadsheet to find the solution.

A brief revision of the relevant mathematics is set out below.

Maclaurin's series

Assuming that $f(x)$ can be expanded as a power series:

$$y = f(x) = a_0 + a_1x + a_2x^2 + a_3x^3 + \cdots + a_nx^n$$

$$\frac{dy}{dx} = f'(x) = a_1 + 2a_2x + 3a_3x^2 + 4a_4x^3 + \cdots + na_nx^{n-1}$$

$$\frac{d^2y}{dx^2} = f''(x) = 2a_2 + 2.3a_3x + 3.4a_4x^2 + \cdots + n(n-1)a_nx^{n-2}$$

$$\frac{d^3y}{dx^3} = f'''(x) = 2.3a_3 + 2.3.4a_4x + \cdots + n(n-1)(n-2)a_nx^{n-3}$$

If we put $x = 0$ in each of the above:

$$a_0 = f(0); \quad a_1 = f'(0); \quad a_2 = \frac{f''(0)}{2!}; \quad a_3 = \frac{f'''(0)}{3!}, \text{etc.}$$

Generally,

$$a_n = \frac{f^n(0)}{n!}$$

Substituting these values:

$$f(x) = f(0) + xf'(0) + \frac{x^2 f''(0)}{2!} + \frac{x^3 f'''(0)}{3!} + \cdots + \frac{x^n f^n(0)}{n!} \qquad (13.14)$$

This is the Maclaurin's series for the expansion of $f(x)$.

Taylor's series

If a curve $y = f(x)$ cuts the y-axis above the origin O at a point A (Fig. 13.11), we can interpret Maclaurin's expression as follows:

Let P be a point on the curve with abscissa x.
Let the values of $f(x)$, $f'(x)$, $f''(x)$, etc., at A be: y_0, y'_0, y''_0, etc.
Let the value of $f(x)$ at P be y_p. Then:

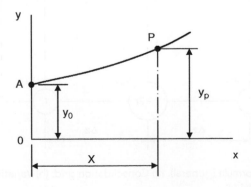

Fig. 13.11   Taylor's series.

$$f(x) \text{ at } P = y_p = y_0 + xy_0' + \frac{x^2 y_0''}{2!} + \frac{x^3 y_0'''}{3!} + \cdots \qquad (13.15)$$

This is a Taylor's series and gives us the value of the coordinate of P in terms of the ordinate gradient, etc., at A and the distance x between A and P.

Explicit finite difference equation

Gibson and Lumb (1953) first illustrated how the numerical solution of consolidation problems can be obtained using the *explicit finite difference equation*. The differential equation for one-dimensional consolidation may be established from Equation (13.13):

$$c_v \frac{\partial^2 u}{\partial z^2} = \frac{\partial u}{\partial t} \qquad (13.16)$$

Consider part of a grid drawn on to a consolidating layer (Fig. 13.12a). The variation of the excess pore pressure, u, with the depth, z, at a certain time, k, is shown in Fig. 13.12b, and the variation of u at the point O during a time increment from k to k + 1 is illustrated by Fig. 13.12c.

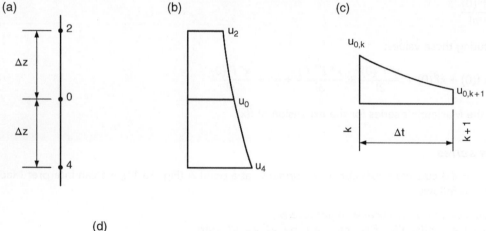

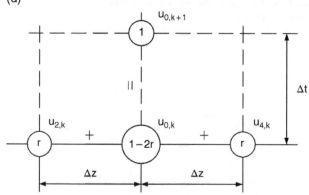

Fig. 13.12   Explicit recurrence formula (general). (a) Consolidation grid. (b) Variation of u with z. (c) Variation of u with t. (d) Schematic form of equation.

In Fig. 13.12b, from Taylor's theorem:

$$u_{2,k} = u_{0,k} - \Delta z u'_{0,k} + \frac{\Delta z^2}{2!} u''_{0,k} - \frac{\Delta z^3}{3!} u'''_{0,k} + \cdots$$

$$u_{4,k} = u_{0,k} + \Delta z u'_{0,k} + \frac{\Delta z^2}{2!} u''_{0,k} + \frac{\Delta z^3}{3!} u'''_{0,k} + \cdots$$

Adding and ignoring terms greater than second order:

$$u_{2,k} + u_{4,k} = 2u_{0,k} + \Delta z^2 u''_{0,k} \tag{13.17}$$

$$\Rightarrow \frac{\partial^2 u}{\partial z^2} = u''_{0,k} = \frac{u_{2,k} + u_{4,k} - 2u_{0,k}}{\Delta z^2} \tag{13.18}$$

In Fig. 13.12c:

$\frac{\partial u}{\partial t}$ is a function $u = f(t)$

By Taylor's theorem:

$$u_{0,k+1} = u_{0,k} + \Delta t u'_{0,k} + \frac{\Delta t^2}{2!} u''_{0,k} + \cdots$$

Ignoring second derivatives and above:

$$\frac{\partial u}{\partial t} = u'_{0,k} = \frac{u_{0,k+1} - u_{0,k}}{\Delta t} \tag{13.19}$$

$$\Rightarrow c_v \left(\frac{u_{2,k} + u_{4,k} - 2u_{0,k}}{\Delta z^2} \right) = \frac{u_{0,k+1} - u_{0,k}}{\Delta t} \tag{13.20}$$

$$\Rightarrow u_{0,k+1} = r(u_{2,k} + u_{4,k} - 2u_{0,k}) + u_{0,k} \tag{13.21}$$

where

$$r = \frac{c_v \Delta t}{\Delta z^2} \tag{13.22}$$

The schematic form of this expression is shown in Fig. 13.12d. Hence, if a series of points in a consolidating layer are established, Δz apart, it is possible by numerical iteration to work out the values of u at any time interval after consolidation has commenced if the initial excess values, u_i, are known.

Impermeable boundary conditions

Figure 13.13a illustrates this case in which conditions at the boundary are represented by:

$$\frac{\partial u}{\partial z} = 0$$

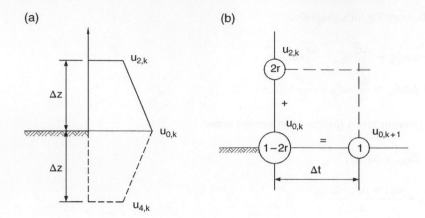

Fig. 13.13 Explicit recurrence formula: treatment for an impermeable boundary. (a) Variation of u with z. (b) Schematic form of equation.

Hence between the points 2_k and 4_k:

$$\frac{\partial u}{\partial z} = \frac{u_{2,k} - u_{4,k}}{2\Delta z} = 0 \tag{13.23}$$

i.e.

$$u_{2,k} = u_{4,k}$$

The equation therefore becomes

$$u_{0,k+1} = 2r(u_{2,k} - u_{0,k}) + u_{0,k} \tag{13.24}$$

and is shown in schematic form in Fig. 13.13b.

The boundary equation can also be used at the centre of a double-drained layer with a symmetrical initial pore pressure distribution. In this case, values for only half the layer need to be evaluated.

Errors associated with the explicit equation

Errors fall into two main groups: truncation errors (due to ignoring the higher derivatives) and rounding-off errors (due to working to only a certain number of decimal places). The size of the space increment, Δz, affects both these errors but in different ways: the smaller the Δz is, the less the truncation error that arises but the greater the round-off error tends to become.

The value of r is also important. For stability, r must not be greater than 0.5 and, for minimum truncation errors, it should be 1/6. The usual practice is to take r as near as possible to 0.5. This restriction means that the time interval must be short, and a considerable number of iterations become necessary to obtain the solution for a large time interval. With present software this is not a problem, but if necessary, use can be made of alternative numerical techniques (e.g. either the *implicit finite difference equation* or the *relaxation method*) details of which can be readily obtained from any numerical modelling textbook.

Example 13.5: Degree of consolidation by finite difference method

A 4 m thick clay layer is drained on its top surface and has a uniform initial excess pore pressure distribution. The consolidation coefficient of the clay is 0.1 m^2/month. Using a numerical method, determine the degree of consolidation that the layer will have undergone 24 months after the commencement of consolidation. Check your answer by the theoretical curves of Fig. 13.4.

Solution:

In a numerical solution, the grid must first be established: for this example, the layer has been split into four layers each of $\Delta z = 1.0$ m. (It is important to consider that since Simpson's rule will be applied to determine the degree of consolidation, the layer should be divided into an even number of strips.) The initial excess pore pressure values have been taken everywhere throughout the layer as equal to 100 units.

In 24 months:

$$r = \frac{c_v \Delta t}{\Delta z^2} = \frac{0.1 \times 24}{1.0} = 2.4$$

For the finite difference equation r must not be greater than 0.5, so use five time increments, i.e. $\Delta t = 24/5 = 4.8$ months and

$$r = \frac{0.1 \times 4.8}{1.0} = 0.48$$

Owing to the instantaneous dissipation at the drained surface, the excess pore pressure distribution at time = 0 can be taken as that shown in Fig. 13.14. (The values obtained during the iteration process are also given.) The finite difference formula is applied to each point of the grid, except at the drained surface:

$$u_{0,k+1} = r(u_{2,k} + u_{4,k} - 2u_{0,k}) + u_{0,k}$$

For example, with the first time increment, the point next to the drained surface has u = 0.48(0 + 100 − 2 × 100) + 100 = 52.0. Note that at the undrained surface the finite difference equation alters.

Degree of consolidation

Area of initial excess pore pressure distribution diagram = 4 × 100 = 400.
Area under final isochrone is obtained by Simpson's rule:

$$\frac{1.0}{3}(87.7 + 4(32.4 + 77.3) + 2 \times 62.8) = 217$$

Hence:

$$U = \frac{400 - 217}{400} = 45.7\%$$

Checking by the theoretical curve: Total time = 24 months, H = 4 m:

$$T = \frac{c_v t}{H^2} = \frac{0.1 \times 24}{16} = 0.15$$

From Fig. 13.4: U = 45%

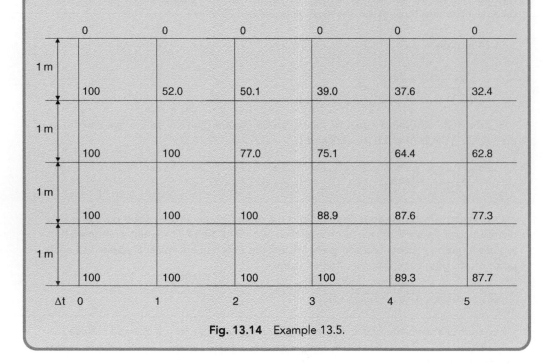

Fig. 13.14   Example 13.5.

13.13   Construction pore pressures in an earth dam

A knowledge of the induced pore pressures occurring during the construction of an earth dam or embankment is necessary, so that stability analyses can be carried out and a suitable construction rate determined. Such a problem is best solved by numerical methods. During the construction of an earth dam (or an embankment), the placing of material above that already in position increases the pore water pressure, whilst consolidation has the effect of decreasing it. The problem is one of a layer of soil that is consolidating as it is increasing in thickness. Gibson (1958) examined this condition. If it is assumed that the water in the soil will experience vertical drainage only, the finite difference equation becomes

$$u_{0,k+1} = r(u_{2,k} + u_{4,k} - 2u_{0,k}) + u_{0,k} + \overline{B}\gamma\Delta z \tag{13.25}$$

where

Δz = the grid spacing, and also the increment of dam thickness placed in time Δt

γ = unit weight of dam material

$\overline{B}$ = pore pressure coefficient

$$r = \frac{c_v \Delta t}{\Delta z^2}$$

In order that Δz is constant throughout the full height of the dam, all construction periods must be approximated to the same linear relationship and then transformed into a series of steps. The formula can only be applied to a layer that has some finite thickness, and as the layer does not exist initially, it is necessary to obtain a solution by some other method for the early stages of construction when the dam is insufficiently thick for the formula to be applicable. Smith (1968b) has shown how a relaxation procedure can be used for this initial stage.

Example 13.6: Excess pore pressure distribution by numerical method

At a stage in its construction, an earth embankment has attained a height of 9.12 m and has the excess pore water pressure distribution shown in Fig. 13.15a. A proposal has been made that further construction will be at the rate of 1.52 m thickness of material placed in one month, the unit weight of the placed material to be 19.2 kN/m^3 and its $\overline{B}$ value about 0.85. Determine approximate values for the excess pore pressures that will exist within the embankment three months after further construction is commenced. The c_v for the soil = 0.558 m^2/month.

Solution:

Check the r value with Δz taken as equal to 1.52 m.
   For Δz = 1.52 m, t = 1.0 month:

$$r = \frac{0.558 \times 1}{1.52^2} = 0.241$$

This value of r is satisfactory and has been used in the solution (if r had been greater than 0.5, then Δt and Δz would have had to be varied until r was less than 0.5).
   A 1.52 m deposit of the soil will induce an excess pressure, throughout the whole embankment, of $1.52 \times 19.2 \times \overline{B}$ = 24.8 kPa. This pressure value must be added to the value at each grid point for each time increment. The pore pressure increase is in fact applied gradually over a month, but for a numerical solution we must assume that it is applied in a series of steps, i.e. 24.8 kPa at t = 1 month, at t = 2 months, and at t = 3 months. From t = 0 to t = 1, no increment is assumed to be added and the initial pore pressures will have dissipated further before they are increased.
   The numerical iteration is shown in Fig. 13.15b.

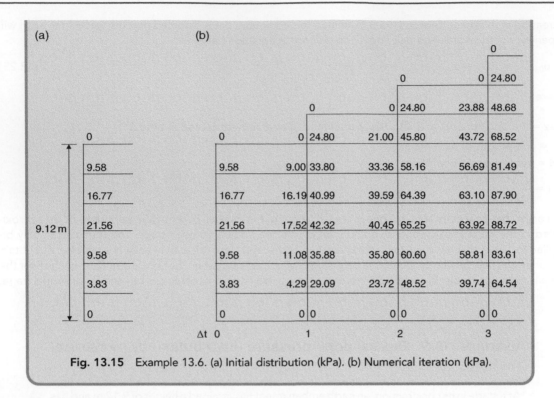

Fig. 13.15 Example 13.6. (a) Initial distribution (kPa). (b) Numerical iteration (kPa).

13.14 Numerical solutions for two- and three- dimensional consolidation

13.14.1 Two-dimensional consolidation

The differential equation for two-dimensional consolidation is derived from Equation (13.13)):

$$c_v\left(\frac{\partial^2 u}{\partial x^2} + \frac{\partial^2 u}{\partial y^2}\right) = \frac{\partial u}{\partial t} \tag{13.26}$$

Part of a consolidation grid is shown in Fig. 13.16a.

From the previous discussion of the finite difference equation (Equations (13.19)–(13.21)), we can write

$$\frac{\partial u}{\partial t} = \frac{u_{0,k+1} - u_{0,k}}{\Delta t} \tag{13.27}$$

$$\frac{\partial^2 u}{\partial y^2} = \frac{c_v}{h^2}\left(u_{2,k} + u_{4,k} - 2u_{0,k}\right) \tag{13.28}$$

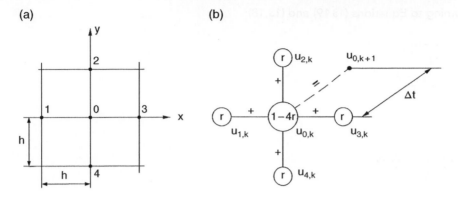

Fig. 13.16 Schematic form of the finite difference equation (two-dimensional). (a) Consolidation grid. (b) Schematic form of equation.

$$\frac{\partial^2 u}{\partial x^2} = \frac{c_v}{h^2}(u_{1,k} + u_{3,k} - 2u_{0,k})$$ (13.29)

Hence, the explicit finite difference equation is

$$u_{0,k+1} = r(u_{1,k} + u_{2,k} + u_{3,k} + u_{4,k}) + u_{0,k}(1 - 4r)$$ (13.30)

where

$$r = \frac{c_v \Delta t}{h^2}$$ (13.31)

The schematic form of this equation is illustrated in Fig. 13.16b.

Impermeable boundary condition

Impermeable boundaries are treated as for the one-dimensional case.

13.14.2 Three-dimensional consolidation

For instances of radial symmetry, the differential equation can be expressed in polar coordinates:

$$c_v\left(\frac{\partial^2 u}{\partial R^2} + \frac{1}{R}\frac{\partial u}{\partial r} + \frac{\partial^2 u}{\partial z^2}\right) = \frac{\partial u}{\partial t}$$ (13.32)

Then, returning to Equations (13.19) and (13.18):

$$\frac{\partial u}{\partial t} = \frac{u_{0,k+1} - u_{0,k}}{\Delta t}$$

$$\frac{\partial^2 u}{\partial z^2} = \frac{u_{2,k} + u_{4,k} - 2u_{0,k}}{\Delta z^2}$$

we have

$$\frac{\partial^2 u}{\partial R^2} = \frac{u_{1,k} + u_{3,k} - 2u_{0,k}}{\Delta R^2} \tag{13.33}$$

$$\frac{1}{R}\frac{\partial u}{\partial R} = \frac{1}{R}\left(\frac{u_{3,k} - u_{1,k}}{2\Delta R}\right) \tag{13.34}$$

If we put $\Delta z = \Delta R = h$, the finite difference equation becomes

$$u_{0,k+1} = r(u_{2,k} + u_{4,k}) + u_{0,k}(1 - 4r) + ru_{1,k}\left(1 - \frac{h}{2R}\right) + ru_{3,k}\left(1 + \frac{h}{2R}\right) \tag{13.35}$$

where

$$r = \frac{c_v \Delta t}{h^2} \tag{13.36}$$

At the origin, where $R = 0$,

$$\frac{1}{R}\frac{\partial u}{\partial R} \rightarrow \frac{\partial^2 u}{\partial R^2}$$

and the equation becomes

$$u_{0,k+1} = ru_{2,k} + 4ru_{3,k} + ru_{4,k} + u_{0,k}(1 - 6r) \tag{13.37}$$

Using the convention $R = mh$, the schematic form for the explicit equation is shown in Fig. 13.17a (for a point at the origin) and Fig. 13.17b (for other interior points).

For drainage in the vertical direction, the procedure is the same, but for radial drainage the expression for $u_{0,k+1}$ at a boundary point, where $\partial u/\partial R = 0$ is given by:

$$u_{0,k+1} = r(u_{2,k} + u_{4,k}) + 2ru_{1,k} + u_{0,k}(1 - 4r) \tag{13.38}$$

Value of r

In three-dimensional work, the explicit recurrence formula is stable if r is either equal to or less than 1/6. This is not so severe a restriction as it would at first appear, since with three-dimensional drainage the time required to reach a high degree of consolidation is much less than for one-dimensional drainage. For two-dimensional work, r should not exceed 0.5.

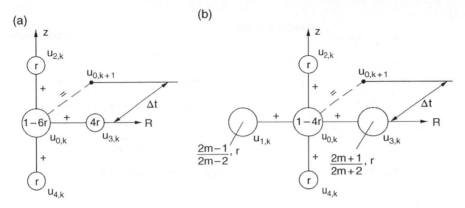

Fig. 13.17 Schematic form of the finite difference equation (three-dimensional). (a) At origin (m = 0). (b) Interior grid point.

13.14.3   Determination of initial excess pore water pressure values

For one-dimensional consolidation problems, u_i can at any point be taken as equal to the increment of the total major principal stress at that point. For two- and three-dimensional problems, u_i must be obtained from Equation (4.13):

$$u_i = B[\Delta\sigma_3 + A(\Delta\sigma_1 - \Delta\sigma_3)] \tag{13.39}$$

As the clay is assumed saturated, B = 1.0.

13.15   Wick, or prefabricated vertical, drains

Sometimes, the natural rate of consolidation of a particular soil is too slow, especially when the layer overlies an impermeable material and, in order that the structure may carry out its intended purpose, the rate of consolidation must be increased. An example of where this type of problem can occur is an embankment designed to carry road traffic. It is essential that most of the settlement has taken place before the pavement is constructed if excessive pavement cracking is to be avoided.

From the model law of consolidation, it is known that the rate of consolidation is proportional to the square of the drainage path length. It is plain therefore, that the consolidation rate is increased if horizontal, as well as vertical, drainage paths are made available to the pore water. This can be achieved by the installation of a system of *wick drains*, or *prefabricated vertical drains (PVD)*. Wick drains are also known as *band drains*.

A PVD is a prefabricated grooved plastic rectangular strip (approximately 75–100 mm wide and 5 mm thick) wrapped in a geotextile filter which is installed vertically into the ground using a special mandrel device attached to an excavator arm. The mandrel drives the PVD to the required depth and is then withdrawn back to the surface, leaving the drain in the ground. A series of these drains installed at regular spacing in both directions makes up the overall drainage system. These provide vertical drainage paths to accelerate the consolidation of the ground into which they are installed. The excess pore water pressures are dissipated by the groundwater flowing through the filter material and into the grooves of the plastic strip. Thereafter, vertical flow of the water to a permeable layer is rapid, enabling in turn, accelerated pore pressure dissipation in the soil.

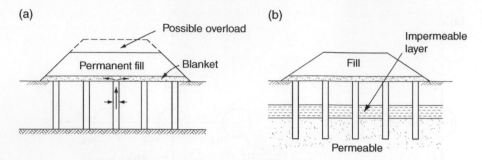

Fig. 13.18   Typical prefabricated vertical drains arrangements.

A typical spatial arrangement is shown in Fig. 13.18. There are occasions when the drains are made to puncture through an impermeable layer when there is a pervious layer beneath it. This creates two-way vertical drainage, as well as lateral, and can result in a considerable speeding up of construction.

The original wick drain design theory was based on simple sand drains (columns of sand poured into pre-drilled boreholes of diameter, d, but the theory holds good for modern PVDs too, if we assume that the wetted perimeter of the borehole ($=\pi d$ per unit length) is equal to the wetted perimeter of the, flatter in shape, PVD.

13.15.1   Design of a prefabricated vertical drain system

Key aspects

The key aspects to consider are listed below:

Spacing of drains: depends upon the type of soil in which they are placed. Spacings vary between 1.0 and approximately 3.0 m. PVDs are effective if the spacing, a, is less than the thickness of the consolidating layer, 2H.

Arrangement of grid: PVDs are laid out in either square (Fig. 13.19a) or triangular (Fig. 13.19b) patterns. Triangular arrangements are more efficient in terms of accelerating consolidation, and these grids form a series of equilateral triangles the sides of which are equal to the drain spacing.

Depth of PVDs: dictated by subsoil conditions. PVDs can be installed to depths of up to 45 m.

Drainage blanket: after drains are installed, a blanket of gravel and sand from 0.33 to 1.0 m thick, is spread over the entire area to provide lateral drainage at the base of the fill.

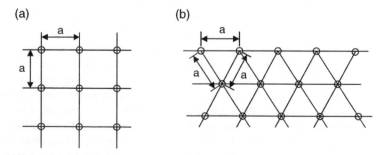

Fig. 13.19   Popular arrangements of prefabricated vertical drains. (a) Square. (b) Triangular.

Overfill or surcharge: often used in conjunction with PVDs. It consists of extra fill material placed above the permanent fill to accelerate consolidation. Once piezometer measurements indicate that consolidation has become slow, this surcharge is removed.

Strain effects: although there is lateral drainage, lateral strain effects are assumed to be negligible. Hence, the consolidation of a soil layer in which PVDs are placed is still obtained from Equation (12.14):

$$\rho_c = m_v \, dp \, 2H$$

Consolidation theory

The three-dimensional consolidation equation, from Equation (13.32), is

$$\frac{\partial u}{\partial t} = c_h \left(\frac{\partial^2 u}{\partial r^2} + \frac{1}{r} \frac{\partial u}{\partial r} \right) + c_v \frac{\partial^2 u}{\partial z^2} \qquad (13.40)$$

where

c_h = coefficient of consolidation for horizontal drainage (when it can be measured: otherwise use c_v).

The various coordinate directions of the equation are shown in Fig. 13.20. The equation can be solved by finite differences.

Equivalent radius, R

The effect of each vertical drain extends to the end of its *equivalent radius*, which differs for square and triangular arrangements (see Figure 13.19).

For a square system:

Area of square enclosed by grid = a^2
Area of equivalent circle of radius R = a^2

i.e. $\pi R^2 = a^2$ or,

$$R = 0.564a \qquad (13.41)$$

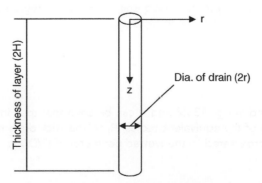

Fig. 13.20 Coordinate directions.

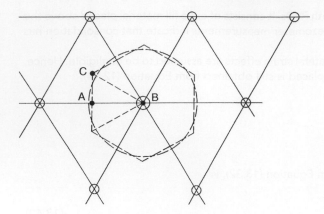

Fig. 13.21   Equivalent radius: triangular system.

For a triangular system:
A hexagon is formed by bisecting the various grid lines joining adjacent drains (Fig. 13.21). A typical hexagon is shown in the figure from which it is seen that the base of triangle ABC, i.e. the line AB, = a/2.
Now,

$$AC = AB \tan \angle ABC = \frac{a}{2} \tan 30° = \frac{a}{2\sqrt{3}}$$

hence:

$$\text{Area of triangle ABC} = \frac{1}{2} \times \frac{a}{2} \times \frac{a}{2\sqrt{3}} = \frac{a^2}{8\sqrt{3}}$$

So that:

$$\text{Total area of the hexagon} = 12 \times \frac{a^2}{8\sqrt{3}} = 0.865a^2$$

Radius of the equivalent circle, R = 0.525a   (13.42)

Determination of consolidation rates from curves

Barron (1948) produced curves which give the relationship between the degree of consolidation due to radial flow only, U_r, and the corresponding radial time factor, T_r:

$$T_r = \frac{c_h t}{4R^2}$$ (13.43)

where t = time considered.
These curves are reproduced in Fig. 13.22 and it can be seen that they involve the use of a factor n. This factor is simply the ratio of the equivalent radius, R, to the wick drain radius, r (for circular drains). For non-circular PVDs, r is approximated to the wetted perimeter of PVD/2π:

$$n = \frac{R}{r} \text{ and should lie between 5 and 100}$$

Design procedure

To determine U (for both radial and vertical drainage) for a particular time, t the procedure becomes:

(i) Determine U_z from the normal consolidation curves of U_z against T_z (Fig. 13.4):

$$T_z = \frac{c_v t}{H^2} \qquad (13.44)$$

where H = vertical drainage path
(ii) Determine T_r (Equation 13.43) based on proposed PVD dimensions and spacing.
(iii) Determine U_r from Barron's curves of U_r against T_r (Fig. 13.22).
(iv) Determine the resultant percentage consolidation, U, from:

$$U = 100 - \frac{1}{100}(100 - U_z)(100 - U_r) \qquad (13.45)$$

and check $U \geq U_r$. If U is not at least as large as Ur, then redesign the spacing and repeat from step (ii).

Smear effects

The curves in Fig. 13.22 are for idealised drains, perfectly installed, clean, and working correctly. During PVDs installation in varved clays (clays with sandwich-type layers of silt and sand within them), the finer and more impervious layers can be dragged down and smear over the more pervious layers to create a zone of reduced permeability around the perimeter of the drain. This smeared zone reduces the rate of consolidation, and *in situ* measurements to check on the estimated settlement rate are necessary on all but the smallest of jobs.

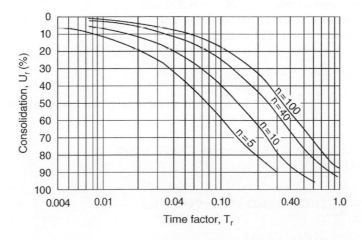

Fig. 13.22 Radial consolidation rates (after Barron, 1948).

Effectiveness of prefabricated vertical drains

PVDs are particularly suitable for soft clays but have little effect on soils with small primary, but large secondary effects, such as peat.

Example 13.7: Prefabricated vertical drain system

A soft clay layer, $m_v = 2.5 \times 10^{-4} m^2/kN$; $c_v = 0.187\ m^2/month$, is 9.2 m thick and overlies impervious shale. An embankment, to be constructed in six months, will subject the centre of the layer to a pressure increase of 100 kPa. It is scheduled that a highway pavement will be constructed on top of the embankment one year after the start of construction and the maximum allowable settlement after this is to be 25 mm.

Determine a suitable PVD system to achieve the requirements.

Solution:

$\rho_c = m_v\ dp\ 2H = 2.5 \times 10^{-4} \times 100 \times 9.2 = 0.230\ m = 230\ mm$

Therefore, the minimum settlement that must have occurred by the time the pavement is constructed = 230 − 25 = 205 mm. i.e.

$$U = \frac{205}{230} = 90\%$$

Assume that settlement commences at half the construction time for the embankment. Then, the time to reach U = 90% = $12 - \frac{6}{2} = 9$ months.

$$T_z = \frac{c_v t}{H^2} = \frac{0.187 \times 9}{9.2^2} = 0.020$$

From Fig. 13.4 $U_z = 16\%$

Try PVDs, of width 100 mm and thickness 5 mm, in a triangular pattern at spacing, a = 1.5 m.

Wetted perimeter of drain = 2(100 + 5) = 210 mm
Equivalent drain radius, r = 210/2π = 33.4 mm
Equivalent radius, R = 0.525a = 0.525 × 1.5 m = 0.788 m

n = R/r = 788/33.4 = 23.6

$$T_r = \frac{c_h t}{4R^2} = \frac{0.187 \times 9}{4 \times 0.788^2} = 0.68 \text{ (Note that no value for } c_h \text{ was given so } c_v \text{ must be used.)}$$

From Fig. 13.22, with $T_r = 0.68$ and n = 23.6 : $U_r \approx 93\%$
Check U > U_r:

$$U = 100 - \frac{1}{100}(100 - 16)(100 - 93) = 94\% (> U_r)$$

The arrangement is therefore satisfactory.

In practice, no PVD system could be designed as quickly as this. The object of the example is simply to illustrate the method. The question of installation costs must be considered, and several schemes would have to be closely examined before a final arrangement could be selected.

13.16   Preconsolidation by surface loading

Total settlements of foundations or embankments on soft clays can be reduced by preloading the area of the structure with mounds of soil or rubble. These impose a bearing pressure on the ground greater than that of the proposed structure or of the completed embankment.

As we have seen, clays will consolidate through time as they reduce in volume due to the expulsion of water in the voids under the applied loading. By preloading the area, that process can commence, and be accelerated, ahead of the foundations being cast or the embankment being constructed. The adoption of preloading can only really be effectively adopted if (i) there is adequate soil/rubble available on the site and (ii) there is adequate lead time ahead of the formation of the foundations to make adoption of the technique worthwhile. The technique is most applicable to soft and/or organic clays and peats.

Exercises

Exercise 13.1

A soil sample of thickness 19.1 mm in an oedometer test experienced 30% primary consolidation after 10 minutes. How long would it take the sample to reach 80% consolidation?

Answer   80 minutes

Exercise 13.2

A 5 m thick clay layer has an average c_v value of 5.0×10^{-2} mm^2/min. If the layer is subjected to a uniform initial excess pore pressure distribution, determine the time it will take to reach 90% consolidation (i) if drained on both surfaces and (ii) if drained on its upper surface only.

Answer   (i) 200 years, (ii) 800 years

Exercise 13.3

In a consolidation test, the following readings were obtained for a pressure increment:

Sample thickness (mm)	Time (min)
16.97	0
16.84	0.25
16.76	1
16.61	4
16.46	9
16.31	16
16.15	25
16.08	36
16.03	49
15.98	64
15.95	81

(i) Determine the coefficient of consolidation of the sample.
(ii) From the point for U = 90% on the test curve, establish the point for U = 50% and hence obtain the test value for t_{50}. Check your value from the formula:

$$t_{50} = \frac{T_{50}H^2}{c_v}$$

Answer $c_v = 1.39$ mm^2/ min , $t_{50} = 10$ minutes

Exercise 13.4

A sample in a consolidation test had a mean thickness of 18.1 mm during a pressure increment of 150–290 kPa. The sample achieved 50% consolidation in 12.5 minutes.
   If the initial and final void ratios for the increment were 1.03 and 0.97, respectively, determine values for the coefficient of volume compressibility and of consolidation of the sample.

Answer $m_v = 0.2113 \times 10^3$ mm^2/kN; $c_v = 1.29$ mm^2/min

Exercise 13.5

For the sample of Exercise 13.4, determine a value for the coefficient of permeability of the soil.

Answer $k = 2.71 \times 10^{-6}$ mm/min

Exercise 13.6

A 2 m thick layer of clay, drained at its upper surface only, is subjected to a triangular distribution of initial excess pore water pressure varying from 1000 kPa at the upper surface to 0.0 at the base. The c_v value of the clay is 1.8×10^{-3} m²/month. By dividing the layer into four equal slices, determine, numerically, the degree of consolidation after four years.

Note: If the total time is split into seven increments, r = 0.049.

Answer   U = 45%

Chapter 14

Stability of Slopes

Learning objectives:

By the end of this chapter, you will have been introduced to:
- planar failure mechanisms in granular slopes and the method of analysis used to assess stability;
- rotational failure mechanisms in cohesive slopes and the various methods of analysis available to assess stability;
- total and effective stress methods of analysis for the short- and long-term analyses of slopes;
- using Taylor's charts for the rapid approximation of the factor of safety of homogenous slopes;
- the use of both the first and second generations of Eurocode 7 to verify the limit state requirements of slope stability.

Landslides, or landslips, are movement of soil or rock down a natural or engineered slope. They are caused by changes in the loading conditions of the slope, changes in the groundwater conditions within the slope, or by a reduction in the shear strength of the soil. There are several forms of landslides including planar slides, rotational slips, soil flows and rockslides and topples. The effect of a landslide can range from the completely insignificant to the catastrophic depending on the location of the slope and the proximity of buildings or transport infrastructure to it.

When considering the stability of slopes, we use an analysis appropriate to the soil type (i.e. granular or cohesive) as the failure mechanism for both differ. Cohesive soils display *rotational failures* and granular soils display *planar failures*.

Methods of analysis

Note For each method of analysis described in the following sections, we consider the three-dimensional problem of a slope failure, in only two dimensions. This follows from the fact that long slopes act in a state of plane strain, as described in Section 3.1.6. Thus, if we consider the 'long' dimension to be into and out of the page, the end effects can be ignored, and the resulting analysis can be considered as a two-dimensional problem. The calculations and expressions involved in the two-dimensional analysis, therefore, are for a section of slope of unit thickness. This explains why the units of some calculated values are *per metre run of slope*.

Smith's Elements of Soil Mechanics, 10th Edition. Ian Smith.
© 2021 John Wiley & Sons Ltd. Published 2021 by John Wiley & Sons Ltd.
Companion website: www.wiley.com/go/smith/soilmechanics10e

14.1   Planar failures

As we saw in Chapter 4, soils such as gravel and sand are collectively referred to as granular soils and normally exhibit only a frictional component of strength. A potential slip surface in a slope of granular material will be planar and the analysis of the slope is relatively simple. However, most soils exhibit both cohesive and frictional strength and purely granular soils are infrequent. Nevertheless, a study of granular soils affords a useful introduction to the later treatment of soil slopes that exhibit both cohesive and frictional strength.

Figure 14.1 illustrates an embankment of dry granular material with an angle of shearing resistance, ϕ', and with its surface sloping at angle β to the horizontal.

Consider an element of the embankment of weight W:

For stability, the forces causing sliding, F_s must not exceed the force resisting the sliding, F_r which in the main comes from the shear strength of the soil along the failure plane. From Equation (4.1) we recall that the shear strength of a purely frictional material is:

$$\tau' = \sigma'_n \tan \phi' \tag{14.1}$$

and from Fig. 14.1b, we see that:

the force perpendicular to slope, $N = W \cos \beta$
the force parallel to slope, $T = W \sin \beta$

Thus,

$$F_r = W \cos \beta \tan \phi'$$

$$F_s = T = W \sin \beta$$

If we express the magnitude of the force resisting sliding, F_r to the magnitude of the forces causing the sliding, F_s we can derive an expression for the factor of safety, F. The higher the factor of safety, the less likely the slope is to fail.

$$F = \frac{\text{Restraining force, } F_r}{\text{Sliding forces, } F_s} \tag{14.2}$$

$$F = \frac{W \cos \beta \tan \phi'}{W \sin \beta}$$

$$F = \frac{\tan \phi'}{\tan \beta} \tag{14.3}$$

For limiting equilibrium, i.e. when the slope is on the point of failure and F = 1, $\tan \beta = \tan \phi'$, i.e. $\beta = \phi'$.

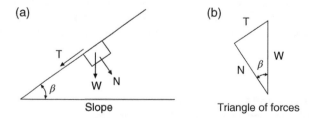

Fig. 14.1   Forces involved in a slope of granular material.

From this it is seen that (a) the weight of a material does not affect the stability of the slope, (b) the safe angle for the slope is the same whether the soil is dry or submerged, and (c) the embankment can be of any height.

Failure of a submerged sloping granular dam can occur however, if the water level of the retained water falls rapidly while the water level in the slope lags behind, as seepage forces are set up in this situation (refer to Section 2.14).

14.1.1 Seepage forces in a granular slope subjected to rapid drawdown

In Fig. 14.2a, the level of the river has dropped suddenly due to tidal effects. The permeability of the soil in the slope is such that the water in it cannot follow the water level changes as rapidly as the river, with the result that seepage occurs from the high water level in the slope to the lower water level of the river.

A flow net can be established for this condition and the excess hydraulic head for any point within the slope can be determined.

Assume that a potential failure plane, parallel to the surface of the slope, occurs at a depth of z and consider an element within the slope of weight W. Let the excess pore water pressure induced by seepage be u at the mid-point of the base of the element.

Normal reaction force, $N = W\cos\beta$

$\Rightarrow$ Normal stress, $\quad \sigma = \dfrac{W\cos\beta}{l} = \dfrac{W\cos^2\beta}{b} \left(\text{since } l = \dfrac{b}{\cos\beta}\right)$

Normal effective stress, $\quad \sigma' = \dfrac{W\cos^2\beta}{b} - u = \dfrac{\gamma z b \cos^2\beta}{b} - u = \gamma z \cos^2\beta - u$

where γ is the average unit weight of the whole slice. It is usually taken that the whole slice is saturated.

Tangential force, $T = W\sin\beta$

$\Rightarrow$ Tangential shear stress, $\quad \tau = \dfrac{W\sin\beta}{l} = \gamma z \sin\beta \cos\beta$

From Equations (14.1) and (14.2), ultimate shear strength of soil $= \sigma'_n \tan\phi' = \tau F$.

$\Rightarrow \gamma z \sin\beta \cos\beta = \left(\gamma z \cos^2\beta - u\right)\dfrac{\tan\phi'}{F}$

$\Rightarrow F = \left(\dfrac{\cos\beta}{\sin\beta} - \dfrac{u}{\gamma z \sin\beta \cos\beta}\right)\tan\phi'$

(a) (b)

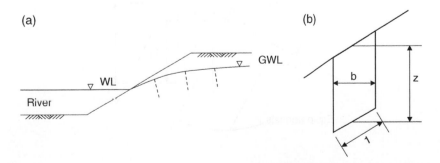

Fig. 14.2 Seepage due to rapid drawdown.

$$F = \left(1 - \frac{u}{\gamma z \cos^2 \beta}\right) \frac{\tan \phi'}{\tan \beta} \tag{14.4}$$

This expression may be written as:

$$F = \left(1 - \frac{r_u}{\cos^2 \beta}\right) \frac{\tan \phi'}{\tan \beta} \tag{14.5}$$

where r_u is known as the *pore pressure ratio* (see Section 14.2.4) and is defined as the ratio of the pore water pressure, to the overburden pressure,

$$r_u = \frac{u}{\gamma z} \tag{14.6}$$

14.1.2   Flow parallel to the surface and at the surface

The flow net for these special conditions is illustrated in Fig. 14.3.

If we consider the same element as before, the excess pore water head, at the centre of the base of the element, is represented by the height h_w in Fig. 14.3. In the figure, $AB = z \cos \beta$ and $h_w = AB \cos \beta$.

Hence, $h_w = z \cos^2 \beta$, so that the excess pore water pressure at the base of the element $= \gamma_w z \cos^2 \beta$.

$$r_u = \frac{u}{\gamma z} = \frac{\gamma_w z \cos^2 \beta}{\gamma z} = \frac{\gamma_w}{\gamma} \cos^2 \beta \tag{14.7}$$

The equation for F becomes:

$$F = \left(1 - \frac{\gamma_w}{\gamma}\right) \frac{\tan \phi'}{\tan \beta} = \left(\frac{\gamma - \gamma_w}{\gamma}\right) \frac{\tan \phi'}{\tan \beta} = \frac{\gamma' \tan \phi'}{\gamma_{sat} \tan \beta} \tag{14.8}$$

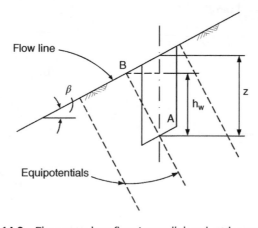

Fig. 14.3   Flow net when flow is parallel and at the surface.

Example 14.1: Safe angle of slope

A granular soil has a saturated unit weight of 18.0 kN/m^3 and an effective angle of shearing resistance of 30°. A slope is to be made of this material. If the factor of safety is to be 1.25, determine the safe angle of the slope (i) when the slope is dry or submerged and (ii) if seepage occurs at and parallel to the surface of the slope.

Solution:

(i) When dry or submerged:

$$F = \frac{\tan \phi'}{\tan \beta} \quad \Rightarrow \quad \tan \beta = \frac{0.5774}{1.25} = 0.462$$
$$\Rightarrow \quad \beta = 25°$$

(ii) When flow occurs at and parallel to the surface:

$$F = \frac{\gamma' \tan \phi'}{\gamma_{sat} \tan \beta} \Rightarrow \quad \tan \beta = \frac{(18 - 9.81) \times 0.5774}{1.25 \times 18} = 0.210$$
$$\Rightarrow \quad \beta = 12°$$

Seepage more than halves the safe angle of the slope in this particular example.

14.1.3   Planar translational slip

Quite often the surface of an existing slope is underlain by a plane of weakness lying parallel to it. This potential failure surface (often caused by downstream creep under alternating winter–summer conditions) generally lies at a depth below the surface that is small when compared with the length of the slope.

Owing to the comparative length of the slope and the depth to the failure surface we can generally assume that the end effects are negligible, and that the factor of safety of the slope against slip can be determined from the analysis of a wedge or slice of the material, as for the granular slope.

Consider Fig. 14.4. The GWL is parallel to the surface and at constant height, h above the failure plane. Thus, the pore water pressure along the slip plane is constant and conditions of steady seepage exist parallel to the ground surface.

Angle of slope = β, depth to failure surface = z, width of slice = b, and weight of slice, W = γzb /m run of slope. Then the excess hydrostatic head at the midpoint of the base of the slice,

$$h_w = h \cos^2\beta \tag{14.9}$$

$\Rightarrow$ pore pressure on slip plane, u = $\gamma_w hb \cos^2\beta$

Forces acting:

Weight of slice, W = $\gamma z(b \cos\beta)$
Normal force, N = $W\cos\beta$
Tangential (sliding) force, T = $W\sin\beta$
Resistance force, R = $\tau' b$

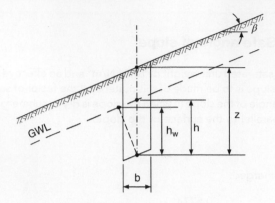

Fig. 14.4 Planar translational slip.

Under drained conditions, τ' *(per unit length of b)* $= \dfrac{c'b + N' \tan \phi'}{b} = \dfrac{c'b + (N-u) \tan \phi'}{b}$

Now,

$$(N - u) = W \cos \beta - u$$

$$= \gamma z b \cos^2 \beta - \gamma_w h \, b \cos^2 \beta$$

$$\Rightarrow \tau'(/\text{metre}) = \frac{c'b + (\gamma z b \cos^2 \beta - \gamma_w h b \cos^2 \beta) \tan \phi'}{b}$$

$$= c' + (\gamma z \cos^2 \beta - \gamma_w h \cos^2 \beta) \tan \phi'$$

$$\Rightarrow R = c'b + (\gamma z - \gamma_w h) b \cos^2 \beta \tan \phi' \quad (R \text{ is resistance along entire length of } b)$$

At limiting equilibrium, the total force resisting shear, R must equal T $(=W \sin \beta = \gamma z b \cos \beta \sin \beta)$.
Therefore, $\quad F = R/T$

$$\Rightarrow F = \frac{c'b + (\gamma z - \gamma_w h) b \cos^2 \beta \tan \phi'}{\gamma z b \cos \beta \sin \beta}$$

$$= \frac{c'}{\gamma z \cos \beta \sin \beta} + \frac{(\gamma z - \gamma_w h) \tan \phi'}{\gamma z \tan \beta} \tag{14.10}$$

Planar failures are most likely to occur when $c' = 0$. In this condition,

$$F = \frac{(\gamma z - \gamma_w h) \tan \phi'}{\gamma z \tan \beta} \tag{14.11}$$

Note: When the GWL is at the ground surface (i.e. $h = z$), we obtain the same expression as derived for a granular slope (Equation 14.8):

$$F = \left(1 - \frac{\gamma_w}{\gamma}\right) \frac{\tan \phi'}{\tan \beta}$$

14.2 Rotational failures

Failures in slopes made from soils that possess both cohesive and frictional strength components tend to be rotational, the actual slip surface approximating to the arc of a circle (Fig. 14.5).

Contemporary methods of investigating the stability of such slopes are based on (a) assuming a slip surface and a centre about which it rotates, (b) studying the equilibrium of the forces acting on this surface, and (c) repeating the process until the worst slip surface is found as illustrated in Fig. 14.6. The worst slip surface is the surface which yields the lowest factor of safety, F, where F is the ratio of the restoring moment to the disturbing moment, each moment considered about the centre of rotation.

The methods of assessing stability using this moment equilibrium approach are described in the next few sections. Further, if stability assessment is to be performed in accordance with Eurocode 7 (either the first or second generation), the procedures described in Section 14.5 are followed.

Regardless of the approach, the critical slip circle is found by considering several trial circles, each differing by the location of their centre, and identifying the one that returns the lowest measure of safety. This is achieved nowadays by using specific slope stability software that can perform repeated analyses in seconds and rapidly find the location of the centre of the critical slip circle.

In the case of soils with angles of shearing resistance that are not less than 3°, the critical slip circle is invariably through the toe – as it is for any soil (no matter what its ϕ' value) if the angle of slope exceeds 53° (Fig. 14.7a). An exception to this rule occurs when there is a layer of relatively stiff material at the base of the slope, which will cause the circle to be tangential to this layer (Fig. 14.7b).

For cohesive soils with a small angle of shearing resistance, the slip circle tends to be deeper and usually extends in front of the toe (Fig. 14.7c). This type of circle can of course be tangential to a layer of stiff material below the embankment which limits the depth to which it would have extended (Fig. 14.7d).

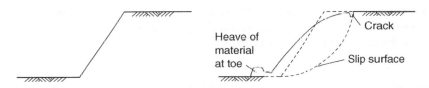

Fig. 14.5 Typical rotational slip in a cohesive soil.

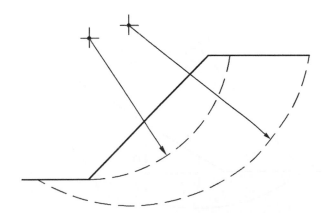

Fig. 14.6 Example of two possible slip surfaces.

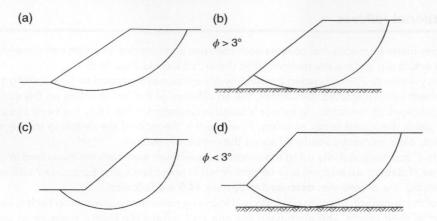

Fig. 14.7 Types of slip failures. (a) Toe failure. (b) Circle tangential to base. (c) Deep slip circle. (d) Circle tangential to deep stiff layer.

14.2.1 Total stress analysis

This analysis, also referred to as an *undrained analysis*, is intended to give the stability of an embankment or cutting immediately after its construction. At this stage it is assumed that the soil in the slope has had no time to drain (i.e. the soil is considered to be in an undrained state) and the strength parameter used in the analysis is the undrained cohesion.

Consider in Fig. 14.8 the sector of soil cut off by arc AB of radius R. Let W equal the weight of the sector and G the position of its centre of gravity. The shear strength of the soil is c_u. The mass of the soil tends to rotate clockwise but is resisted by the shear strength acting along the failure surface.

Taking moments about O, the point of rotation:

disturbing moment, $M_d = We$
restoring moment, $M_r = c_u LR = c_u R\theta R = c_u R^2 \theta$ (θ in radians)

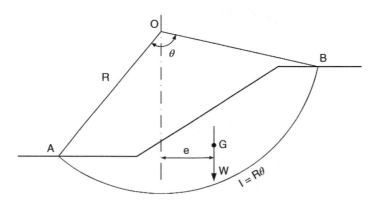

Fig. 14.8 Total stress analysis.

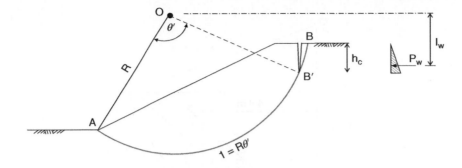

Fig. 14.9 Tension crack in a cohesive soil.

Therefore, the factor of safety of the slope is defined:

$$F = \frac{M_r}{M_d} = \frac{c_u R^2 \theta}{We}$$

Note: the term $R\theta$ is the length of the slip circle through the soil (length of an arc of a circle) where θ is measured in radians.

The position of G is not needed, and it is only necessary to ascertain where the line of action of W is. This can be obtained by dividing the sector into a set of vertical slices and taking moments of area of these slices about a convenient vertical axis.

14.2.2 Effect of tension cracks

With a slip in a cohesive soil, there will be a tension crack at the top of the slope (Fig. 14.9), along which no shear resistance can develop. In the undrained state the depth of the crack, h_c, is given by the following formula (from Equation 8.9):

$$h_c = \frac{2c_u}{\gamma} \tag{14.12}$$

The effect of the tension crack is to shorten the arc AB to AB'. If the crack is to be allowed for, the angle θ' must be used instead of θ in the formula for F, and the full weight W of the sector is still used in order to compensate for any water pressures that may be exerted if the crack fills with rainwater.

If the crack does fill with rainwater, a hydrostatic water pressure will occur over the length of the crack, resulting in an additional component to the disturbing moment, as illustrated in Example 14.3.

Example 14.2: Factor of safety against sliding

Figure 14.10 gives details of an embankment to be made of cohesive soil with $c_u = 20$ kPa. The unit weight of the soil is 19 kN/m³.

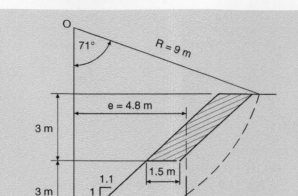

Fig. 14.10   Example 14.2.

(a) For the trial circle shown, determine the factor of safety against sliding soon after construction. The weight of the sliding sector is 329 kN acting at an eccentricity of 4.8 m from the centre of rotation.

(b) What would the factor of safety be if the shaded portion of the embankment were removed?

In both cases assume that no tension crack develops.

Solution:

(a) Disturbing moment = $329 \times 4.8 = 1579$ kNm

Restoring moment = $c_u R^2 \theta = 20 \times 9^2 \times \dfrac{71}{180} \times \pi = 2007$ kNm

$\Rightarrow F = \dfrac{2007}{1579} = 1.27$

(b) Area of portion removed = $1.5 \times 3 = 4.5\ \text{m}^2$
Weight of portion removed = $4.5 \times 19 = 85.5$ kN

Eccentricity from O = $3.3 + \dfrac{3.3 + 1.5}{2} = 5.7$ m

Relief of disturbing moment = $5.7 \times 85.5 = 487$ kNm

$\Rightarrow F = \dfrac{2007}{1579 - 487} = 1.84$

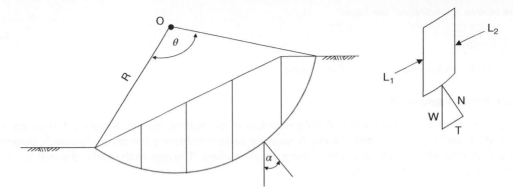

Fig. 14.11   The Swedish method of slices.

14.2.3   The Swedish, or Fellenius, method of slices analysis

A more accurate assessment of the factor of safety can be gained using this method. In this method the sliding section is divided into a suitable number of vertical slices, the stability of one such slice being considered in Fig. 14.11 (the lateral reactions on the two vertical sides of the wedge, L_1 and L_2, are assumed to be equal). By analysing the equilibrium of each slice and then adding up the totals for all slices, we can establish the factor of safety of the slope.

   The solution is solved graphically using a scale drawing, or more commonly, analytically using a software package. For a graphical approach, at the base of each slice set off its weight to some scale. The weight of the slice is equal to the mid-height of the slice times its width times the unit weight. Next draw the direction of its normal component, N, and by completing the triangle of forces determine its magnitude, together with the magnitude of the tangential component T. Repeat for all slices.

   N and T for each slice can also be determined using trigonometry (see Example 14.4).

Total stress analysis

The factor of safety is established by considering the moment equilibrium about the centre of rotation, O:

Disturbing moment, $M_d = R\Sigma T$
Restoring moment, $M_r = c_u R^2 \theta$

Hence

$$F = \frac{c_u R^2 \theta}{R \sum T} = \frac{c_u R \theta}{\sum T} \tag{14.13}$$

The same result can be obtained using a part analytical approach. We can see from the triangle of forces, that:

   $N = W\cos\alpha$
   $T = W\sin\alpha$

where α is the angle between the normal, N and the vertical.

From the above relationship, we have:

$$F = \frac{c_u R\theta}{\sum W \sin \alpha} \tag{14.14}$$

A tension crack can be allowed for in the analysis.

Effective stress analysis

As time passes after the construction of an embankment or cutting, the slope will no longer be in the undrained state and any slope stability analysis must be done considering the drained strength parameters c' and $\tan \phi'$ and the pore pressure, u acting along the slip surface. The approach used in the undrained analysis can be adapted to cover this case:

Taking moments about the centre of rotation, O:

Disturbing moment = $R\sum T$
Restoring moment = $R(c'R\theta + \sum N' \tan\phi')$

Hence,

$$F = \frac{c'R\theta + \sum N' \tan \phi'}{\sum T} = \frac{c'R\theta + \sum(N - ul) \tan \phi'}{\sum T} \tag{14.15}$$

where ul = pore pressure along base of slice.

The effect of a tension crack can again be allowed for, and in this case:

$$h_c = \frac{2c'}{\gamma} \tan \left(45 + \frac{\phi'}{2} \right) \tag{14.16}$$

Example 14.3: Swedish method of slices: undrained state

Figure 14.12 shows how a proposed 12.0 m high compacted clay embankment is to be constructed on the ground surface. An undrained slope stability analysis is to be carried out and the centre of the critical slip circle has been established as occurring at point O.

The completed embankment can be assumed to be homogenous and will possess constant density and undrained shear strength throughout its mass.

Determine the factor of safety of the slope in the short-term (undrained state) if;

(a) no tension crack develops;
(b) a tension crack develops and fills with rainwater.

Solution:

(a) No tension crack
Draw the slope to scale using CAD software or on graph paper (Fig. 14.13) and split the sliding section up into a suitable number of slices. Between four and seven slices is a typical amount.

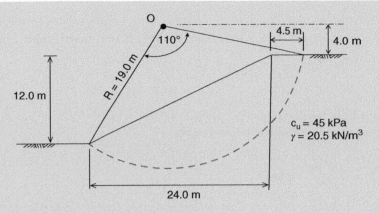

Fig. 14.12 The problem.

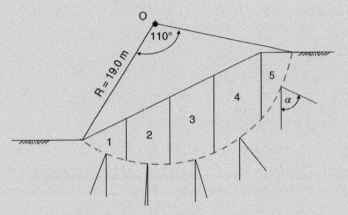

Fig. 14.13 Choice of slices.

Read the mid-height, z and width, b of each slice from the scale drawing and thus calculate the weight, W of each slice ($W = \gamma bz$). At the mid-point of the base of each slice, draw a vertical line and the normal line. Since the normal line must touch the circle at 90°, its direction is established by the knowledge that it passes through the centre of the circle and the mid-point of the base of the slice. The angle, α between these two lines is measured.

The calculations are then set out in a table.

Slice	b (m)	z (m)	α (°)	Area (m²)	Weight (kN)	Sliding force, T (kN/m run) (=Wsinα)
1	6.00	3.20	−21	19.2	394	−141
2	6.00	7.40	−3	44.4	910	−48
3	6.00	9.80	16	58.8	1205	332
4	6.00	10.00	36	60.0	1230	723
5	4.50	5.40	60	24.3	498	431
						Σ 1297

$$c_u R\theta = 45 \times 19 \times 110/180 \times \pi = 1641 \text{ kN/m run of slope}$$
$$F = \frac{c_u R\theta}{\sum T} = \frac{1641}{1297} = 1.27$$

(b) With rain-filled tension crack
   Determine the depth of the tension crack:

$$h_c = \frac{2c_u}{\gamma} = \frac{2 \times 45}{20.5} = 4.4 \text{ m}$$

The effect of the tension crack is to reduce the angle subtended at the centre of the circle from 110° to 95°. Further, the rainwater filled crack gives rise to a hydrostatic water pressure, P_w, as shown on Fig. 14.14, which contributes to the disturbing moment. The lever arm of P_w is the distance l_w shown on Fig. 14.14.

$$P_w = \tfrac{1}{2} \times 4.4^2 \times 9.81 = 95 \text{ kN}$$
$$M_w = P_w \times l_w = 95 \times 6.9 = 656 \text{ kNm}$$
$$c_u R\theta = 45 \times 19 \times 95/180 \times \pi = 1418 \text{ kN/m run of slope}$$
$$F = \frac{c_u R\theta}{\sum T + M_w} = \frac{1418}{1297 + 656} = 0.73$$

It is clear that the presence of the tension crack reduces the safety of the slope. If the tension crack does not fill with rainwater (i.e. $P_w = 0$), F increases to 1.09.

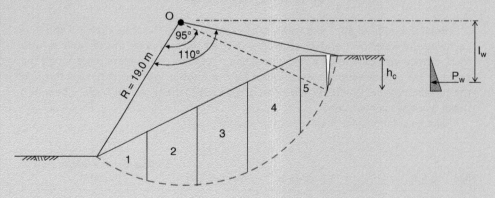

Fig. 14.14 Effect of tension crack.

Example 14.4: Swedish method of slices: drained state

The 6.1 m high homogeneous clay embankment shown in Fig. 14.15 is constructed on the surface of a hard layer of clay, which the slip circle touches tangentially. The slope angle, $\beta = 35°$ and the slip circle cuts through the upper surface, 3.2 m behind the crest of the slope. The long-term pore pressure is estimated to be constant throughout the embankment at 50 kPa.

Determine the long-term factor of safety of the trial slip circle shown.

Solution:

Draw the slope to scale on graph paper or using CAD and split the sliding section up into a suitable number of slices (Fig. 4.16).

For each slice, determine W and measure the angle α. Thereafter determine N ($=W\cos\alpha$) and T ($=W\sin\alpha$).

Slice	b (m)	z (m)	α (°)	Area (m²)	Weight (kN)	Normal, N (kN/m run)	Tangential, T (kN/m run)
1	3.1	1.2	−4	3.7	72	72	−5
2	3.0	2.9	12	8.7	168	164	35
3	2.8	3.7	30	10.4	200	173	100
4	3.2	2.3	50	7.4	142	91	109
						Σ 500	Σ 239

$$SN' \tan \phi' = \left(\sum N - u\right) \times \tan \phi' = [500 - (4 \times 50)] \times 0.364 = 109 \text{ kN}$$

$$c'R\theta = 20 \times 10.7 \times 76/180 \times \pi = 284 \text{ kN/m run of slope}$$

$$F = \frac{c'R\theta + \sum N' \tan \phi'}{\sum T} = \frac{109 + 284}{239} = 1.64$$

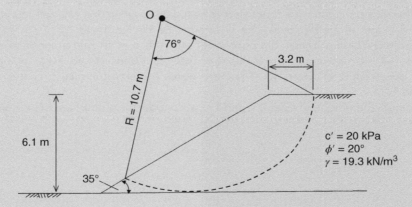

Fig. 14.15 Example 14.4.

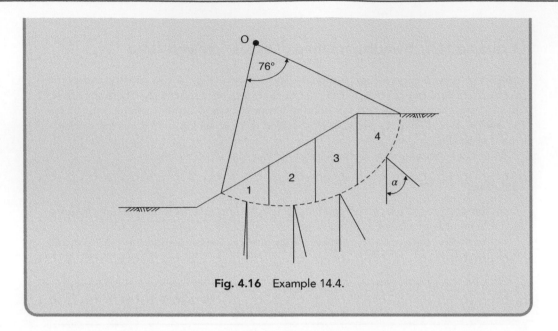

Fig. 4.16   Example 14.4.

Example 14.5: Swedish method of slices, two soils

Consider again the embankment of Example 14.4. Another possible failure mechanism for the embankment is for a deep-seated slip circle to develop cutting through the hard layer as shown in Fig. 14.17. The horizontal GWL is coincident with the original ground surface.

Determine the factor of safety for this trial slip circle.

Solution:

As two soils are present, we must apply the relevant shear strength to the base of the slices. The slice boundaries are thus selected based on ensuring only one soil acts wholly along the length of the slice base. The slices are shown in Fig. 14.18.

Since the GWL is horizontal, the pore water pressure, u at the mid-point of the base of slices 1–4 is simply equal to the depth of that point beneath the GWL, multiplied by the unit weight of water, γ_w. The pore water pressure for slice 5 comes from Example 14.4.

In addition, the weights of slices 2, 3 and 4 must consider the proportions of both soils in the slice.

The measurements taken from Fig. 14.18 enable N' and T to be established for each slice:

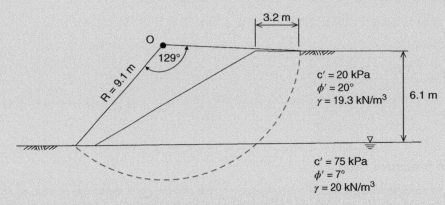

Fig. 14.17 Example 14.5.

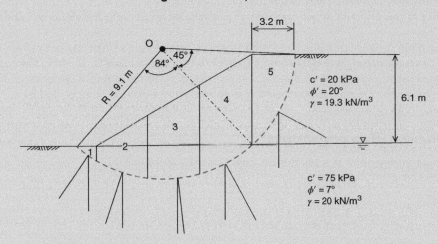

Fig. 14.18 Choice of slices – ensure base of each slice is only in one soil.

Slice	z (m)	b (m)	Area (m²)	W (kN)	α (°)	u (kPa)	N = Wcos α (kN/m run)	N' (=N − u) (kN/m run)	T (=Wsin α) (kN/m run)
1	0.8	1.3	1.0	20	−35	7.8	16.4	8.6	−12
2 Upper	1.1	3.6	4.0	213	−18	18.6	202.8	184.2	−66
2 Lower	1.9		6.8						
3 Upper	3.1	3.6	11.2	381	6	22.6	378.9	356.3	40
3 Lower	2.3		8.3						
4 Upper	5.3	3.6	19.1	455	30	11.8	393.7	382.0	227
4 Lower	1.2		4.3						
5	4.6	3.2	14.7	284	57	50.0	154.7	104.7	238
								Σ	428

$\Sigma N'$ upper layer = 104.7 kN/m run

$\Sigma N'$ lower layer = 8.6 + 184.2 + 356.3 + 382.0 = 931.0 kN/m run

$\Sigma N' \tan \phi' = (104.7 \times \tan 20°) + (931.0 \times \tan 7°) = 152.4$ kN/m run

$c'R\theta = (20 \times 9.1 \times 45° \times \pi/180) + (75 \times 9.1 \times 84° \times \pi/180) = 1144$ kN/m run

$$F = \frac{152.4 + 1144}{428} = 3.03$$

14.2.4   Pore pressure ratio, r_u

As mentioned in the previous section, if the long-term factor of safety of a slope is required, an analysis must be carried out in terms of effective stress. Such an analysis can be used in fact for any intermediate value of pore pressure between undrained and drained.

Before looking at the effective stress methods of analysis, let us consider the determination of the pore pressure ratio, r_u.

There are two main types of problem in considering pore pressures in a slope: those in which the value of the pore water pressure depends upon the magnitude of the applied stresses (e.g. during the rapid construction of an embankment), and those where the value of the pore water pressure depends upon either the groundwater level within the embankment or the seepage pattern of water impounded by it.

Rapid construction of an embankment

The pore pressure at any point in a soil mass is given by the expression:

$$u = u_0 + \Delta u \tag{14.17}$$

where

$u_0 =$ initial value of pore pressure before any stress change, and
$\Delta u =$ change in pore pressure due to change in stress.

From Equation (4.13):

$$\Delta u = B[\Delta \sigma_3 + A(\Delta \sigma_1 - \Delta \sigma_3)] \tag{14.18}$$

Skempton (1954) showed that the ratio of the pore pressure change to the change in the total major principal stress gives another pore pressure coefficient, $\overline{B}$:

$$\overline{B} = \frac{\Delta u}{\Delta \sigma_1} = B\left[\frac{\Delta \sigma_3}{\Delta \sigma_1} + A\left(1 - \frac{\Delta \sigma_3}{\Delta \sigma_1}\right)\right] \tag{14.19}$$

The coefficient $\overline{B}$ can be used to determine the magnitude of pore pressures set up at any point in an embankment if it is assumed that no drainage occurs during construction (a fairly reasonable thesis if the construction rate is rapid). Now, from Equation (14.6),

$$r_u = \frac{u}{\gamma z} = \left(\frac{u_0}{\gamma z} + \frac{\overline{B}\Delta\sigma_1}{\gamma z}\right) \qquad (14.20)$$

A reasonable assumption to make for the value of the major principal stress is that it equals the weight of the material above the point considered.

Hence,

$$\Delta\sigma_1 = \gamma z$$

and

$$r_u = \frac{u_0}{\gamma z} + \overline{B} \qquad (14.21)$$

For soils placed at or below optimum water content (see Chapter 15), u_0 is small and can even be negative. Its effect is of little consequence and may be ignored so that the analysis for stability at the end of construction is often determined from the relationship $r_u = \overline{B}$.

The pore pressure coefficient $\overline{B}$ is determined from a special stress path test known as a *dissipation test*. Briefly, a sample of the soil is placed in a triaxial cell and subjected to increases in the principal stresses $\Delta\sigma_1$ and $\Delta\sigma_1$ of magnitudes approximating to those expected in the field. The resulting pore pressure is measured and B obtained.

Steady seepage

It is easy to determine r_u from a study of the flow net (Fig. 14.19). The procedure is to trace the equipotential through the point considered up to the top of the flow net, so that the height to which water would rise in a standpipe inserted at the point is h_w.

Since $u = \gamma_w h_w$,

$$r_u = \frac{h_w \gamma_w}{\gamma z} \qquad (14.22)$$

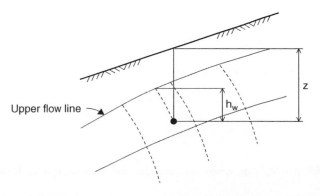

Fig. 14.19   Determination of excess head at a point on a flow net.

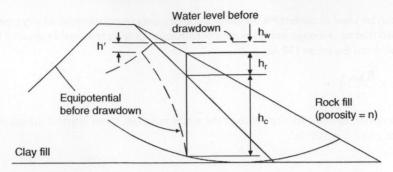

Fig. 14.20 Upstream dam face subjected to sudden drawdown.

Rapid drawdown

In the case of lagoons, a sudden drawdown in the level of the slurry is unlikely, but the problem is important in the case of a normal earth dam. Bishop (1954) considered the case of the upstream face of a dam subjected to this effect, the slope having a rock fill protection as shown in Fig. 14.20.

A simplified expression for u under these conditions is obtained by the following calculation:

$$u = u_0 + \Delta u \tag{14.23}$$

where

$$u_0 = \gamma_w \left(h_w + h_r + h_c - h' \right)$$

If it is assumed that the major principal stress equals the weight of material, then the initial total major principal stress, σ_{10} is given by the expression:

$$\sigma_{10} = \gamma_c h_c + \gamma_r h_r + \gamma_w h_w \tag{14.24}$$

where γ_c and γ_r are the saturated unit weights of the clay and the rock and h_c, h_r and h' are as indicated in the figure.

The final total major principal stress, after drawdown, will be:

$$\sigma_{1F} = \gamma_c h_c + \gamma_{dr} h_r \tag{14.25}$$

where γ_{dr} equals the drained unit weight of the rock fill.

Change in major principal stress $= \sigma_{1F} - \sigma_{10} = h_r(\gamma_{dr} - \gamma_r) - \gamma_w h_w$

i.e.

$$\Delta\sigma_1 = -\gamma_w n h_r - \gamma_w h_w \tag{14.26}$$

Note: Porosity of rock fill, $n = V_v/V$ or, when we consider unit volume, $n = V_v$. Hence $(\gamma_{dr} - \gamma_r) = -\gamma_w n$.

$$\Rightarrow \Delta u = -\overline{B}(\gamma_w n h_r + \gamma_w h_w) \tag{14.27}$$

The pore pressure coefficient $\overline{B}$ can be obtained from a laboratory test, but standard practice is to assume, conservatively, that $\overline{B} = 1.0$. In this case,

$$\Delta u = -\gamma_w(nh_r + h_w) \qquad (14.28)$$

and the expression for u becomes,

$$u = \gamma_w\left[h_c + h_r(1-n) - h'\right] \qquad (14.29)$$

The measurement of *in situ* pore water pressures is described in Section 7.5.

14.2.5 Effective stress analysis by Bishop's method

The effective stress methods of analysis now in general use were evolved by Bishop (1955). Figure 14.21 illustrates a circular failure arc, ABCD, and shows the forces on a vertical slice through the sliding segment. Let L_n and L_{n+1} equal the lateral reactions acting on sections n and n + 1 respectively. The difference between L_n and L_{n+1} is small, and the effect of these forces can be ignored with little loss in accuracy. Let the other forces acting on the slice be:

W = weight of slice
P = total normal force acting on base of slice
T = shear force acting on base of slice ($=\tau l$)

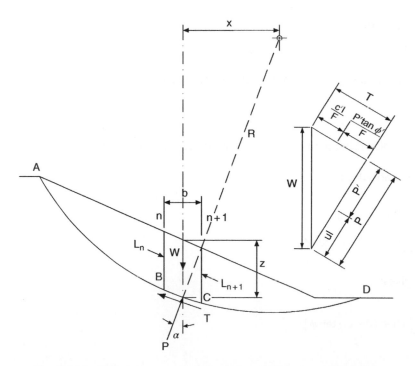

Fig. 14.21 Effective stress analysis: forces acting on a vertical slice.

and the other notation is:

z = height of slice
b = breadth of slice
l = length of BC (assume straight line)
α = angle between P and the vertical
x = horizontal distance from centre of slice to centre of rotation, O.

In terms of effective stress, we have seen from Equation (4.8) that the shear strength of the soil, τ is given by:

$$\tau = c' + (\sigma_n - u) \tan \phi'$$

The shear strength actually mobilised is:

$$\tau_m = \frac{c' + (\sigma_n - u) \tan \phi'}{F} \qquad (14.30)$$

Now, the total normal stress on the base of the slice is:

$$\sigma_n = \frac{P}{l}$$

Therefore,

$$\tau_m = \frac{c' + \left(\dfrac{P}{l} - u\right) \tan \phi'}{F} \qquad (14.31)$$

The total shear force resisting sliding on the base of the slice is:

$$T = \tau_m l$$

For equilibrium, the disturbing moment must equal the restoring moment:

i.e. taking moments about O;

$$\sum Wx = \sum TR = \sum \tau_m lR$$
$$= \frac{R}{F} \sum [c'l + (P - ul) \tan \phi']$$

and rearranging gives:

$$F = \frac{R}{\sum Wx} \sum [c'l + (P - ul) \tan \phi'] \qquad (14.32)$$

From the earlier observation that the difference between L_n and L_{n+1} is small, we see that the only vertical force acting on the slice is W. Hence,

$$P = W \cos \alpha$$

$$\Rightarrow F = \frac{R}{\sum Wx} \sum [c'l + (W \cos \alpha - ul) \tan \phi']$$

Now, by putting $x = R \sin \alpha$, we get

$$F = \frac{1}{\sum W \sin \alpha} \sum [c'l + (W \cos \alpha - ul) \tan \phi'] \tag{14.33}$$

We can now express u in terms of the pore pressure ratio, r_u:

$$u = r_u \gamma z = r_u \frac{W}{b}$$

Now,

$$b = l \cos \alpha$$

$$\Rightarrow u = \frac{r_u W}{l \cos \alpha} = \frac{r_u W}{l} \sec \alpha$$

$$\Rightarrow F = \frac{1}{\sum W \sin \alpha} \sum [c'l + W(\cos \alpha - r_u \sec \alpha) \tan \phi'] \tag{14.34}$$

This formula gives a solution generally known as the *conventional method* which allows rapid determination of F when sufficient slip circles are available to permit the determination of the most critical. For analysing the stability of an existing spoil heap, it should prove perfectly adequate.

Example 14.6: Bishop's conventional method

Determine the factor of safety of the embankment shown in Fig. 14.22 for the trial slip circle indicated.

The soil properties are:

Unit weight of soil = 18 kN/m³
Effective angle of shearing resistance = 35°
Effective cohesion = 15 kPa
Pore pressure ratio = 0.45

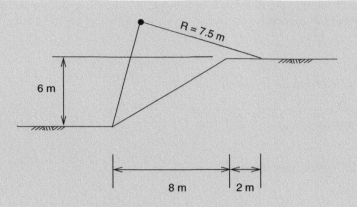

Fig. 14.22   Example 14.6.

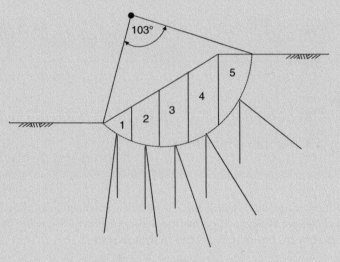

Fig. 14.23   Choice of slices.

Solution:

Divide the slip into a suitable number of slices, say 5, as indicated in Fig. 14.23.

(b) Set out calculations in a table to aid the determination of the factor of safety, F:

Slice	z (m)	b (m)	W (kN)	α (°)	$\cos \alpha$	$\sec \alpha$	$\cos \alpha$ $-r_u$ $\sec \alpha$	$W(\cos \alpha - r_u \sec \alpha) \tan \phi$	$\sin \alpha$	$W \sin \alpha$
1	1.0	2.0	36.0	−12	0.978	1.022	0.518	13.1	-0.21	-7.5
2	2.7	"	97.2	4	0.998	1.002	0.547	37.2	0.07	6.8
3	3.8	"	136.8	20	0.940	1.064	0.461	44.2	0.34	46.8
4	4.2	"	151.2	36	0.809	1.236	0.253	26.8	0.59	88.9
5	2.9	"	104.4	60	0.5	2.0	-0.40	~~-29.3~~* 0	0.87	90.4
								$\Sigma 121.3$		$\Sigma\, 225.4$

*When <0, set to = 0 (i.e. suggests negative friction, which is impossible).

$$\sum c'l = c'R\theta = 15 \times 7.5 \times \frac{\pi}{180} \times 103° = 202.2 \text{ kN/m}$$

$$F = \frac{1}{\sum W \sin \alpha}\sum[c'l + W(\cos\alpha - r_u \sec\alpha)\tan\phi']$$

$$= \frac{1}{225.4} \times (202.2 + 121.3)$$

$$F = 1.44$$

Bishop's routine, or rigorous, method

As demonstrated by slice 5 in Example 14.6, the conventional method has some errors in its solution. In addition, the method was found to give errors of up to 15% in the value of F obtained, although the error is on the safe side since it gives a lower value than is the case. In the construction of new embankments and earth dams, however, this error would lead to unnecessarily high costs, and it becomes particularly pronounced with a deep slip circle where the variations of α over the slip length are large.

Therefore, *Bishop's rigorous method* is normally adopted in preference. The derivation of the solution is presented below.

Return to Equation (14.32):

$$F = \frac{R}{\sum Wx}\sum[c'l + (P - ul)\tan\phi']$$

From Equation (14.31), and recalling that $T = \tau_m l$,

$$T = \frac{1}{F}(c'l + P'\tan\phi') \quad \text{where} \quad P' = P - ul$$

By resolving forces vertically (refer to Fig. 14.21):

$$W = P \cos \alpha + T \sin \alpha$$

$$\Rightarrow W = [ul \cos \alpha + P' \cos \alpha] + \left[\frac{P' \tan \phi'}{F} \sin \alpha + \frac{c'l}{F} \sin \alpha \right] \tag{14.35}$$

$$= l \left[u \cos \alpha + \frac{c' \sin \alpha}{F} \right] + P' \left[\cos \alpha + \frac{\tan \phi' \sin \alpha}{F} \right]$$

which reduces to

$$P' = \frac{W - l \left(u \cos \alpha + \dfrac{c'}{F} \sin \alpha \right)}{\cos \alpha + \dfrac{\tan \phi' \sin \alpha}{F}} \tag{14.36}$$

but $P' = P - ul$

$\Rightarrow$ substituting (14.36) into (14.32) gives

$$F = \frac{R}{\sum Wx} \sum \left[c'l + \frac{\left(W - ul \cos \alpha - \dfrac{c'l}{F} \sin \alpha \right) \tan \phi'}{\cos \alpha + \dfrac{\tan \phi' \sin \alpha}{F}} \right] \tag{14.37}$$

Now substitute $x = R \sin \alpha$, $b = l \cos \alpha$ and $\dfrac{ub}{W} = \dfrac{u}{\gamma z} = r_u$

Therefore:

$$F = \frac{1}{\sum W \sin \alpha} \sum \left[(c'b + W(1 - r_u) \tan \phi') \frac{\sec \alpha}{1 + \dfrac{\tan \phi' \sin \alpha}{F}} \right] \tag{14.38}$$

No unique solution of Equation (14.38) exists since F is on both sides of the equation. The solution, therefore, involves a series of successive approximations. The analysis is best carried out by tabulating the calculations and

Slice no.	z (m)	b (m)	$W = \gamma z b$	$\alpha°$	$\sin \alpha$	$W \sin \alpha$ (1)	$c'b$ (2)	$W(1 - r_u) \times \tan \phi'$ (3)	$2 + 3$ (4)	$\sec \alpha$	$\tan \alpha$	(5) $\dfrac{\sec \alpha}{1 + \dfrac{\tan \phi' \tan \alpha}{F}}$	(4) × (5) (6)
												F =	F =
												F =	F =

Fig. 14.24   Example spreadsheet template for slope stability calculation.

can be greatly simplified through the use of a spreadsheet, as per the example shown in Fig. 14.24. However, in most design offices these days, slope stability problems are solved using slope stability software.

Example 14.7: Bishop's conventional and rigorous methods

The cross-section of an earth dam sitting on an impermeable base is shown in Fig. 14.25. The stability of the downstream slope is to be investigated using the slip circle shown and given the following information:

$\gamma_{sat} = 19.2 \text{ kN/m}^3$
$c' = 12 \text{ kPa}$
$\phi' = 20°$
$R = 9.15 \text{ m}$
Angle subtended by arc of slip circle, $\theta = 89°$

  For this circle, determine the factor of safety (a) by the conventional method and (b) by the rigorous method.

Solution:

The earth dam is drawn to scale by CAD or on graph paper. The first step in the analysis is to divide the sliding sector into a suitable number of slices and determine the pore pressure ratio at the mid-point of the base of each slice.

  The phreatic surface must be drawn, using the method of Casagrande (as described in Section 2.14.3). A rough form of the flow net must then be established, so that the equipotentials through the centre points of each slice can be inserted. Five slices is an appropriate number for this slope (Fig. 4.26).

  The determination of the r_u values is required for both methods and will be considered first.

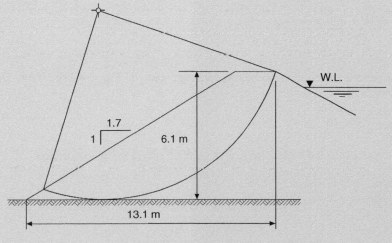

Fig. 14.25 Example 14.7.

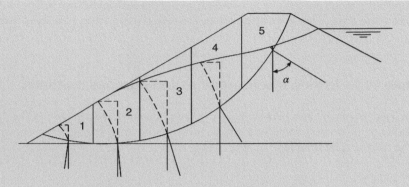

Fig. 4.26   Flow net and choice of slices.

Slice	h_w (m)	u (kPa)	z (m)	r_u
1	0.654	6.42	0.95	0.352
2	1.958	19.21	2.44	0.410
3	2.440	23.90	3.32	0.376
4	2.020	19.82	3.50	0.295
5	0.246	2.41	1.74	0.072

(a) The calculations for the conventional method are set out in Fig. 14.27a.

$\theta = 89°$

$$c'R\theta = 12 \times 9.15 \times \frac{\pi}{180} \times 89° = 170.6 \, \text{kN/m}$$

$$F = \frac{100.9 + 170.6}{207.9} = 1.31$$

(b) The rigorous method calculations are set out in Fig. 14.27b. With the first approximation:

$$F = \frac{297.5}{207.9} = 1.43$$

This value was obtained by assuming a value for F of 1.5 in the expression:

$$\frac{\sec \alpha}{1 + \frac{\tan \phi' \sin \alpha}{F}}$$

of column (5).

(a) Conventional method

Slice	z (m)	b (m)	W (kN)	α (°)	cos α	sec α	h_w (m)	r_u	cos α − r_u sec α	W(cos α − r_u sec α) × tan φ'	sin α	W(sin α)
1	0.95	2.35	42.9	−10	0.985	1.015	0.654	0.352	0.628	9.8	−0.174	−7.4
2	2.44	2.35	110.1	4	0.998	1.002	1.958	0.410	0.587	23.5	0.070	7.7
3	3.32	2.35	149.8	20	0.940	1.064	2.440	0.376	0.540	29.4	0.342	51.2
4	3.50	2.35	157.9	35	0.819	1.221	2.020	0.295	0.459	26.4	0.574	90.6
5	1.74	2.35	78.5	57	0.545	1.836	0.246	0.072	0.412	11.8	0.839	65.8
										Σ100.9		Σ207.9

(b) Rigorous method

Slice	z (m)	b (m)	W (kN)	α (°)	(1) sin α	(2) W sin α	(3) c'b	(4) W(1 − r_u) × tan φ'	(2)+(3)	sec α	tan α	(5) $\dfrac{\sec α}{1 + \dfrac{\tan φ' \tan α}{F}}$ F=1.5	(5) F=1.43	(6) (4) × (5) F=1.5	(6) F=1.43
1	0.95	2.35	42.9	−10	−0.17	−7.4	22.6	10.1	38.3	1.015	−0.18	1.061	1.063	40.6	40.7
2	2.44	2.35	110.1	4	0.07	7.7	22.6	23.6	51.8	1.002	0.07	0.986	0.985	51.1	51.1
3	3.32	2.35	149.8	20	0.34	51.2	22.6	34.1	62.2	1.064	0.36	0.978	0.974	60.9	60.6
4	3.50	2.35	157.9	35	0.57	90.6	22.6	40.5	68.7	1.221	0.70	1.043	1.036	71.7	71.2
5	1.74	2.35	78.5	57	0.84	65.8	22.6	26.5	54.7	1.836	1.54	1.337	1.319	73.1	72.2
						Σ207.9								Σ297.5	295.8

Fig. 14.27 Spreadsheet solutions.

Columns (5) and (6) are now recalculated using F = 1.43 and a revised value of F is obtained:

$$F = \frac{295.8}{207.9} = 1.42$$

This is approximately equal to the assumed value of 1.43 and is taken as correct. Thus, the factor of safety of the slope is 1.42.

Had the assumed and derived values of F not been approximately equal, the iterative procedure could have been repeated once again to find an improved value of F, as can be demonstrated through the example_14.7.xls spreadsheet.

Example 14.8: Bishop's rigorous method

Figure 14.28 gives details of the cross-section of an embankment. The soil has the following properties:

$$\phi' = 35°, c' = 10 \, \text{kPa}, \gamma = 16 \, \text{kN/m}^3$$

For the slip circle shown, determine the factor of safety for the following values of r_u: 0.2, 0.4 and 0.6.

Plot the variation of F with r_u.

Solution:

The calculations are based on the rigorous method and are shown in Figs 14.29 and 14.30.

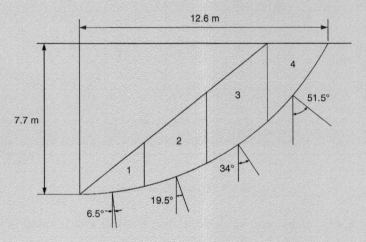

Fig. 14.28 Example 14.8.

$r_u = 0.2$

						(1)	(2)	(3)	(4)			(5)		(6)	
Slice	z (m)	b (m)	W (kN)	α (°)	$\sin \alpha$	$W \sin \alpha$	$c'b$	$W(1-r_u) \times \tan \phi'$	$(2)+(3)$	$\sec \alpha$	$\tan \alpha$	$\dfrac{\sec \alpha}{1 + \dfrac{\tan \phi' \tan \alpha}{F}}$		$(4) \times (5)$	
												$F = 1.5$	$F = 1.47$	$F = 1.5$	$F = 1.47$
1	1.00	3.15	50.4	7	0.113	5.7	31.5	28.2	59.7	1.006	0.114	0.956	0.955	57.1	57.0
2	3.08	3.15	155.2	20	0.334	51.8	31.5	87.0	118.5	1.061	0.354	0.910	0.908	107.8	107.5
3	4.00	3.15	201.6	34	0.559	112.7	31.5	112.9	144.4	1.206	0.675	0.917	0.913	132.5	131.8
4	2.70	3.15	136.1	52	0.783	106.5	31.5	76.2	107.7	1.606	1.257	1.012	1.004	109.1	108.2
						$\Sigma 276.8$								$\Sigma 406.5$	404.6

$$F = \frac{406.5}{276.8} = 1.47 \qquad F = \frac{404.6}{276.8} = 1.46$$

$r_u = 0.4$

(3)	(4)	(5)		(6)	
$W(1-r_u) \times \tan \phi'$	$(2)+(3)$	$\dfrac{\sec \alpha}{1 + \dfrac{\tan \phi' \tan \alpha}{F}}$		$(4) \times (5)$	
		$F = 1.3$	$F = 1.17$	$F = 1.3$	$F = 1.17$
21.2	52.7	0.948	0.942	49.9	49.6
65.2	96.7	0.891	0.875	86.2	84.7
84.7	116.2	0.885	0.859	102.8	99.9
57.2	88.7	0.958	0.917	84.9	81.3
				$\Sigma 323.9$	315.4

$$F = \frac{323.9}{276.8} = 1.17 \qquad F = \frac{315.4}{276.8} = 1.14$$

$r_u = 0.6$

(3)	(4)	(5)		(6)	
$W(1-r_u) \times \tan \phi'$	$(2)+(3)$	$\dfrac{\sec \alpha}{1 + \dfrac{\tan \phi' \tan \alpha}{F}}$		$(4) \times (5)$	
		$F = 1.0$	$F = 0.86$	$F = 1.0$	$F = 0.86$
14.1	45.6	0.932	0.921	42.5	42.0
43.5	75.0	0.850	0.823	63.7	61.7
56.5	88.0	0.819	0.778	72.1	68.5
38.1	69.6	0.854	0.793	59.5	55.2
				$\Sigma 237.8$	227.4

$$F = \frac{237.8}{276.8} = 0.86 \qquad F = \frac{227.4}{276.8} = 0.82$$

Fig. 14.29 Calculations for each value of r_u.

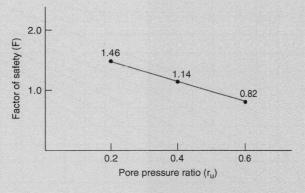

Fig. 14.30 Variation of F with r_u.

14.3   Slope stability design charts

Design charts offer a means of rapidly determining an approximate factor of safety of a homogeneous, regular slope. These slopes rarely exist in reality but nonetheless many embankments and cuttings can be considered as being of this form, if only to offer a simple and rapid, yet very approximate solution to assessing their stability. Once an approximate factor of safety is found from a design chart approach, it would be normal practice for a more rigorous analysis to then be performed to establish a more accurate and reliable measure of the safety of the slope.

14.3.1   Rapid determination of F for a homogeneous, regular slope

It can be shown that for two similar slopes made from two different soils the ratio $c_m/\gamma H$ is the same for each slope provided that the two soils have the same angle of shearing resistance. The ratio $c_m/\gamma H$ is known as the stability number and is given the symbol N, where c_m = mobilised cohesion, γ = unit weight of soil, and H = vertical height of slope.

For any type of soil, the critical circle always passes through the toe when $\beta > 53°$. In theory, when $\phi = 0°$ (in practice when $\phi < 3°$) and $\beta > 53°$, the critical slip circle can extend to a considerable depth (Fig. 14.7c).

Taylor (1948) prepared two set of curves that relate the stability number to the angle of the slope: the first (Fig. 14.31) is for the general case of a $c' - \phi'$ soil whilst the second (Fig. 14.32) is for a soil with $\phi = 0°$, a slope angle of less than 53° and with a layer of stiff material or rock at a depth DH below the top of the embankment. D is known as the depth factor and, depending upon its value, the slip circle will either emerge at a distance nH in front of the toe or pass through the toe (using the dashed lines, the value of n can be obtained from the curves).

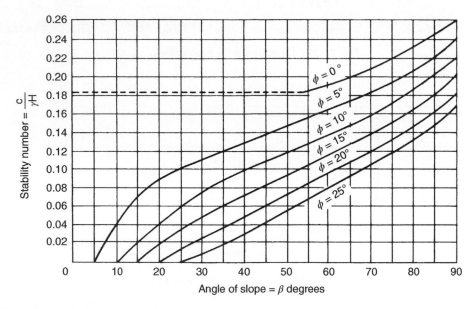

Fig. 14.31   Taylor's curves for rapid assessment of F. (For $\phi = 0°$ and $\beta < 53°$, use Fig. 14.32.) Based on Taylor (1948).

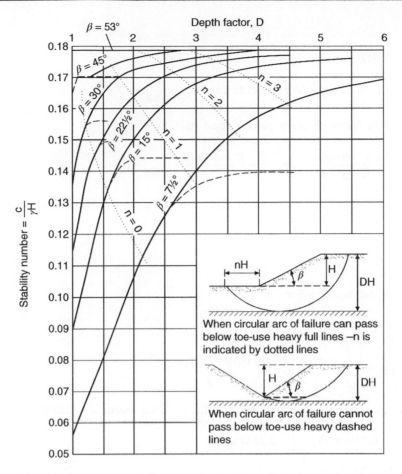

Fig. 14.32 Effect of depth limitation on Taylor's curves (for $\beta < 53°$ and $\phi' > 0$, use Fig. 14.31). *Note*: $c'/\gamma H = 0.181$ at D = ∞ for all β values. Based on Taylor (1948).

Example 14.9: Taylor's charts (i)

An embankment has a slope of one vertical to two horizontal. The properties of the slope are: $c' = 5$ kPa, $\phi' = 20°$, $\gamma = 18$ kN/m³ and H = 11 m. Using Taylor's charts, determine an approximation for the factor of safety of the slope.

Solution:

From the charts it is seen that a slope with $\phi' = 20°$ and a slope angle, β of 26.6° has a stability number of 0.017. This means that, if the factor of safety for shearing resistance was unity, c_m, the cohesion which must be mobilised, would be found from the expression:

$\dfrac{c_m}{\gamma H} = 0.017$ i.e. $c_m = 18 \times 11 \times 0.017 = 3.37$ kPa

$\Rightarrow$ Factor of safety, with respect to cohesion $= \dfrac{5}{3.37} = 1.48$

This is not the factor of safety used in slope stability, which is:

$$F = \dfrac{\text{Shear strength}}{\text{Disturbing shear}} \quad i.e. \quad F = \dfrac{c' + \sigma' \tan \phi'}{\tau}$$

This safety factor applies equally to cohesion and to friction. F can be found by successive approximations:

$$F = \dfrac{c'}{F} + \dfrac{\sigma' \tan \phi'}{F}$$

As an initial estimate, try F = 1.1:

$$F = \dfrac{c'}{F} + \dfrac{\sigma' \tan \phi'}{F}$$

$$\phi' = \tan^{-1}\left(\dfrac{\tan \phi'}{F}\right) = \tan^{-1}\left(\dfrac{\tan 20°}{1.1}\right) = 18.3°$$

Use this value of ϕ' to establish a new N value from the charts: N = 0.019

$\Rightarrow c_m = 0.019 \times 18 \times 11 = 3.76$ kPa

$\Rightarrow F$ (for c') $= F_{c'} = \dfrac{5}{3.76} = 1.33$

$F_{c'} \neq F_{\phi'}$, so revise the estimate of F.

Try F = 1.2:

$$\phi' = \tan^{-1}\left(\dfrac{\tan \phi'}{F}\right) = \tan^{-1}\left(\dfrac{\tan 20°}{1.2}\right) = 16.9°$$

From the charts N = 0.024 $\Rightarrow F_{c'} = \dfrac{5}{4.75} = 1.05$

Try F = 1.15:

$$\phi' = \tan^{-1}\left(\dfrac{\tan \phi'}{F}\right) = \tan^{-1}\left(\dfrac{\tan 20°}{1.15}\right) = 17.6°$$

From the charts N = 0.022 $\Rightarrow F_{c'} = \dfrac{5}{4.36} = 1.15 \left(= F_{\phi'}\right)$

i.e. Factor of safety of slope = 1.15

Example 14.10: Taylor's charts (ii)

Slope = 1 vertical to four horizontal, $c' = 12.5\,kPa$, $H = 31\,m$, $\phi' = 20°$, $\gamma = 16\,kN/m^3$. Estimate the F value of the slope.

Solution:

Angle of slope = 14°, so the slope is safe as it is less than the angle of shearing resistance, ϕ'. With this case, N from the charts = 0.

The procedure is identical with Example 14.9.

Try F = 1.5:

$$\phi' = \tan^{-1}\left(\frac{\tan\phi'}{F}\right) = \tan^{-1}\left(\frac{\tan 20°}{1.5}\right) = 13.5°$$

From the charts, $N = 0.005 \quad \Rightarrow \quad F_{c'} = \frac{12.5}{0.005 \times 31 \times 16} = 5.04$

Try F = 2.0:

$$\phi' = \tan^{-1}\left(\frac{\tan\phi'}{F}\right) = \tan^{-1}\left(\frac{\tan 20°}{2.0}\right) = 10.3°$$

From the charts $N = 0.016 \quad \Rightarrow \quad F_{c'} = \frac{12.5}{7.95} = 1.57$

Try F = 1.9:

$$\phi' = \tan^{-1}\left(\frac{\tan\phi'}{F}\right) = \tan^{-1}\left(\frac{\tan 20°}{1.9}\right) = 11°$$

From the charts $N = 0.013 \quad \Rightarrow \quad F_{c'} = \frac{12.5}{6.45} = 1.94$ (acceptable)

Factor of safety for slope = 1.9

14.3.2 Homogeneous slope with a constant pore pressure ratio

If on a trial slip circle the value of F is determined for various values of r_u and the results plotted, a linear relationship is found between F and r_u (see Example 14.8). The usual values of r_u encountered in practice range from 0.0 to 0.7 and it has been established that this linear relationship between F and r_u applies over this range. The factor of safety, F may therefore be determined from the expression:

$$F = m - nr_u \tag{14.39}$$

in which m is the factor of safety with respect to total stresses (i.e. when no pore pressures are assumed) and n is the coefficient which represents the effect of the pore pressures on the factor of safety. These terms m and n are known as stability coefficients and were evolved by Bishop and Morgenstern (1960) and can be used to give a rapid approximation of F using specially prepared design charts. The process involved, however, is a bit lengthy and time consuming and since F is only approximate, the use of the charts these days is seen as dated.

14.4   Wedge failure

When the potential sliding mass of the soil is bounded by two or three straight lines, we have a wedge failure. Wedge failures can be brought about by a variety of geological conditions, and one example is shown in Fig. 14.33a with the design approximation illustrated in Fig. 14.33b.

The form of construction within an earth structure can also dictate that any stability failure will be of a wedge type. One example is that of an earth dam with a sloping impermeable core (Fig. 14.33c). Various wedge failure patterns could be assumed for the purpose of analysis. One such form is illustrated in Fig. 14.33d.

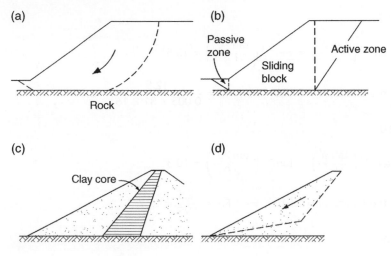

Fig. 14.33   Sliding block or wedge failure.

Example 14.11: Wedge failure

The cross-section of a sloping core dam is shown in Fig. 14.34a and a probable form of wedge failure. Using the suggested failure shape determine the factor of safety of the dam.

Relevant properties: rock fill: $\phi' = 40°$, $\gamma = 18$ kN/m³; core: $c_u = 80$ kPa.

Solution:

The forces acting on wedges (1) and (2) are shown in Fig. 14.34b. In the diagram ϕ_m = angle of shearing resistance mobilised in the rock fill and c_m = cohesion mobilised in the core.

$$T_1 = c_m BE \text{ kN/m run of dam}; T_2 = N_2 \tan \phi_m \text{ kN/m run of dam}$$

The procedure starts by selecting suitable F values and evaluating the corresponding values for ϕ_m and c_m:

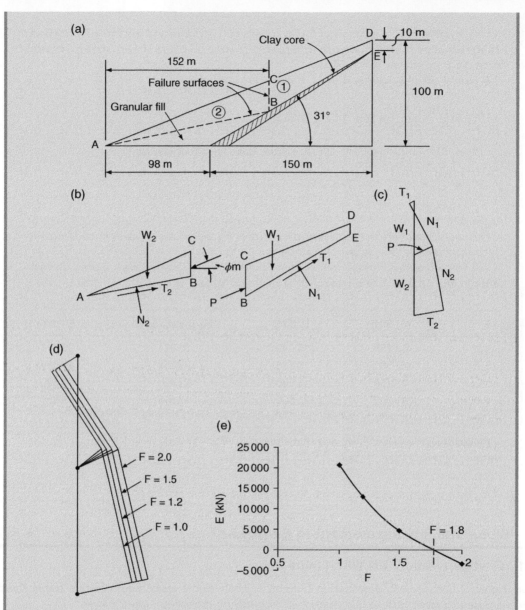

Fig. 14.34 Example 14.11: (a) Cross-section of dam. (b) Wedge forces. (c) Polygon of forces. (d) Full force diagram. (e) Determination of F.

$F = 1.0$ $c_m = 80$ kPa; $\phi_m = 40°$

$F = 1.2$ cm $= 66.7$ kPa; $\phi_m = \tan^{-1}\left(\dfrac{\tan\phi}{1.2}\right) = 35°$

$F = 1.5$ $c_m = 53.3$ kPa; $\phi_m = 29°$

$F = 2.0$ $c_m = 40$ kPa; $\phi_m = 23°$

For each value of F, a polygon of forces for slice (1) is drawn and the force P obtained. Using this value for P the polygon of forces for wedge (2) is now drawn to give the total force diagram shown in Fig. 14.34c.

A typical set of calculations (for F = 2.0) is set out below.

$$W_1 = \left(\frac{10 + 29}{2}\right) \times 96 \times 18 = 33\,696 \text{ kN}$$

$$T_1 = 40 \times 112 = 4480 \text{ kN} \quad (112\,\text{m is the base length of wedge 1})$$

$$W_2 = \frac{29}{2} \times 152 \times 18 = 39\,672 \text{ kN}; \quad \phi_m = 23°$$

As the directions of P and T_2 are known, the value obtained for N_2 will be correct. The error of closure, E, can therefore be assessed by comparing the value of T_2 determined from the force diagram with the value calculated from $N_2 \tan \phi_m$.

The force diagrams for the four chosen values for F are shown superimposed on each other in Fig. 14.34d. The corresponding values of N_2 and T_2 are set out below.

F	N_2 (kN)	T_2 (kN)	$N_2 \tan \phi_m$ (kN)	E (kN)
1.0	42 750	15 080	35 871	20 791
1.2	42 500	16 700	29 759	13 059
1.5	42 250	18 800	23 420	4 620
2.0	41 500	21 000	17 616	−3 384

By plotting F against E, we see that the value of E = 0 when F = 1.8. Hence, for the wedge failure surfaces chosen, F = 1.8 (Fig. 14.34e).

14.5   Slope stability assessment to Eurocode 7

14.5.1   First generation: EN 1997-1:2004

The principles of Eurocode 7, described in Chapter 6, apply to the assessment of slope stability and the reader is advised to refer back to that chapter whilst studying the following few pages.

The overall stability of slopes is covered in Section 11 of Eurocode 7, and the GEO limit state is the principal state that is considered. The procedure to check overall stability uses the methods described earlier in this chapter and applies to both the undrained and drained states. Partial factors are applied to the characteristic values of the soils' shear strength parameters, and to the representative values of the actions (e.g. weights of slices) to obtain the design values.

For circular failure surfaces, during a method of slices analysis, the weight of a single slice can contribute to the disturbing moment or it may contribute to the restoring moment, as illustrated by Example 14.7 (Fig. 4.26). In that example, slices 2–5 contribute to the disturbing moment and slice 1 contributes to the restoring moment. This follows from the particular choice of position of the centre of the slip circle: had the centre of the slip circle been in a different location, the directions of the moment of each slice might have been different.

As explained in Chapter 6, an action is considered as either *favourable* or *unfavourable*. However, as we have just seen, the choice of the position of the centre of the circle influences whether the weight of a slice would be favourable or unfavourable and, because of this, it is impossible to know from the outset whether an action will be favourable or unfavourable. When using Design Approach 1 Combination 1, the partial factors (see Table 6.1) are different for favourable and unfavourable actions, whereas for Combination 2 the partial factors are the same (=1.0). To this end, when using Design Approach 1 for circular failure surfaces, Combination 2 will almost always be used to check the overall stability.

The GEO limit state requirement is satisfied if the design effect of the actions, E_d is less than or equal to the design resistance, R_d, i.e. $E_d \leq R_d$. Here E_d is the design value of the disturbing moment (or force, for planar failure surfaces) and R_d is the design value of the restoring moment (or force). By representing the ratio of the restoring moment (or force) to the disturbing moment (or force) as the over-design factor, Γ, it is seen that the limit state requirement is satisfied if $\Gamma \geq 1$.

In addition to the ultimate limit states checks, serviceability limit states, such as excessive deformation of the embankment, or settlement of the compressible soil beneath, are also checked.

14.5.2   Second generation: EN 1997-1:202x and EN 1997-3:202x

In the second generation, the design of slopes, cuttings and embankments is covered in Section 4 of EN 1997-3:202x. In accordance with the procedure for establishing the Geotechnical Category (described in Chapter 6), the Geotechnical Complexity Class must be established at an early stage. Most routine cuttings and embankments design work will fall into Geotechnical Complexity Class 2. Higher complexity earth structures will fall into GCC3, and lower complexity ones will fall into GCC1.

GCC3 slopes include situations where: the ground includes soils susceptible to excessive deformation, the ground is experiencing ongoing settlement or is unstable, significant seepage can cause internal erosion or piping, or where significant dynamic or cyclic loading might undermine the slope stability.

GCC1 slopes cover situations where there is negligible risk of the occurrence of an ultimate or serviceability limit state, where the cutting depth is shallow or where the embankment height is low or where the slope angle is very low: these all being related too to the soil strength properties.

The limit state check for the overall stability of a slope is checked using the material factor approach (MFA) – specifically, partial factor combination Design Case 3 and Materials set M2. The partial factors for DC3 are given in Table 6.6 and for M2 are given in Table 6.7.

As with the first generation, serviceability limit states are also checked.

Example 14.12: Taylor's charts to Eurocode 7 (first generation)

Using Taylor's charts, determine the margin of safety of a proposed motorway cutting slope, as detailed below, using the procedures of Eurocode 7 GEO limit state Design Approaches 1 and 2 for both the short- and the long-term conditions.

The slope will be constructed from a regular, homogeneous clay deposit and will have the following properties:

Height of slope, H = 15 m
Angle of slope, $\beta = 21°$
Weight density, $\gamma_k = 20\,\text{kN/m}^3$
Undrained shear strength, $c_{u;k} = 90\,\text{kPa}$

Effective cohesion, $c'_k = 5$ kPa

Effective angle of shearing resistance $\phi'_k = 25°$

Solution:

As seen in Section 14.3.1, when using Taylor's chart, the factor of safety applied to each component of the shear strength (c' and $\tan\phi'$) is the same. However, when applying partial factors of safety in a Eurocode 7 design, different factors are applied to these two components and thus the use of Taylor's charts in a Eurocode 7 design is slightly more complex. The stability number is calculated using the design values of shear strength, together with the design weight of the slope (i.e. the permanent, unfavourable design action acting).

(a) Design Approach 1

1. *Combination 1 (partial factor sets A1 + M1 + R1)*
    From Table 6.1: $\gamma_{G;\,unfav} = 1.35$; $\gamma_{cu} = 1.0$; $\gamma_{c'} = 1.0$; $\gamma_{\phi'} = 1.0$

    Design material properties:

$$c_{u;d} = \frac{c_u}{\gamma_{cu}} = \frac{90}{1.0} = 90 \text{ kPa}$$

$$c'_d = \frac{c'}{\gamma_{c'}} = \frac{5}{1.0} = 5 \text{ kPa}$$

$$\phi'_d = \tan^{-1}\left(\frac{\tan\phi'}{\gamma_{\phi'}}\right) = 25°$$

Short-term : $N_d = \dfrac{c_{u;d}}{H \times \gamma \times \gamma_G} = \dfrac{90}{15 \times 20 \times 1.35} = 0.222$

From Fig. 14.31: For N = 0.222, $\phi_{u;d} = 0°$: maximum achievable slope angle, $\beta = 77°$ i.e. the short-term GEO limit state requirement is satisfied since 77° > 21°

Long-term $N_d = \dfrac{c'_d}{H \times \gamma \times \gamma_G} = \dfrac{5}{15 \times 20 \times 1.35} = 0.012$

From Fig. 14.31: For N = 0.012, $\phi'_d = 25°$: maximum achievable slope angle, $\beta = 32°$ i.e. the long-term GEO limit state requirement is satisfied since 32° > 21°

2. *Combination 2 (partial factor sets A2 + M2 + R1)*
    From Table 6.1: $\gamma_{G;\,unfav} = 1.0$; $\gamma_{cu} = 1.4$; $\gamma_{c'} = 1.25$; $\gamma_{\phi'} = 1.25$.

$$c_{u;d} = \frac{90}{1.4} = 64.3\ kPa;\ c'_d = \frac{5}{1.25} = 4\ kPa;\ \phi'_d = \tan^{-1}\left(\frac{\tan 25°}{1.25}\right) = 20.5°$$

$$Short\text{-}term \quad N_d = \frac{c_{u;d}}{H \times \gamma \times \gamma_G} = \frac{64.3}{15 \times 20 \times 1.0} = 0.214$$

From Fig. 14.31: For N = 0.214, $\phi_{u;d}$ = 0°: maximum achievable slope angle, β = 72° i.e. the short-term GEO limit state requirement is satisfied since 72° > 21°

$$Long\text{-}term \quad N_d = \frac{c'_d}{H \times \gamma \times \gamma_G} = \frac{4}{15 \times 20 \times 1.0} = 0.013$$

From Fig. 14.31: For N = 0.013, ϕ'_d = 20.5°: maximum achievable slope angle, β = 24° i.e. the long-term GEO limit state requirement is satisfied since 24° > 21°

(b) Design Approach 2
   (Partial factor sets A1 + M1 + R2)
      From Table 6.1: $\gamma_{G;\ unfav}$ = 1.35; γ_{cu} = 1.0; $\gamma_{c'}$ = 1.0; $\gamma_{\phi'}$ = 1.0; γ_{Rh} = 1.1.
      Design material properties:

$$c_{u;d} = \frac{90}{1.0} = 90\ kPa;\ c'_d = \frac{5}{1.0} = 5\ kPa;\ \phi'_d = \tan^{-1}\left(\frac{\tan 25°}{1.0}\right) = 25°$$

In Design Approach 2, the design resistance is obtained by reducing the characteristic resistance by the relevant partial factor on resistance (see Section 6.4.7 and EN 1977-1:2004, Annex B). In this case, it is clear that the partial factor on sliding resistance is to be used.

$$Short\text{-}term \quad N_d = \frac{c_{u;d}}{H \times \gamma \times \gamma_G \times \gamma_{Rh}} = \frac{90}{15 \times 20 \times 1.35 \times 1.1} = 0.202$$

From Fig. 14.31: For N = 0.202, $\phi_{u;d}$ = 0°: maximum achievable slope angle, β = 65° i.e. the short-term GEO limit state requirement is satisfied since 65° > 21°

$$Long\text{-}term \quad N_d = \frac{c'_d}{H \times \gamma \times \gamma_G \times \gamma_{Rh}} = \frac{5}{15 \times 20 \times 1.35 \times 1.1} = 0.011$$

From Fig. 14.31: For N = 0.011, ϕ'_d = 25°: maximum achievable slope angle, β = 32° i.e. the long-term GEO limit state requirement is satisfied since 32° > 21°
It is seen that the proposed design satisfies the requirements of both Design Approach 1 and Design Approach 2. As would be expected for a cutting, the long-term state is the governing condition, i.e. maximum safe slope angle is lowest in the long-term.

Example 14.13: Taylor's charts to Eurocode 7 (second generation)

Repeat Example 14.12 using the MFA of the second generation of Eurocode 7, assuming that the slope is in Consequence Class 2.

Solution:

Material Factor Approach: DC3 and M2
    From Table 6.6: $\gamma_G = 1.0$
    From Table 6.7: $\gamma_{cu} = 1.4\ K_M$; $\gamma_{c'} = 1.25\ K_M$; $\gamma_{\tan\phi} = 1.25\ K_M$ ($K_M = 1.0$, since CC2)
    Design material properties:

$$c_{u;d} = \frac{90}{1.4} = 64.3\,\text{kPa}; c'_d = \frac{5}{1.25} = 4\,\text{kPa}; \phi'_d = \tan^{-1}\left(\frac{\tan 25°}{1.25}\right) = 20.5°$$

$$\text{Short-term:} \quad N_d = \frac{c_{u;d}}{H \times \gamma \times \gamma_G} = \frac{64.3}{15 \times 20 \times 1.0} = 0.214$$

From Fig. 14.31: For N = 0.214, $\phi_{u;d} = 0°$: maximum achievable slope angle, $\beta = 72°$

$$\text{Long-term:} \quad N_d = \frac{c'_d}{H \times \gamma \times \gamma_G} = \frac{4}{15 \times 20 \times 1.0} = 0.013$$

From Fig. 14.31: For N = 0.013, $\phi'_d = 20.5°$: maximum achievable slope angle, $\beta = 24°$

i.e. the limit state requirement is satisfied since the lowest safest slope angle ($\beta = 24°$) > 21°

Example 14.14: Rotational failure to Eurocode 7 (first generation)

Check the GEO limit state for the earth dam described in Example 14.7 using Design Approach 1, Combination 2.

Solution:

From Table 6.1, the relevant partial factors are: $\gamma_{c'} = 1.25$; $\gamma_{\phi'} = 1.25$.
    The design shear strength parameters c'_d and ϕ'_d are determined:

$$c'_d = \frac{c'}{\gamma_{c'}} = \frac{12}{1.25} = 9.6\,\text{kPa}$$

$$\phi'_d = \tan^{-1}\left(\frac{\tan\phi'}{\gamma_{\phi'}}\right) = \tan^{-1}\left(\frac{\tan 20°}{1.25}\right) = 16.2°$$

Slice	z (m)	b (m)	W (kN)	α (°)	$\cos\alpha$	$\sec\alpha$	h_w (m)	r_u	$\cos\alpha - r_u\sec\alpha$	$W(\cos\alpha - r_u\sec\alpha)\times\tan\phi'$	$\sin\alpha$	$W(\sin\alpha)$
1	0.95	2.35	42.9	−10	0.985	1.015	0.654	0.352	0.628	7.8	−0.174	−7.4
2	2.44	2.35	110.1	4	0.998	1.002	1.958	0.410	0.587	18.8	0.070	7.7
3	3.32	2.35	149.8	20	0.940	1.064	2.440	0.376	0.540	23.6	0.342	51.2
4	3.50	2.35	157.9	35	0.819	1.221	2.020	0.295	0.459	21.1	0.574	90.6
5	1.74	2.35	78.5	57	0.545	1.836	0.246	0.072	0.412	9.4	0.839	65.8
										$\Sigma80.7$		$\Sigma207.9$

Fig. 14.35 Example 14.14: conventional method.

(a) *Conventional method*: using c_d' and ϕ_d' the conventional analysis is performed and the calculations are set out in Fig. 14.35.

$$c_d'R\theta = 9.6 \times 9.15 \times \frac{\pi}{180} \times 89° = 136.4\ \text{kN/m}$$

Over-design factor, $\quad \Gamma = \dfrac{80.7 + 136.4}{207.9} = 1.04$

Since $\Gamma > 1$, the GEO limit state requirement is satisfied.

(b) *Rigorous method*: the calculations are set out in Fig. 14.36.

						(1)	(2)	(3)	(4)			(5)		(6)	
Slice	z (m)	b (m)	W (kN)	α (°)	$\sin\alpha$	$W\sin\alpha$	$c'b$	$W(1-r_u)\times\tan\phi'$	$(2)+(3)$	$\sec\alpha$	$\tan\alpha$	$\dfrac{\sec\alpha}{1+\dfrac{\tan\phi'\tan\alpha}{F}}$		$(4)\times(5)$	
												$F=1.5$	$F=1.17$	$F=1.5$	$F=1.17$
1	0.95	2.35	42.9	−10	−0.17	−7.4	22.6	8.1	30.7	1.015	−0.18	1.051	1.062	32.2	32.5
2	2.44	2.35	110.1	4	0.07	7.7	22.6	18.9	41.5	1.002	0.07	0.989	0.985	41.0	40.9
3	3.32	2.35	149.8	20	0.34	51.2	22.6	27.2	49.8	1.064	0.36	0.994	0.976	49.5	48.6
4	3.50	2.35	157.9	35	0.57	90.6	22.6	32.4	55.0	1.221	0.70	1.075	1.040	59.1	57.2
5	1.74	2.35	78.5	57	0.84	65.8	22.6	21.2	43.8	1.836	1.54	1.414	1.328	61.9	58.1
						$\Sigma207.9$								$\Sigma243.7$	237.3

Fig. 14.36 Example 14.14: rigorous method.

Over-design factor, $\quad \Gamma = \dfrac{\sum(6)}{\sum(1)} = \dfrac{237.3}{207.9} = 1.14$

that is, the GEO limit state requirement is satisfied.

Example 14.15: Rotational failure to Eurocode 7 (second generation)

Repeat Example 14.14 using the MFA of the second generation of Eurocode 7, assuming that the slope is in Consequence Class 2.

Solution:

Material Factor Approach: DC3 and M2
   From Table 6.6: $\gamma_G = 1.0$
   From Table 6.7: $\gamma_{c'} = 1.25\,K_M$; $\gamma_{\tan\phi} = 1.25\,K_M$ ($K_M = 1.0$, since CC2)
   Since the partial factors have the same magnitude as those used in Example 14.14, all calculations will be the same as Example 14.14.
   The rotational failure limit state requirement is thus satisfied.

Example 14.16: Planar failure to Eurocode 7 (first generation)

Consider an infinite slope, constructed at an angle of $\beta = 20°$ to the horizontal using the granular soil described in Example 14.1. Check the GEO limit state against failure along a plane parallel to and 2 m beneath the surface of the slope, using Design Approach 1.

Solution:

(i) *Combination 1 (partial factor sets A1 + M1 + R1)*
   From Table 6.1, the relevant partial factors are: $\gamma_{G;\,unfav} = 1.35$; $\gamma_{\phi'} = 1.0$.
   From Section 14.1 it was seen that the sliding force, per unit area $= W\sin\beta$

$$
\begin{aligned}
\text{Design sliding force per unit area, } G_d &= \gamma \times z \times \sin\beta \times \gamma_{G;\,unfav} \\
&= 18 \times 2 \times \sin 20° \times 1.35 \\
&= 16.6\,\text{kPa}
\end{aligned}
$$

No other actions contribute to the sliding force so $E_d = G_d = 16.6\,kPa$

$$\phi'_d = \tan^{-1}\left(\frac{\tan\phi'}{\gamma_{\phi'}}\right) = \tan^{-1}\left(\frac{\tan 30°}{1.0}\right) = 30°$$

Design resisting force per unit area, $R_d = \gamma \times z \times \cos\beta \times \tan\phi'_d$
$$= 18 \times 2 \times \cos 20° \times \tan 30°$$
$$= 19.5\,kPa$$

Therefore, the limit state requirement is satisfied since $E_d \le R_d$.

(ii) *Combination 2 (partial factor sets A2 + M2 + R1)*
From Table 6.1: $\gamma_{G;\,unfav} = 1.0$; $\gamma_{\phi'} = 1.25$.

Design sliding force per unit area, $G_d = \gamma \times z \times \sin\beta \times \gamma_{G;\,unfav}$
$$= 18 \times 2 \times \sin 20° \times 1.0$$
$$= 12.3\,kPa$$

that is, $E_d = 12.3\,kPa$.

$$\phi'_d = \tan^{-1}\left(\frac{\tan\phi'}{\gamma_{\phi'}}\right) = \tan^{-1}\left(\frac{\tan 30°}{1.25}\right) = 24.8°$$

Design resisting force per unit area, $R_d = \gamma \times z \times \cos\beta \times \tan\phi'_d$
$$= 18 \times 2 \times \cos 20° \times \tan 24.8°$$
$$= 15.6\,kPa$$

Once again, the limit state requirement is satisfied since $E_d \le R_d$.

Example 14.17: Planar failure to Eurocode 7 (second generation)

Repeat Example 14.16 using the MFA of the second generation of Eurocode 7, assuming that the slope is in Consequence Class 2.

Solution:

Material actor Approach: DC3 and M2
  From Table 6.6: $\gamma_G = 1.0$
  From Table 6.7: $\gamma_{\tan\phi} = 1.25\,K_M$ ($K_M = 1.0$, since CC2)

Design sliding force per unit area, $G_d = \gamma \times z \times \sin\beta \times \gamma_G$
$$= 18 \times 2 \times \sin 20° \times 1.0$$
$$= 12.3 \text{ kPa}$$

that is, $E_d = 12.3$ kPa.

$$\phi'_d = \tan^{-1}\left(\frac{\tan\phi'}{\gamma_{\tan\phi'}}\right) = \tan^{-1}\left(\frac{\tan 30°}{1.25}\right) = 24.8°$$

Design resisting force per unit area, $R_d = \gamma \times z \times \cos\beta \times \tan\phi'_d$
$$= 18 \times 2 \times \cos 20° \times \tan 24.8°$$
$$= 15.6 \text{ kPa}$$

Therefore, the limit state requirement is satisfied since $E_d \leq R_d$.

Exercises

Exercise 14.1

A proposed cutting is to have the dimensions shown in Fig. 14.37. The soil has the following properties: $\phi' = 15°$, $c' = 13.5$ kPa, $\gamma = 19.3$ kN/m³.

Determine the factor of safety against slipping for the slip circle shown (i) ignoring tension cracks and (ii) allowing for a tension crack.

Answer   (i) 1.7, (ii) 1.6

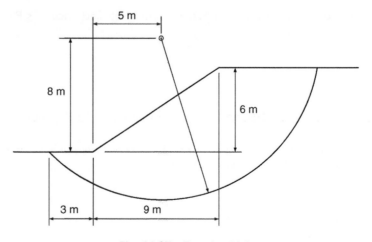

Fig. 14.37   Exercise 14.1.

Exercise 14.2

Investigate the stability of the embankment shown in Fig. 14.38. The embankment consists of granular fill placed upon a deep deposit of clay. Both soils have a bulk unit weight
of 19.3 kN/m³. The fill has $c' = 7.2$ kPa and $\phi' = 30°$, whilst the clay soil has $c_u = 32.5$ kPa.
   Analyse the slip circle shown (ignore tension cracks).

Answer F = 1.2

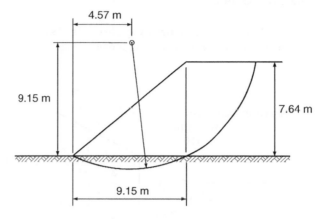

4.57 m

9.15 m

7.64 m

9.15 m

Fig. 14.38 Exercise 14.2.

Exercise 14.3

The surface of a granular soil mass is inclined at 25° to the horizontal. The soil is saturated throughout with a water content of 15.8%, particle specific gravity of 2.65 and
an effective angle of shearing resistance of 38°. At a depth of 1.83 m, water is seeping
through the soil parallel to the surface.
   Determine the factor of safety against slipping on a plane parallel to the surface of
the soil at a vertical depth of 3.05 m below the surface.

Answer 1.37

What would be the factor of safety for the same plane if the level of the seeping water
rose to the surface of the soil?

Answer 0.90

Exercise 14.4

Using Taylor's curves, determine the factor of safety for the following slopes
(assume D = 1.0):

   $H = 30.5$ m, $\beta = 40°$, $c' = 10.8$ kPa, $\phi' = 25°$, $\gamma = 14.4$ kN/m³

Answer ≈0.95

$$H = 15.25\,\text{m}, \beta = 20°, c_u = 24.0\,\text{kPa}, \gamma = 19.3\,\text{kN/m}^3$$

Answer   0.8

$$H = 22.8\,\text{m}, \beta = 30°, c' = 9.6\,\text{kPa}, \phi' = 25°, \gamma = 16.1\,\text{kN/m}^3$$

Answer   1.2

Exercise 14.5

During the analysis of a trial slip circle on a soil slope, the rotating mass of the soil was divided into five vertical slices as shown in Fig. 14.39. The position of the GWT is also shown.

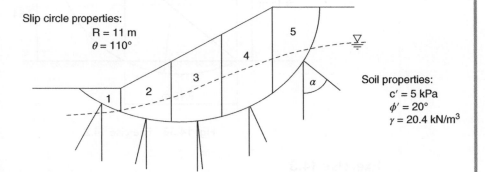

Fig. 14.39   Exercise 14.5.

The following measurements were made:

Slice	b (m)	z (m)	h_w (m)	α (°)
1	2.8	1.0	0	−20
2	3.4	3.0	0.8	−12
3	3.4	5.0	2.0	4
4	3.4	6.0	2.4	18
5	3.4	3.4	0.6	48

Determine the factor of safety of the slope using Bishop's conventional method of analysis.

Answer   1.70

Exercise 14.6

In the stability analysis of an earth embankment, the slip circle shown in Fig. 14.40 was used and the following figures obtained:

Slice	Breadth, b (m)	Weight, W (kN)	α (°)
1	5.65	372	−26
2	5.65	656	−7
3	5.65	1070	12
4	5.65	1220	30
5	5.65	686	54

With the values above, and using the conventional method, determine the safety factor of the slope assuming the pore pressure ratio to be 0.4 and the cohesion of soil and the angle of shearing resistance (with regard to effective stresses) to be 10 kPa and 30°, respectively. The radius of the slip circle was 18.8 m and the angle subtended at the centre of the circle was 100°.

Answer 1.12

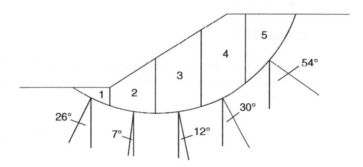

Fig. 14.40 Exercise 14.6.

Exercise 14.7

Using the slip circle shown in Fig. 14.41, determine the F values for $r_u = 0.4$, 0.6 and 0.8 using Bishop's rigorous method if the slope has the following properties:

$$\gamma = 21 \text{ kN/m}^3, c' = 5 \text{ kPa}, \phi' = 37.5°$$

Plot the relationship of r_u against F.

Answer
r_u	0.4	0.6	0.8
F	1.5	1.0	0.4

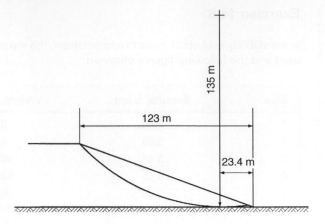

Fig. 14.41 Exercise 14.7.

Exercise 14.8

Check the GEO limit state requirement for stability for the embankment described in Exercise 14.6 using (a) Eurocode 7 (1st generation) Design Approach 1, Combination 2, and (b) the MFA of the 2nd generation (assume slope is in Consequence Class 2).

Answer $\Gamma = 0.9$

Exercise 14.9

Details of an earth embankment are shown in Fig. 14.42. Using the material factor approach of EN 1997-3:202x, determine the over-design factor for the slope, given the following data:

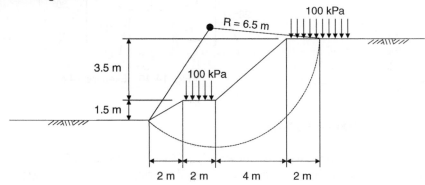

Fig. 14.42 Exercise 14.9.

Unit weight of soil = 20 kN/m³
Effective angle of shearing resistance = 30°
Effective cohesion = 15 kPa
Pore pressure ratio = 0.3
Use five slices, each 2 m wide.

Answer $\Gamma = 1.4$

Chapter 15

Soil Compaction, Highway Foundation Design and Ground Improvement

> **Learning objectives:**
>
> By the end of this chapter, you will have been introduced to:
> - the various plant used to compact soils in the field;
> - the laboratory compaction tests used to ascertain the maximum dry density and optimum moisture content of soils;
> - the specification and the measurement of field density;
> - the principles of highway foundation design in accordance with the DMRB;
> - the CBR and MCV tests;
> - various ground improvement techniques.

Soil compaction is extensively employed in the construction of embankments and in strengthening the *subgrades* of roads and runways. The subgrade is simply the soil immediately beneath the road or runway. The purpose of compaction is to reduce the volume of the soil by mechanical means thus reducing its permeability and compressibility and increasing its shear strength. By applying mechanical energy to the soil, the solid particles are packed together, by the expulsion of air. This results in a reduction of void ratio and an increase in the soil density.

The compaction of a soil, measured in terms of dry density, is a function of water content, compactive effort and the nature of the soil. Laboratory compaction tests are used to determine the relationship between the dry density, ρ_d and water content, w. Often, in the subject of the compaction of soils, the term *moisture content* is used rather than *water content*. For a given compactive effort, the maximum dry density for a soil occurs at the soil's optimum moisture content.

15.1 Field compaction of soils

The compaction plant used to compact wide areas of subgrade or placed fill are almost always rollers, of which there are several forms. These can be categorised in terms of total mass and the effective depth of soil which they can compact in a single passing. Obviously the greater the mass of a roller, the greater its compactive effort, and the static mass per metre width of roller is defined as the total mass on the roller divided by its width. Where a roller has more than one axle the machine's mass category is determined using the roller which gives the greatest value of mass per unit width.

Smith's Elements of Soil Mechanics, 10th Edition. Ian Smith.
© 2021 John Wiley & Sons Ltd. Published 2021 by John Wiley & Sons Ltd.
Companion website: www.wiley.com/go/smith/soilmechanics10e

Several of the following pieces of plant are now fitted with advanced on-board compaction control systems which allow the operator to see at a glance, the density and state of compaction of the soil as work progresses. This enables an extremely efficient compaction process, whereby the operator knows when the soil is at the required degree of compaction, and therefore knows that they can move the equipment on to the next section to be compacted.

Smooth wheeled roller

Probably the most commonly used roller in the world is the smooth wheeled roller. It consists of hollow steel drums so that its weight distribution can be altered by the addition of ballast (sand or water) to the rollers. These rollers vary in mass/m width from about 2100 kg to over 54 000 kg and are suitable for the compaction of most soils except uniform, and silty, sands. They are usually self-propelled although towable units are also available.

The successful operation of smooth wheeled rollers is difficult (and often impossible) when site conditions are wet, and in these circumstances rollers that can be towed by either track-laying or wheeled tractors are used. Both dead weight and vibratory units are available.

Vibratory roller

The smooth wheeled vibratory roller is either self-propelled or towed and has a mass/m width value from 270 kg to over 5000 kg. The compactive effort is raised by vibration, generally in the form of a rotating shaft (powered by the propulsion unit) that carries out-of-balance weights. Tests have shown that the best results on both heavy clays and granular soils are obtained when the frequency of vibration is in the range 2200–2400 cycles per minute. Vibration is obviously more effective in granular soil but the effect of vibration on a 200 mm layer of cohesive soil can effectively double the compactive effort. A vibrating roller operating without the vibration activated is regarded as being a smooth wheeled roller.

Pneumatic-tyred roller

In its usual form, the pneumatic-tyred roller comprises a cab and drive unit housed above and between two axles, the rear axle generally having one less wheel than the forward axle (so arranged that they track in with the rear wheels). Some models have five wheels at the front and four at the back, others have four at the front and three at the back. The dead load of the vehicle gives a mass per wheel in the range of 1000 kg to over 12 000 kg. The mass per wheel is simply the total mass of the unit divided by the number of wheels. A certain amount of vertical movement of the wheels is provided for so that the roller can exert a steady pressure on uneven ground – a useful facility in the initial stages of compacting a fill.

This type of roller originated as a towed unit but is now widely available as a self-propelled vehicle. It is suitable for most types of soil and has particular advantages on wet cohesive materials.

Sheepsfoot roller

This roller consists of a hollow steel drum from which the feet project, dead weight being provided by placing water or wet sand inside the drum. It is generally used as a towed assembly (although self-propelled units are available), with the drums mounted either singly or in pairs.

The feet are usually either club-shaped or tapered and the number on a 5000 kg roller varies between 64 and 240. The dimensions of the feet vary greatly between models, in order to provide a range of alternative compaction options to cover different soil conditions. Variations in the shape of the feet have been used in regions where soil conditions are not suited to the above two shapes.

The sheepsfoot roller is only satisfactory on cohesive soils, but at low water contents the resulting compaction of such soils is probably better than can be obtained with other forms of plant. Their use in the UK was not common in the 20th century because of the generally wet soil conditions, but this is steadily changing as drier soil conditions are encountered more often these days.

Deep impact rollers

Recent advances in the development of compaction equipment has seen the introduction of rollers that can compact significant depths of soil in a single pass. High weight polygonal drum rollers can achieve much greater compaction efficiency and depths than equivalent weight smooth wheeled rollers. Weighing around 32 000 kg these machines can be used to compact large extents of subgrade on highway construction projects in about half the time of alternative dead weight rollers. This method of compaction is known as *rolling dynamic compaction*.

The grid roller

This is a towed unit consisting of rolls made up from 38 mm diameter steel bars at 130 mm centres, giving spaces of 90 mm square. The usual mass of the roller is about 5500 kg which can be increased to around 11 000 kg by the addition of dead weights; there are generally two rolls, but a third can be added to give greater coverage. The grid roller is suitable for many soil types, but wet clays tend to adhere to the grid and convert it into a form of smooth roller.

Rammers and vibrators

Manually controlled power rammers can be used for all soil types and are useful when rolling is impractical due to restricted site conditions. Vibrating plates produce high dry densities at low water content in sand and gravels and are particularly useful when other plant cannot be used.

15.2 Laboratory compaction of soils

15.2.1 British standard compaction tests

Three different compaction tests are specified in BS 1377-4:1990 (BSI, 1990) and these are briefly described below.

The 2.5 kg rammer method

An air-dried representative sample of the soil under test is passed through a 20 mm sieve and 5 kg is collected. This soil is then thoroughly mixed with enough water to give a fairly low value of water content. For sands and gravelly soils, the commencing value of w should be about 5% but for cohesive soils it should be about 8–10% less than the plastic limit of the soil tested.

The soil is then compacted in a metal mould of internal diameter 105 mm using a 2.5 kg rammer, of 50 mm diameter, free falling from 300 mm above the top of the soil: see Fig. 15.1. Compaction is effected in three layers, of approximately equal depth. Each layer is given 27 blows, which are spread evenly over the surface of the soil.

The compaction can be considered as satisfactory when the compacted soil is not more than 6 mm above the top of the mould (otherwise the test results become inaccurate and should be discarded). The top of the compacted soil is trimmed level with the top of the mould. The base of the mould is removed and the mould and the test sample it encloses are weighed.

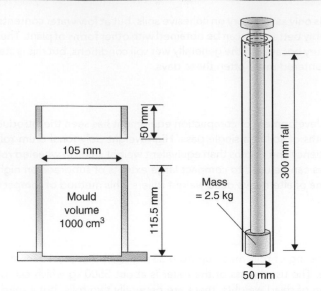

Fig. 15.1 Equipment for the 2.5 kg rammer compaction test.

Samples for water content determination are then taken from the top and the base of the soil sample, the rest of the soil being removed from the mould and broken down and mixed with the remainder of the original sample that passed the 20 mm sieve. A suitable increment of water (to give about 2% increase in water content) is thoroughly mixed into the soil and the compaction is repeated. The test should involve not less than five sets of compaction but it is usually continued until the weight of the wet soil in the mould passes some maximum value and begins to decrease.

Eventually, when the test has been completed, the values of water content corresponding to each volume of compacted soil are determined and it becomes possible to plot the dry density to water content relationship.

The 4.5 kg rammer method

In this compaction test, the mould and the amount of dry soil used are the same as for the 2.5 kg rammer method but a heavier compactive effort is applied to the test sample. The rammer has a mass of 4.5 kg with a free fall of 450 mm above the surface of the soil. The number of blows per layer remains the same, 27, but the number of layers compacted is increased to five.

The vibrating hammer method

It is possible to obtain the dry density/water content relationship for a granular soil with the use of a heavy electric vibrating hammer, such as the Kango. A suitable hammer, according to BS 1377-4:1990 (BSI, 1990), would have a frequency between 25 and 45 Hz, and a power consumption of 600–750 W. It should be in good condition and have been correctly maintained. The hammer is fitted with a special tamper (see Fig. 15.2) and for gravels and sands is considered to give more reliable results than the dynamic compaction techniques just described.

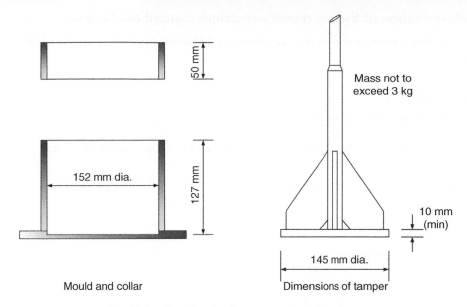

Fig. 15.2   The vibrating hammer compaction test.

The British Standard vibrating hammer test is carried out on soil in the 152 mm diameter mould, with a mould volume of 2305 cm^3, the soil having passed the 37.5 mm sieve. The soil is mixed with water, as for any compaction test, and is compacted in the mould in three approximately equal layers by pushing the tamper firmly down on to the soil and operating the hammer for 60 seconds, per layer.

The vibrating hammer method should only be used for fine-grained granular soils and for the fraction of medium- and coarse-grained granular soils passing the 37.5 mm sieve. For highly permeable soils, such as clean gravels and uniformly graded coarse sands, compaction by the vibrating hammer usually gives more dependable results than compaction by either the 2.5 kg rammer or the 4.5 kg rammer.

15.2.2   Soils susceptible to crushing during compaction

The procedure for each of the three compaction tests just described is based on the assumption that the soils tested are not susceptible to crushing during compaction. Because of this, each newly compacted specimen can be broken out of the mould and mixed with the remaining soil for the next compaction.

This technique cannot be used for soils that crush when compacted. With these soils, at least five separate 2.5 kg air dried samples of soil are prepared at different water contents and each sample is compacted, once only, and then discarded.

Preparation of the soil at different water contents

It is important, when water is added to a soil sample, that it is mixed thoroughly to give a uniform dispersion. Inadequate mixing can lead to varying test results and some form of mechanical mixer should be used. Adequate mixing is particularly important with cohesive soils and with highly plastic soils it may be necessary to place the mixed sample in an airtight container for at least 16 hours, in order to allow the moisture to migrate throughout the soil.

15.2.3   Determination of the dry density–moisture content relationship

For each of the three compaction tests described the following readings must be obtained for each compaction:

M_1 = Mass of mould
M_2 = Mass of mould + soil
w = water/moisture content (as a decimal)

The bulk density and the dry density values for each compaction can now be obtained:

$$\rho_b = \frac{M_2 - M_1}{1000} \text{ Mg/m}^3 \text{ (for the 2.5 kg and 4.5 kg rammers)}$$

$$\rho_b = \frac{M_2 - M_1}{2305} \text{ Mg/m}^3 \text{ (for the vibrating hammer test)}$$

From Equation (1.13), we have:

$$\rho_d = \frac{\rho_b}{1 + w}$$

When the values of dry density and moisture content are plotted, the resulting curve has a peak value of dry density. The corresponding moisture content is known as the *optimum moisture content* (omc). The reason for this is that at low w values the soil is stiff and difficult to compact, resulting in a low dry density with a high void ratio. As w is increased however, the water lubricates the soil, increasing the workability and producing high dry density and low void ratio. However, beyond omc, pore water pressures begin to develop and the water tends to keep the soil particles apart resulting in low dry densities and high void ratios.

With all soils, an increase in the compactive effort results in an increase in the maximum dry density and a decrease in the optimum moisture content (Fig. 15.3).

15.2.4   Percentage air voids, V_a

Saturation line

Figure 15.3 illustrates the saturation line, or *zero air voids line* as it is often called. It represents the dry densities that would be obtained if all the air in the soil could be expelled, so that after compaction the sample

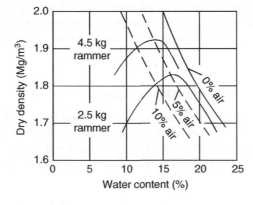

Fig. 15.3   Typical compaction test results.

would be fully saturated. This state is impossible to achieve by compaction either in the laboratory or in the field, but with the compactive efforts now available, it is quite common for a soil to have as little as 5% air voids after compaction.

The equation of the zero air voids line is given by:

$$\rho_d = \frac{\rho_w}{\dfrac{1}{G_s} + w} \tag{15.1}$$

where all symbols have their usual meaning, as defined in Chapter 1.

BS 1377-4:1990 (BSI, 1990) expresses the percentage air voids as the volume of air in the soil expressed as a percentage of the total volume, rather than as a percentage of the void volume. Hence 5% air voids does not mean the same as 95% degree of saturation.

Air voids line

Just as for the saturation line, it is possible to draw a line showing the dry density to water content relationship for a particular air voids percentage, V_a (see Fig. 15.3.)

$$V_a = \frac{\text{Volume of air}}{\text{Total volume}} (\%) \tag{15.2}$$

Referring to Fig. 1.12 and Equation (1.10), we have:

$$V - V_a = V_s - V_w$$

$$V\left[1 - \frac{V_a}{V}\right] = \frac{M_s}{G_s\rho_w} + \frac{M_w}{\rho_w}$$

$$= \frac{M_s}{G_s\rho_w}\left[1 + G_s\frac{M_w}{M_s}\right]$$

$$\frac{M_s}{V} = \frac{G_s\rho_w\left[1 - \dfrac{V_a}{V}\right]}{\left[1 + G_s\dfrac{M_w}{M_s}\right]}$$

From Equations (15.2) and (1.1)

$$\Rightarrow \rho_d = \frac{G_s\rho_w(1 - V_a)}{[1 + G_s w]}$$

$$\rho_d = \frac{\rho_w(1 - V_a)}{\left[\dfrac{1}{G_s} + w\right]} \tag{15.3}$$

where w is the water content expressed as a decimal, and
V_a is the percentage of air voids (%)
This is the equation of the air voids line, for all percentages.

It is seen that by putting $V_a = 0$ in Equation (15.3) we obtain the relationship between dry density and water content for zero air voids, i.e. the saturation line (Equation (15.1)).

15.2.5 Correction for gravel content

As noted, the standard compaction test is carried out on soil that has passed the 20 mm sieve. Any coarse gravel or cobbles in the soil is removed.

Provided that the percentage of excluded material, X, does not exceed 5% there will be little effect on the derived maximum dry density and optimum moisture content. If X is much greater than 5% then the test values will be affected and may be quite different to those pertaining to the natural soil. There is no generally accepted method to allow for this problem but for X up to 30%, some form of correction to allow for the removal of the gravel can be applied as set out below.

The percentage of material retained on the 20 mm sieve, X, can be obtained either by weighing the amount on the sieve (if the soil is oven dried) or else from the particle size distribution curve.

In the compacted soil, this percentage of coarse gravel and cobbles has been replaced by an equal mass of soil of a smaller size. Generally, the particle specific gravity of the gravel will be greater than that of the soil of smaller size so the volume of the gravel excluded would occupy a smaller volume than that of the soil which replaced it.

If ρ_d = maximum dry density obtained from the test, then:

$$\text{mass of gravel excluded} = \frac{X}{100}\rho_d$$

Considering unit volume, the volume of soil that replaced the gravel = X, so that the volume of gravel omitted,

$$= \frac{X}{100}\frac{\rho_d}{\rho_w G_s}$$

where G_s = particle specific gravity of excluded gravel.

$$\Rightarrow \text{Difference in volume} = \frac{X}{100} - \frac{X}{100}\frac{\rho_d}{\rho_w G_s} = \frac{X}{100}\left(1 - \frac{\rho_d}{\rho_w G_s}\right)$$

Corrected maximum dry density,

$$\rho_{d,max} = \frac{\rho_d}{1 - \dfrac{X}{100}\left(1 - \dfrac{\rho_d}{\rho_w G_s}\right)} = \frac{\rho_d}{1 + \dfrac{X}{100}\left(\dfrac{\rho_d}{\rho_w G_s} - 1\right)} \tag{15.4}$$

Similarly, corrected optimum moisture content,

$$w_{omc,\ corr} = \frac{100 - X}{100} w_{omc} \tag{15.5}$$

where w_{omc} = the optimum moisture content obtained from the test.

Even when a gravel correction is applied, compaction test results are not representative for a soil with X > 30%. BS 1377 suggests that in such cases the CBR mould (a mould 152 mm in diameter and 127 mm high) can be used. The compaction procedure is similar to that for the standard test except that the soil is compacted in the bigger mould in three equal layers with the same 2.5 kg rammer falling 300 mm but the number of blows per layer is increased to 62.

Correction for the excluded gravel (i.e. the particles greater than 37.5 mm) can be carried out in the manner proposed for the 20 mm size.

Example 15.1: 2.5 kg compaction test

The following results were obtained from a compaction test using the 2.5 kg rammer and a proctor mould of internal volume 1 l.

Test number	1	2	3	4	5	6
Mass of mould + wet soil (g)	2783	3057	3224	3281	3250	3196
Water content (%)	8.1	9.9	12.0	14.3	16.1	18.2

The weight of the mould, less its collar and base, was 1130 g and the soil had a particle specific gravity of 2.70.

Plot the curve of dry density against water content and determine the optimum water content. On your diagram plot the lines for 5 and 0% air voids.

Solution:

The calculations are best tabulated:

w (%)	Mass of mould + wet soil (g)	Mass of wet soil, M (g)	Bulk density, ρ (Mg/m^3)	Dry density, ρ_d (Mg/m^3)
8.1	2783	1653	1.65	1.53
9.9	3057	1927	1.93	1.76
12.0	3224	2094	2.09	1.87
14.3	3281	2151	2.15	1.88
16.1	3250	2120	2.12	1.83
18.2	3196	2066	2.07	1.75

The relationship between ρ_d and V_a is given by the expression:

$$\rho_d = \frac{\rho_w(1 - V_a)}{\left[\dfrac{1}{G_s} + w\right]}$$

Values for the 0 and 5% air voids lines can be obtained by substituting $V_a = 0$ and $V_a = 0.05$ in the above formula along with different values for w.

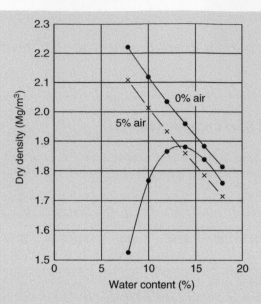

Fig. 15.4   Example 15.1.

w (%)	8	10	12	14	16	18
ρ_d (Mg/m³)						
$V_a = 0$	2.22	2.13	2.04	1.96	1.89	1.81
$V_a = 0.05$	2.11	2.02	1.94	1.86	1.79	1.73

The complete plot is shown in Fig. 15.4.
   From the plot,

$\rho_{d,max} = 1.88$ Mg/m³
omc = 14%

Example 15.2: Corrected ρ_{dmax} and omc for gravel content

A 2.5 kg rammer compaction test on a soil sample that had been passed through a 20 mm sieve gave a maximum dry density of 1.91 Mg/m³ and an optimum moisture content of 13.7%. If the percentage mass of the soil retained on the sieve was 20%, determine the gravel-corrected values for $\rho_{d,max}$ and the omc. The particle specific gravity of the retained particles was 2.78.

Solution:

$$\rho_{d,max} = \frac{\rho_d}{1 + \dfrac{X}{100}\left(\dfrac{\rho_d}{\rho_w G_s} - 1\right)}$$

$$= \frac{1.91}{1 + 0.2\left(\dfrac{1.91}{1.0 \times 2.78} - 1\right)}$$

$$= 2.04 \; \text{Mg/m}^3$$

$$w_{omc,corr} = \frac{100 - X}{100} w_{omc} = \frac{80}{100} \times 0.137 = 11\%$$

15.3   Specification of the field compacted density

With the main types of compaction plant, the optimum moisture contents of many soils are fairly close to their natural water contents and in such cases, compaction can be carried out without variation of water content.

In the United Kingdom, the *Design manual for roads and bridges* (DMRB) (Highways England et al., 2020a) is used as the basis for highway design and is generally referred to in parallel with the *Specification for highway works* (SHW) (Highways England et al., 2016a). The DMRB emphasises the importance of placing and compacting soils whose natural water contents render them in an *acceptable* state. Soils in an acceptable state will generally achieve the degree of compaction required and will be able to be excavated, transported and placed satisfactorily. The DMRB gives guidance on the limits of acceptability for different types of soil, and the methods of test by which acceptability should be assessed for each. In both the SHW and DMRB, different types of soil (e.g. cohesive, granular) together with their anticipated use (e.g. general fill, landscape fill) are identified apart by a *Class* type. The DMRB recommends different types of test (e.g. water content, compaction, MCV (see Section 15.7)) and appropriate limits of acceptability for each class, and if the soil is placed at a water content within the limits, a satisfactory degree of compaction be achieved.

Overcompaction

Care should be taken to ensure that the compactive effort in the field does not place the soil into the range beyond optimum moisture content. In Fig. 15.5, the point A represents the maximum dry density corresponding to the optimum moisture content. If the soil being compacted has a water content just below the optimum value the dry density attained will be point B, but if compaction is continued after this stage the optimum moisture content decreases (to point D) and the soil will reach the density shown by point C. Although the dry density is higher, point C is well past the optimum value and the soil will therefore be much softer than if compaction had been stopped once point B had been reached.

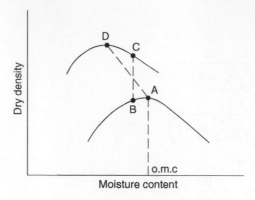

Fig. 15.5   Overcompaction.

15.3.1   Compactive effort in the field

The amount of compactive effort delivered to a point in a soil during compaction depends upon both the mass of the compacting unit and the number of times that it runs over the point (i.e. the number of passes). Obviously the greater the number of passes the greater the compactive effort but, as discussed in the previous section, this greater compactive effort will not necessarily achieve a higher dry density. The number of passes must correspond to the compactive effort required to achieve the maximum dry density when the actual water content of the soil equals the optimum moisture content. Specification by *method compaction* (see Section 15.3.3) is when the contractor is instructed to use particular compaction plant and a fixed number of passes.

15.3.2   Relative compaction

Laboratory compaction tests use different compactive efforts from those of many of the large pieces of compaction plant so results from these tests cannot always be used directly to predict the maximum dry density values that will be achieved in the field. What can be used is the *relative compaction*, which is the percentage ratio of the *in situ* maximum dry density of the compacted fill material to the maximum dry density obtained with the relevant laboratory compaction test. Specification by *end product compaction* (see Section 15.3.4) is when the contractor is directed to achieve a certain minimum value relative compaction and is allowed to select their own plant.

15.3.3   Method compaction

The *Specification for Highway Works* (2016a), gives two Tables (6.1 and 6.4) from which, knowing the classification characteristics of the soil, it is possible to decide upon the most suitable compaction equipment together with the number of passes that it must use. With this information a specification by method compaction can be prepared. The tables give an instruction for the method of compaction to be used: e.g. *using a vibrating roller of mass 1000 kg, the compaction should take place by compacting layers of soil of maximum thickness 100 mm by a number of 12 passes of the roller.*

15.3.4   End-product compaction

The values of the maximum dry density and the optimum moisture content are obtained using the 2.5 kg rammer or the vibrating hammer test, depending upon which is more relevant to the expected field compaction. The value of the required *relative compaction* should be equal to or greater than the value quoted in Table 6.1 of the SHW (2016a), usually 90–95%. The soil or fill in the field should then be compacted to that density and the compacted density verified through *in situ* density measurement by, for example, the sand replacement test.

15.3.5   Air voids percentage

Another form of end-product specification is to instruct that a certain minimum value of air voids percentage is to be obtained in the compacted soil. This value of V_a was for many years taken as between 5 and 10%, but later research suggested that a value of less than 5% is required to remove the risk of collapse on inundation sometime after construction (Trenter and Charles, 1996; Charles et al. 1998).

15.4   Field measurement tests

During construction, tests for bulk density and water content should be carried out at regular intervals if proper control of the compaction is to be achieved. The number and spacing of tests is governed by the nature of the particular project and by the type of the compaction activities taking place.

15.4.1   Bulk density determination

These tests are described in BS 1377-9:1990 (BSI, 1990) and BS EN ISO 11272:2017 (BSI, 2017).

Core-cutter method

Details of the core-cutter apparatus, which is suitable for cohesive soils, are given in Fig. 15.6. After the cutter has been first pressed into the soil and then dug out, the soil is trimmed to the size of the cutter and both

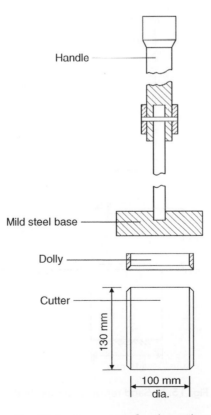

Fig. 15.6   Core cutter for clay soil.

cutter and soil are weighed. Given the weight and dimensions of the cutter, the bulk density of the soil can be obtained.

Sand replacement test

For granular soils, the apparatus shown in Fig. 15.7 is used. The test procedure is described in BS 1377-9:1990 (BSI, 1990). A small round hole (about 100 mm diameter and 150 mm deep) is dug, and the mass of the excavated material is carefully determined. The volume of the hole thus formed is obtained by pouring into it sand of known density from a special graduated container. Given the weight of sand in the container before and after the test, the weight of sand in the hole, and hence the volume of the hole, can be determined.

The apparatus shown in Fig. 15.7 is suitable for fine to medium grained soils and is known as the small pouring cylinder method. For coarse grained soils a larger pouring cylinder is used. This cylinder has an internal diameter of 215 mm and a height of 170 mm to the valve or shutter. The excavated hole in this case should be about 200 mm in diameter and about 250 mm deep. This larger pouring cylinder can also be used for fine to medium grained soils.

A variation on the procedure just described is given in BS EN ISO 11272:2017 (BSI, 2017). In this procedure the dug hole is lined with polyethylene film and the sand that fills the hole and levelled off at the surface is retained within the film removed from the hole and weighed. As with the above procedure, knowing the mass of the excavated soil and the density of the dry sand, the bulk density of the soil is established in the same way.

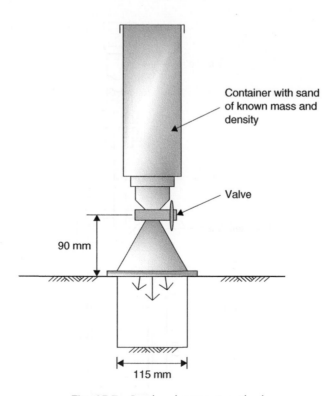

Container with sand of known mass and density

Valve

90 mm

115 mm

Fig. 15.7   Sand replacement method.

Example 15.3: Sand replacement test

During construction of a highway, a sand replacement test was performed to determine the *in situ* density of the subgrade. The following results were obtained:

Total mass of sand used in test = 8500 g
Mass of sand retained in cylinder at end of test = 2500 g
Mass of soil removed from hole = 5036 g

Previously the following had been established in the laboratory:

Mass of sand held in cone beneath cylinder = 1760 g
Density of sand = 1.6 Mg/m^3

Determine the bulk density of the subgrade soil.

Solution:

Mass of sand in hole, M_{sand} = 8500 − 2500 − 1760 = 4240 g

$$\Rightarrow \quad \text{Volume of hole, V} = \frac{M_{sand}}{\rho_{sand}} = \frac{4240 \times 10^{-6}}{1.6} = 2.65 \times 10^{-3} \text{ m}^3$$

$$\Rightarrow \rho_b = \frac{M_{soil}}{V} = \frac{5036 \times 10^{-6}}{2.65 \times 10^{-3}} = 1.90 \text{ Mg/m}^3$$

Nuclear density gauge

This instrument consists of an aluminium probe which is pushed into the soil, connected to a detection and measuring unit placed on the soil surface. Neutrons emitted from a source in the probe lose their energy by collision with soil and water particles. A detector fitted to the instrument then picks up the number of slow neutrons deflected back to the instrument. Since this number depends upon the amount of water present in the soil, measurements of the water content, the bulk density and the dry density of the soil can be obtained directly. The apparatus gives a rapid and dependable reading and thus the test has become much more widely used than the sand replacement method test. However, as the apparatus uses a nuclear source, strict legislation exists as to its use which can render it unapproved for use on some projects.

Non-nuclear density gauge

Because of the health and safety aspects associated with the nuclear density gauge, a non-nuclear version of the apparatus has been developed in recent years. This probe uses advanced electrical impedance spectroscopy rather than nuclear scatter to establish the density and water content of the soil. It gives rapid results and is likely to become more commonplace than the nuclear density gauge.

15.4.2 Water content determination

Quick water content determinations are essential if compaction work is to proceed smoothly, and the *Manual of contract documents for highway works*, *MCHW 2* (2016b) permits the use of quicker drying techniques

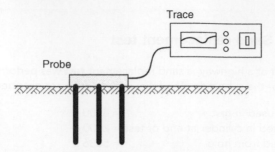

Fig. 15.8   Time domain reflectometry probe.

than are specified in BS 1377. The most common of these tests are described below. Nevertheless, it is standard practice to calibrate these results occasionally by collecting suitable soil samples and determining their water content values accurately by oven drying.

Nuclear and non-nuclear density gauges

These have been discussed in the previous section. Both gauges are capable of determining the water content of the *in situ* soil. The test results are both accurate and rapid, and the gauges are reasonably commonplace on highways projects.

Time domain reflectometry (TDR)

The TDR method for the measurement of soil water content is relatively new and an early review of the method was given by Topp and Davis (1985). The technique involves the determination of the propagation velocity of an electromagnetic pulse sent down a fork-like probe installed in the soil (Fig. 15.8). The velocity is determined by measuring the time taken for the pulse to travel down the probe and be reflected back from its end. The propagation velocity depends on the dielectric constant of the material in contact with the probe (i.e. the soil). The dielectric constant of free water is 80 and that of a soil matrix is typically three to six. Hence as the water content of the soil changes, there is a measurable change in the dielectric constant of the system, which affects the velocity of the pulse. Therefore, by measuring the time taken by the pulse, the water content of the soil around the probe can be established.

Soil moisture capacitance probe

Descriptions of the capacitance technique for determining soil water content are given by Dean *et al.* (1987) and Bell *et al.* (1987). The probe utilises the same principle as TDR, i.e. that the capacitance of a material depends on its dielectric constant. The probe is used inside a PVC access tube installed vertically into the soil. Indirect measurements of the capacitance of the surrounding soil are determined at the required depth and these are translated into soil water content using a simple formula. A description of the use of a capacitance probe for measuring soil water content is given by Smith *et al.* (1997).

15.5   Highway design

As mentioned earlier, highway design in the UK follows the *Design manual for roads and bridges* (DMRB) (Highways England *et al.*, 2020a). The main concern in the design of a highway pavement structure is that its overall thickness from soil to running surface is such that the soil beneath the road will not be overstressed by the imposed traffic loads. A useful reference document for readers interested in this subject area is the *ICE Manual of highway design and management* (Walsh, 2011).

Pavements can be either *flexible* or *rigid*:

- *Flexible*: These structures can flex under loading. They are constructed using asphaltic, cement-bound granular or hydraulically bound granular bases, all overlain with layers of asphalt materials.
- *Rigid*: These may be *continuously reinforced concrete pavement* (CRCP) normally with an asphalt overlay of at least 30 mm, or *continuously reinforced concrete base* (CRCB) with a 100 mm thick asphalt overlay.

The choice of which depends largely upon the national or local authority requirements, as well as on economic considerations. As most new highways are flexible, the required construction plant is readily available and therefore flexible highway construction is cheaper and tends to dominate new highway construction. However, rigid pavements do occasionally exist, though the cost associated with these structures is often considered excessive.

15.5.1 Components of a flexible road

Both a road and an airfield runway consist of two basic parts: the *pavement* and the *foundation*, both of which comprise different layers of compacted materials as shown in Fig. 15.9.

- *Pavement*: the layers of bound and bituminous materials above *formation* level. The pavement distributes wheel loads over an area so that the bearing capacity of the subgrade is not exceeded. It usually consists of two or more layers of material: a top layer or wearing/surface course which is durable and waterproof with good skid resistance, and a base material. For economic reasons, the base material is sometimes split into two layers, a base and a subbase.
- *Foundation*: comprises the subgrade and the layers above, up to and including the subbase. It is the platform on which the more expensive and structurally significant layers are placed. It must satisfy certain minimum standard requirements, regardless of the underlying soil conditions.

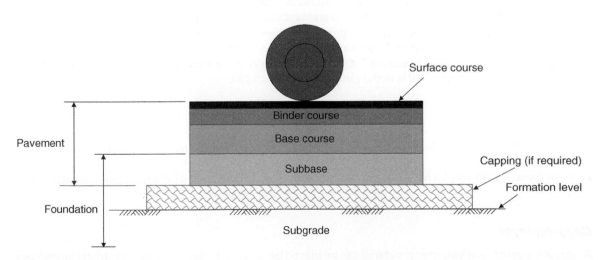

Fig. 15.9 Flexible pavement construction.

15.5.2 Highway design in the UK

The subgrade is seldom stiff or strong enough to withstand applied stresses from traffic loads placed on a thin, non-engineered running surface above, so two aspects of design are considered:

- improve the strength of the subgrade and thereby reduce the required pavement thickness;
- design and construct a sufficiently thick pavement to suit the subgrade.

For the trunk road network in the UK, pavement foundations are designed in accordance with the DMRB design standard, CD 225 *Design for new pavement foundations*. This document is used alongside CD 226 *Design for new pavement construction*. Pavement foundations are classified using four classes. The classification depends on the surface modulus at the top of the foundation level, and the amount of anticipated traffic.

The procedures involved require an assessment of the subgrade strength and stiffness to be made to inform the foundation design. For weak subgrades, strengthening or stabilisation techniques (see Section 15.6) may be employed to improve the ground conditions to ensure that the pavement does not experience unacceptable settlement during service.

The foundation design determines if a capping layer is required (and if so, its thickness) and the thickness of the subbase to ensure the applied stress placed on the subgrade does not exceed the acceptable limit. During the design process, both the short-term and long-term conditions are assessed. As we have seen throughout this book, undrained and drained states must be considered in all geotechnical design work. In the case of highway design, it should also be borne in mind that heavy construction plant will apply high stresses onto the subgrade when the capping and/or subbase are being placed. Thus, the short-term applied loadings will be considerably higher than the long-term where traffic loads are distributed through the pavement layers and less stress is applied to the subgrade than during construction.

Subgrade stiffness

The assessment of the subgrade involves establishing the subgrade surface stiffness modulus, E. Both the short-term modulus and the long-term modulus are established. The foundation design depends on the magnitude of E, and a specific highway strength test – the California bearing ratio (CBR) test – is routinely used to estimate the magnitude of E. The CBR test is described in Section 15.5.3.

Traditionally, the DMRB based the design of pavement foundations on the subgrade CBR value. The latest version of DMRB (2020a) however, guides highway designers to design foundations based on the subgrade surface stiffness modulus, E instead. Determination of the soil stiffness can be made from field penetration tests (such as those described in Section 7.6), or an estimate of E from the CBR may be made using Equation (15.6) (where the *in situ* CBR is in the range 2% < CBR < 12%).

$$E = 17.6(CBR)^{0.64}(MPa) \tag{15.6}$$

- If E < 30 MPa, the subgrade is too weak to support the construction of a pavement, and ground improvement becomes necessary.
- If E is between 30 and 50 MPa, the provision of a capping layer as part of the foundation is required.

Capping layer

A *capping layer* of relatively cheap material can be placed between the subgrade and the subbase to improve the strength of the structure – see Fig. 15.9. The design procedure is followed to establish the required thickness of capping and subbase layers: the thickness of each depends on the subgrade surface stiffness modulus. The method given in the manual permits alternative combinations of capping and subbase layer thickness. These thicknesses depend on the particular materials used for each layer. Increasingly, recycled concrete and other recycled construction waste materials are used as capping. The suitability of materials for use as capping or subbase is validated through the design method being followed: *restricted design* or *performance design*.

Restricted and performance design methods

The DMRB permits two methods of pavement foundation design: *restricted design* or *performance design*.

- *Restricted design*: based on a limited selection of materials linked to an assumed performance which does not require verification via performance testing of the foundation. This approach tends to be used more on smaller schemes, where less material testing is required.
- *Performance design*: offers economic and environmental benefits through innovation and/or the use of materials not permitted within restricted foundation designs. The assurance of material performance is verified through materials testing. Used for major works and provides greater flexibility in design and range of materials than restricted design.

15.5.3   The California bearing ratio test

As we have seen, the strength and stiffness of the subgrade are the main factors in determining the thickness of the pavement, although the susceptibility to frost must also be considered. Subgrade strength is expressed in terms of its California Bearing Ratio (CBR) value. The CBR value is measured by an empirical test devised by the California State Highway Association and is simply the resistance to a penetration of 2.5 mm of a standard cylindrical plunger of 49.65 mm diameter, expressed as a percentage of the known resistance of the plunger to various penetrations in crushed aggregate, notably 13.2 kN at 2.5 mm penetration and 20.0 kN at 5.0 mm penetration.

The reason for the odd dimension of 49.65 mm for the diameter of the plunger is that the test was originally devised in the USA and used a plunger with a cross-sectional area of 3.0 square inches. This area translates as 1935 mm^2 and, as the test is international, it is impossible to vary this area and therefore the plunger has a diameter of 49.65 mm.

Laboratory CBR test

The laboratory CBR test is generally carried out on remoulded samples of the subgrade and is described in BS 1377-4:1990 (BSI, 1990). The usual form of the apparatus is illustrated in Fig. 15.10. The sample must be compacted to the expected field dry density at the appropriate water content. The appropriate water content is the *in situ* value used for the field compaction. However, if the final strength of the subgrade is required a further CBR test must be carried out with the soil at the same dry density but with the water content adjusted to the value that will eventually be reached in the subgrade after construction. This value of water content is called the *equilibrium water content*.

The dry density value can only be truly determined from full-sized field tests using the compaction equipment that will eventually be used for the road construction. Where this is impracticable, the dry density can be taken as that corresponding to 5% air voids at a moisture content corresponding to the omc of the standard compaction test. In some soils this will not be satisfactory – it is impossible to give a general rule; for example, in silts, a spongy condition may well be achieved if a compaction to 5% air voids is attempted. This state of the soil would not be allowed to happen *in situ* and the laboratory tester must therefore increase the air voids percentage until the condition disappears.

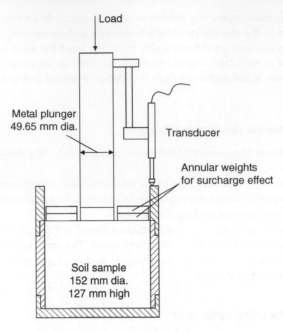

Fig. 15.10 The California bearing ratio test.

With its collar, the mould into which the soil is placed has a diameter of 152 mm and a depth of 177 mm. The soil is broken down, passed through a 20 mm sieve and adjusted to have the appropriate water content. The final compacted sample has dimensions 152 mm diameter and 127 mm height.

Two approaches to compacting the soil to the required density are available:

(i) *Static compaction*: sufficient wet soil to fill the mould when compacted is weighed out and placed in the mould. The soil is now compressed into the mould in a compression machine to the required height dimension. This is the most satisfactory method.

(ii) *Dynamic compaction*: the weighed wet soil is compacted into the mould in five layers using either the 2.5 kg rammer, the 4.5 kg rammer or the vibrating hammer. The amount of compaction for each layer is determined by experience, several trial runs being necessary before the amount of compactive effort required to leave the soil less than 6 mm proud of the mould top is determined. The soil is now trimmed, and the mould weighed so that the density can be determined.

After compaction the plunger is seated on to the top of the sample under a specific load: 50 N for CBR values up to 30% and 250 N for soils with CBR values above 30%.

The plunger is made to penetrate into the soil at the rate of 1.0 mm/minute and the plunger load is recorded for each 0.25 mm penetration up to a maximum of 7.5 mm. Test results are plotted in the form of a load–penetration diagram by drawing a curve through the experimental points. Usually, the curve will be convex upwards (curve Test 1 in Fig. 15.11), but sometimes the initial part of the curve is concave upwards, and over this section, a correction becomes necessary. The correction consists of drawing a tangent to the curve at its steepest slope and producing it back to cut the penetration axis. This point is regarded as the origin of the penetration scale for the corrected curve.

The plunger resistance at 2.5 mm is expressed as a percentage of 13.2 kN and the plunger resistance at 5.00 mm is expressed as a percentage of 20.0 kN. The higher of these two percentages is taken as the CBR value of the soil tested.

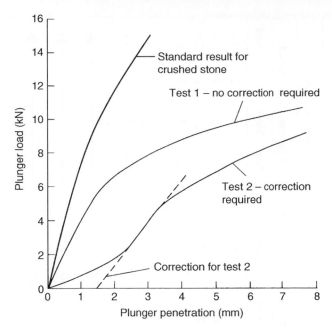

Fig. 15.11 Typical CBR results.

Surcharge effect

When a subgrade or a subbase material is to be tested, it is possible to allow for the increase in strength due to the road construction material placed above. Surcharge weights, in the form of annular discs with a mass of 2 kg can be placed on top of the soil test sample. The plunger penetrates through a hole in the disc to reach the soil. Each 2 kg disc is roughly equivalent to 75 mm of surcharge material.

In situ CBR test

When an existing road is to be reconstructed, the value of the subgrade strength must be determined. For such a situation the CBR test can be carried out *in situ* but it must be remembered that the water content of the soil should be the equilibrium value so that the test should be carried out in a newly excavated pit, not less than 1 m deep, on the freshly exposed soil. The test procedure for the determination of the *in situ* CBR is described in BS 1377-9:1990 (BSI, 1990). *In situ* CBR testing can also be employed on new highway constructions.

Estimation of CBR values

Table 15.1 gives estimates of equilibrium CBR values for the most common soils in the United Kingdom for various conditions of construction, groundwater and pavement thickness. The data in Table 15.1 comes from the work of Powell *et al.* (1984). A high water table is taken as one that is 300 mm below foundation level whilst a low water table is one that is 1000 mm below foundation level. For water tables at depths between these limits, the CBR value may be found by interpolation. In Table 15.1 a thick pavement has a total thickness of 1200 mm including any capping layer (used for motorways) and a thin pavement has a total thickness of 300 mm. For pavement thicknesses between these limits the value of the CBR may be interpolated.

Table 15.1 Equilibrium CBR values for common soils in UK.

Type of soil	Plasticity index	High water table — Poor — Thin	Poor Thick	Average Thin	Average Thick	Good Thin	Good Thick	Low water table — Poor — Thin	Poor Thick	Average Thin	Average Thick	Good Thin	Good Thick
Heavy clay	70	1.5	2	2	2	2	2	1.5	2	2	2	2	2.5
	60	1.5	2	2	2	2	2.5	1.5	2	2	2	2	2.5
	50	1.5	2	2	2.5	2	2.5	2	2	2	2.5	2	2.5
	40	2	2.5	2.5	3	2.5	3	2.5	2.5	3	3	3	3.5
Silty clay	30	2.5	3.5	3	4	3.5	5	3	3.5	4	4	4	6
Sandy clay	20	2.5	4	4	5	4.5	7	3	4	5	6	6	8
	10	1.5	3.5	3	6	3.5	7	2.5	4	4.5	7	6	>8
Silt*	–	1	1	1	1	2	2	1	1	2	2	2	2
Sand (poorly graded)	–	← 20 →											
Sand (well graded)	–	← 40 →											
Sandy gravel (well graded)	–	← 60 →											

*Estimated, assuming some probability of material saturating.
Data from Powell *et al.* (1984).

Example 15.4: CBR test

A CBR test on a sample of subgrade gave the following results:

Plunger penetration (mm)	Plunger load (kN)	Plunger penetration (mm)	Plunger load (kN)
0.25	1.0	3.25	10.1
0.50	1.6	3.50	10.7
0.75	2.4	3.75	11.2
1.00	3.6	4.00	11.7
1.25	4.5	4.25	12.2
1.50	5.3	4.50	12.7
1.75	6.0	4.75	13.0
2.00	6.8	5.00	13.5
2.25	7.5	5.25	14.1
2.50	8.3	5.50	14.4
2.75	9.0	5.75	14.6
3.00	9.4	6.00	14.9

The standard force penetration curve, corresponding to 100% CBR has the following values:

Plunger penetration (mm)	2	4	6	8
Plunger load (kN)	11.5	17.6	22.2	26.3

Determine the CBR value of the subgrade.

Solution:

The standard penetration curve is shown drawn in Fig. 15.12.

The test points are plotted, and a smooth curve drawn through them. In this case there is no need for the correction procedure as the curve is concave upwards in its initial stages.

From the test curve it is seen that at 2.5 mm the plunger load is 8.3 kN and at 5.0 mm the penetration is 13.5 kN.

The CBR value is therefore either

$$\frac{8.3}{13.2} \times 100 \quad \text{or} \quad \frac{13.5}{20.0} \times 100$$

i.e. 63 or 67%

The higher value is taken as the CBR, i.e. CBR = 67%.

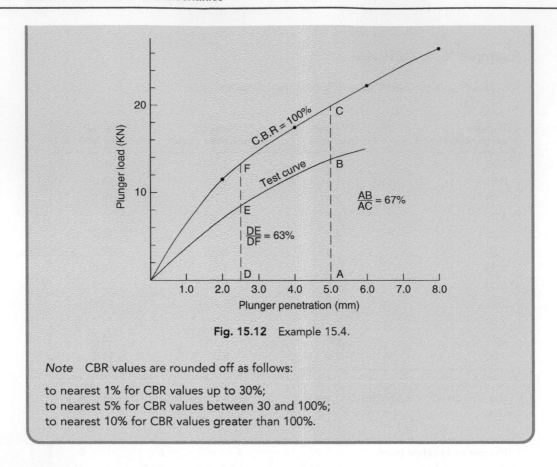

Fig. 15.12   Example 15.4.

Note   CBR values are rounded off as follows:

to nearest 1% for CBR values up to 30%;
to nearest 5% for CBR values between 30 and 100%;
to nearest 10% for CBR values greater than 100%.

15.5.4   Drainage and weather protection

Whenever practical, the water table should be prevented from rising above a depth of 300 mm below the foundation level. This is usually achieved by the installation of a drainage system to intercept a rising ground water table.

15.5.5   Frost susceptibility of subgrades and base materials

Cohesive soils: can be regarded as non-frost susceptible when I_p is greater than 15% for well drained soils and 20% for poorly drained soils (i.e. water table within 600 mm of formation level).

   Non-cohesive soils: except for limestone gravels, may be regarded as non-frost susceptible if with less than 10% fines.

   Limestone gravels are likely to be frost susceptible if the average saturation moisture content of the limestone aggregate exceeds 2%.

   Chalks: all crushed chalks are frost susceptible. Magnitude of heave increases linearly with the saturation water content of the chalk aggregate.

Limestone: all oolitic and magnesian limestones with an average saturation water content of 3% or more must be regarded as frost susceptible.

All hard limestones with less than 2% of average saturation water content within the aggregate and with 10% or less fines can be regarded as non-frost susceptible.

Granites: crushed granites with less than 10% fines can be regarded as non-frost susceptible.

Slags: crushed, graded slags are not liable to frost heave if they have less than 10% fines. Slags are a by-product of steel making, specifically the wate product of smelting in kilns.

Pulverised fuel ash: coarse fuel ashes with less than 40% fines are unlikely to be frost susceptible. Fine ashes may be frost susceptible and tests should be carried out before such materials are used in the top 450 mm.

15.5.6   Traffic assessment

Within the DMRB, traffic assessment is carried out in accordance with the procedures set out in document CD 224 *Traffic assessment*. The traffic assessment serves two functions: to estimate the *design traffic* for new roads, and to estimate past and future *design traffic* for the maintenance of existing roads. The traffic loads imposed on the running surface stress the layers that comprise the pavement and in turn stress the subgrade beneath. Considering new roads, it is important that the traffic assessment is performed so that the correct required thickness of pavement is established such that the subgrade will not be overstressed by the traffic loadings.

The flow of commercial vehicles (i.e. those greater than 15 kN unladen vehicle weight) is established in order to determine the cumulative design traffic. Commercial vehicles are placed into categories depending on the type of the vehicle (e.g. coach or lorry), the number of axles and the vehicle rigidity. Buses and coaches are placed in the *public service vehicle* (PSV) category, and lorries are placed into one of two *other goods vehicles* categories (OGV1 and OGV2) depending on their size. Traffic flow data should be determined from traffic studies and, from these, the percentage of OGV2 vehicles in the total flow is determined. To establish the cumulative design traffic, use is made of existing data and growth factors for the particular design period required. CD224 is used alongside CD 225 and CD226.

15.5.7   Design life of a road

An important part in the design of a road is the decision as to the number of years of life for which the road can be economically built. The DMRB suggests a design life of 20 years for bituminous roads, because their life may be extended by a strengthening overlay. However, during the design of streets in residential areas, a design life of 40 years may be considered.

15.6   Subgrade improvement through soil stabilisation

If the subgrade is too wet, weak or compressible to permit highway construction, additives may be mixed with the soil to reduce the water content and increase stiffness and strength, and reduce permeability. The most common additive used is lime powder, but cement, pulverised fuel ash (PFA), bitumen emulsion and certain non-hazardous waste materials and chemicals are also used.

In lime stabilisation, the powdered lime is spread over the surface of the soil from a spreader unit towed by a tractor or tracked vehicle. The lime is then ploughed into the soil using a towed unit containing a series of ploughs. This results in a good mixing of the soil and lime. Hydration of the lime with the water in the soil commences during mixing and it is not uncommon to see clouds of water vapour rising from the soil during this process.

Cement stabilisation follows the same process as for lime. The methods for mixing the other, alternative additives with the soil are also similar but require specialist equipment, such as bowsers, to spread the liquid additives onto the surface, or to inject the liquid beneath the surface.

15.7   The moisture condition value, MCV

The selection of a satisfactory soil for earthworks in road construction involves the visual identification of unsuitable soils, the classification tests described in Chapter 1 together with the use of at least one of the three compaction tests described in this chapter. The compaction test chosen is the one that uses a compactive effort nearest to the expected construction compactive effort and is used to determine the optimum moisture content value, i.e. the upper value of water content beyond which the soil becomes unworkable. The system can give good results, particularly with experienced engineers, but there are occasions when the assessment of a particular soil's suitability for earthworks is still difficult.

The moisture condition test is an attempt to remove some of the selection difficulties and was proposed by Parsons in 1976. It is essentially a strength test in which the compactive effort necessary to achieve near full compaction of the test sample is determined. The moisture condition value, MCV, is a measure of this compactive effort and is correlated with the undrained shear strength, c_u or the CBR value, that the soil will attain when subjected to the same level of compaction (Parsons and Boden, 1979).

The test procedure is given in BS 1377-4:1990 (BSI, 1990) and details of the apparatus are shown in Fig. 15.13. The test consists of placing a 1.5 kg sample of soil that has passed through a 20 mm sieve into a cylindrical mould of internal diameter 100 mm. The sample is then compacted to maximum bulk density with blows from a 7 kg rammer, 97 mm in diameter and falling 250 mm. After a selected number of blows (see Example 15.5), the penetration of the rammer into the mould is measured by a vernier attachment and noted. The test is terminated when no further significant penetration is noted or as soon as water is seen to extrude from the base of the mould. The latter requirement is essential if the water content of the sample is not to change.

MCV tests are intended to replace the previous techniques of defining an upper limit of moisture content for soil acceptability. Generally speaking, a soil with an MCV of not less than 8.5, which is comparable to the *in situ* compactive effort of present day plant, will be acceptable for earthworks. Although no difficulty will be experienced with testing the majority of soils, particularly those of a cohesive nature, problems can arise with granular soils with a low fines content.

15.7.1   Determination of MCV

The difference in penetration for a given number of blows, B, and a further three times as many blows (i.e. 4B blows) is calculated and is plotted against the logarithm of B. The calculation is repeated for all relevant B values so that a plot similar to Fig. 15.14 of Example 15.5 is obtained. The MCV is taken to be $10 \log_{10} B$ where B = the number of blows at which the change of penetration = 5 mm. As is seen from Fig. 15.14 the chart can be prepared so that the value of $10 \log_{10} B$, to the nearest 0.1, can be read directly from the plot. The value of 5 mm for the change in penetration value was arbitrarily selected as the point at which no further significant increase in density can occur. This avoids having to extrapolate the point at which zero penetration change occurs.

The MCV for a soil at its natural water content can be obtained with one test, the wet soil being first passed through a 20 mm sieve and a 1.5 kg sample collected.

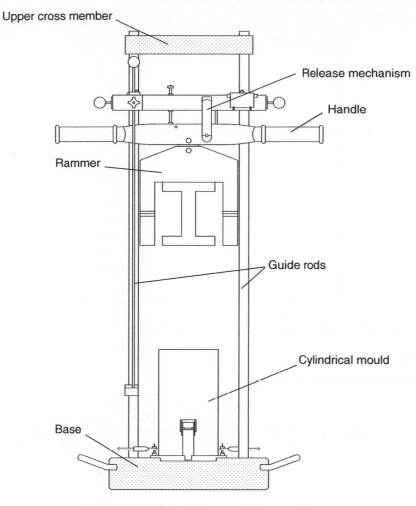

Fig. 15.13 The moisture content apparatus.

Example 15.5: MCV test

A sample of silty clay was subjected to a moisture condition test at its natural water content. The results obtained are set out below.

No. of blows, B	Penetration (mm)
1	45.6
2	55.3
3	62.7
4	67.1
6	74.5
8	79.5
12	86.1
16	90.7
24	96.6
32	99.4
48	101.9
64	102.9
96	103.5
128	103.8

Determine the MCV for the soil.

Solution:

The calculations are best tabulated:

No. of blows, B	Penetration (mm)	Change in penetration with additional 3B blows (mm)
1	45.6	21.5
2	55.3	24.2
3	62.7	23.4
4	67.1	23.6
6	74.5	22.1
8	79.5	19.9
12	86.1	15.8
16	90.7	12.2
24	96.6	6.9
32	99.4	4.4
48	101.9	
64	102.9	
96	103.5	
128	103.8	

The plot of change in penetration to number of blows is shown in Fig. 15.14. The plot crosses the line at B = 28.0. Hence the MCV of the soil is 10 = $\log_{10} 28.0$ = 14.5. The value of MCV can also be read directly on the bottom axis.

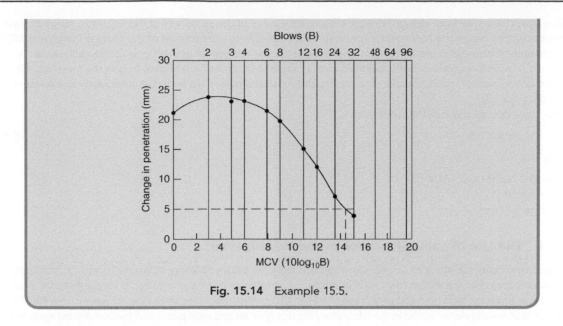

Fig. 15.14 Example 15.5.

15.7.2 The calibration line of a soil

If the relationship of MCV to water content is required, then a sample (approximately 25 kg) of the soil is air dried and passed through the 20 mm BS sieve. At least four sub-samples are taken and mixed thoroughly with different quantities of water to give a suitable range of water contents. Ideally, the range of water contents should be such as to yield MCVs between 3 and 15.

Test samples of mass 1.5 kg are prepared from each sub-sample and tested in the same manner as for the natural test and a series of MCVs are thus obtained. The water contents of the test samples are determined and a plot of the series of water contents against MCVs is produced. The best-fit plot is linear and is referred to as the *MCV calibration line*. An example of a calibration line is shown in Fig. 15.15.

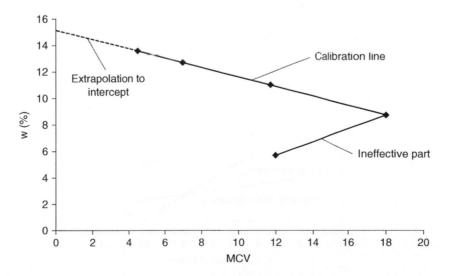

Fig. 15.15 Typical MCV calibration line.

The purpose of the best straight-line plot is to gain a value of the *sensitivity* of the soil and a range of water contents within which the MCV test is applicable. The sensitivity is a measure of the change in value of MCV for a specified change in water content and is equal to the reciprocal of the slope of the calibration line. If required, the calibration line is extrapolated back to the water content axis (y-axis) to give the *intercept*. Often an ineffective part of the calibration line can be detected if samples are tested at sufficiently low water contents (Fig. 15.15).

The equation of the calibration line is:

$$w = a - b(MCV) \tag{15.7}$$

where

w = water content at MCV (%)
a = intercept (%)
b = slope of line

15.7.3   The use of calibration lines in site investigation

To determine the equation of a calibration line it is only necessary to know its slope, b, and its intercept, a. This information can be extremely useful in deciding whether or not a soil is suitable for earthworks. Typical calibration lines for different soils are shown in Fig. 15.16. It is important to realise, however, that MCV calibration test results determined in the laboratory do not give an accurate picture of the conditions likely to be experienced in the field. As a result, a degree of care and judgement is required when using values of sensitivity and intercept to judge a soil's *in situ* acceptability for earthworking.

The value of the intercept gives a measure of the potential of the soil to retain moisture when in a state of low compaction. This potential increases with the value of the intercept.

The slope of the line gives a measure of the soil's change in MCV value, i.e. in strength, with variation in water content. The steeper the slope the more acceptable the soil. A soil with a very low slope will suffer significant changes in its MCV value over a small range of water content and could well be an ideal soil for compaction on a fine day yet utterly unsuitable in rainy or misty conditions. Such a soil has a very high sensitivity to water content changes and would therefore be considered unacceptable for earthworks.

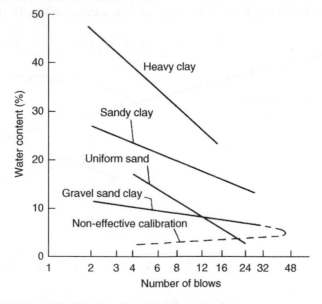

Fig. 15.16   Typical calibration line for different soils. Based on Parsons (1976).

15.8   Ground improvement techniques

Various processes can be used to strengthen weak and compressible soils. Some of these techniques, described in this section, are specifically useful for improving the strength and reducing the compressibility and permeability at depth. Surface soils can usually be improved through normal compaction methods as described in Section 15.1.

The subject of *ground improvement* is wide and varied and a thorough description of the range of techniques is given by Han (2015). This section provides a summary of only a few of the techniques available.

15.8.1   Consolidation by preloading

Total settlements of foundations on soft clays can be reduced by preloading the area of the structure with mounds of soil or rubble. These impose a bearing pressure on the ground greater than that of the proposed structure.

Clays will consolidate through time as they reduce in volume due to the expulsion of water in the voids under the applied loading. By preloading the area, that process can commence, and be accelerated, ahead of the foundations being cast. The adoption of preloading can only really be effectively adopted if (a) there is adequate soil/rubble available on the site, and (b) there is adequate lead-time ahead of the formation of the foundations to make adoption of the technique worthwhile. The technique is most applicable to soft and/or organic clays and peats.

15.8.2   Deep compaction techniques

As seen in Section 15.1, normal surface compaction processes can only compact the upper 1 or 2 m of soil. To enable compaction of the ground at greater depths, specialist equipment and techniques are required.

Vibro-compaction

Heavy vibrating pokers are lowered through loose granular soils and raised to permit sand to be poured down the resulting hole. The poker penetrates downwards under its own weight and compacts the surrounding soil up to a distance of about 2.5 m from the probe by virtue of the temporary reduction in effective stress caused by the vibration. The process is repeated leaving a column of compacted soil in the ground and can be assisted with jets of water (vibro-flotation). Vibro-compaction densifies the ground and thereafter increases strength and improves the bearing capacity. The method cannot be used in cohesive soils although soils with up to about 25% silt can also be treated. Pokers are normally spaced at 1.5–3 m and can compact suitable soils to a depth of about 12 m.

Vibro-flotation

In order to assist the penetration of the vibrating poker into the ground, water jets can be fitted at the top and bottom of the poker. In this case the process is referred to as *vibro-floatation* and the poker is called a *vibrofloat*.

Vibro-replacement (stone columns)

This technique can be used to improve the load carrying capacity and accelerate consolidation of soft silts and clays. The soil is reinforced by the insertion of stone columns from the surface to the required depth. This is achieved with the use of a vibrating probe similar to the technique used in vibro-compaction. The probe is allowed to penetrate the soil and does so by displacing the soil radially. Once the required depth has been reached, the probe is withdrawn, and the hole created by the probe backfilled with graded aggregate up to 75 mm in size. The probe is then reintroduced to both compact and radially displace the aggregate. The process is repeated until the required stone column has been created. With soft clays, soil is removed not displaced by means of water jets fitted to the probe. The method is really only suitable for light foundation loads as heavy loads can cause excessive settlement.

Dynamic consolidation

This method involves the dropping of a large weight 100–400 kN from a height of 5–30 m onto the surface of the soil. It is seen that the energy delivered to the soil per blow can be as high as 12 000 kNm although the energy values normally used lie between 1500 and 5000 kNm. The impact of the weight with the soil creates shock waves that can penetrate to a depth of 10 m. In cohesionless soils, these shock waves create liquefaction immediately followed by compaction of the soil, whereas in cohesive soils they create excessive pour water pressures which are followed by the consolidation of the soil.

Before the work is commenced, the area to be treated is covered with a layer of granular material of thickness between 0.5 and 1.0 m. The layer acts as a working platform for the equipment and helps to prevent excessive penetration of the weight. It also provides a preload surcharge of about 10–20 kPa and helps to drain away water as it is driven out of the soil.

The tamper is dropped a set number of times on a grid pattern over the site to form a pass. Two or three passes on a site, dependent on soil type and condition, may be required. The imprints formed at each drop position are infilled with granular material after each pass. In the case of cohesive soils not all the blows are delivered at once as it is necessary to have pauses in order that full consolidation for a particular treated area is first achieved. These pauses can extend to weeks in some cases.

The benefit of this process over earlier ones is that this method does not introduce any material deep into the ground. It simply compacts what is already there.

15.8.3   Grouting

The engineering properties of a soil can be improved by the injection of chemical fluids known as *grouts* which solidify in the soil mass and hence strengthen the soil structure. It should be fairly obvious that the system is only effective if the voids of the soil can be penetrated by the grout and it therefore has little application for cohesive soils, except when fissures require to be sealed. Grouting is mainly restricted to granular soil and weathered rocks. The procedure is expensive and is mainly used when other methods of soil improvement are not applicable.

Grouting has two main applications:

Compensation grouting: where the grout provides an increase in strength and stiffness to the soil to stop or control vertical settlement, movement or collapse of the soil above;
Permeation grouting: where the grout provides an impermeable barrier to flow of water through the soil.

In the main, cementitious grouts are used in geotechnical applications, however bentonite clay and bituminous based slurries can also be used as the grout mixture. Chemical grouts, comprising various compounds (e.g. sodium silicate and calcium chloride) mixed with hardening agents, can also be used in permeation grouting applications. The use of chemicals allows an element of time control of the grout-hardening process.

15.8.4   Prefabricated vertical drains

These have been described in detail in Section 13.15. Vertical drains are used in order to accelerate the consolidation of soft ground. They are suitable for very soft and soft soils with a high water content and have particular application beneath new embankments. Prefabricated plastic drains are inserted into the ground using a custom-built machine called a stitcher. A regular grid of drains leads to a controlled consolidation process once the surcharge loading (e.g. the embankment) is placed on top.

15.8.5 Geosynthetics

Geosynthetics are materials manufactured from various types of polymers used to enhance, augment and make possible, cost effective geotechnical, transportation and environmental engineering construction projects. They are used to provide one or more of the following functions:

- separation
- reinforcement
- filtration
- drainage
- liquid barrier

Separation

The base of a pavement construction may be subjected to separation if it is placed directly onto the surface of a soft subgrade. Separation is the upward migration of particles of the fine subgrade soil accompanied by the downward movement of the denser base particles. Such intermixing of soil particles can create a weak zone at the interface between the two materials resulting in considerable reduction in bearing capacity strength. The placing of a relatively weak strength geotextile fabric on the surface of a soft subgrade prior to constructing the base, is all that is necessary to provide a permanent solution to separation between the two materials.

Reinforcement

The use of plastic reinforcement in reinforced soil retaining walls is now well established. The technique is mentioned in Section 9.8. In the construction of an earth embankment on top of a soft foundation soil, a layer of geotextile fabric placed on the surface of the soft soil can give enough tensile strength to allow it to support an incremental layer of the embankment without spreading or edge failure during consolidation, and thus permit stage construction to be carried out. The subbases of roads supported by soft subgrades can be strengthened by the inclusion of layers of a geotextile fabric.

Filtration

Where a cohesive soil is subjected to seepage, a suitable geotextile can be used to prevent the migration of the fine soil particles in exactly the same way as the granular filters described in Chapter 2. A geotextile filter placed at the end of the seepage path operates in a different manner to a granular filter. Soil particles tend to collect at the boundary between the soil and the geotextile, and this appears to induce a self-filtration effect within the soil.

Drainage

Special types of permeable geotextile fabrics can be used to form drainage layers in basements and behind retaining walls in exactly the same manner as the layers of granular material illustrated in Fig. 8.25.

Liquid barrier

To prevent contaminants leaching, for example from a landfill site, into the groundwater, impermeable barriers can be created where required through the use of geomembranes.

The most common types of geosynthetics are:

Geotextiles: flexible, textile-like fabrics of controlled permeability, used to provide all of the above functions (except liquid barrier) in soil, rock and waste materials. It should be noted that natural fibre geotextiles (e.g. using jute) are manufactured in some parts of the world and these products are also considered to fall within the geotextile classification;
geomembranes: impermeable polymeric sheets used as barriers for liquid or solid waste containment;
geogrids: stiff or flexible polymer grid-like sheets with large apertures used primarily as reinforcement of unstable soil and waste masses;
geonets: stiff polymer net-like sheets with in-plane openings used primarily as a drainage material within landfills or in soil and rock masses;
geosynthetic clay liners: prefabricated bentonite clay layers incorporated between geotextiles and/or geomembranes and used as a barrier for liquid or solid waste containment;
geopipes: perforated or solid wall polymeric pipes used for the drainage of various liquids;
geocomposites: hybrid systems of any, or all, of the above geosynthetic types which can function as specifically designed. For use in soil, rock, waste and liquid related problems.

The geosynthetics industry has exhibited much innovation that has led to other speciality products. These include threaded soil masses, polymeric anchors, and encapsulated soil cells. As with geocomposites, their primary function is product-dependent and can be any of the five major functions of geosynthetics.

15.9 Environmental geotechnics

Environmental geotechnics brings together the principles of geotechnical engineering with the concerns for the protection of the environment and the subject has become increasingly important to the geotechnical engineer in recent years. Applications of environmental geotechnics include contaminated land (both its control and reclamation) containment of toxic wastes, design of landfill sites and the management of mining wastes. These applications have a number of common features which epitomise environmental geotechnics problems: soil water flow problems, soil chemistry, and local and national government legislation. Many of the environmental geotechnics issues concern the leaching of toxins into the soil and groundwater supplies, and so the soil properties which are of greatest significance are permeability, void ratio and plasticity. The study of environmental geotechnics is a subject in its own right and is beyond the scope of this book. Readers interested in this aspect of geotechnics should refer to the texts of Attewell (1993), Cairney (1998), and Harris (1994) for a description of the subject.

Exercises

Exercise 15.1

The results of a compaction test on a soil are set out below.

Water content (%)	9.0	10.2	12.5	13.4	14.8	16.0
Bulk density (Mg/m^3)	1.923	2.051	2.220	2.220	2.179	2.096

Plot the dry density to moisture content curve and determine the maximum dry density and the optimum moisture content.

If the particle specific gravity of the soil was 2.68, determine the air void percentage at maximum dry density.

Answer $\rho_{d,max} = 1.97 \text{ Mg/m}^3$
          omc = 13%
          Percentage air voids = 1.0%

Exercise 15.2

A 2.5 kg rammer compaction test was carried out in a 105 mm diameter mould of volume 1000 cm^3 and a mass of 1125 g.
   Test results were:

Water content (%)	10.0	11.0	12.0	13.0	14.0
Mass of wet soil and mould (g)	3168	3300	3334	3350	3320

Plot the curve of dry density against water content and determine the test value for $\rho_{d,max}$ and omc.
   On your diagram plot the zero and 5% air voids lines (take G_s as 2.65). If the percentage of gravel omitted from the test (particle specific gravity = 2.73) was 10%, determine corrected values for $\rho_{d,max}$ and the omc.

Answer   From test: $\rho_{d,max} = 1.97 \text{ Mg/m}^3$;   omc = 12%
         Corrected values: $\rho_{d,max} = 2.03 \text{ Mg/m}^3$;   omc = 11%

Exercise 15.3

An MCV test carried out on a soil sample gave the following results:

No. of blows, B	Penetration (mm)
1	38.0
2	45.1
3	64.0
4	72.3
6	78.1
8	80.1
12	81.0
16	81.5
24	81.7
32	81.7
48	81.7

Determine the MCV of the soil.

Answer MCV = 7.5

Exercise 15.4

An MCV test carried out on a glacial till gave the results set out below:

w (%)	5.6	6.8	7.8	8.0	8.5	9.0	9.5
MCV	13.3	13.5	13.7	13.1	11.1	8.5	6.0

Plot the MCV/w relationship and hence determine the equation of the calibration line and determine its sensitivity.

Show that the MCV values obtained for the two lowest water content values form the ineffective part of the calibration line.

Answer w = 10.7 − 0.21 MCV
Sensitivity = 1/0.21 = 4.8

Chapter 16
An Introduction to Constitutive Modelling in Geomechanics

Learning objectives:

By the end of this chapter, you will have been introduced to the theory of stress–strain behaviours of materials and the different constitutive models used in geotechnics which include:
- the principles of linear elasticity theory and its capabilities and limitations;
- the principles of rigid plasticity theory and its capabilities and limitations;
- the principles of elastoplasticity theory;

and have encountered examples of application of the theories of constitutive modelling in geomechanics.

16.1 Introduction

Soils are complex materials that, in comparison with other engineering materials, are one of the most difficult to characterise. As explained in Chapters 1 and 3, soils consist of a solid skeleton of grains in contact with each other and voids filled with air and/or water or other fluids. The soil skeleton transmits normal and shear forces at the grain contacts, and it behaves in a very complex manner that depends on a large number of factors such as the void ratio and the confining pressure.

Understanding the likely behaviour of the soil is essential for successful geotechnical engineering, and as we have seen in earlier chapters, this requires data and information obtained from a range of sources including field observations and measurements, and laboratory and field testing. Equally, developing models, from simple conceptual models to complex numerical models of the anticipated response of the soil behaviour, is also an essential component of modern geotechnical practice.

With the development of numerical methods such as the finite element and finite difference methods, it has become possible to analyse and predict soil behaviour. Such analyses depend mainly on the representation of the relations between stresses and strains for the various materials involved in a geotechnical structure.

In numerical computations, the relations between stresses and strains in a given material are represented by a *constitutive model* which consists of mathematical expressions that model the behaviour of the soil in a single element. In mathematical terms, they are required in the solution of *boundary value problems*. Such models are an appropriate simplification of reality. However, it is important to identify the appropriate level

Smith's Elements of Soil Mechanics, 10th Edition. Ian Smith.
© 2021 John Wiley & Sons Ltd. Published 2021 by John Wiley & Sons Ltd.
Companion website: www.wiley.com/go/smith/soilmechanics10e

of simplification: if engineers are unaware of the simplifications that they have made, problems may arise if the assumptions are inappropriate for the particular application.

Thus, the reliability of the constitutive model is best checked or validated by comparing its results with laboratory experiments which subject soil samples to similar general stress or strain changes. Triaxial testing, for example, provides the bulk of the available data on the mechanical behaviour of soil elements and therefore provides a useful background against which to introduce ideas of constitutive modelling of soils.

So, the purpose of a constitutive model is to simulate the soil behaviour, with sufficient accuracy under all loading conditions in the numerical computations. Many of the commercially available finite element and finite difference programs, such as PLAXIS and FLAC, allow implementation of simple, as well as advanced, constitutive models.

Constitutive models have gone through substantial improvements since their introduction several decades ago, and the development of new and improved models is ongoing. An appropriate choice of the mechanical constitutive model for soils is the key issue for successful engineering modelling of geomechanical problems. Amongst other factors, the choice of constitutive model also depends on the specific application and requirements of the problem.

This chapter introduces the development of constitutive models used to represent the mechanical behaviour of soils and provides an overview of the principles and the main features and components of existing constitutive models. However, this chapter is not intended to provide in-depth detail, but rather provides some fundamentals of constitutive models, their capabilities and their limitations, which are important for geotechnical engineers.

16.2 Stress-strain behaviour

Consider a typical stress–strain curve obtained from a uniaxial test of most materials (metal, soil, concrete, etc.) as shown in Fig. 16.1. Only the axial component of stress and strain is recorded. In this example, cycles of loading–reloading are performed. In order to describe the behaviour of the material mathematically, this complex non-linear and irreversible behaviour must be approximated without losing significant features.

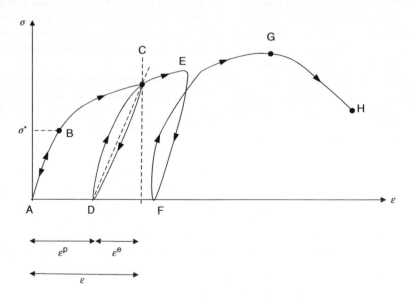

Fig. 16.1 Typical stress–strain curve of a uniaxial test.

Such a schematic material response may be described as follows:

1. Section AB: the response is reversible around the origin; here deformations are reversible, i.e. if the material is unloaded, the initial shape of the material will be recovered. Thus, the behaviour of the material can be approximated by an *isotropic linear elastic model*;
2. At point B: a threshold for reversibility is reached when the stress exceeds a *yield shear* or *elastic limit*. Beyond this limit, denoted σ^*, irreversibility is characterised by permanent or unrecoverable deformation once the applied stress is removed;
3. Section BC: beyond the yield point, non-linear behaviour is observed: the stress–strain curve bends towards the strain axis.
4. During unloading (CD and EF) and reloading (DC and FE), the response is almost parallel to the initial response about the origin (AB) (if the hysteretic phenomenon is ignored); thus, this section can be modelled by an isotropic linear elastic model;
5. At the end of a reloading phase (at C for example), the strain, ε, is the sum of an unrecoverable (plastic) strain ε^p and a recoverable (elastic) strain ε^e:

$$\varepsilon = \varepsilon^e + \varepsilon^p \tag{16.1}$$

6. Sections CE and EG: the stress in the reloading phase exceeds a new yield stress: the previous yielding stress has increased its value to the highest value taken by the stress state during the loading history. The material looks harder than it was originally; this phenomenon is known as *strain hardening*;
7. Section GH: in contrast to most materials, which strain harden in uniaxial tests, soils may present a more complex behaviour under some circumstances (e.g. case of a dense or medium dense soil) known as *strain softening* where the stress decreases as the strain exceeds a certain limit. It is important to note that, although the stress decreases, this behaviour is different from unloading. During unloading, both strain and stress increments are negative, whereas during strain softening these increments have opposite signs;
8. At point H: the material then fails when the stress reaches a final failure stress.

Constitutive relations act to describe, in terms of phenomenological laws, such a stress-strain behaviour of soil. Different constitutive models exist allowing us to describe the stress-strain behaviour including elastic, rigid plastic, and elastoplastic models as shown in Fig. 16.2:

16.3   Selecting the most appropriate constitutive model

Choosing an appropriate constitutive model is an essential step in the analysis of a geotechnical problem. As seen above, different models exist. However, it is known that the theory of elasticity is most often used to predict the response of a structure under serviceability loads (e.g. immediate settlement as described in Chapter 12), while the theory of plasticity is most often used to predict the ultimate load (e.g. bearing capacity as described in Chapter 10). In order to predict the complete response, we need a single constitutive model which allows us to do both. This normally means choosing a constitutive model from the theory of *elastoplasticity*.

Consider an example of a boundary value problem to predict the response of a rigid shallow footing subjected to a vertical load (Fig. 16.3). First, we would try to predict the immediate settlement. This can be done using the theory of elasticity where the behaviour is almost linear and reversible. The second step would be to separately try to predict the maximum bearing pressure that can be carried by the soil under this foundation. Beyond this value, the foundation will collapse. In this case, we would typically use the theory of plasticity to predict the ultimate bearing pressure which could be applied.

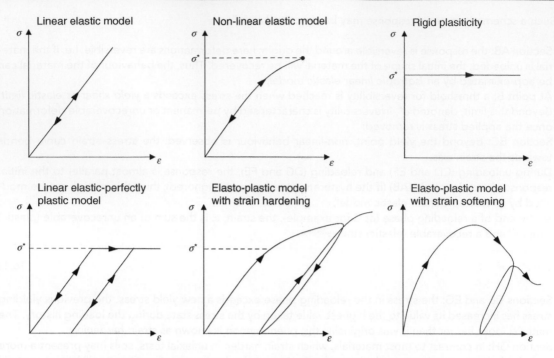

Fig. 16.2   Examples of constitutive models to describe the stress–strain behaviour.

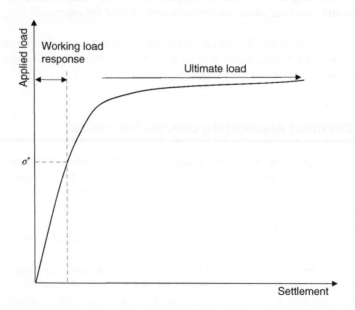

Fig. 16.3   Load-settlement curve of a rigid shallow foundation subjected to a vertical load.

Traditionally, two separate theories and constitutive models were used to describe the soil behaviour: one to describe the response under working/serviceability loading conditions and one to describe the response at the ultimate limit state. However, nowadays we can use one constitutive model, the elastoplastic model, to predict the entire load deformation curve or the load settlement curve for the footing problem.

16.4 Linear elasticity theory

16.4.1 Linear isotropic elasticity

In the theory of elasticity, the constitutive law required to describe material behaviour is called *Hooke's law*. Consider a hanging weight on a wire of length, L as shown in Fig. 16.4. The action of the weight is to cause an elongation in the wire of magnitude x. According to Hooke's law, the force, F applied on the wire is given by:

$$F = kx \tag{16.2}$$

where x is the elongation of the wire and k is the wire's stiffness.

For many materials, there is a range of loads for which the elongation varies linearly with the applied load, F and is recovered when the load is removed, as shown in Fig. 16.5a. There is no permanent deformation of the wire, thus representing an elastic reversible behaviour. The slope of the linear relationship is related to the uniaxial stiffness, k of the material of the wire.

One of the properties or independent constants usually used to describe an elastic response is known as Young's modulus, E which was defined in Section 3.1.3 and, with reference to Fig. 16.4, is expressed as:

$$E = \frac{F/A}{x/L} \tag{16.3}$$

where A is the cross-sectional area of the wire.

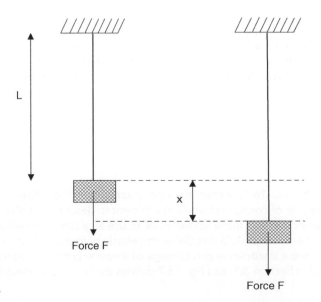

Fig. 16.4 Hooke's law application.

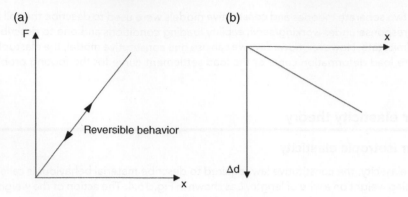

Fig. 16.5 Elastic behaviour of a wire for a finite range of applied loads.

Another elasticity constant is the Poisson's ratio, ν. This constant was also introduced in Section 3.1.3. It is recognised that most materials subjected to tension would not only extend in the direction of the applied force, but they would also shrink laterally. The lateral contraction (or expansion in case of compression loads) is characterised by the Poisson's ratio, ν. Thus, as the wire becomes longer, its diameter becomes smaller.

The ratio of the magnitude of the induced diameter strain to the imposed longitudinal strain is defined as Poisson's ratio:

$$\nu = \frac{\Delta d/d}{x/L} \tag{16.4}$$

where

d is the diameter of the wire, and
Δd is the reduction in diameter.

The Poisson's ratio relationship is depicted in Fig. 16.5b.

Now consider a material subjected to a vertical stress increment, $\delta\sigma_1$ as shown in Fig. 16.6a. This would result in a vertical strain increment, $\delta\varepsilon_1$, and a lateral strain increment, $\delta\varepsilon_3$.

In incremental stress–strain form, the elastic behaviour can then be written mathematically as:

$$\delta\sigma_1 = E\,\delta\varepsilon_1 \tag{16.5}$$

and

$$\delta\varepsilon_3 = \nu\,\delta\varepsilon_1 \tag{16.6}$$

Equations (16.5) and (16.6) are represented graphically in Fig. 16.6b.

Note that, the pair of constants E and ν is sufficient to describe the elastic response of isotropic materials. However, in many cases it is more fundamental to use an alternative pair of elastic constants: the bulk modulus, K and the shear modulus, G that divide the elastic deformation into a volumetric part (change of size at constant shape) and a distortional part (change of shape at constant volume) respectively. These parameters were introduced in Section 3.1, and Fig. 16.7 defines the behaviours described by each elastic constant: E, ν, K and G.

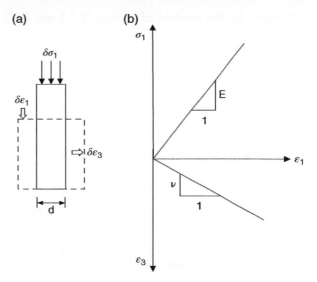

Fig. 16.6   Stress-strain behaviour of an elastic material.

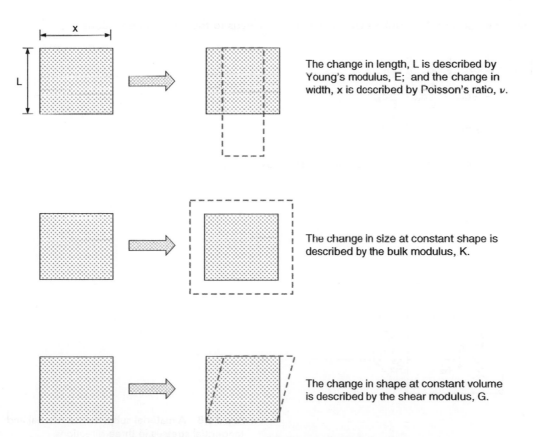

The change in length, L is described by Young's modulus, E; and the change in width, x is described by Poisson's ratio, ν.

The change in size at constant shape is described by the bulk modulus, K.

The change in shape at constant volume is described by the shear modulus, G.

Fig. 16.7   Material constants defining an elastic behaviour.

Drawing on the content of Sections 3.1.4 and 3.1.5, we see that K and G are given by the following expressions:

$$K = \frac{E}{3(1 - 2\upsilon)} \tag{16.7}$$

and

$$G = \frac{E}{2(1 + \upsilon)} \tag{16.8}$$

16.4.2 Stress, strain and stiffness tensors

Figure 16.8 duplicates Fig. 3.2a. The stresses and strains can be represented in tensor form to allow analysis of the behaviour of the element of soil in the general 3D case.

In this case, the *stress tensor*, σ is proportional to the *strain tensor*, ε through the *stiffness tensor*, **D** such that:

$$\sigma = \mathbf{D}\varepsilon \tag{16.9}$$

A generalization of the Hooke's law in all directions leads to the following equations:

$$\begin{cases} \varepsilon_{xx} = \dfrac{1}{E}\sigma_{xx} - \dfrac{\upsilon}{E}\sigma_{yy} - \dfrac{\upsilon}{E}\sigma_{zz} \\[2mm] \varepsilon_{yy} = -\dfrac{\upsilon}{E}\sigma_{xx} + \dfrac{1}{E}\sigma_{yy} - \dfrac{\upsilon}{E}\sigma_{zz} \\[2mm] \varepsilon_{zz} = -\dfrac{\upsilon}{E}\sigma_{xx} - \dfrac{\upsilon}{E}\sigma_{yy} + \dfrac{1}{E}\sigma_{zz} \end{cases} \tag{16.10}$$

where σ_{xx}, σ_{yy} and σ_{zz} are the applied normal stresses.

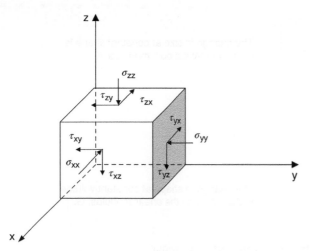

Fig. 16.8 A material subjected to normal and tangential stresses in three directions.

In addition, shear strains are proportional to the applied shear stress through the shear modulus, G such that:

$$
\begin{cases}
2\gamma_{xy} = \dfrac{1}{G}\tau_{xy} \\[2mm]
2\gamma_{xz} = \dfrac{1}{G}\tau_{xz} \\[2mm]
2\gamma_{yz} = \dfrac{1}{G}\tau_{yz}
\end{cases}
\tag{16.11}
$$

where γ_{xy}, γ_{xz} and γ_{yz} are the shear strains and τ_{xy}, τ_{xz} and τ_{yz} are the shear stresses.

Putting together these equations in a matrix form gives:

$$
\begin{bmatrix}
\varepsilon_{xx} \\ \varepsilon_{yy} \\ \varepsilon_{zz} \\ 2\gamma_{xy} \\ 2\gamma_{xz} \\ 2\gamma_{yz}
\end{bmatrix}
=
\begin{bmatrix}
\dfrac{1}{E} & -\dfrac{\nu}{E} & -\dfrac{\nu}{E} & 0 & 0 & 0 \\[2mm]
-\dfrac{\nu}{E} & \dfrac{1}{E} & -\dfrac{\nu}{E} & 0 & 0 & 0 \\[2mm]
-\dfrac{\nu}{E} & -\dfrac{\nu}{E} & \dfrac{1}{E} & 0 & 0 & 0 \\[2mm]
0 & 0 & 0 & \dfrac{2(1+\nu)}{E} & 0 & 0 \\[2mm]
0 & 0 & 0 & 0 & \dfrac{2(1+\nu)}{E} & 0 \\[2mm]
0 & 0 & 0 & 0 & 0 & \dfrac{2(1+\nu)}{E}
\end{bmatrix}
\begin{bmatrix}
\sigma_{xx} \\ \sigma_{yy} \\ \sigma_{zz} \\ \tau_{xy} \\ \tau_{xz} \\ \tau_{yz}
\end{bmatrix}
\tag{16.12}
$$

or

$$
\begin{bmatrix}
\sigma_{xx} \\ \sigma_{yy} \\ \sigma_{zz} \\ \tau_{xy} \\ \tau_{xz} \\ \tau_{yz}
\end{bmatrix}
=
\dfrac{E}{(1+\nu)(1-2\nu)}
\begin{bmatrix}
1-\nu & \nu & \nu & 0 & 0 & 0 \\[2mm]
\nu & 1-\nu & \nu & 0 & 0 & 0 \\[2mm]
\nu & \nu & 1-\nu & 0 & 0 & 0 \\[2mm]
0 & 0 & 0 & \dfrac{(1-2\nu)}{2} & 0 & 0 \\[2mm]
0 & 0 & 0 & 0 & \dfrac{(1-2\nu)}{2} & 0 \\[2mm]
0 & 0 & 0 & 0 & 0 & \dfrac{(1-2\nu)}{2}
\end{bmatrix}
\begin{bmatrix}
\varepsilon_{xx} \\ \varepsilon_{yy} \\ \varepsilon_{zz} \\ 2\gamma_{xy} \\ 2\gamma_{xz} \\ 2\gamma_{yz}
\end{bmatrix}
\tag{16.13}
$$

or we can write:

$$\sigma = D\varepsilon$$

where

$$
D = \dfrac{E}{(1+\nu)(1-2\nu)}
\begin{bmatrix}
1-\nu & \nu & \nu & 0 & 0 & 0 \\[2mm]
\nu & 1-\nu & \nu & 0 & 0 & 0 \\[2mm]
\nu & \nu & 1-\nu & 0 & 0 & 0 \\[2mm]
0 & 0 & 0 & \dfrac{(1-2\nu)}{2} & 0 & 0 \\[2mm]
0 & 0 & 0 & 0 & \dfrac{(1-2\nu)}{2} & 0 \\[2mm]
0 & 0 & 0 & 0 & 0 & \dfrac{(1-2\nu)}{2}
\end{bmatrix}
$$

is the *elastic constitutive tensor*, or *stiffness tensor*.

If we consider a typical triaxial drained test on a soil sample, the response of a soil specimen can be described in terms of the mean effective stress, p′, and the deviatoric stress, q. Thus, drawing on Equations (3.11)–(3.16), the elastic response can be written as follows:

$$p' = K\varepsilon_v \tag{16.14}$$

$$q = 3G\varepsilon_q \tag{16.15}$$

where

ε_v is the volumetric strain, defined as $\varepsilon_v = \varepsilon_{xx} + \varepsilon_{yy} + \varepsilon_{zz}$
ε_q is the deviatoric strain, defined as $\varepsilon_q = \frac{2}{3}(\varepsilon_{zz} - \varepsilon_{xx})$ considering ε_{zz} as the axial strain.

Over the years, numerous geotechnical problems have been solved by modelling soil responses using the theory of elasticity. One successful application of elasticity theory is in soil–structure interaction problems for foundations.

Figure 16.9 shows experimental results from a test on dense silica sand conducted by Poon (2005) and the prediction of a linear elastic model (Carter, 2006). This figure presents the measured relationship between the vertical load applied to a circular footing, and the resulting vertical displacement (settlement) of the footing. It is seen that part of the curve is approximately linear over a range of loading. Therefore, in this range of applied loads it is reasonable to consider an elastic soil behaviour which can be modelled quite accurately for this section of the load–displacement curve.

Countless other examples exist where constitutive models have been proved to accurately predict geomechanical behaviour, verified through experimental measurements and observations.

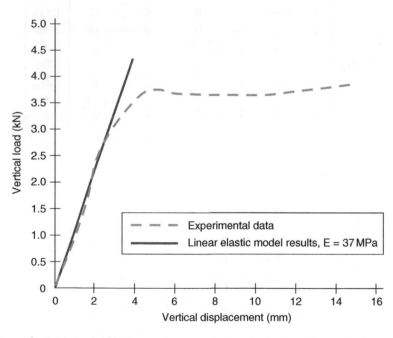

Fig. 16.9   Behaviour of a rigid circular footing on dense sand: experimental and linear elastic model data. Based on Carter (2006).

16.4.3 Limitations of linear elastic constitutive models

The behaviour of soils may be predicted with reasonable accuracy by elastic models in cases of stress states not approaching failure. However, phenomena such as irreversibility of a portion of the strains, shear-dilatancy, and most soil behaviour near and beyond failure cannot be described by the elasticity theory. Therefore, it is important to take into consideration some of the important limitations of using elasticity to predict soil behaviour:

- soils are not truly linear, nor are they generally isotropic;
- elasticity does not allow the coupling of the shear and volumetric modes of soil response;
- with elasticity we cannot estimate the strength of the soil and we cannot describe any hardening or softening that might be exhibited by real soil.

The shortcomings of Hooke's law for elastic behaviour may be demonstrated by using two examples presented by Lade (2005).

Equation (6.12) shows that in the elastic matrix that relates the strain increments to stress increments, zeros occupy the upper-right and lower-left sub-matrices. The zeros in the upper-right sub-matrix signify that there are no relations between normal strains and shear stresses. However, Example (1) in Fig. 16.10 shows a dense sand specimen subjected to a normal stress and a shear stress. When the shear stress increases, increments in shear strains will result. However, observations of real soil indicates that the soil will also dilate such that the specimen's height increases too, as witnessed for dense soils in the shear box test (see Section 4.2.2).

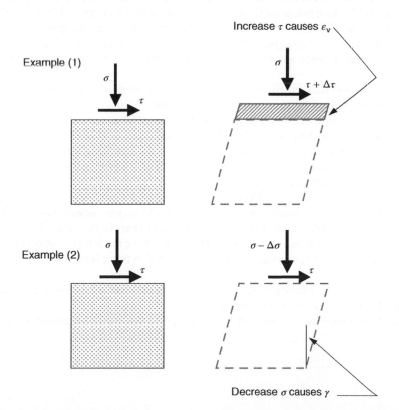

Fig. 16.10 Examples to indicate the limitations of Hooke's law for elastic behaviour. Based on Lade (2005).

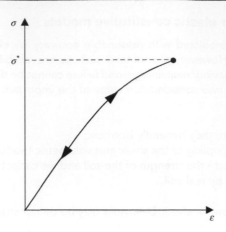

Fig. 16.11   Non-linear stress–strain behaviour.

Thus, normal strain increments result from increments in shear stress. Since zeros are present in the upper-right submatrix of the elastic matrix, no normal strain increments can be predicted from increments in shear stress. Therefore, Hooke's law is not able to model this behaviour of soils.

Also, the zeros in the lower-left sub-matrix of the elastic material matrix may be checked by the experiment in Example (2) in Fig. 16.10. Here the initial condition is the same as in Example (1), but now the normal stress on the specimen is reduced. Observations of real soil behaviour shows that the specimen will produce shear strain increments for such a reduction in normal stress, and sufficient reduction will result in failure. However, the zeros in this sub-matrix prevent any prediction of shear strain increments due to increments in normal stresses.

These two examples clearly demonstrate some serious shortcomings of the elasticity theory for modelling the behaviour of soils for stress conditions approaching failure.

In addition to these limitations, as mentioned before the elastic behaviour is not necessarily linear. In other words, if a load with a value less than the yield stress is applied in a model, and then the model is unloaded, the model may effectively come back to its original shape, but the stress and strain might not follow a linear relation of proportionality as shown in Fig. 16.11. For such non-linear elastic materials (such as rubber), we say that the slope of the stress-strain curve is *strain-dependent*.

Elastic deformation continues until the yield stress of the material. Both linear and non-linear elastic materials will elastically return to an 'unloaded' state after loading without permanent deformations, but the relationship between stress and strain will be different. It is linear for linear elastic material and more complex in a non-linear case. Thus, a logical first step to improving the linear elastic models described above is to make the material parameters depend on stress and/or strain level. By doing this it is possible to satisfy several of the requirements.

For materials with a non-linear elastic portion, either the *tangent modulus* or the *secant modulus* is used in design calculations. These moduli are shown in Fig. 16.12.

The secant modulus is given by the line joining the origin to the current stress–strain point (σ, ε) and is defined:

$$E_{sec} = \frac{\sigma}{\varepsilon} \tag{16.16}$$

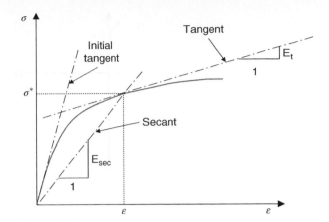

Fig. 16.12 Tangent and secant modulus of a non-linear elastic model.

The *tangent modulus*, which represents the slope of the stress–strain curve at the point under consideration, is defined as the rate of change of stress with strain. In this case:

$$E_t = \frac{\delta\sigma}{\delta\varepsilon} \tag{16.17}$$

Similarly, the *secant shear modulus* is given by:

$$G_{sec} = \frac{\tau}{\gamma} \tag{16.18}$$

and the *tangent shear modulus* is given by:

$$G_t = \frac{\delta\tau}{\delta\gamma} \tag{16.19}$$

16.5   Rigid plasticity theory

16.5.1   Rigid plastic model

Some of the development that has been made in constitutive modelling is by considering the *rigid plastic* soil model which can be used in the theory of plasticity to predict ultimate loads that may be applied to soil. Consider an ideal material that does not deform until its strength is fully mobilised as illustrated in Fig. 16.13. For soils, that strength is usually the available shear strength. Using a constitutive model of this form allows us to handle one 'end' of the boundary value problem. For example, in the case of a shallow footing, it allows us to predict the ultimate or collapse load.

In order to define such a constitutive model, we need a *failure criterion*, defining the strength of the material, such as the Mohr–Coulomb criterion which expresses the combination of stresses acting on soil that will cause it to fail and unable to carry further load, e.g.

$$\sigma_1' = \frac{1 + \sin\phi'}{1 - \sin\phi'}\sigma_3' + 2c'\sqrt{\frac{1 + \sin\phi'}{1 - \sin\phi'}} \tag{16.20}$$

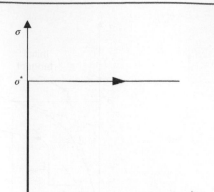

Fig. 16.13   Rigid plastic model.

Equation (16.20) is an alternative representation of Equation (8.7), where σ'_1 and σ'_3 are the principal stresses.

The two constants involved in this linear equation, are of course the effective cohesion c' and the effective angle of shearing resistance ϕ'.

The Mohr–Coulomb criterion can be used in the theory of plasticity to predict the collapse loads of engineering structures. For example, it can be used in the determination of the ultimate bearing capacity (see Section 10.3) to establish the maximum load that can be placed on a cohesive-frictional soil without collapse.

16.5.2   Limitations of the rigid plasticity theory

As is the case with all theories, the rigid plastic model has its limitations. Soils are not generally rigid before collapse as some deformation of the soil will normally occur before the ultimate collapse load is reached. However, the theory for a rigid plastic material can tell us nothing about the deformations which may precede collapse. Furthermore, the classical theory assumes what is known as *associated plastic flow* which will be addressed later in this chapter. For a frictional soil this generally means that the volume change that the soil undergoes as it is failing is over-estimated and this can have major consequences, particularly for the prediction of the mechanisms of collapse within the soil and therefore, potentially, the predicted collapse load itself.

16.6   Elastoplasticity theory

In order to overcome the different limitations of the elastic and the rigid plastic soil models, a different category of models, which combines the different features of these two separate constitutive models, has been developed over the past years. Soil models that fit into this category are termed *elastoplastic constitutive models*.

The elastoplastic models are based on three fundamental notions: the *yield criterion*, the *plastic flow rule* and the *hardening rule*.

16.6.1 The yield criterion

This criterion defines the elastic limit, σ^* beyond which the material becomes plastic. The yield point, σ^* defined previously, is generalised into a hypersurface in the multi-dimensional stress space, called the *yield surface* (see Fig. 16.14). By definition, the yield surface encloses a domain in stress space in which any infinitesimal stress change produces only recoverable strain.

The yield surface divides the stress space into two parts: the interior of the yield surface corresponds to the state of reversible elastic deformations, and at the yield surface, the deformations consist of a reversible elastic part and an irreversible plastic part. The boundary is characterised by a scalar yield function, f of the domain, called the yield function, and given by:

$$f\left(\sigma_{xx}, \sigma_{yy}, \sigma_{zz}, \tau_{xy}, \tau_{yz}, \tau_{zx}, \alpha_1, \alpha_2, \dots, \alpha_n\right) = 0 \qquad (16.21)$$

where σ_{xx}, σ_{yy}, σ_{zz}, τ_{xy}, τ_{yz}, τ_{zx} are the stress components and α_1, α_2, ..., α_n are specific material properties. Alternatively, we may write:

$$f(\sigma, \alpha) = 0 \qquad (16.22)$$

When defining a yield criterion, care must be taken to ensure that a rotation of the coordinate system does not influence the conditions at which the material yields. Thus, it is convenient to define a yield criterion in terms of certain *invariants* which are not affected by a rotation of the coordinate system such as the principal stresses. So, it is more convenient to write the yield criterion as:

$$f(\sigma_1, \sigma_2, \sigma_3, \alpha_1, \dots, \alpha_n) = 0 \qquad (16.23)$$

where σ_1, σ_2, σ_3 are the principal stresses.

In many cases, it may be more convenient to use another set of invariants, such as for example:

- The invariants of *stress tensor*: I_1, I_2, I_3
- The invariants of *deviatoric stress tensor*: J_1, J_2, J_3

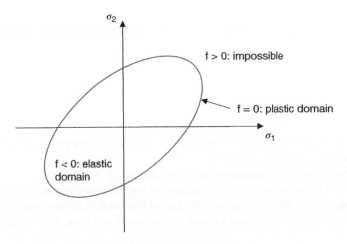

Fig. 16.14 Yield surface.

Details of the derivations of these stress invariants are given by Smith (1971) for readers interested in study-ing this subject deeper. I_1 and J_2 are the most useful of these invariants.

With reference to Fig. 16.4, we may summarise the following states:

- $f(\sigma_{ij}) < 0$ (interior of the domain): the stress state, σ_{ij}, lies inside the yield surface; any increment of stress creates only incremental elastic strain $d\varepsilon^e$; i.e. the incremental plastic strain, $d\varepsilon^P = 0$;
- $f(\sigma_{ij}) = 0$ (boundary of the domain): the stress state, σ_{ij}, lies on the yield surface. If subjected to loading, irreversible strain results, i.e. $d\varepsilon^P \neq 0$. In the case of unloading or if the stress is maintained on the yield surface, only elastic strain develops, i.e. $d\varepsilon^P = 0$;
- $f(\sigma_{ij}) > 0$ (outside the domain): This state is impossible. If a state of stress, σ_{ij}, is allowed to exist outside the yield surface, then for any increment of stress from this state, irreversible deformation would occur. No unloading with reversible response would be possible, which is contradictory to the previous uniaxial con-cept defined from uniaxial tests shown in Fig. 16.1. Therefore, the yield surface must change its position, or deform, in order to follow the stress state when plastic flow occurs, so that the stress state lies on or within the yield surface. This condition is known as the *Consistency condition*.

(a) Consistency condition

When the stress state reaches the yield surface $f = 0$, two cases of elastoplastic behaviour are possible:

1. the surface f doesn't evolve (elastic-perfectly plastic model);
2. the surface f evolves during loading (hardening elastoplastic model).

In plastic loading with perfect plasticity, the state of stress can only be altered by a redistribution between the different stress components such that the stress point can be imagined as sliding along the yield surface. In other words, plastic deformations can occur as long as the stress point is located on the yield surface. For perfect plasticity, a stress point may remain in one fixed position on the yield surface, or slide along it with redistribution of stresses among the different components.

Therefore, from a mathematical point of view, this plastic loading condition can be translated by:

$$f(\sigma + d\sigma) = f(\sigma) + \nabla f^T d\sigma = 0 \tag{16.24}$$

where $d\sigma$ is a stress increment, and ∇f is the normal to the yield surface as shown in Fig. 16.15.

∇f^T is defined:

$$\nabla f^T = \left[\frac{\partial f}{\partial \sigma_x}, ..., \frac{\partial f}{\partial \tau_{zx}} \right]^T \tag{16.25}$$

Since $f(\sigma) = 0$ (yield criterion), then $df = \nabla f^T d\sigma = 0$

Thus, during plastic loading, the change in stress occurs tangential to the yield surface in the case of perfect plasticity. This is called the *consistency condition*.

In unloading, the state of stress immediately becomes elastic: $df = \nabla f^T d\sigma < 0$.

If hardening is present, the stress point will still remain on the boundary defined by $f = 0$ but this boundary will be shifted according to the relevant hardening rules as the loading progresses. In unloading, the stress point moves from the boundary of the yield surface to the inside and thus, immediately recovers its elastic properties.

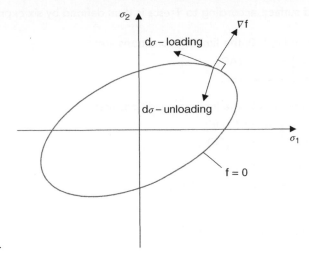

Fig. 16.15   Plastic loading and unloading.

(b) Examples of yield criteria

(i) Tresca criterion

In the 19th century Tresca subjected metal specimens to a combination of stresses. After these experiments, he proposed that yielding occurs as a result of the maximum shear stress reaching a critical value given by:

$$\tau_{max} = \max\left(\frac{1}{2}|\sigma_1 - \sigma_2|, \frac{1}{2}|\sigma_2 - \sigma_3|, \frac{1}{2}|\sigma_3 - \sigma_1|\right) = \max\left(\tau_{xy}, \tau_{yz}, \tau_{zx}\right) = k^* \qquad (16.26)$$

where k^* can be obtained from a simple tension test, $k^* = \frac{1}{2}\sigma^*$

In plane stress, the yield surface may then be plotted as shown in Fig. 16.16.

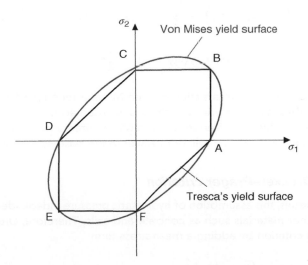

Fig. 16.16   Tresca and Von Mises yield criteria.

The yield surface according to Tresca is then defined by six expressions:

- If $\sigma_1 > 0$ and $\sigma_2 > 0$, the limiting conditions are:
    - $\frac{1}{2}|\sigma_1| = \frac{1}{2}\sigma_1 = k^*$ (AB)
    - $\frac{1}{2}|\sigma_2| = \frac{1}{2}\sigma_2 = k^*$ (BC)

- If $\sigma_1 < 0$ and $\sigma_2 < 0$, the limiting conditions are:
    - $\frac{1}{2}|\sigma_1| = -\frac{1}{2}\sigma_1 = k^*$ (DE)
    - $\frac{1}{2}|\sigma_2| = -\frac{1}{2}\sigma_2 = k^*$ (EF)

- If $\sigma_1 < 0$ and $\sigma_2 > 0$:
    - $\frac{1}{2}|\sigma_1 - \sigma_2| = \frac{1}{2}(\sigma_2 - \sigma_1) = k^*$ (CD)

- If $\sigma_1 > 0$ and $\sigma_2 < 0$:
    - $\frac{1}{2}|\sigma_1 - \sigma_2| = \frac{1}{2}(\sigma_1 - \sigma_2) = k^*$ (AF)

(ii) Von-Mises criterion

The weakness of Tresca's yield criterion is that only the maximum shear stress is considered and the influence of the two smaller shear stresses is ignored. A more accurate criterion, taking all three principal shear stresses into account, is that of Von Mises.

In this case, the yield function is given by:

$$f(J_2) = \sqrt{J_2} - k^* \tag{16.27}$$

or

$$f(\sigma) = \sigma_e - \sigma^* \tag{16.28}$$

where σ_e is an equivalent stress defined as:

$$\sigma_e = \left(\tau_{xy}^2 + \tau_{yz}^2 + \tau_{zx}^2\right)^{\frac{1}{2}} = \left[\frac{1}{2}(\sigma_1 - \sigma_2)^2 + \frac{1}{2}(\sigma_2 - \sigma_3)^2 + \frac{1}{2}(\sigma_3 - \sigma_1)^2\right]^{\frac{1}{2}} \tag{16.29}$$

$$\sigma_e = \left[\frac{1}{2}(\sigma_{xx} - \sigma_{yy})^2 + \frac{1}{2}(\sigma_{yy} - \sigma_{zz})^2 + \frac{1}{2}(\sigma_{zz} - \sigma_{xx})^2 + 3\tau_{xy}^2 + 3\tau_{yz}^2 + 3\tau_{zx}^2\right]^{\frac{1}{2}} \tag{16.30}$$

In contrast to Tresca's criterion, Von Mises criterion predicts that the principal stress may exceed σ^* provided that the other principal stress is adjusted accordingly (Fig. 16.16).

As with Tresca's criterion, the Von Mises criterion is independent of the hydrostatic pressure. In 3D, the yield surface depicts a cylinder parallel to the hydrostatic axis, $\sigma_1 = \sigma_2 = \sigma_3$ (Fig. 16.17).

(iii) Drucker–Prager criterion

For metals, the assumption of hydrostatic pressure independence is realistic. However, this assumption fails for other materials such as concrete and soils. Therefore, Drucker and Prager formulated a modified Von Mises criterion by adding a mean stress term:

$$f(I_1, J_2) = \sqrt{J_2} + \alpha I_1 - k^* \tag{16.31}$$

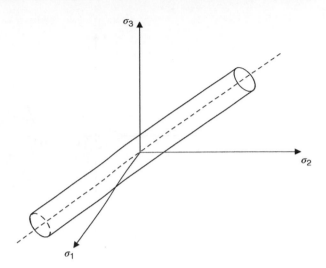

Fig. 16.17   Von Mises yield criterion in principal stress space.

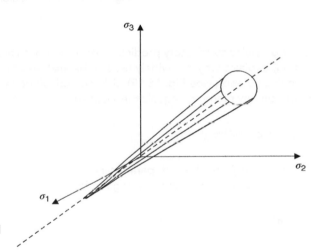

Fig. 16.18   Drucker–Prager yield criterion in principal stress space.

or

$$f(\boldsymbol{\sigma}) = \sigma_e + \alpha\sigma_m - \sigma^* \tag{16.32}$$

where $\sigma_m = \frac{1}{3}I_1 = \frac{1}{3}(\sigma_1 + \sigma_2 + \sigma_3)$ is the mean stress.

In 3D, the yield surface has a conical shape as shown in Fig. 16.18. As can be seen, there is a limit for negative (tensile) mean stresses.

(iv) Mohr–Coulomb criterion

The Mohr–Coulomb yield (failure) criterion is similar to the Tresca's criterion, with additional modifications for materials with different tensile and compressive yield strengths. This model has been extensively described in Chapter 4 and used elsewhere in this book, as it is a widely used and simple to apply in many situations. The

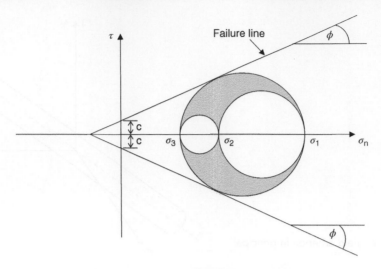

Fig. 16.19 Mohr–Coulomb failure criterion.

criterion yields satisfactory predictions of soil strength and, together with its relatively straightforward application, explains why it is widely used in the analysis of most practical problems which involve soil strength.

In this criterion (see Fig. 16.19), yielding and failure takes place in the soil mass when the mobilised shear
stress on any plane, τ_m equals the shear strength τ_f which is given by:

$$\tau_m = c + \sigma_n \tan \phi = \tau_f \tag{16.33}$$

where c and ϕ are strength parameters.

The yield function, therefore, is given by:

$$f(\boldsymbol{\sigma}) = \tau - c - \sigma_n \tan \phi = 0 \tag{16.34}$$

In terms of principal stresses:

$$\tau = \frac{\sigma_1 - \sigma_3}{2} \cos \phi \tag{16.35}$$

$$\sigma_n = \frac{\sigma_1 + \sigma_3}{2} + \frac{\sigma_1 - \sigma_3}{2} \sin \phi \tag{16.36}$$

Thus,

$$f(\boldsymbol{\sigma}) = (\sigma_1 - \sigma_3) - (\sigma_1 + \sigma_3) \sin \phi - 2c \cos \phi = 0 \tag{16.37}$$

The Mohr Coulomb failure surface is then an irregular hexagonal pyramid in the principal stress space as
illustrated in Fig. 16.20.

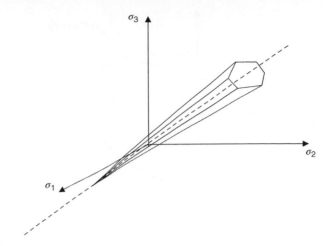

Fig. 16.20 Mohr–Coulomb criterion in principal stress space.

16.6.2   The plastic flow rule

When a structure is loaded from a neutral state, elastic strains first appear. These strains are obtained from elastic models such as Hooke's law, described in Section 16.4.1. However, as the loading increases, the material begins to yield, and plastic strains will develop. At some point the load carrying capacity of the structure becomes exhausted, and plastic strains become infinitely large leading to the collapse of the structure.

It should be emphasised that when examining the strains in a plastic material, plastic increments of strain are used rather than a total accumulated strain. One reason for this is that when the material is subjected to a certain stress state, the corresponding strain state could be one of many. Similarly, the strain state could correspond to many different stress states as can be observed from Fig. 16.21.

It's not possible therefore to make use of stress-strain relations in plastic regions, since there is no unique relationship between the current stress and the current strain, i.e. there is no unique relation between stresses

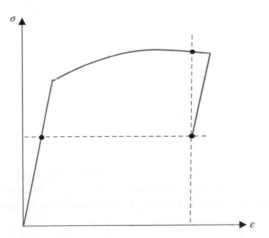

Fig. 16.21   Stress–strain curve showing: different strain states for a given stress state, and different stress states for a given strain state.

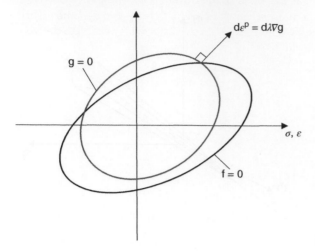

Fig. 16.22   Plastic strain increment.

and total strains in the plastic range. Thus, it is convenient to consider only plastic strain increments, $d\varepsilon^P$. So, we can relate the current stress to the current strain increment, and these are the *stress–strain laws* which are used in plasticity theory. The total strain can be obtained by summing, or integrating, the strain increments.

(a) Computation of strain increments

After the initiation of plasticity, the constitutive elastic equations become invalid. Plastic strains depend on the overall loading history of the material and the stress-strain relations are given in terms of strain increments. Exploiting Equation (16.1), the total strain increment may be decomposed into its elastic and plastic parts:

$$d\varepsilon = d\varepsilon^e + d\varepsilon^P \tag{16.38}$$

$d\varepsilon^e$ is obtained from an elastic model whereas for the plastic increment, $d\varepsilon^P$, the direction and magnitude of the strain increment vector has to be calculated. This is done with the flow rule which defines the plastic potential function g (Fig. 16.22).
Then,

$$d\varepsilon^P = d\lambda\nabla g \tag{16.39}$$

where $\delta\lambda$ is a positive scalar of proportionality;
$g = 0$ describes a surface in stress space, and
∇g is a vector normal to the potential surface $g = 0$.

Thus, $d\varepsilon^P$ is a vector normal to the surface with a length determined by $\delta\lambda$.
The expression for g must be determined experimentally similarly to the way the yield function, f, is established. But this is quite a demanding task. Therefore, as a first estimate, we can take g = f.
Then,

$$d\varepsilon^P = d\lambda\frac{\partial f}{\partial\sigma} = d\lambda\nabla f \tag{16.40}$$

In this case, we define an *associated flow rule* i.e. when the potential function, g, is the same as the yield function, f. Thus, in this case the strains are related to the yield function. The use of the yield criterion to derive plastic strain increments is also referred to as the *normality rule*.

In contrast to the associated flow rule, in situations where the strains are not connected directly to the yield surface, we must use the non-associated flow rule.

(b) Examples of plastic flow rules

i) Von Mises criterion

The plastic strain increments, using the yield function defined in Equation (16.28), are given by:

$$
d\varepsilon^P = \begin{bmatrix} d\varepsilon^P_{xx} \\ d\varepsilon^P_{yy} \\ d\varepsilon^P_{zz} \\ d\gamma^P_{xy} \\ d\gamma^P_{yz} \\ d\gamma^P_{zx} \end{bmatrix} = d\lambda \times \frac{1}{2\sigma_e} \begin{bmatrix} 2\sigma_{xx} - \sigma_{yy} - \sigma_{zz} \\ 2\sigma_{yy} - \sigma_{zz} - \sigma_{xx} \\ 2\sigma_{zz} - \sigma_{xx} - \sigma_{yy} \\ 6\tau_{xy} \\ 6\tau_{yz} \\ 6\tau_{zx} \end{bmatrix} \tag{16.41}
$$

Given a particular state of strain, the relative volume change can be determined as:

$$
\frac{\Delta V}{V} = (1 + \varepsilon_{xx})(1 + \varepsilon_{yy})(1 + \varepsilon_{zz}) - 1 \approx \varepsilon_{xx} + \varepsilon_{yy} + \varepsilon_{zz} = 0 \tag{16.42}
$$

Thus, the associated Von Mises flow rule predicts that no volumetric changes occur as a result of plastic straining, which for metals is in agreement with what can be observed experimentally. However, soils, concrete and other granular materials do exhibit a volumetric dilation during plastic flow which is reflected by Drucker–Prager criterion.

ii) Drucker–Prager criterion

In this criterion and using the yield function defined by Equation (16.32), the plastic strain increments are given by:

$$
d\varepsilon^P = \begin{bmatrix} d\varepsilon^P_{xx} \\ d\varepsilon^P_{yy} \\ d\varepsilon^P_{zz} \\ d\gamma^P_{xy} \\ d\gamma^P_{yz} \\ d\gamma^P_{zx} \end{bmatrix} = d\lambda \times \left(\frac{1}{2\sigma_e} \begin{bmatrix} 2\sigma_{xx} - \sigma_{yy} - \sigma_{zz} \\ 2\sigma_{yy} - \sigma_{zz} - \sigma_{xx} \\ 2\sigma_{zz} - \sigma_{xx} - \sigma_{yy} \\ 6\tau_{xy} \\ 6\tau_{yz} \\ 6\tau_{zx} \end{bmatrix} + \begin{bmatrix} \frac{1}{3}\alpha\sigma_{xx} \\ \frac{1}{3}\alpha\sigma_{yy} \\ \frac{1}{3}\alpha\sigma_{zz} \\ 0 \\ 0 \\ 0 \end{bmatrix} \right) \tag{16.43}
$$

Then the relative change in volume is:

$$
\frac{\Delta V}{V} = d\varepsilon^P_{xx} + d\varepsilon^P_{yy} + d\varepsilon^P_{zz} = d\lambda \times \frac{1}{3} \times \alpha \times (\sigma_{xx} + \sigma_{yy} + \sigma_{zz}) \tag{16.44}
$$

which is not necessarily equal to zero.

However, the volumetric dilation predicted by the associated Drucker–Prager rule is often somewhat larger than what can be verified experimentally. Therefore, a *non-associated flow rule* can be used where the flow rule is defined by some other function. For example, the yield and potential functions may be given by:

$$f(\sigma) = \sqrt{J_2} + \alpha I_1 - K \tag{16.45}$$

$$g(\sigma) = \sqrt{J_2} + \beta I_1 - K \tag{16.46}$$

with $\beta < \alpha$

(c) Establishing an elastic-perfectly plastic relation

In computational elastoplastic analysis, the process proceeds by applying a load increment which produces a displacement increment and thus a total strain increment. The stress increment corresponding to the total strain increment can be obtained by a constitutive relation similar to the one from elasticity.

$$d\sigma = \mathbf{D}^{ep} d\varepsilon \tag{16.47}$$

where $\mathbf{D}^{ep}$ is the elastoplastic constitutive matrix

The development of an elastoplastic relationship consists of the following sequence of steps:

1. We have $d\varepsilon = d\varepsilon^e + d\varepsilon^p$ and from Hooke's law we know that:

$$d\sigma = \mathbf{D}d\varepsilon^e = \mathbf{D}(d\varepsilon - d\varepsilon^p) \tag{16.48}$$

where $\mathbf{D}$ is the elastic constitutive matrix.

2. The flow rule gives:

$$d\varepsilon^p = d\lambda \nabla g \tag{16.49}$$

so

$$d\sigma = \mathbf{D}d\varepsilon - d\lambda\, \mathbf{D}\nabla g \tag{16.50}$$

3. From the consistency condition:

$$\nabla f^{\mathsf{T}} d\sigma = 0 \tag{16.51}$$

so

$$\left(\frac{\partial f}{\partial \sigma}\right)^{\mathsf{T}} (\mathbf{D}d\varepsilon - d\lambda\, \mathbf{D}\nabla g) = 0 \tag{16.52}$$

$$d\lambda = \frac{\left(\frac{\partial f}{\partial \sigma}\right)^{T} \mathbf{D}\, d\varepsilon}{\left(\frac{\partial f}{\partial \sigma}\right)^{T} \mathbf{D} \frac{\partial g}{\partial \sigma}} \tag{16.53}$$

Substituting Equation 16.53 into 16.50,

$$d\sigma = \mathbf{D}\left(d\varepsilon - \frac{\left(\frac{\partial f}{\partial \sigma}\right)^{T} \mathbf{D}\, d\varepsilon\, \frac{\partial g}{\partial \sigma}}{\left(\frac{\partial f}{\partial \sigma}\right)^{T} \mathbf{D} \frac{\partial g}{\partial \sigma}} \right) \tag{16.54}$$

Thus, an *Elastoplastic constitutive relation* can be written as follows:

$$d\sigma = \left(\mathbf{D} - \frac{\mathbf{D} \times \frac{\partial g}{\partial \sigma} \left(\frac{\partial f}{\partial \sigma}\right)^{T} \mathbf{D}}{\left(\frac{\partial f}{\partial \sigma}\right)^{T} \mathbf{D} \frac{\partial g}{\partial \sigma}} \right) d\varepsilon \tag{16.55}$$

with

$$\mathbf{D}^{ep} = \mathbf{D} - \frac{\mathbf{D} \frac{\partial g}{\partial \sigma} \left(\frac{\partial f}{\partial \sigma}\right)^{T} \mathbf{D}}{\left(\frac{\partial f}{\partial \sigma}\right)^{T} \mathbf{D} \frac{\partial g}{\partial \sigma}}$$

This defines the stress increment uniquely once the total strain increment and the current state of stress are known.

Application: Mohr–Coulomb constitutive model

- Yield (failure) condition:

$$f(\sigma) = (\tau - \sigma_n \tan\phi - c) = (\sigma_1 - \sigma_3) - (\sigma_1 + \sigma_3)\sin\phi - 2\,c\,\cos\phi = 0 \tag{16.56}$$

In p-q space (as introduced in Chapter 5) this condition can be written as:

$$\sigma_1 = 3p - 2\sigma_3,\ \sigma_3 = \sigma_1 - q,\ \text{so } \sigma_1 = 3p - 2\sigma_1 + 2q \tag{16.57}$$

$$\sigma_1 = \frac{3p + 2q}{3} \tag{16.58}$$

$$\sigma_3 = 3p - 2\sigma_3 - q = \frac{3p - q}{3} \tag{16.59}$$

$$\sigma_1 + \sigma_3 = \frac{6p + q}{3} \tag{16.60}$$

$$q = \eta p + c^{*} \tag{16.61}$$

where $c^{*} = \dfrac{6c\cos\phi}{3 - \sin\phi}$ and $\eta = \dfrac{6\sin\phi}{3 - \sin\phi}$

$$f = q - \eta p - c^* = 0 \tag{16.62}$$

- Elastic law:

$$\left\{ \begin{array}{c} \Delta p \\ \Delta q \end{array} \right\} = \left[\begin{array}{cc} K & 0 \\ 0 & 3G \end{array} \right] \left\{ \begin{array}{c} \Delta \varepsilon_v^e \\ \Delta \varepsilon_q^e \end{array} \right\} \tag{16.63}$$

- Let's assume an associated flow rule and perfect plasticity:

$$\Rightarrow g = f$$

- Consistency condition:

$$\frac{\partial f}{\partial p} dp + \frac{\partial f}{\partial q} dq = 0 \tag{16.64}$$

$$dp = K(d\varepsilon_v - d\varepsilon_v^p) \tag{16.65}$$

$$dq = 3G\left(d\varepsilon_q - d\varepsilon_q^p\right) \tag{16.66}$$

$$d\varepsilon_v^p = d\lambda \frac{\partial f}{\partial p} \tag{16.67}$$

$$d\varepsilon_q^p = d\lambda \frac{\partial f}{\partial q} \tag{16.68}$$

Substituting into Equation (16.64):

$$\frac{\partial f}{\partial p} K d\varepsilon_v - \frac{\partial f}{\partial p} K d\lambda \frac{\partial f}{\partial p} + \frac{\partial f}{\partial q} 3G d\varepsilon_q - \frac{\partial f}{\partial q} 3G d\lambda \frac{\partial f}{\partial q} = 0 \tag{16.69}$$

$$d\lambda = \frac{\left[\dfrac{\partial f}{\partial p}, \dfrac{\partial f}{\partial q} \right] \left[\begin{array}{cc} K & 0 \\ 0 & 3G \end{array} \right] \left\{ \begin{array}{c} d\varepsilon_v \\ d\varepsilon_q \end{array} \right\}}{\left[\dfrac{\partial f}{\partial p}, \dfrac{\partial f}{\partial q} \right] \left[\begin{array}{cc} K & 0 \\ 0 & 3G \end{array} \right] \left\{ \begin{array}{c} \dfrac{\partial f}{\partial p} \\ \dfrac{\partial f}{\partial q} \end{array} \right\}} \tag{16.70}$$

$$\frac{\partial f}{\partial p} = -\eta$$

$$\frac{\partial f}{\partial q} = 1$$

$$d\lambda = \frac{-\eta K d\varepsilon_v + 3G d\varepsilon_q}{\eta^2 K + 3G} \tag{16.71}$$

$$d\varepsilon_v^p = -d\lambda \eta, \, d\varepsilon_q^p = d\lambda$$

$$\left\{ \begin{array}{c} dp \\ dq \end{array} \right\} = \left[\begin{array}{cc} K & 0 \\ 0 & 3G \end{array} \right] \left\{ \begin{array}{c} d\varepsilon_v - d\varepsilon_v^p \\ d\varepsilon_q - d\varepsilon_q^p \end{array} \right\} \tag{16.72}$$

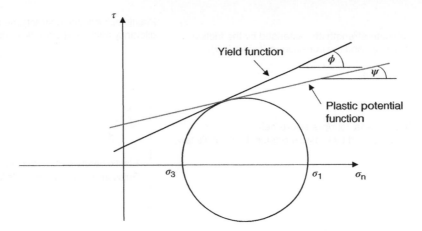

Fig. 16.23 Non-associated Mohr–Coulomb flow rule.

$$\left\{ \begin{array}{c} dp \\ dq \end{array} \right\} = \left[\begin{array}{cc} K & 0 \\ 0 & 3G \end{array} \right] \left\{ \begin{array}{c} d\varepsilon_v \\ d\varepsilon_q \end{array} \right\} - \left[\begin{array}{cc} K & 0 \\ 0 & 3G \end{array} \right] d\lambda \left\{ \begin{array}{c} -\eta \\ 1 \end{array} \right\} \tag{16.73}$$

$$\left\{ \begin{array}{c} dp \\ dq \end{array} \right\} = \left[\begin{array}{cc} K - \dfrac{\eta^2 K^2}{\eta^2 K + 3G} & \dfrac{3GK\eta}{\eta^2 K + 3G} \\[3ex] \dfrac{3GK\eta}{\eta^2 K + 3G} & 3G - \dfrac{9G^2}{\eta^2 K + 3G} \end{array} \right] \left\{ \begin{array}{c} d\varepsilon_v \\ d\varepsilon_q \end{array} \right\} \tag{16.74}$$

with

$$\mathbf{D}^{ep} = \left[\begin{array}{cc} K - \dfrac{\eta^2 K^2}{\eta^2 K + 3G} & \dfrac{3GK\eta}{\eta^2 K + 3G} \\[3ex] \dfrac{3GK\eta}{\eta^2 K + 3G} & 3G - \dfrac{9G^2}{\eta^2 K + 3G} \end{array} \right]$$

For a non-associated flow rule, the Mohr–Coulomb flow rule is defined through the *dilatancy angle, ψ,* of the soil (Fig. 16.23). This angle is the constant of the Mohr-Coulomb model that defines the plastic volumetric strain.

The potential function can be defined as:

$$g(\sigma) = \tau - \sigma_n \tan \psi - c = 0 \tag{16.75}$$

If $\psi = \phi$, then it's an associated flow.

As can be noticed from the previous example of Mohr-Coulomb model, such elastic-perfectly plastic models assume no plastic hardening or softening, i.e. the process of plastically deforming the material does not cause its intrinsic strength to alter. This last assumption simplifies the mathematical treatment, but at a cost. Another assumption made in the simpler models of this type is that if volume change or dilation accompanies plastic straining, it does so at a constant rate. This also simplifies the mathematical treatment. By far the most

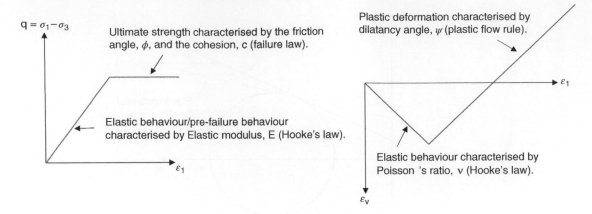

Fig. 16.24   Mohr–Coulomb elastic-perfectly plastic model.

commonly used form of elastoplastic constitutive model is the Mohr–Coulomb model. It is found in most commercially available numerical programs and in most design codes used in geotechnical engineering today.

Figure 16.24 represents schematically the Mohr–Coulomb model for a sample subjected to biaxial or triaxial compression, in which the minimum principal stress, σ_3, is maintained constant. As the shear stress is applied, the response of the model is initially linear elastic. If the stress is increased, shear failure will be initiated in the material. The soil then deforms at constant shear stress and while deforming plastically, it will dilate. In the simplest form of the model, the dilation continues indefinitely as the sample continues to be sheared. Hooke's law of elasticity defines the pre-failure behaviour. The strength parameters and the confining stress level determine the ultimate strength of the soil, while the plastic flow rule, and particularly the dilatancy angle, ψ, determines the volume changes that occur as the soil deforms plastically.

(d)   Limitations of elastic-perfectly plastic models

Within the framework of an elastoplastic theory of soil behaviour, the major limitations with the simpler models are related to the assumptions made about the plastic flow rule and the nature of any hardening or softening. The simpler models predict never-ending dilation, violating the concept of critical state behaviour. Moreover, more accurate pre-failure predictions are needed which will often require the inclusion of some plastic hardening into the model.

Therefore, the more general class of such models is that consisting of elastoplastic soil models allowing the possibility of pre-peak hardening and/or of post-peak softening which improve our understanding and our predictions of the pre-failure behaviour of soils. In other words, hardening soil models were introduced particularly to improve predictions of soil response in the working load range. The best-known soil model in this category is the *Modified Cam Clay*. This model is now available in most software packages used in geotechnical practice.

Consider, as before, the biaxial or triaxial compression of a dense soil sample subjected to a constant confining pressure, σ_3. A constitutive model which features plastic pre-peak hardening and post-peak softening will exhibit a typical stress–strain curve as illustrated schematically in Fig. 16.25.

During this deformation the soil may initially contract, then dilate, and ultimately it should reach a critical state of deformation. The shape of the stress-strain response depends on the assumptions concerning

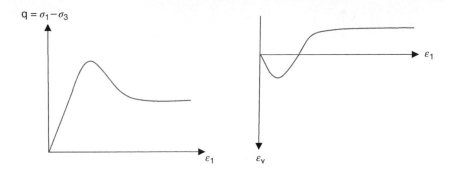

Fig. 16.25 Typical stress-strain curve with pre-peak hardening and post-peak softening.

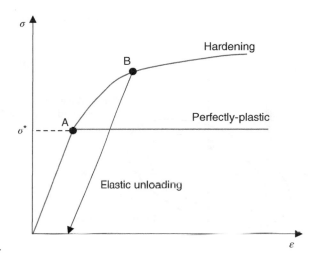

Fig. 16.26 Stress–strain curve with a hardening rule.

elasticity and the hardening law. The mobilised strength depends on the relationship between friction and dilation and, finally, the volumetric response depends on the plastic flow rule.

16.6.3 The hardening rule

As already discussed, most materials exhibit some degree of hardening accompanying plastic straining. In the perfectly plastic case, once the stress reaches the yield point A as shown in Fig. 16.26, plastic deformation starts, so long as the stress is maintained at A. If the stress is reduced, elastic unloading occurs. In the hardening case, once yield occurs, the stress needs to be continually increased in order to drive the plastic deformation. If the stress is held constant, for example at B, no further plastic deformation will occur; at the same time, no elastic unloading will occur. Note that this condition cannot occur in the perfectly-plastic case, where there is either plastic deformation or elastic unloading.

This means that the shape and size of the yield surface changes during plastic loading. This change may be rather arbitrary and extremely difficult to describe accurately. The initial yield surface will be of the form $f(\sigma_{ij}) = 0$ as discussed previously. In the perfectly plastic case, the yield surface remains unchanged. In the more general case (taking into account hardening), the yield surface may change size, shape and position, and can be described by:

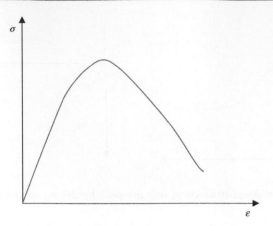

Fig. 16.27 Stress–strain curve with softening.

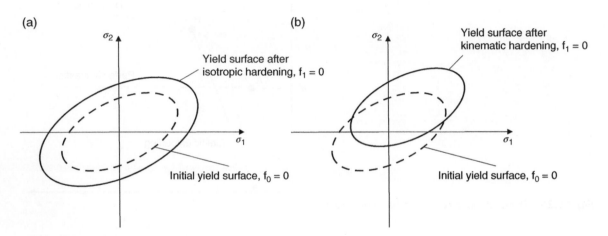

Fig. 16.28 Hardening types: (a) isotropic hardening, (b) kinematic hardening.

$$f(\sigma_{ij}, h_i) = 0 \tag{16.76}$$

Here, h_i represents one or more hardening parameters, which change during plastic deformation and determine the evolution of the yield surface. They may be scalars or higher-order tensors. At first yield, the hardening parameters are zero, and $f(\sigma_{ij}, 0) = f_0(\sigma_{ij})$. The description of how the yield surface changes with plastic deformation is called the *hardening rule*.

Materials can also strain soften, as is the case with dense soils. In this case, the stress-strain curve 'turns down', as shown in Fig. 16.27. The yield surface for such a material will in general decrease in size with further straining.

Therefore, hardening is often described by a combination of two specific types of hardening, namely *isotropic hardening* and *kinematic hardening*, as illustrated in Fig. 16.28.

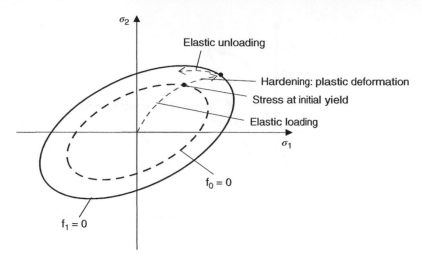

Fig. 16.29   Isotropic hardening.

(a) Isotropic hardening

Isotropic hardening is where the yield surface remains the same shape but expands with increasing stress as illustrated in Fig. 16.29.

In particular, the yield function takes the form:

$$f_1 = f(\sigma_{ij}, h_i) = f_0(\sigma_{ij}) - \sigma^*(\alpha) = 0 \tag{16.77}$$

where α is a hardening variable.

The shape of the yield function is specified by the initial yield function and its size changes with hardening.

(b) Kinematic hardening

The isotropic model implies that, if the yield strength in tension and compression are initially the same, i.e. the yield surface is symmetric about the stress axes, they remain equal as the yield surface develops with plastic strain.

However, in the case of *Bauschinger effect* (Fig. 16.30), a hardening in tension will lead to a softening in a subsequent compression. In this case, the kinematic hardening rule can be used. This is where the yield surface remains the same shape and size but translates in stress space (Fig. 16.30). In the figure, σ_{y1} is defined as the yield stress corresponding to the new yield surface after hardening.

The yield function now takes the general form:

$$f_1 = f(\sigma_{ij}, h_i) = f_0(\sigma_{ij} - \alpha_{ij}) = 0 \tag{16.78}$$

The hardening parameter here is the stress α_{ij}, known as the *back-stress* or *shift stress*; the yield surface is shifted relative to the stress-space axes by α_{ij} (Fig. 16.31).

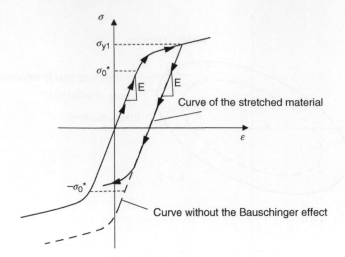

Fig. 16.30 Bauschinger effect.

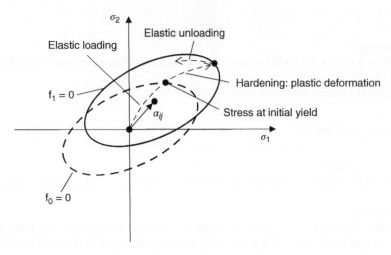

Fig. 16.31 Kinematic hardening.

(c) Establishing elastoplastic constitutive relations with hardening
Uniaxial stress–strain

Consider an example of a uniaxial stress–strain curve for an elastoplastic model as shown in Fig. 16.32.

To characterize the behaviour of the material, the following steps are considered:

- Elastic regime: $d\sigma = E\,d\varepsilon$
- Elastoplastic regime in unloading: $d\sigma = E\,d\varepsilon$
- Elastoplastic regime in loading: $d\sigma = E^{ep}\,d\varepsilon$ where E^{ep} is the elastoplastic tangent modulus to be determined

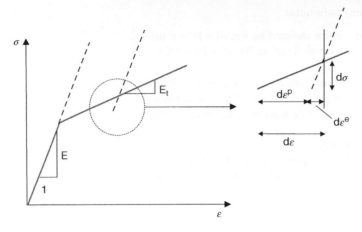

Fig. 16.32   Uniaxial stress-strain curve with hardening.

i) the elastic regime:
The elastic domain in the stress space is defined as the interior of the domain enclosed by the surface $f(\sigma, \alpha) = 0$ which defines the yield surface.
In an elastic domain: $f(\sigma, \alpha) < 0$
For the initial elastic domain:

$$\varepsilon^P = 0$$

But $d\alpha$ in 1D is defined as $d\alpha = \text{sign}(\sigma)d\varepsilon^P$
and $d\alpha \geq 0$
So

$$d\alpha = |d\varepsilon^P|$$

and

if $\varepsilon^P = 0$, then $\alpha = 0$ and $f(\sigma 0) < 0$

An additional requirement of the initial elastic domain is that it should contain the null stress state: $f(0, 0) < 0$
Therefore, a yield function can be defined as follows:

$$f(\sigma, \alpha) = |\sigma| - \sigma^*(\alpha)$$

$\sigma^*(\alpha)$ is the yield stress and is positive and $\sigma^*(0) = \sigma_0^*$
$\sigma^*(\alpha)$ is also named the hardening law. σ_0^* is the initial yield stress.

The hardening law provides the evolution of the yield stress $\sigma^*(\alpha)$ in terms of the hardening variable α.
It's common to consider a linear hardening law:

$$\sigma^* = \sigma_0^* + H'\alpha$$

$$d\sigma^* = H'd\alpha$$

where H′ is a hardening parameter.

So the yield surface can be defined as: $f(\sigma, \alpha) = |\sigma| - \sigma^*(\alpha) = 0$
The elastic domain can be defined as: $f(\sigma, \alpha) = |\sigma| - \sigma^*(\alpha) \leq 0$

- $f(\sigma, \alpha) < 0$ so $|\sigma| < \sigma^*(\alpha)$, then σ is in the elastic domain
- $f(\sigma, \alpha) = 0$ so $|\sigma| = \sigma^*(\alpha)$, then σ is on the yield surface
- $f(\sigma, \alpha) > 0$ so $|\sigma| > \sigma^*(\alpha)$, this is an impossible stress state

ii) the elastoplastic regime:
Consider an elastoplastic regime in plastic loading:
Then σ is on the yield surface and $f(\sigma, \alpha) = |\sigma| - \sigma^*(\alpha) = 0$

$$df(\sigma, \alpha) = 0$$

$$d|\sigma| - d\sigma^*(\alpha) = 0$$

$$\text{sign}(\sigma)d\sigma - H'd\alpha = 0$$

$$\text{sign}(\sigma)d\sigma - H'\text{sign}(\sigma)d\varepsilon^p = 0$$

So $d\varepsilon^p = \dfrac{1}{H'}d\sigma$

$$d\varepsilon = d\varepsilon^e + d\varepsilon^p$$

$$d\varepsilon^e = \dfrac{d\sigma}{E}$$

Then

$$d\varepsilon = \frac{1}{E}d\sigma + \frac{1}{H'}d\sigma = \left(\frac{1}{E} + \frac{1}{H'}\right)d\sigma$$

or

$$d\sigma = \frac{1}{\left(\frac{1}{E} + \frac{1}{H'}\right)}d\varepsilon = E^{ep}d\varepsilon$$

so

$$E^{ep} = E\frac{H'}{E + H'}$$

The hardening parameter H′ plays an important role in the definition of the slope E^{ep} of the elastoplastic domain.

Depending on the value of H′, different situations arise:

- $H' > 0$, then $E^{ep} > 0 \Rightarrow$ Plasticity with strain hardening
- $H' = 0$, then $E^{ep} = 0 \Rightarrow$ Perfect plasticity
- $H' < 0$, then $E^{ep} < 0 \Rightarrow$ Plasticity with strain softening

Triaxial stress-strain

- *Elastic regime:* σ is in the elastic domain, so

  $$d\sigma = D d\varepsilon$$

  where **D** is the elastic constitutive tensor
- *Elastoplastic regime in unloading:* σ is on the yield surface but $f < 0$, so

  $$d\sigma = D d\varepsilon$$

- *Elastoplastic regime in plastic loading:*
    σ is on the yield surface and $f = 0$, so

  $$d\sigma = D^{ep} d\varepsilon$$

  where D^{ep} is the elastoplastic constitutive tensor to be determined.
- *Determination of* D^{ep}:

  $$d\varepsilon = d\varepsilon^e + d\varepsilon^p,$$

  – Hooke's law: $d\sigma = D d\varepsilon^e = D(d\varepsilon - d\varepsilon^p)$

  – Flow rule:

  $$d\varepsilon^p = d\lambda \frac{\partial g}{\partial \sigma}$$

  $$d\sigma = D d\varepsilon - d\lambda D \frac{\partial g}{\partial \sigma}$$

  – Hardening variable: $d\alpha = d\lambda$
  – Consistency condition:

  $$\frac{\partial f}{\partial \sigma} d\sigma - \frac{\partial \sigma^*}{\partial \alpha} d\alpha = 0$$

  $$\frac{\partial f}{\partial \sigma} d\sigma - H' d\lambda = 0$$

  $$\frac{\partial f}{\partial \sigma} \left(D d\varepsilon - d\lambda D \frac{\partial g}{\partial \sigma} \right) - H' d\lambda = 0$$

$$d\lambda = \frac{\left(\frac{\partial f}{\partial \sigma}\right)^{T} D d\varepsilon}{H' + \left(\frac{\partial f}{\partial \sigma}\right)^{T} D \frac{\partial g}{\partial \sigma}}$$

Then

$$d\sigma = \left(D - \frac{D \frac{\partial g}{\partial \sigma} \left(\frac{\partial f}{\partial \sigma}\right)^{T} D}{H' + \left(\frac{\partial f}{\partial \sigma}\right)^{T} D \frac{\partial g}{\partial \sigma}} \right) d\varepsilon$$

$$D^{ep} = D - \frac{D \frac{\partial g}{\partial \sigma} \left(\frac{\partial f}{\partial \sigma}\right)^{T} D}{H' + \left(\frac{\partial f}{\partial \sigma}\right)^{T} D \frac{\partial g}{\partial \sigma}}$$

Notice that H' plays the same role as in the uniaxial case. It determines the expansion or the contraction of the elastic domain in the stress space.

- $H' > 0$: Expansion of the elastic domain $\Rightarrow$ Plasticity with hardening
- $H' < 0$: Contraction of the elastic domain $\Rightarrow$ Plasticity with softening
- $H' = 0$: Constant elastic domain $\Rightarrow$ Perfect plasticity

References

American Society for Testing and Materials (2017) *Standard Practice for Classification of Soils for Engineering Purposes (Unified Soil Classification System)*. D2487-17. ASTM International. Philadelphia, Pennsylvania.

Atkinson, J.H. (2007) *The Mechanics of Soils and Foundations*, 2nd edn. Taylor & Francis, Oxon.

Atkinson, J.H and Bransby, P.L. (1978) *The Mechanics of Soils*. McGraw-Hill, London.

Atterberg, A. (1911) *Die Plastizitat der Tone*. Int. Mitt. für Bodenkunden, I, (1), 10–43, Berlin.

Attewell, P. (1993) *Ground Pollution: Environment, Geology, Engineering and Law*. E & FN Spon, London.

Barden, L. (1974) *Sheet pile wall design based on Rowe's method*. Part III of CIRIA Technical Report No. 54 – A comparison of quay wall design methods, London.

Barron, R.A. (1948) Consolidation of fine grained soils by drain wells *Trans. ASCE*, **113**: 718–742.

Bell, A.L. (1915) Lateral pressure and resistance of clay and the supporting power of clay foundations. *Minutes Proc. Inst. Civil. Engrs*, 199, (1915), pp. 233–272, London.

Bell, J.P., Dean, T.J. and Hodnett, M.G. (1987) Soil moisture measurement by an improved capacitance technique, Part II. Field techniques, evaluation and calibration. *J. Hydrol.*, **93**, pp. 79–90.

Berezantzev, V.G., Khristorforov, V. and Golubkov, V. (1961) Load bearing capacity and deformation of piled foundations. *Proc. 5th Int. Conf. ISSMFE*, 2, pp. 11–15, Paris (1961).

Bishop, A.W. (1954) The use of pore-pressure coefficients in practice. *Géotechnique*, **4**, (2), pp. 148–152.

Bishop, A.W. (1955) The use of the slip circle in the stability analysis of slopes. *Géotechnique*, **5**, (1), pp. 7–17.

Bishop, A.W. and Morgenstern, N. (1960) Stability coefficients for earth slopes. *Géotechnique*, **10**, (4), pp. 129–153.

Bishop, A.W., Green, G.E., Garga, V.K., Andresen, A. and Brown, J.D. (1971) A new ring shear apparatus and its application to the measurement of residual strength. *Géotechnique*, **21**, (4), pp. 273–328.

Bond, A.J. (2011) A procedure for determining the characteristic value of a geotechnical parameter. *Proc. 3rd Int Symp on Geotechnical Safety and Risk (ISGSR)*, pp. 537–547, Munich, Germany (2–3 June 2011).

Bond A. and Harris A. (2008) *Decoding Eurocode 7*. Taylor and Francis, London.

Bond, A.J., Formichi, P., Spehl, P. and van Seters, A.J. (2019a) Tomorrow's geotechnical toolbox: EN 1990:202x, basis of structural and geotechnical design, *Proc. XVII Eur. Conf. Soil Mech. Geotech. Eng.*, Reykjavik (1–6 September 2019).

Bond, A.J., Jenner, C. and Moormann, C. (2019b) Tomorrow's geotechnical toolbox: EN 1997-3:202x, Geotechnical structures. *Proc. XVII Eur. Conf. Soil Mech. Geotech. Eng.*, Reykjavik (1–6 September 2019).

Boussinesq, J. (1885) *Application des potentiels à l'étude de l'équilibre et de mouvement des solides élastiques*. Gauthier–Villard, Paris.

Bowles, J.E. (1996) *Foundation Analysis and Design*, 5th edn. McGraw-Hill Book Co., New York.

British Standards Institution (1990) *British Standard Methods of Test for Soils for Civil Engineering Purposes*. BS 1377, Parts 1–9. British Standards Institution, London.

Smith's Elements of Soil Mechanics, 10th Edition. Ian Smith.
© 2021 John Wiley & Sons Ltd. Published 2021 by John Wiley & Sons Ltd.
Companion website: www.wiley.com/go/smith/soilmechanics10e

British Standards Institution (1995) *Code of Practice for Strengthened/Reinforced Soils and Other Fills*. BS 8006. British Standards Institution, London.

British Standards Institution (2002a) *Eurocode – Basis of Structural Design*. BS EN 1990. British Standards Institution, London.

British Standards Institution (2002b) BS EN 1991-1-1 *Eurocode 1. Actions on Structures. General Actions*. British Standards Institution, London.

British Standards Institution (2003) BS EN ISO 14686. *Hydrometric Determinations. Pumping Tests for Water Wells. Considerations and Guidelines for Design, Performance and Use*. British Standards Institution, London.

British Standards Institution (2004) *Eurocode 7: Geotechnical Design – Part 1: General Rules*. BS EN 1997-1. British Standards Institution, London.

British Standards Institution (2007a) *Eurocode 7: Geotechnical Design – Part 2: Ground Investigation and Testing*. BS EN 1997-2. British Standards Institution, London.

British Standards Institution (2007b): *UK National Annex to Eurocode7 – Geotechnical Design – Part 1: General Rules*. UK NA to BS EN 1997-1:2004. British Standards Institution, London.

British Standards Institution (2009): *UK National Annex to Eurocode7 – Geotechnical Design – Part 2: Ground Investigation and Testing*. UK NA to BS EN 1997-2:2007. British Standards Institution, London.

British Standards Institution (2009–2020) BS EN ISO 22476 *Geotechnical Investigation and Testing – Field Testing, Parts 1–15*. British Standards Institution, London.

British Standards Institution (2010) BS 8006-1:2010 *Code of Practice for Strengthened/Reinforced Soils and Other Fills*. British Standards Institution, London.

British Standards Institution (2011) BS 8006-2:2011 Code of Practice for Strengthened/Reinforced Soils. *Soil Nail Design*. British Standards Institution, London.

British Standards Institution (2012) BS EN ISO 22282 *Geotechnical Investigation and Testing – Geohydraulic Testing, Parts 1–6*. British Standards Institution, London.

British Standards Institution (2014–2019) BS EN ISO 17892 *Geotechnical Investigation and Testing – Laboratory Testing of Soil*. Parts 1 – 12. London.

British Standards Institution (2015) *Code of Practice for Site Investigations*. BS 5930, London.

British Standards Institution (2016–2018) BS EN ISO 22477 *Geotechnical Investigation and Testing – Testing of Geotechnical Structures. Parts 1–4*. British Standards Institution, London.

British Standards Institution (2017) *BS EN ISO 11272 Soil Quality – Determination of Dry Bulk Density*. British Standards Institution London.

British Standards Institution (2018a) BS EN ISO 14688-1 *Geotechnical Investigation and Testing – Identification and Classification of Soil – Part 1: Identification and Description*. British Standards Institution, London.

British Standards Institution (2018b) BS EN ISO 14689 *Geotechnical Investigation and Testing – Identification and Classification of Rock*. British Standards Institution, London.

British Standards Institution (2018c) BS EN ISO 14688-2 *Geotechnical Investigation and Testing – Identification and Classification of Soil – Part 2: Principles for a Classification*. British Standards Institution, London.

British Standards Institution (2006–2011) BS EN ISO 22475 *Geotechnical Investigation and Testing – Sampling Methods and Groundwater Measurements. Parts 1–3*. British Standards Institution, London.

British Standards Institution (2020) *Code of Practice for Foundations*. BS 8004:2015 + A1:2020. British Standards Institution, London.

Broms, B.B. (1966) Methods of calculating the ultimate bearing capacity of piles, a summary. *Sols–Soils*, **5**, (18 and 19), pp. 21–31.

Brown, J.D. and Meyerhof, G.C. (1969) Experimental study of bearing capacity in layered clays. *Proc. 7th Int. Conf. ISSMFE*, 2, pp. 45–51, Mexico City (1969).

Cairney, T. (1998) *Contaminated Land: Problems and Solutions*. CRC Press, London.

Campbell, G.S. and Gee, G.W. (1986) Water potential: miscellaneous methods. In *Methods of Soil Analysis. Part 1. Physical and Mineralogical Methods* (A. Klute). American Society of Agronomy–Soil Science Society of America.

Carter, J.P. (2006). Who needs constitutive models? *Aust. Geomech. J.*, **41**, pp 1–27

Casagrande, A. (1937) Seepage through earth dams. *J. New Eng. Water Works Assoc.*, **51**, (2), pp. 131–172.

Casagrande, A. (1947) Classification and identification of soils. *Proc. Am. Soc. Civ. Eng.*, **73**, pp. 783–810.

CEN (2020) *prEN1990 Eurocode – Basis of Structural and Geotechnical Design*. CEN, Brussels.

CEN (2018a) *prEN 1997-1 Geotechnical Design – General Rules*. CEN, Brussels.

CEN (2018b) *prEN 1997-2 Geotechnical Design – Ground Properties*. CEN, Brussels.

CEN (2018c) *prEN 1997-3 Geotechnical Design – Geotechnical Structures*. CEN, Brussels.

Charles, J.A., Skinner, H.D. and Watts, K.S. (1998) The specification of fills to support buildings on shallow foundations: the "95% fixation". *Ground Eng.*, January, Londonpp. 29–33.

Chen, W.F. (1975) *Limit Analysis and Soil Plasticity*. Elsevier.

Clayton, C.R.I. (1995) The standard penetration test (SPT): methods and use. CIRIA Report 143, London.

Clayton, C.R.I. and Symons, I.F. (1992) The pressure of compacted fill on retaining walls. (Technical Note). *Géotechnique*, **42**, (1), pp. 127–130.

Coulomb, C.A. (1776) Essais sur une application des règles des maxims et minimis à quelques problèmes de statique relatifs à l'architecture. *Mem. Acad. Roy. Pres. Divers, Sav.* 5,7, Paris.

Darcy, H. (1856) *Les fontaines publiques de la ville de Dijon*. Dalmont, Paris.

Dean, T.J., Bell, J.P. and Baty, A.J.B. (1987) Soil moisture measurement by an improved capacitance technique, part 1. Sensor design and performance. *J. Hydrol.*, **93**, pp. 67–78.

De Beer, E.E. (1963) The scale effect in the transposition of the results of deep sounding tests on the ultimate bearing capacity of piles and caisson foundations. *Géotechnique*, **13**, (1)pp. 39–75.

De Beer, E.E. and Martens, A. (1957) Method of computation of an upper limit for the influence of the homogenity of sand layers in the settlement of bridges. *Proc. 4th Int. Conf. ISSMFE*, 1, London (1957).

Dunnicliff, J. (1993) *Geotechnical Instrumentation for Monitoring Field Performance*. John Wiley & Sons, Chichester.

Estaire, J., Arroyo, M., Scarpelli, G. and Bond, A.J. (2019). Tomorrow's geotechnical toolbox: design of geotechnical structures to EN 1997:202x, *Proc. XVII Eur. Conf. Soil Mech. Geotech. Eng.*, Reykjavik (1–6 September 2019).

Fadum, R.E. (1948) Influence values for estimating stresses in elastic foundations. *Proc. 2nd Int. Conf. ISSMFE*, 3, Rotterdam (1948).

Fellenius, W. (1927) *Erdstatische Berechnungen*. Ernst, Berlin.

Franzen, G., Arroyo, M., Lees, A., Kavvadas, M., van Seters, A. and Bond, A.J. (2019) Tomorrow's geotechnical toolbox: EN 1997-1:202x, general rules. *Proc. XVII Eur. Conf. Soil Mech. Geotech. Eng.*, Reykjavik (1–6 September 2019).

Fredlund, D.G., Rahardjo, H. and Fredlund, M.D. (2012) *Unsaturated Soil Mechanics in Engineering Practice*. Wiley.

Gibson, R.E. (1958) The progress of consolidation in a clay layer increasing in thickness with time. *Géotechnique*, **8**, pp. 171–182.

Gibson, R.E. and Lumb, P. (1953) Numerical solution for some problems in the consolidation of clay. *Proc. Inst. Civ. Engrs*, Part 1, pp. 182–198, London.

Gibbs, H.J. and Holtz, W.G. (1957). Research on determining the density of sands by spoon penetration test. *Proc. 4th Int. Conf. ISSMFE*, 1, London (1957).

Han, J. (2015) *Principles and Practices of Ground Improvement*. Wiley, New Jersey.

Hansen, J.B. (1957) Foundation of structures – general report. *Proc. 4th Int. Conf. ISSMFE*, London (1957).

Hansen, J.B. (1970) A revised and extended formula for bearing capacity. *Danish Geotech. Inst.*, Bulletin 28, Copenhagen.

Harris, M. (1994) *Contaminated Land: Investigation, Assessment and Remediation*. Thomas Telford, London.

Head, K.H. (2006) *Manual of Soil Laboratory Testing*. Vol 1 Whittles, Dunbeath.

Head, K.H. and Epps R.J. (2011) *Manual of Soil Laboratory Testing*. Vol 2. Whittles, Dunbeath.

Head, K.H. and Epps R.J. (2013) *Manual of Soil Laboratory Testing*. Vol 3. Whittles, Dunbeath.

Hicks M. (2013). An explanation of characteristic values of soil properties in Eurocode 7, in *Modern Geotechnical Design Codes of Practice* (Eds P. Arnold et al.). IOS Press, Amsterdam.

Highways England, Transport Scotland, Welsh Assembly Government and The Department for Infrastructure Northern Ireland (2020a) *Design Manual for Roads and Bridges*. Standards for Highways, London.

Highways England, Transport Scotland, Welsh Assembly Government and The Department for Infrastructure Northern Ireland (2020b) *CD224 Traffic Assessment*. Standards for Highways, London.

Highways England, Transport Scotland, Welsh Assembly Government and The Department for Infrastructure Northern Ireland (2020c) *CD225 Design for New Pavement Foundations*. Standards for Highways, London.

Highways England, Transport Scotland, Welsh Assembly Government and The Department for Infrastructure Northern Ireland (2020d) *CD226 Design for New Pavement Construction*. Standards for Highways, London.

Highways England, Transport Scotland, Welsh Assembly Government and The Department for Infrastructure Northern Ireland (2016a) *Manual of Contract Documents for Highway Works*, MCHW 1; Specification for highway works: Series 600. Earthworks, London.

Highways England, Transport Scotland, Welsh Assembly Government and The Department for Infrastructure Northern Ireland (2016b) *Manual of Contract Documents for Highway Works*, MCHW 2; Notes for guidance on the Specification for highway works: Series NG 600. Earthworks, London.

Ingold, T.S. (1979) The effects of compaction on retaining walls. *Geotechnique*, **29**, (3), pp. 265–283.

Jaky, J. (1944) The coefficient of earth pressure at rest. *J. Soc. Hung. Archit. Eng.*, **78**, (22)pp. 355–358.

Janbu, N. (1957) Earth pressure and bearing capacity calculations by generalised procedure of slices. *Proc. 4th Int. Conf. ISSMFE*, London (1957).

Kerisel, J. and Absi, E. (1990) *Active and Passive Earth Pressure Tables*, 3rd edn. Balkema, Rotterdam.

Kjellman, W., Kallstenius, T. and Wager, O. (1950) Soil sampler with metal foils. *Proc. Royal Swed. Geot. Inst.*, No. 1, Stockholm (January 1950).

Lade, P.V. (2005) *Overview of Constitutive Models for Soils*. ASCE Geo-Frontiers Congress, Austin, USA.

Lade P.V. (2016) *Triaxial Testing of Soils*. Wiley.

Lunne, T., Robertson, P.K. and Powell, J.J.M. (1997) *Cone Penetration Testing in Geotechnical Practice*. E & FN Spon, London.

Meigh, A.C. and Nixon, I.K. (1961) Comparison of in situ tests for granular soils. *Proc. 5th Int. Conf. ISSMFE*, 1, pp. 449–507, Paris (1961).

Meyerhof, G.G. (1953) The bearing capacity of foundations under eccentric and inclined loads. *Proc. 3rd Int. Conf. ISSMFE*, Zurich (1953).

Meyerhof, G.G. (1956) Penetration tests and bearing capacity of cohesionless soils. *Proc. ASCE, J. Soil. Mech. Found. Div.*, **85**, (SM6), pp. 1–19.

Meyerhof, G.G. (1959) Compaction of sands and bearing capacity of piles. *Proc. ASCE, J. Soil Mech. Found. Div.*, **85**, (SM6), pp. 1–30.

Meyerhof, G.G. (1963) Some recent research on the bearing capacity of foundations. *Can. Geotech. J.*, **1**, (1), pp. 16–23.

Meyerhof, G.G. (1974) State-of-the-art of penetration testing in countries outside Europe. *Proc. 1st Euro. Symp. on Penetration Testing*, 2, pp. 40–48, Stockholm (5 June 1974).

Meyerhof, G.G. (1976) Bearing capacity and settlement of piled foundations. *J. Geotech. Eng. Div., ASCE*, **102**, (GT3), pp. 195–228.

Mikkelsen, P.E. and Green, G.E. (2003) Piezometers in fully grouted boreholes. *Proc. Symp Field Measurements in Geomech.*, Oslo, Norway (23–26 September 2003).

Muir Wood, D. (1991) *Soil Behaviour and Critical State Soil Mechanics*. Cambridge University Press.

Murray, H.H. (2006) *Applied Clay Mineralogy*. Elsevier, Amsterdam.

Newmark, N.M. (1942) Influence charts for computation of stresses in elastic foundations. University of Illinois Engng Exp. Stn., Bull. No. 338.

Norbury, D.R., Arroyo, M., Foti, S., Garin, H., Reiffsteck, P. and Bond, A.J. (2019). Tomorrow's geotechnical toolbox: EN 1997-2:202x, Ground investigation. *Proc. XVII Eur. Conf. Soil Mech. Geotech. Eng.*, Reykjavik (1–6 September 2019).

Ng, C. W. W. and Menzies, J. (2007) *Advanced Unsaturated Soil Mechanics in Engineering Practice*. CRC Press.

Orr, T. (2012) How Eurocode 7 has affected geotechnical design: a review. *Proc. ICE – Geotech. Eng.*, **165**, (6), 337–350.

Padfield, C.J. and Mair, R.J. (1984) Design of retaining walls embedded in stiff clay. CIRIA Report 104, London.

Parry, R.H.G. (1960) Triaxial compression and extension tests on remoulded saturated clay. *Géotechnique*, **10**, pp. 166–180.

Parsons, A.W. (1976) The rapid determination of the moisture condition of earthwork material. Department of Environment, Department of Transport, TRRL Report LR750, Crowthorne, Berks.

Parsons, A.W. and Boden, J.B. (1979) The moisture condition test and its potential applications in earthworks. Department of Environment, Department of Transport, TRRL Report SR522, Crowthorne, Berks.

Poon, M.S.B. (2005) The Behaviour of Shallow Foundations on Silica Sand Subjected to Inclined Load. PhD Thesis, The University of Sydney.

Potts, D.M. and Fourie, A.B. (1984) The behaviour of a propped retaining wall: results of a numerical experiment. *Géotechnique*, **34**, pp. 383–404.

Powell, W.D., Potter, J.F., Mayhew, H.C. and Nunn, M.E. (1984) The structural design of bituminous roads. Department of Environment, Department of Transport, TRRL Report LR1132, Crowthorne, Berks.

Prandtl, L. (1921) Uber die Eindringungsfestigkeit plastischer Baustoffe und die Festigkeit von Schneiden. *Zeitschrift fur Angewandte Mathematik*, **1**, (1), pp. 15–20.

Rankine, W.J.M. (1857) On the stability of loose earth. *Philos. Trans. R. Soc.*, **147**, (Part 1), pp. 9–27.

Ridley, A.M. (2015) Soil suction — what it is and how to successfully measure it. *Proc. Field Meas. Geomech.*, Sydney, Australia (9–11 September 2015).

Rowe, P.W. (1952) Anchored sheet-pile walls. *Proc. Inst. Civ. Engrs.*, Part 1, 1, pp. 27–70, London (January 1952).

Rowe, P.W. (1957) Sheet-pile walls in clay. *Proc. Inst. Civ. Engrs.*, 7, pp. 629–654, London (July 1957).

Rowe, P.W. (1958) Measurements in sheet-pile walls driven into clay. *Proc. Brussels Conf. 58 on Earth Pressure Problems. Belgian group of ISSMFE*, Brussels (1958).

Rowe, P.W. and Barden, L. (1966) A new consolidation cell. *Géotechnique*, **16**, (2), pp. 162–170.

Schmertmann, J.H. (1953) Estimating the true consolidation behaviour of clay from laboratory test results. *Proc. ASCE*, 120 (1953).

Schmertmann, J.H. (1970) Static cone to compute settlement over sand. *Proc. ASCE, J. Soil Mech. Found. Div.*, **96**, (SM3), pp. 1001–1043.

Schmertmann, J.H., Hartman, I.P. and Brown, P.R. (1978) Improved strain influence factor diagrams. *Proc. ASCE, J. Geotech. Engng. Div.*, **104**, (GT8), pp. 1131–1135.

Schneider H. and Schneider M. (2013) Dealing with uncertainties in EC7 with emphasis on determination of characteristic soil properties, in *Modern Geotechnical Design Codes of Practice* (Eds P. Arnold et al.). IOS Press, Amsterdam.

Schuppener, B. (2008) Personal communication, Eurocodes Background and applications: "Dissemination of information for training" workshop, Brussels (18–20 February 2008).

Sherard, J.L., Dunnigan, L.P. and Talbot, J.R. (1984a) Basic properties of sand and gravel filters. *ASCE J. Geotech. Eng.*, **110**, (6), pp. 684–700.

Sherard, J.L., Dunnigan, L.P. and Talbot, J.R. (1984b) Filters for silts and clays. *ASCE J. Geotech. Eng.*, **110**, (6), pp. 701–718.

Simpson, B. (2011) *Concise Eurocodes: Geotechnical Design, BS EN 1997-1: Eurocode 7, Part 1.* BSI, London.

Site Investigation Steering Group (2013) *Effective Site Investigations.* ICE Publishing, London.

Skempton, A.W. (1951) The bearing capacity of clays. *Proc. Building Research Congress*, pp. 180–189, London (1951).

Skempton, A.W. (1953) The colloidal activity of clays. *Proc. 3rd Int. Conf. ISSMFE*, 1, Zurich (1953).

Skempton, A.W. (1954) The pore pressure coefficient A and B. *Géotechnique*, **4**, (4), pp. 143–147.

Skempton, A.W. and Bjerrum, L. (1957) A contribution to settlement analysis of foundations on clay. *Géotechnique*, **7**, (4), pp. 168–178.

Smith, G.N. (1968a) *Determining the Settlements of Embankments on Soft Clay.* Highways and Public Works, London.

Smith, G.N. (1968b) *Construction Pore Pressures in an Earth Dam.* Civil Engineering Public Works Review, London.

Smith, G.N. (1971) *An Introduction to Matrix and Finite Element Methods in Civil Engineering.* Applied Science Publishers, London.

Smith, I., Oliphant, J. and Wallis, S.G. (1997) Field validation of a computer model for forecasting mean weekly in situ moisture condition value. *Electron. J. Geotech. Eng.*, **2**, http://www.ejge.com/.

Sokolovski, V.V. (1960) *Statics of Soils Media.* (Trans. by D.H. Jones and A.N. Schofield). Butterworth & Co. Ltd., London.

Stannard, D.I. (1992) Tensiometers – theory, construction and use. *Geotech. Test. J.*, **15**, (1), pp. 48–58.

Steinbrenner, W. (1934) Tafeln zur Setzungsberechnung. *Die Strasse*, **1**, pp. 121–124.

Taylor, D.W. (1948) *Fundamentals of Soil Mechanics.* John Wiley & Sons Inc., New York; Chapman & Hall, London.

Terzaghi, K. (1925) *Erdbeaumechanik auf bodenphysikalischer grundlage.* Deuticke, Vienna.

Terzaghi, K. (1943) *Theoretical Soil Mechanics.* John Wiley, London and New York.

Terzaghi, K. and Peck, R.B. (1948) *Soil Mechanics in Engineering Practice.* Chapman and Hall, London; John Wiley & Sons Inc., New York.

Terzaghi, K. and Peck, R. (1967) *Soil Mechanics in Engineering Practice*, 2nd edn. John Wiley, New York.

Terzaghi, K., Peck, R.B. and Mesri, G. (1996) *Soil Mechanics in Engineering Practice*, 3rd edn. John Wiley, New York.

Thenault, L. (2012) Design of sheet pile walls to Eurocode 7: the effect of passive pressure. MEng Disseration, Edinburgh Napier University.

Topp, G.C. and Davis, J.L. (1985) Measurement of soil water content using time-domain reflectometry (TDR): a field evaluation. *Soil Sci. Soc. Am. J.*, **49**, pp. 19–24.

Trenter, N.A. and Charles, J.A. (1996) A model specification for engineered fills for building purposes. *Proc. Inst. Civ. Eng., Geotech. Eng.*, **119**, (4), pp. 219–230.

van Seters, A.J. and Franzen, G. (2019) Tomorrow's geotechnical toolbox: EN 1997, Overview. *Proc. XVII Eur. Conf. Soil Mech. Geotechn. Eng.*, Reykjavik (1–6 September 2019).

Vesic, A.S. (1963) Bearing capacity of deep foundations in sand. Soil Mechanics Lab. Report, Georgia Inst. of Technology.

Vesic, A.S. (1970) Tests on instrumented piles, Ogeechee River site. *J. Soil Mech., Found. Div., ASCE*, **96**, (SM2), pp. 561–584.

Vesic, A.S. (1973) Analysis of ultimate loads of shallow foundations. *J. Soil Mech., Found. Div., ASCE*, **99**, (SMI), pp. 45–73.

Vesic, A.S. (1975) Bearing capacity of shallow foundations. Chapter 3. In *Foundation Engineering Handbook* (Eds H.F. Winterkorn & H.Y. Fang). Van Nostrand, Reinhold Co., New York.

Vidal, H. (1966) *La terre armée.* Annales de l'Institut Technique du Bâtiment et des Travaux Publics, 19, (223 and 224), France.

Walsh, I.D. (2011) *Manual of Highway Design and Management*, ICE manuals (Ed. I.D. Walsh. Institution of Civil Engineers, London.

Wilson, G. (1941) The calculation of the bearing capacity of footings on clay. *J. Inst. Civ. Eng.*, London, **17**, (1), pp. 87–96.

Wise, W.R. (1992) A new insight on pore structure and permeability. *Water Resour. Res.*, **28**, (2), pp. 189–198.

Index

Smith's Elements of Soil Mechanics, 10th Edition. Ian Smith.
© 2021 John Wiley & Sons Ltd. Published 2021 by John Wiley & Sons Ltd.
Companion website: www.wiley.com/go/smith/soilmechanics10e

Printed and bound by CPI Group (UK) Ltd, Croydon, CR0 4YY

27/10/2024

14580163-0002